CW00350297

MANY WORLDS?

Many Worlds?

Everett, Quantum Theory, and Reality

Edited by
SIMON SAUNDERS, JONATHAN BARRETT,
ADRIAN KENT, AND DAVID WALLACE

OXFORD
UNIVERSITY PRESS

*This book has been printed digitally and produced in a standard specification
in order to ensure its continuing availability*

OXFORD
UNIVERSITY PRESS

Great Clarendon Street, Oxford OX2 6DP
United Kingdom

Oxford University Press is a department of the University of Oxford.
It furthers the University's objective of excellence in research, scholarship,
and education by publishing worldwide.

Oxford is a registered trade mark of Oxford University Press in the UK
and in certain other countries

British Library Cataloguing in Publication Data
Data available

Library of Congress Cataloging in Publication Data
Data available

ISBN 978-0-19-956056-1

To my parents (JB)
To Adriana Kent and Patrick Rabbitt, with gratitude and love (AK)
To Kalypso, Ari, and Daphne, with love and appreciation; to Itamar
Pitowsky, in remembrance (SS)
To Lynne and Phil Wallace (DW)

Preface

This paper seeks to clarify the foundations of quantum mechanics . . . The aim is not to deny or contradict the conventional formulation of quantum theory, which has demonstrated its usefulness in an overwhelming variety of problems, but rather to supply a new, more general and complete formulation, from which the conventional interpretation can be *deduced*.

So wrote Hugh Everett III in 1957, ambitiously and yet understatedly. His 'relative state formulation of quantum mechanics' was and is still more radical: it claims that the formalism of quantum mechanics, taken completely literally, describes a reality where every macroscopic superposition of quantum states is really a splitting of the universe into parallel copies. Hence the more standard, and more nakedly radical, name for Everett's proposal: *many-worlds quantum theory.*

It sounds like science fiction: suggest to a non-physicist that this is what our best theory of the world tells us and the response is often incredulity. Yet over the fifty years since Everett's proposal, the many-worlds concept has become enormously influential in theoretical physics. It remains highly controversial. Many physicists accept some version of many worlds as the correct reading of quantum theory; many others take the idea very seriously; many, too, firmly reject it. Meanwhile, the last two decades have seen a renaissance in the conceptual and technical development of Everett's ideas: work in areas as disparate as decoherence theory and decision theory has helped to clarify exactly what a commitment to many worlds might—or perhaps must—involve.

Fifty years on, the time seems ripe for an assessment of many-worlds quantum theory. The four of us do not agree on its merits. Two of us (Saunders and Wallace) argue that Everett's proposal is the only coherent way to understand unitary quantum theory, and so—in the absence of viable alternatives to unitary quantum theory—current physics makes many-worlds quantum theory compulsory. One of us (Kent) argues that the conventional formulation of quantum theory and the experimental predictions that support it cannot in fact be deduced from Everett's proposal: in other words, that Everett failed in his stated aim. One of us (Barrett) remains agnostic. But we do agree about the profound challenge posed by Everett's ideas and the work that has come from them, and the concomitant importance of a critical assessment.

In the summer of 2007—to mark the 50th anniversary of Everett's original paper—we organized two conferences, to bring together the best advocacy of the

Everett interpretation and engage it with the sharpest criticism. The venues and attendees of these conferences illustrate the interdisciplinary nature of research on many-worlds quantum theory and its wide-ranging importance. The first was held in Oxford, organized by Saunders (a philosopher of physics) and under the aegis of the Philosophy Faculty of Oxford University; the second was held in the Perimeter Institute for Theoretical Physics in Waterloo, Canada, and was organized by Barrett and Kent (theoretical physicists) and by Wallace (a physicist-turned-philosopher).

This book grew out of the contributions to those conferences, but it is more than a 'simple' conference proceedings. Instead, we have drawn on the expertise present at the Oxford and Perimeter meetings to put together a collection of papers which illustrates, as clearly as possible, the state of play of many-worlds quantum theory fifty years on. The four of us agree that there are powerful arguments in its favour, and serious counter-arguments that need to be overcome for it to succeed. We believe that the best of each were on display at the Oxford and PI conferences, and we hope that the best of each can be found within this volume.

The book begins with an introduction to many-worlds quantum theory, and a summary of the main arguments that follow. They are arranged in six parts. The first and second centre on whether and how the formalism of quantum theory, when applied to macroscopic systems, already contains within it a description of many worlds. Part 1 gives arguments in favour of that claim; Part 2 gives arguments against. Part 3 is about probability, and the case for recovering or enforcing the probability interpretation of standard quantum theory given the branching structure of reality as argued for in Part 1. Part 4 is also largely on probability and the relationship between many-worlds and standard quantum theory, but its authors aim to refute those claims of Part 3, and particularly those resting on decision theory and Bayesian methods, or on any likely variant of them. Part 5 is given over to some of the main contemporary rivals to many worlds—to proposed alternative realist versions or variants of quantum theory. In contrast Part 6 considers, from the perspective of many worlds, old and new questions: about Everett's original ideas and their early reception; about time-symmetric quantum mechanics; and about the multiverse concept as it independently arises in contemporary cosmology.

Neither this book nor the conferences from which it emerged would have been possible without support from the Fundamental Questions Institute Foundation (FQXi), and from the Perimeter Institute (PI). In connection with the PI meeting, we would particularly like to thank Mike Lazaridis, Perimeter Institute's Founder and Board Chair, for making everything possible, Howard Burton, PI's then Executive Director, and all the faculty, researchers and staff of Perimeter Institute for their support of and many contributions to the meeting. Special thanks go to Kate Gillespie and Dawn Bombay, for their cheerfully efficient administration.

The Oxford meeting is indebted to the support of the Philosophy Faculty and its administrators, headed by Thomas Moore, Secretary to the Faculty, to Andrew Davies, IT Manager, for his help with filming the conference and building the conference website, and to all the staff at 10 Merton Street for their cooperation and hospitality. Special thanks go to Peter Taylor, who gave freely and unstintingly of time and energy in managing the conference, and to Paul Tappenden, for preparing transcripts of discussions.

Jonathan Barrett
Adrian Kent
Simon Saunders
David Wallace

November 2009

Contents

6. NOT ONLY MANY WORLDS

List of Contributors

David Albert is Professor of Philosophy at Columbia University.

Harvey R. Brown is Professor of Philosophy at the University of Oxford and a Fellow of Wolfson College, Oxford.

Jeffrey Bub is Distinguished University Professor in the Department of Philosophy and Institute for Physical Science and Technology at the University of Maryland.

Peter Byrne is an Investigative Reporter.

David Deutsch is Visiting Professor of Physics at the University of Oxford.

Hilary Greaves is a Tutorial Fellow in Philosophy at Somerville College, Oxford, and a Lecturer in Philosophy of Physics at the University of Oxford.

Jonathan Halliwell is Professor of Theoretical Physics at Imperial College, London.

Jim Hartle is Research Professor and Professor Emeritus at the University of California, Santa Barbara, and an External Faculty Member of the Santa Fe Institute.

John Hawthorne is Waynflete Professor of Metaphysical Philosophy at the University of Oxford and a Fellow of Magdalen College, Oxford.

Adrian Kent is Reader in Quantum Physics at DAMTP, University of Cambridge, a Fellow of Wolfson College, Cambridge, and an Associate Member of Perimeter Institute for Theoretical Physics.

James Ladyman is Professor of Philosophy at the University of Bristol.

Tim Maudlin is Professor of Philosophy at Rutgers University.

Wayne Myrvold is Associate Professor of Philosophy at the University of Western Ontario.

David Papineau is Professor of Philosophy at King's College, London.

Itamar Pitowsky was Professor of Philosophy at the Hebrew University of Jerusalem.

Huw Price is an ARC Federation Fellow and Challis Professor of Philosophy at the University of Sydney.

Simon Saunders is Professor of Philosophy of Physics at the University of Oxford, and a Fellow of Linacre College, Oxford.

Rüdiger Schack is Professor of Mathematics at Royal Holloway, University of London.

Max Tegmark is Professor of Physics at M.I.T.

Christopher Timpson is a Tutorial Fellow in Philosophy at Brasenose College, Oxford, and a Lecturer in Philosophy of Physics at the University of Oxford.

Lev Vaidman is Professor of Physics at the University of Tel Aviv.

Antony Valentini is a Research Fellow at Imperial College, London.

David Wallace is a Tutorial Fellow in Philosophy at Balliol College, Oxford, and a Lecturer in Philosophy of Physics at the University of Oxford.

Wojciech Zurek is a Laboratory Fellow at Los Alamos National Laboratory.

Many Worlds? An Introduction

Simon Saunders

This problem of getting the interpretation proved to be rather more difficult than just working out the equation.

P.A.M. Dirac

Ask not if quantum mechanics is true, ask rather what the theory implies. What does realism about the quantum state imply? What follows then, when quantum theory is applied without restriction, if need be to the whole universe?

This is the question that this book addresses. The answers vary widely. According to one view, 'what follows' is a detailed and realistic picture of reality that provides a unified description of micro- and macroworlds. But according to another, the result is *nonsense*—there is no physically meaningful theory at all, or not in the sense of a realist theory, a theory supposed to give an intelligible picture of a reality existing independently of our thoughts and beliefs. According to the latter view, the formalism of quantum mechanics, if applied unrestrictedly, is at best a fragment of such a theory, in need of substantive additional assumptions and equations.

So sharp a division about what appears to be a reasonably well-defined question is all the more striking given how much agreement there is otherwise, for all parties to the debate in this book are agreed on realism, and on the need, or the aspiration, for a theory that unites micro- and macroworlds, at least in principle. They all see it as legitimate—obligatory even—to ask whether the fundamental equations of quantum mechanics, principally the Schrödinger equation, *already* constitute such a system. They all agree that such equations, if they are to be truly fundamental, must ultimately apply to the entire universe. And most of the authors also agree that the quantum state *should* be treated as something physically real. But now disagreements set in.

For the further claim argued by some is that if you allow the Schrödinger equation unrestricted application, supposing the quantum state to be something physically real, then without making any additional hypotheses, there follows a conservative picture of the small macroscopic, consistent with standard applications of quantum mechanics to the special sciences, a picture that extends to the biological sciences, to people, planets, galaxies, and ultimately the entire universe, but only insofar as this universe is one of *countlessly many others, constantly branching in time, all of which are real.* The result is the *many worlds theory,* also known as the *Everett interpretation* of quantum mechanics.

But contrary claims say this picture of many worlds is in no sense inherent in quantum mechanics, even when the latter is allowed unrestricted scope and even given that the quantum state is physically real. And if such a picture *were* implied by the Schrödinger equation, that would only go to show that this equation must be restricted or supplemented or changed. For (run contrary claims) this picture of branching worlds fails to make physical sense. The stuff of these worlds, what they are made of, is never adequately explained, nor are the worlds precisely defined; ordinary ideas about time and identity over time are compromised; the concept of probability itself is in question. This picture of many branching worlds is inchoate. There are realist alternatives to many worlds, among them theories that leave the Schrödinger equation unchanged.

These are the claims and counterclaims argued in this book. This introduction is in three parts. The first is partisan, making the case for many worlds in the light of recent discoveries; the crucial new datum, absent from earlier discussions, is decoherence theory, which in this book takes centre stage. Section 2 is even-handed, and sketches the main arguments of the book: on ontology, the existence of worlds; on probability, as reduced to the branching structure of the quantum state; and on alternatives to many worlds, realist proposals that leave the Schrödinger equation unchanged. The third and final section summarizes some of the mathematical ideas, including the consistent histories formalism.

1 THE CASE FOR MANY WORLDS

1.1 Realism and Quantum Mechanics

As Popper once said, physics has always been in crisis, but there was a special kind of crisis that set in with quantum mechanics. For despite all its obvious empirical success and fecundity, the theory was based on rules or prescriptions that seemed inherently contradictory. There never was any real agreement on these matters among the founding fathers of the theory. Bohr and later Heisenberg in their more philosophical writings provided little more than a fig-leaf; the emperor, to the eyes of realists, wore no clothes. Textbook accounts of quantum mechanics in the past half-century have by and large been operationalist. They say as little as possible about Bohr and Heisenberg's philosophy or about realism.

In what sense are the rules of quantum mechanics contradictory? They break down into two parts. One is the *unitary* formalism, notably the Schrödinger equation, governing the evolution of the quantum state. It is deterministic and encodes spacetime and dynamical symmetries. Whether for a particle system or a system of fields, the Schrödinger equation is linear: the sum of two solutions to the equation is also a solution (the superposition principle). This gives the solution space of the Schrödinger equation the structure of a vector space (Hilbert space).

However, there are also rules for another kind of dynamical evolution for the state, which is—well, *none* of the above. These rules govern the *collapse* of the wavefunction. They are indeterministic and non-linear, respecting none of the spacetime or dynamical symmetries. And unlike the unitary evolution, there is no obvious route to investigating the collapse process empirically.

Understanding state collapse, and its relationship to the unitary formalism, is the *measurement problem* of quantum mechanics. There are other conceptual questions in physics, but few if any of them are genuinely paradoxical. None, for their depth, breadth, and longevity, can hold a candle to the measurement problem.

Why not say that the collapse is simply irreducible, 'the quantum jump', something primitive, inevitable in a theory which is fundamentally a theory of chance? Because it isn't only the collapse process itself that is under-specified: the time of the collapse, within relatively wide limits, is undefined, and the criteria for the kind of collapse, linking the set of possible outcomes of the experiment to the wavefunction, are strange. They either refer to another theory entirely—classical mechanics—or worse, they refer to our 'intentions', to the 'purpose' of the experiment. They are the *measurement postulates*—('probability postulates' would be better, as this is the only place where probabilities enter into quantum mechanics). One is the *Born rule*, assigning probabilities (as determined by the quantum state) to macroscopic outcomes; the other is the *projection postulate*, assigning a new microscopic state to the system measured, depending on the macroscopic outcome. True, the latter is only needed when the measurement apparatus is functioning as a state-preparation device, but there is no doubt that *something* happens to the microscopic system on triggering a macroscopic outcome.

Whether or not the projection postulate is needed in a particular experiment, the Born rule is essential. It provides the link between the possible macroscopic outcomes and the antecedent state of the microscopic system. As such it is usually specified by giving a choice of vector basis—a set of orthogonal unit vectors in the state space—whereupon the state is written as a superposition of these. The modulus square of the amplitude of each term in the superposition, thus defined, is the probability of the associated macroscopic outcome (see Section 3 p.37). But what dictates the choice of basis? What determines the time at which this outcome happens? How does the measurement apparatus interact with the microscopic system to produce these effects?

From the point of view of the realist the answer seems obvious. The apparatus itself should be modelled in quantum mechanics, then its interaction with the microscopic system can be studied dynamically. But if this description is entirely quantum mechanical, if the dynamics is unitary, it is *deterministic*. Probabilities only enter the conventional theory explicitly with the measurement postulates. The straightforwardly physicalistic strategy seems *bound* to fail.

How are realists to make sense of this? The various solutions that have been proposed down the years run into scores, but they fall into two broadly recognizable classes. One concludes that the wavefunction describes not the microscopic system itself, but our *knowledge* of it, or the information we have available of it (perhaps 'ideal' or 'maximal' knowledge or information). No wonder modelling the apparatus in the wavefunction is no solution: that only shifts the problem further back, ultimately to 'the observer' and to questions about the mind, or consciousness, or information—all ultimately philosophical questions. Anti-realists welcome this conclusion; according to them, we neglect our special status as the knowing subject at our peril. But from a realist point of view this just leaves open the question of what the goings-on at the microscopic level, thus revealed, actually are. By all means constrain the spatiotemporal description (by the uncertainty relations or information-theoretic analogues), but still *some* spatiotemporal description must be found, down to the lengthscales of cells and complex molecules at least, even if not all the way to atomic processes. That leads to the demand for equations for variables that do not involve the wavefunction, or, if none is to be had in quantum mechanics, to something entirely new, glimpsed hitherto only with regard to its statistical behaviour. This was essentially Einstein's settled view on the matter.

The only other serious alternative (to realists) is *quantum state realism*, the view that the quantum state *is* physically real, changing in time according to the unitary equations and, somehow, *also* in accordance with the measurement postulates.

How so? Here differences in views set in. Some advocate that the Schrödinger equation itself must be changed (so as to give, in the right circumstances, collapse as a fundamental process). They are for a *collapse* theory.

Others argue that the Schrödinger equation can be left alone if only it is supplemented by additional equations, governing 'hidden' variables. These, despite their name, constitute the real ontology, the stuff of tables and chairs and so forth, but their behaviour is governed by the wavefunction. This is the *pilot-wave* theory. Collapse in a theory like this is only 'effective', as reflecting the sudden irrelevance (in the right circumstances) of some part of the wavefunction in its influence on these variables. And once irrelevant in this way, always irrelevant: such parts of the wavefunction can simply be discarded. This explains the appearance of collapse.

But for others again, no such additional variables are needed. The collapse is indeed only 'effective', but that reflects, not a change in the influence of one part of the quantum state on some hidden or 'real' ontology, but rather the change in dynamical influence of *one part of the wavefunction over another*—the *decoherence* of one part from the other. The result is a branching structure to the wavefunction, and again, collapse only in a phenomenological, effective sense. But then, if our world is just one of these branches, all these branches must be worlds. Thus the *many worlds* theory—worlds not spatially, but *dynamically* separated.

This concept of decoherence played only a shadowy role in the first 50 years of quantum mechanics, but in the past three decades it has been developed more systematically. As applied to the wavefunction to derive a structure of branching worlds it is more recent still. It changes the nature of the argument about the existence of worlds. The claim is that the worlds are dynamically robust patterns in the wavefunction, obeying approximately classical equations. They are genuine discoveries, the outcome of theoretical investigations into the unitary formalism, not from posits or hypotheses. And *if* this is so, it puts collapse theories and pilot-wave theories in a different light. It shows them as modifying or supplementing quantum mechanics not in the name of realism, and still less because of any conflict with experiment, but because reality as depicted by quantum mechanics is in conflict with a priori standards of acceptability—at the fundamental level, as too strange, at the emergent level of worlds, as insufficiently precise or as ontologically too profligate.

But if decoherence theory makes a difference to our understanding of the quantum state, as applied to sufficiently complex many-particle systems, it is not so clear that it touches that other ground on which the Everett interpretation has been rejected—that if quantum mechanics is a purely deterministic theory of many worlds, the idea of objective probability is *simply no longer applicable*. Failing a solution to this, the 'incoherence problem', the Everett interpretation does not provide an empirical theory at all. And with that the argument against modifying quantum mechanics on realist grounds completely collapses.

We shall see how this argument played out in the first three decades of the Everett interpretation in a moment. In fact, it is just here, on the interpretation of probability, that inattention to the concept of decoherence was most damaging. And looking ahead, there is a further important sense in which the argument over many worlds has been changed. In recent years, with the development of decision-theory methods for quantifying subjective probability in quantum mechanics, the link between probability in the subjective sense and an objective counterpart has been greatly clarified. Specifically, it can be shown that agents who are rational, in order to achieve their ends, have no option but to use the modulus squared branch amplitudes in weighting their utilities. In this sense the Born rule has been *derived*.

That goes with other arguments about probability. From a philosophical point of view this link with rational belief, or credence, has always been the most important—and baffling—of the roles played by objective probability. If it is shown to be played by these branching structures then they *are* objective probabilities.

Meanwhile other puzzles specific to the picture of branching worlds (particularly to do with personal identity) can be solved or sidelined.

These claims need to be argued one by one. The first step is to get a clearer understanding of the early debates about many worlds, in the first quarter-century since their appearance.

1.2 Early History

The many-worlds interpretation of quantum mechanics was first proposed by H. Everett III in his doctoral ('long') dissertation, written under his supervisor J.A. Wheeler. It was cut down to a quarter of its size on Wheeler's insistence, but in this form it won a PhD. It won him a ringing endorsement by Wheeler too, when it was published shortly thereafter, as ' "Relative State" Formulation of Quantum Mechanics' (Everett [1957], Wheeler [1957]).[1] The main mathematical ideas of this paper are explained in Section 3.

Ten years later it was endorsed again, but this time with rather greater fidelity to Everett's ideas, by B. DeWitt [1967, 1970] (introducing the terminology 'many worlds' and 'many universes' for the first time in print). DeWitt, in collaboration with his PhD student N. Graham, also published Everett's long dissertation, as 'The Theory of the Universal Wave Function'. This and a handful of much smaller articles made up their compilation *The Many Worlds Interpretation of Quantum Mechanics* (DeWitt and Graham [1973]).

But as such it had a notable deficiency. It lacked an account of why the wavefunction must be viewed in terms of one sort of multiplicity rather than another—and why, even, a multiplicity at all. What reason there was to view the quantum state in this way (as a superposition of possible outcomes) came from the measurement postulates—the very postulates that Everett, by his own admission, was intent on doing away with.

This is the 'preferred basis problem' of the Everett interpretation (Ballentine [1973]). If the basis is determined by the 'purpose' of the experiment, then the measurement postulates are blatantly still in play. DeWitt, who was more interested in applying quantum mechanics to gravitating systems (hence, ultimately, to the entire universe) than in getting rid of any special mention of experiments in the definition of the theory, went so far as to postulate the existence of apparatuses as an *axiom*.

But Everett was able to derive at least a fragment of the Born rule. Given that the measure over the space of branches is a function of the branch amplitudes, the question arises: What function? If the measure is to be additive, so that the measure of a sum of branches is the sum of their measures, it follows that it is the modulus square—that was something. The set of branches, complete with

---

[1] This story is told by Peter Byrne in Chapter 17.

additive measure, then constitute a probability space. As such, versions of the Bernouilli and other large number theorems can be derived. They imply that the measure of all the branches exhibiting anomalous statistics (with respect to this measure) is small when the number of trials is sufficiently large, and goes to zero in the limit—that was something more.

It was enough to raise the prospect of a frequentist account of quantum probability—meaning, an account that identifies probabilities with actual relative frequencies—and to raise the hope, more generally, that the probability interpretation of quantum mechanics was itself *derivable* from the unitary formalism. DeWitt was further impressed by a result due to his PhD student N. Graham [1970] (and discovered independently by J.B. Hartle [1968]): for the kth possible outcome of an experiment, it is possible to construct a 'relative-frequency operator', of which the grand superposition of all possible outcomes on N repetitions of the experiment is an eigenstate in the limit as $N \to \infty$. The corresponding eigenvalue, he showed, equals the Born rule probability for the kth outcome in a single trial.

Another way of putting it (DeWitt [1970 pp.162–3]) was that the components of the total superposition representing 'maverick' worlds (recording anomalous statistics at variance with the Born rule) are of measure zero, in the Hilbert space norm, in the limit $N \to \infty$. But whilst formal criteria like these may be required of an acceptable theory of probability, they are hardly sufficient in themselves. They do not explain probability in terms of existing states of affairs, which invariably involve only finitely many trials. To suppose that states whose records of outcomes are large are 'close to' states in the infinite limit, and therefore record the right relative frequencies (or that those that do are more probable than those that do not), is to beg the question.[2]

These are defects of frequentism as a theory of probability; they are hardly specific to the Everett interpretation. But in an important respect frequentism in the context of many worlds seems to fare worse. For assuming (as Everett's original notation suggested) that branches are in one-one correspondence with sequences of experimental outcomes, the set of all worlds recording the results of N trials can be represented by the set of all possible sequences of length N. In that case, there is an obvious rival to the Born rule: this set of sequences has a natural statistical structure independent of the amplitude, anyway invisible, attached to each world. A priori there are many more of them for which the relative frequency of the kth outcome is close to one half, than of those in which it is closer to zero or one.

This is to treat each distinct sequence as equiprobable. Predictions using this rule would have been wildly contradicted by the empirical evidence, true, but that only goes to show that the probability rule in the Everett interpretation

---

[2] Ochs [1977] and Farhi et al. [1989] offered improvements on the rigour of the argument, but not on its physical significance.

is not forced by the equations—that the Born rule, far from being an obvious consequence of the interpretation of the quantum state in terms of many worlds, appears quite unreasonable.

Attempts to patch up this problem only made clearer its extent. The difficulty—call it the 'combinatorics problem'—was first pointed out by Graham [1973]: he suggested that experiments must involve as an intermediary a thermodynamic system of large numbers of degrees of freedom, for which the count of states did reflect the Born rule quantities. But few found this argument persuasive. And the problem highlighted the deficiencies of Everett's derivation of the Born rule: for why assume that the probability measure on the space of branches is a function only of the branch amplitudes? This assumption, given the picture of many worlds, now seemed ad hoc.

Failing a solution to the preferred basis problem, the theory was not even well defined; if the problem is solved by evoking a special status to experiments, the theory was not even a form of realism; however it is solved, the very picture of many worlds suggests a probability measure at odds with the statistical evidence. It is hardly surprising that the Everett interpretation was ignored by J.S. Bell, when he posed his famous dilemma:

Either the wavefunction, as given by the Schrödinger equation, is not everything, or it is not right.   (Bell [1987 p.201])

But the situation looks quite different today.

1.3 Ontology and Decoherence

Decoherence theory has its roots in Ehrenfest's theorem[3] and the very early debates about foundations in quantum mechanics, but its development was slow. It remains to this day more of a heterogeneous collection of techniques than a systematic theory. But these techniques concern a common question: under what circumstances and with respect to what basis and dynamical variables does a superposition of states behave dynamically just as if it were an incoherent mixture of those same states? This question already arises in conventional quantum mechanics using the measurement postulates (the choice of the von Neumann 'cut'): at what point in the unitary evolution can the latter be applied? If too early, interference effects of salience to the actual behaviour of the apparatus will be destroyed.

The 1980s saw a plethora of toy models attempting to answer this question. Many relied on the system–environment distinction and the Schmidt decomposition (see Section 3 pp.43–4). Together, apart from exceptional cases, they defined a unique basis, with respect to which at any time mixtures and superpositions of states were exactly equivalent, under the measurement postulates, for

---

[3] This theorem is explained and improved on by Jim Hartle in Chapter 2.

observables restricted to the system alone or to the environment alone. But not for other observables; and the basis that resulted was in some cases far from well localized, and sensitive to the details of the dynamics and the system-environment distinction.

In contrast, states well localized in phase space—wavepackets—reliably decohere, and even though elements of a superposition, evolve autonomously from each other for a wide class of Hamiltonians. With respect to states like these, Ehrenfest's theorem takes on a greatly strengthened form. But decoherence in this sense is invariably approximate; it is never an all-or-nothing thing.

Very little of this literature made mention of the Everett interpretation: it was hoped that decoherence would solve the measurement problem in a one-world setting. And where Everett's ideas were involved, it was the concept of 'relative state', as formulated (in terms of the decoherence of the environment) by W.H. Zurek [1982]. According to Zurek, such states need not coexist; rather, macroscopic quantities were subject to 'environmental superselection', after the idea of superselection rules, latterly introduced, prohibiting the development of superpositions of states corresponding to different values of super-selected quantities like charge. But if so, surely decoherence, like superselection, has to be exact, returning us to the Schmidt decomposition.[4]

The exception was H.D. Zeh's 'On the interpretation of measurement in quantum mechanics' [1970], which engaged with Everett's proposal more comprehensively. In it, Zeh set out the idea of dynamical decoherence as a stability condition. He gave the example of sugar molecules of definite chirality; in a superposition of left- and right-handed molecules, each term evolves by a dynamical process which, however complicated, is almost completely decoupled from the motion of the other. Dynamical stability, he proposed, was the key to defining the preferred basis problem in the Everett interpretation.

But Zeh's argument was qualitative, and in subsequent publications (Zeh [1973], Kübler and Zeh [1973]), which did give detailed calculations, he used the Schmidt decomposition instead. This still chimed with Everett's idea of the 'relative state' (Section 3.2), but the idea of dynamical stability was marginalized. And an inessential one was added: Zeh spoke of the need for a 'localization of consciousness' not only in space and time but also in 'certain Hilbert-space components' [1970 p.74], an idea he attributed to Everett. That fostered the view that some high-level hypothesis about mentality was needed if the Everett interpretation was to go through.[5] But if the Everett

---

[4] The attempt to define the preferred basis in terms of the Schmidt decomposition was taken to its logical conclusion in the 'modal' interpretation, as developed in the mid 1990s by D. Dieks, G. Bacciagaluppi, and P. Vermas, among others. But this ran into an embarrassment of technical difficulties (see, in particular, Bacciagaluppi [2000]). The extension to the N-body case was restrictive, the basis thus selected was defective, particularly in the case of quantum fields, and no non-trivial Lorentz covariant theory of this kind could be found.

[5] Taken up by M. Lockwood in his *Mind, Brain and Quantum* [1989] and by J. Barrett in *The Quantum Mechanics of Minds and Worlds* [1999] (for commentaries, see my [1996a, 2001]).

interpretation recovers the elements of quantum chemistry, solid state physics, and hydrodynamics, as the effective theories governing each branch, questions of mentality can be left to the biological sciences. They need no more intrude in the Everett interpretation than they do in a hidden-variable or collapse theory.

Add to this mix the consistent histories formalism of R. Griffiths [1984] and R. Omnès [1988], and especially as developed by M. Gell-Mann and J.B. Hartle [1990, 1993]. In this approach a division between subsystem and environment is inessential: the key idea is the coarse-graining of certain dynamical variables and the definition of quantum histories as time-ordered sequences of coarse-grained values of these variables (Section 3.3). The wavefunction of the universe—in the Heisenberg picture—is in effect the superposition of these histories. The choice of variables—of the history space—is equivalent to the choice of preferred basis.

But this is to put the matter in Everettian terms. The consistent histories theory was based rather on a certain formal constraint, required if a space of quantum histories is to have the structure of a probability space: the 'consistency condition' (Section 3.4). Meanwhile the quantum state, in the Heisenberg picture, could be viewed as *no more* than a probability measure on this space, of which only one history, it seemed, need be real.[6]

The goal was once again a one-world interpretation of quantum mechanics. But for that, fairly obviously, the history space had to be fixed once and for all. It was clear from the beginning that there were many consistent history spaces for a fixed initial state and Hamiltonian—which one should we choose? But it was thought that at least the actual history of the world and its history space up to some time, once given, dictated the probabilities for subsequent events unequivocally. Far from it: as F. Dowker and A.P.A. Kent shortly showed, a history space up to some time can be deformed into any one of a continuous infinity of other history spaces for subsequent times, preserving the consistency condition exactly (Dowker and Kent [1996]).

But this difficulty does not apply to the marriage of decoherence theory in the more general sense with the consistent histories formalism, as carried through by Gell-Mann and Hartle and J.J. Halliwell in the 1990s and 2000s.[7] Decohering histories in the latter sense are robustly defined. But decoherence (and the consistency condition) obtained in this way is never exact. On a one-world

Everett made no mention of consciousness, although he did speak of 'experience'. Zeh has continued to insist on the need, in the Everett interpretation, for a special postulate concerning consciousness (see e.g. Zeh [2000]).

[6] This turns the wavefunction of the universe into something more like a law than a physically existing thing. See Antony Valentini in Chapter 16 for criticism of an analogous proposal in the context of pilot-wave theory.

[7] Chapter 2, by Jim Hartle, is a general review; Chapter 3, by Jonathan Halliwell, is a detailed study of the important example of hydrodynamic variables.

interpretation, 'sufficiently small' interference ϵ is then the criterion for probabilistic change at the fundamental level; what value of $\epsilon > 0$, precisely, is
the trigger? Worse, there is no algorithm for extracting even approximately
decohering histories for *any* Hamiltonian and *any* state. How are the latter to be
constrained? This is at best a programme for modifying quantum mechanics, or
replacing it.

It is otherwise if decoherent histories theory is in service of the Everett
interpretation—defining, among other things, the preferred basis. In that case
it hardly matters if, for some states and regimes, decohering histories are simply
absent altogether: the fundamental reality, the wavefunction of the universe, is
still well defined (Zurek [1993], Saunders [1993, 1995a]). Context-dependence
and inexactitude as to how it is to be broken down into recognizable parts is to
be expected.

It is a further and quite distinct question as to what kinds of parts or worlds,
governed by what kinds of equations, are thus identified. The ones so far
discovered are all approximately classical—classical, but with dissipation terms
present that reflect their quantum origins.[8] In the terminology of Gell-Mann
and Hartle [1990], such a set of decoherent histories is a 'quasiclassical domain'
(in their more recent writings, 'realm'). Might there be non-classical realms,
other preferred bases, involving sets of equations for completely alien variables?
Perhaps, but that need not pose any difficulty for the Everett interpretation.
There is no reason to think that the kind of under-determination of consistent
history space by past history, discovered by Dowker and Kent, applies to realms.

And now for the killer observation (Wallace [2003a]): this business of
extracting approximate, effective equations, along with the structures or patterns
that they govern, is *routine* in the physical sciences. Such patterns may be high-
level, 'emergent' ontology; they are fluids, or crystals, or complex hydrocarbon
molecules, ascending to cells, living organisms, planets, and galaxies. Equally
they are atoms and nuclei, as modelled (with great difficulty) in quantum
chromodynamics and electroweak theory, or phonons, or superconductors, or
Bose condensates, in condensed matter physics—the list goes on and on (what
isn't emergent ontology?). It is in this sense—so the claim goes—that worlds
are shown to exist in the wavefunction, rather than be put in by hand. They are
investigated just as is any other emergent ontology in the special sciences.[9]

If so, doesn't it follow that many worlds *also* exist in pilot-wave theory? As
formulated by L. de Broglie in 1927 and by D. Bohm in 1952, the 'pilot

---

[8] Going the other way—given that worlds are defined in terms of states well localized in position
and momentum space—consistency follows trivially, and it is relatively easy to see (from Ehrenfest's
theorem) that states like these obey approximately classical equations (see Hartle, Chapter 2).
Convinced Everettians such as DeWitt, Deutsch, and Vaidman saw no need for decoherence theory
in consequence. (It went almost unmentioned in Deutsch [1997] and in Vaidman [1998], [2002];
when DeWitt did take note of it (De Witt [1993]), he applied it to branching in the absence of
experiments. When induced by experiments 'Everett has already dealt with it'.)

[9] This argument is reprised by David Wallace in Chapter 1.

wave' is just the wavefunction as given by the Schrödinger equation, so it is mathematically, structurally, identical to the universal state in the Everett interpretation. Indeed, decoherence theory plays the same essential role in dictating 'effective' wavepacket collapse in that theory as it does in the Everett interpretation.[10] There follows the gibe: pilot-wave theories are 'parallel universe theories in a state of chronic denial' (Deutsch [1996 p.225]).[11]

A similar consideration applies even to those models of dynamical collapse in which, from an Everettian point of view, the amplitudes of all worlds save one are only suppressed (they remain non-zero—and, 'for all practical purposes', are subject only to the unitary evolution). It applies, uncomfortably, to the only realistic (albeit non-relativistic) collapse theories so far available, those due to G.C. Ghirardi and his co-workers (Ghirardi et al. [1986], [1990]), in which the collapse attenuates but does not eliminate altogether components of the state. In these theories all the structures of the wavefunction of the universe as goes (what on Everettian terms would be) *other* worlds are still there, their *relative* amplitudes all largely unchanged. This is the so-called 'problem of tails'. On any broadly structuralist, functionalist approach to the physical sciences, these structures to the tails are still real.[12]

The distinctively new feature of the Everett interpretation today is not only that the preferred basis problem is solved; it is that the very existence of worlds, of a multiplicity of patterns in the wavefunction, each obeying approximately classical laws, is *derived*. The fact that they make an unwelcome appearance in every other form of quantum-state realism, from which they can be removed, if at all, only with difficulty, proves the point. But there is more.

1.4 Probability and Decision Theory

Decoherence bears on the probability interpretation if only because it explains why there is a plurality at all. It shows that *some* kind of statistical analysis is perfectly reasonable. But it also undercuts at a stroke the combinatorics problem. For decoherence comes in degrees; there is no good answer to the question—How many decohering worlds? Numbers like these can be stipulated, but from the point of view of the dynamical equations, they would amount to arbitrary conventions. There is no longer a statistical structure to the set of

---

[10] As Bohm effectively acknowledged when considering the problem of when 'empty waves' in pilot-wave theory could be ignored: 'It should be noted that exactly the same problem arises in the usual interpretation of the quantum theory, for whenever two packets overlap, then, even in the usual interpretation, the system must be regarded as, in some sense, covering the states corresponding to both packets simultaneously.' (Bohm [1952 p.178 fn.18]). Here Bohm referenced his textbook on quantum mechanics published the previous year, containing an early treatment of decoherence (Bohm [1951 ch.6, 16, sec.25]).

[11] The argument is made in detail by Brown and Wallace [2005]. In Chapter 16, Antony Valentini gives a reply, followed by a commentary by Harvey Brown.

[12] On this point see Tim Maudlin in Chapter 4.

branches independent of the amplitudes. The Born rule no longer has an obvious rival (Saunders [1998]).

At this point, whatever DeWitt's hopes of deriving the probability interpretation of the theory from the equations, one might hope to settle for the probability rule as a *hypothesis*—much as is done, after all, in conventional quantum mechanics. The ontology—the branching structure to the wavefunction—may not force any particular probability measure, very well; then choose one on empirical grounds. The Born rule is a natural candidate and recovers the probabilistic predictions of conventional quantum mechanics. Why not simply postulate it?

But that would be to reinstate a part—a small part, given that decoherence theory is now dictating the basis, but a part nonetheless—of the measurement postulates. And on reflection, it is not a wholly uncontentious part. There is a puzzle, after all, as to whether, being among a superposition of worlds, we are not in some sense in them all. About to perform a quantum experiment, if the Everett interpretation is to be believed, all outcomes are obtained. We know this in advance, so where is there any uncertainty? And this problem can be lumped with the preferred basis problem: why not let the questions of personal identity and preferred basis follow together from a theory of mentality, or a theory of computation, or of quantum information (Albert and Loewer [1988], Lockwood [1989], Barrett [1999])? Or be posited, in terms of new axioms, at the level of worlds (Deutsch [1985])?

The appeal to mentality is in the tradition of Wheeler and Wigner rather than that of Everett and DeWitt, and we have turned our backs on it. But still, it makes it clearer why it is unsatisfactory to simply *posit* a probability interpretation for the theory. If there is chance in the Everett interpretation, it should be identified as some objective physical structure, and that structure should be shown to fill all (or almost all) the chance-roles—including, plausibly, the role of uncertainty. It cannot just pretend to fill it, or fill it by decree (Greaves [2004]). However, that may turn out to be more of a linguistic matter than is commonly thought. As argued by Papineau [1996], the notion of uncertainty appears to play no useful rule in decision theory.

But there is another chance-role, what the philosopher D.K. Lewis has called the 'principal principle', that all are agreed is indispensable. Let S be the statement that the chance of E at t is p, and suppose our background knowledge K is 'admissible' (essentially, that it excludes information as to whether E happened or not): then our credence in E, conditional on S and K, should be p.

Here 'credence' is subjective probability, degrees of belief. It is probability in the tradition of F.K. Ramsey, B. de Finetti, and L. Savage. Credence is what matters to decision theory, statistical inference, and statistical test. So long as the notion of free will and agency is not in question in the Everett interpretation—or no more so than in classical mechanics or pilot-wave theory—the Ramsey–de

Finetti operational characterization of credence in terms of an agent's betting behaviour will still be available. If, indeed, their criterion for consistency among bets—the 'no Dutch book argument'—still makes sense among branching worlds, this all in itself is a solution to the incoherence objection (that the concept of probability simply *makes no sense* in the Everett interpretation): for still, an agent must make reasoned choices among quantum games.

Very well: probability in the sense of credence may still be implicit in our behaviour even in a branching universe. But we have been given no reason as yet as to why it should track some objective counterpart. It seems positively odd that it should track the mod-squared branch amplitudes, anyway invisible, as required by the Everettian version of the principal principle.

But no one-world theory of objective probability does very well when it comes to the principal principle. Of course it can be made to *sound* rather trite—why shouldn't our degree of subjective uncertainty be set equal to the degree of objective uncertainty?—but that is little more than a play on words. The answer must ultimately depend on what, concretely, 'degrees of objective uncertainty' (chances) really are. For physicalists, they had better be fixed by the physical facts—perhaps by the entire sequence of physical facts of a chance process, or even of all chance processes. But then what sort of physical facts or quantities, exactly? How can any normal physical quantity, a 'Humean magnitude'[13] (like mass or relative distance or field intensity), have such a special place in our rational lives? It is hard for that matter to see how a *problematic* quantity like 'potentiality' or 'propensity' can play this role either. But if objective probabilities float free of the physical facts altogether, it is even harder to see why they should matter.

This dilemma was stated by Lewis in a famous passage:

The distinctive thing about chances is their place in the 'Principal Principle', which compellingly demands that we conform our credences about outcomes to our credences about their chances. Roughly, he who is certain the coin is fair must give equal credence to heads and tails . . . I can see, dimly, how it might be rational to conform my credences about outcomes to my credences about history, symmetries, and frequencies. I haven't the faintest notion how it might be rational to conform my credences about outcomes to my credences about some mysterious unHumean magnitude. Don't try to take the mystery away by saying that this unHumean magnitude is none other than chance! I say that I haven't the faintest notion how an unHumean magnitude can possibly do what it must do to deserve that name—namely, fit into the principle about rationality of credences—so don't just stipulate that it bears that name. Don't say: here's chance, now is it Humean or not? Ask: is there any way that any Humean magnitude could fill the chance-role? Is there any way that an unHumean magnitude could? What I fear is that the answer is 'no' both times! Yet how can I reject the very idea of chance, when I

---

[13] After the philosopher D. Hume, who insisted that nothing was available to inspection other than 'matters of fact'. The problem that follows is strikingly similar to Hume's 'problem of induction' (the problem of identifying 'causes' rather than 'chances' in terms of matters of fact).

know full well that each tritium atom has a certain chance of decaying at any moment? (Lewis [1986a pp.xv–xvi]).

Why hasn't Lewis already given the answer—that chances are made out in terms of history, symmetries, and frequencies? According to 'naive frequentism', probabilities *just are* actual relative frequencies of outcomes. But we know the deficiencies of this. Unfortunately, it seems that no *sophisticated* form of frequentism is workable either. Lewis put the problem like this: not even all facts about the actual world, future as well as past, could pin down facts about probability. If a history-to-chance conditional (the chance is thus-and-so given such-and-such a sequence of events) is made true by some pattern of events, past and future, there must be a chance that that pattern happens; there must be a chance it *doesn't* happen. But the pattern that may result instead may yield a completely different history to chance conditional for the original pattern. Chance, if supervenient on the actual history of a single world, is 'self-undermining' (Lewis [1986a]).[14]

Now for the punchline: *none* of this is a problem in Everettian quantum mechanics. The self-undermining problem is fairly easily solved. And much less obviously, the principal principle can be *explained*. For replace Lewis's Humean tapestry of events by an Everettian tapestry of events, connected not only by relations in space and time but also by the new fundamental relations introduced by quantum mechanics (the transition amplitudes), and then (as argued above) it has the structure of a collection of branching, approximately classical histories. Lewis's question, of why, given this, we should think that the branch amplitudes should dictate our rational credences, is answered thus: an agent who arranges his preferences among various branching scenarios—quantum games—in accordance with certain principles of rationality, *must* act as if maximizing his expected utilities, as computed from the Born rule.

This argument was first made by D. Deutsch in his paper 'Quantum theory of Probability and Decisions' [1999]. It was, in essence, a form of Dutch-book argument, strengthened by appeal to certain symmetries of quantum mechanics. But it hinged on a tacit but relatively powerful assumption subsequently identified by Wallace as 'measurement neutrality' (Wallace [2002]). It is the assumption that an agent should be indifferent as to which of several measurement apparatuses is used to measure a system in a given state, so long as they are all instruments designed to measure the same dynamical variable (by the lights of conventional quantum mechanics).

In fact Deutsch [1999] made no explicit mention of the Everett interpretation. If, indeed, experimental procedures are appropriately operationalized, measurement neutrality is effectively built in (Saunders [2004]). But that can

---

[14] For these and other deficiencies of one-world theories of chance, see David Papineau in Chapter 7. Papineau also defends the claim that the notion of uncertainty plays no useful role in decision theory.

hardly be assumed in the context of the present debate, where it is disputed that the Everett interpretation recovers the measurement postulates, FAPP ('for all practical purposes'), or underwrites ordinary operational procedures.

In the face of this, Wallace [2003b, 2007] offered a rather different, two-part argument. The first was a formal derivation of the Born rule from (rather weak) axioms of pure rationality, given only, in place of measurement neutrality, 'equivalence'—the rule, roughly, that an agent should be indifferent between experiments that yield exactly the same amplitudes for exactly the same outcomes. The second part consisted of informal, pragmatic, but still normative arguments for this rule. This paper attracted wide comment, but primarily by way of allegedly rational—bizarre perhaps, but rational—counterexamples to the Born rule. Wallace's arguments for equivalence were not addressed (or not by Baker [2006], Lewis [2007], and Hemmo and Pitowsky [2007]).

The upshot: pragmatic and rational constraints force the compliance of an agent's expected utilities, as computed from his credences, with his expected utilities as computed from the Born rule. The result is as good a solution to Lewis's dilemma as could be desired—as good as those rational and pragmatic constraints are judged reasonable.[15] Obviously they are somewhat idealized; it is the same with the Dutch book arguments of de Finetti and Ramsey, where an agent's utilities are supposedly quantified in terms of (relatively small) financial rewards; but so long as nothing question-begging or underhand is going on, Wallace's result is already a milestone. Nothing similar has been achieved for any one-world physical theory.

Why is that exactly? Is it for want of perseverance or ingenuity? Perhaps; the same general strategy is available to any other physical probability theory that implies meaningful pragmatic constraints, independent of its interpretation in terms of probability (examples that come to mind include classical statistical mechanics and pilot-wave theory). But Everettian quantum mechanics is special in another respect. As Wallace points out, at the heart of his arguments for the equivalence rule (and Deutsch's original argument) is a certain symmetry—the case of the equi-amplitude outcomes—that cannot possibly be respected in any one-world theory. A tossed coin in any one-world theory must land one way or the other. However perfect the symmetry of the coin, this symmetry cannot be respected (not even approximately) by the dynamics governing its motion on any occasion on which it is actually thrown. But it can in Everettian quantum mechanics.[16]

This link with rationality is not all of the meaning of physical probability, however. It is not even the only link needed with credence. The two come

---

[15] In Chapter 8 by David Wallace these constraints are written down as axioms, and the entire argument is formalized.

[16] This point, in the related context of quantum mechanical symmetry-breaking, was earlier recognized by Zeh [1975]. (To ward off any possible confusion: this is *not* Wallace's equivalence rule, but only a very special case of it.)

together also in statistical inference—in inference from observed statistics to objective probabilities, in accordance, say, with Bayesianism. This applies, above all, to confirming or disconfirming quantum theory itself; to confirming or disconfirming statements about probabilities made by the theory on the basis of observed statistics.

The difficulty in the case of the Everett interpretation is that failing an antecedent understanding of branch amplitudes in terms of probability, the predictions of the theory don't speak of probabilities at all: they speak only of branch amplitudes. The theory predicts not that the more probable statistics are such-and-such, but that the statistics in higher amplitude branches are such-and-such. Suppose such-and-such statistics *are* observed; why is that reason to believe the theory is confirmed? Or if not, that it is disconfirmed?

The principal principle normally does this job, converting probabilities, as given by a physical theory, into degrees of belief. But we lack at this point any comparable principle for converting branch amplitudes into degrees of belief. The quantum decision theory argument is in this context unavailable—it forces the principal principle only for an agent who *already accepts* that his pragmatic situation is as dictated by the branching structure of the wavefunction. If you believe that that is true, then you are already halfway to believing that Everettian quantum mechanics is true. And if you don't, then the gap is as wide as ever.

Call this the *evidential problem*. It was a relative newcomer to the debate over the Everett interpretation (Wallace [2002], Myrvold [2005]). However, a general strategy for solving the problem was rather quickly proposed by H. Greaves. Define a more general (Bayesian) confirmation theory in which the principal principle governs, not credence, which necessarily involves the notion of uncertainty, but 'quasicredence'—which, say, quantifies one's concerns (a 'caring measure'), rather than uncertainties—subject to two constraints: conditional on the proposition that E occurs with chance p, it is to be set equal to p; and conditional on the proposition that E occurs on branches with weight p, it is to be set equal to p.

If quasicredences are updated by Bayesian conditionalization, and if degrees of belief in theories (whether branching or non-branching theories) are marginals of this quasicredence function, then they behave just as one would desire in the context of rival theories. That is, the resulting confirmation theory can adjudicate between a chance theory and a weighted-branching theory, and between rival weighted-branching theories (if there are such), and rival chance theories, without prejudice to any (Greaves [2007]). Most important of all, it passes the obvious test: it does not confirm a branching theory come what may, whatever the branch weights.[17]

---

[17] In Chapter 9, Hilary Greaves, in collaboration with Wayne Myrvold, essentially derives this confirmation theory as a Savage-style representation theorem.

Evidently, it also bypasses the question of whether or not there is any uncertainty in the Everett picture—of whether or not Everettian quantum mechanics is a theory of probability at all—a strategy she had introduced earlier (Greaves [2004]). It has been dubbed the 'fission programme'. The evidential problem, in other words, can be solved regardless of one's views on questions of uncertainty in the context of branching. Both Deutsch and Wallace similarly avoided any appeal to the notion of uncertainty.

But now suppose, for the sake of argument, that all these arguments *do* go through. If so many of the chance-roles can be shown to be played by branching and branch amplitudes, can they not be identified with chance (Saunders [2005], Wallace [2006])? Is it any different in the identification of, say, thermal quantities, with certain kinds of particle motions? Or temporal quantities, with certain functions on a spacetime manifold? There remains the question of what sense, if any, attaches to the notion of uncertainty—given, for the sake of argument, complete knowledge of the wavefunction—but contra Greaves and others, a case can be made on the side of uncertainty too. For take a world as a complete history, and ourselves at some time as belonging to a definite world. There are vast numbers of worlds, all exactly alike up to that time. We do not know what the future will bring, because we do not know which of these worlds is our own (Wallace [2006], Saunders and Wallace [2008]).[18]

The issue is in part—perhaps in large part—a matter of how we *talk* about future contingencies. There is already a comparable difficulty in talk of the past and future, and of change, in the 'block universe' picture of four-dimensional spacetime—and plenty of scope there to interpret our ordinary talk in nonsensical terms. Lots of philosophers (and some physicists) have. For most of us, however, in that context, it is more reasonable to make sense of ordinary talk of change in terms of the relations among events ('before', 'after', and 'simultaneous'—or 'spacelike'), treating words like 'now' as we do spatial demonstratives like 'here' (Saunders [1995a, 1996b, 1998]). We make sense of ordinary talk of time and change in terms of the physics, not nonsense.

Or take the example of sensory perception: what do we perceive by the senses, if physics is to be believed? Nothing but our own ideas, according to most philosophers in the 17th and 18th centuries. We directly see only sense data, or retinal stimuli; everything else is inferred. Hence the 'problem of the external world'. Well, it may be a problem of philosophy, but as a proposal for linguistic reform it is a non-starter. That would be another example of bad interpretative practices.

A better practice, by a wide margin, is 'the principle of charity' (Wallace [2005]): interpretation (or translation) that preserves truth (or, this a

---

[18] I argue the case for this account of branching in Chapter 6.

variant, that preserves knowledge). But some may conclude from all this that if all that is at issue is our ordinary use of words, rather less hangs on the question of uncertainty than might have been thought—and on whether branching and branch amplitudes 'really is' probability. The success of the fission programme points to that.

But whether the fission programme can be judged a success, whether, indeed, any of the arguments just summarized really succeed in their aim, is what this book aims to discover.

2 THE ARGUMENTS OF THE BOOK

The book is structured in six parts. Parts 1 and 2 are on ontology in the Everett interpretation, giving constructive and critical arguments respectively. Part 3 is on probability; Part 4, critical of many worlds, is largely focused on this. Part 5 is on alternatives to many worlds, consistent with realism and the unitary formalism of quantum mechanics. Part 6 collects chapters that are friendly to many worlds but essentially concern something other than its defence—the origins of the theory, its reception, its open questions.

Ontology, probability, alternatives, and open questions; we take them each by turn.

2.1 Ontology

A general objection on the grounds of ontology is that decoherence theory does not do what is claimed because decoherence is only approximate and context-dependent. In some regimes it is too slow to give classicality, or it is absent altogether.

These and related arguments are addressed by David Wallace in Chapter 1. There he presents an outlook on realism in general and on 'emergence' in particular. The extraction of quasiclassical equations—a whole class of such, one for each history—is an example of FAPP reasoning as it operates across the board in the special sciences, according to Wallace. The framework is broadly structuralist and functionalist, in roughly the sense of D. Dennett's writings. It may be true that one has to know what one is looking for in advance, by means of which effective, phenomenological equations are obtained, but it is the same for extracting equations for protons, nuclei, and atoms from the field equations of the Standard Model. Likewise for quasiparticles in condensed matter physics, or (a big jump this) living organisms in molecular biology—and from thence to anatomy, evolutionary biology, and the rest. The fact that in certain regimes decoherence is absent altogether—that classicality, branching, and worlds, are absent altogether—is, says Wallace, scarcely a difficulty. It is not as though we need to recover a theory of biology for all possible regimes of molecular physics, in order to have such a theory for some.

If Wallace's reading of the extraction of classicality from the quantum is correct, it had better apply to decoherence theory as it is actually applied. In Chapter 2 Jim Hartle gives an overview of the field of decohering histories, while in Chapter 3 Jonathan Halliwell gives a detailed model in terms of hydrodynamic variables, one of the most realistic models to date. Readers are invited to judge for themselves.

A second objection is at first sight more philosophical, but it can be read as a continuation of the first. How does talk of macroscopic objects so much as get off the ground? What is the 'deep-down ontology in the Everett interpretation? It can't just be wavefunction, argues Tim Maudlin in Chapter 4; it is simply unintelligible to hold that a function on a high-dimensional space represents something physically real, unless and until we are told what it is a function *of* —of what inhabits that space, what the elements of the function's domain are. If they are particle configurations, then there had better *be* particle configurations, in which case not only the wavefunction is real.

Here one can hardly take instruction from the special sciences, where instrumentalism (or at least agnosticism) about ontology at deeper levels is a commonplace. In any case it would be question-begging, by Maudlin's lights, because, failing an account of what exists at the fundamental ontology, we do not have emergent structures either. But on that point Wallace and Maudlin differ profoundly.

But don't the chapters by Hartle and Halliwell prove otherwise? No—not according to Maudlin. They help themselves to resources they are not entitled to in the context of realism. Physicists indifferent to the question of realism in quantum mechanics may well speak of a function over particle configurations;[19] others may speak in the same way—doing the same calculations, even—but with a hidden-variable theory in mind. But when the topic is realism in quantum mechanics, commitments like this have to be made explicitly. Compare the situation in pilot-wave theory and collapse theories, where in recent years the question of fundamental ontology has received a great deal of attention. So, if it is denied that particles or fields exist and that only the wavefunction is real, then the wavefunction is not a function of particle or field configurations. So of what is it a function?

One can try to treat this challenge as only a verbal dispute—very well, let's speak of 'quantum-state realism' or 'structure-of-the-state realism' instead. But the objection is at bottom a request for *clarification*, for an *intelligible* account of the microworld. So what does it consist in, exactly? Or even in outline? (See also Section 3 p.44.)

---

[19] Hartle and Halliwell both steer clear of questions of realism in quantum foundations.

Agreed, this question of fundamental ontology is important. It is a shame that it has been paid so little attention. That is the main complaint made by David Deutsch in Chapter 18—Maudlin finds an unlikely bed-fellow. The difference between them, to put it in Bayesian terms, is in their priors: Maudlin unlike Deutsch is sceptical that any solution is possible. Maudlin sees pilot-wave and collapse theories as examples of how ontological questions should be settled, but he doubts that anything like the methods used there can apply to many worlds. There again, much of that debate has been driven by the challenge that their ontologies too contain many worlds, and devising ways by which they can be eliminated.

Is there some metatheoretic perspective available? Are there general philosophical guidelines for conducting debates like these? John Hawthorne in Chapter 5 tries, with certain caveats, to say what they might be. His is a metaphysical image to counter the naturalistic one given in Wallace's chapter. He reminds us of the long-standing concern in philosophy over how the gap between the 'manifest' image and fundamental ontology—or the 'fundamental book of the world', as Hawthorne puts it—can be bridged. He proposes a demarcation between 'conservative' and 'liberal' strategies, where the former is straightforwardly an identification of macrodescriptions with descriptions at the fundamental level. The latter in contrast involves 'metaphysical generational principles'; these he (rightly) thinks are rejected by Everettians. But if only identifications of the former kind are available, their task, thinks Hawthorne, is much harder. Typical of identificatory projects in science—by means of 'bridge principles', for example, as was popular in logical empiricist philosophy of science—are 'uncloseable explanatory gaps', bridge principles that are claimed to be true 'but you can't see for the life of you, no matter how much you look, why they are true while certain competing principles are false' (p.149). That is particularly familiar in philosophy of mind where the explanatory gap between descriptions in terms of consciousness and physicalistic descriptions is widely acknowledged. According to Hawthorne, the Everett interpretation threatens to bring with it too many new, uncloseable explanatory gaps.

But if the example is the mind–body problem, isn't functionalism precisely an answer to that? Perhaps—but there at least some input–output facts about stimulus and behaviour are uncontroversially in place. Not so in the Everett interpretation, argues Hawthorne, where 'it is hard to know what the take-home message of the functionalist is in a setting where none of the fundamental-to-macro associations are given' (p.150).

The second lesson that emerges from Hawthorne's analysis is the importance of what he calls 'metasemantical' principles—broadly speaking, theories of how semantical rules ought to operate for connecting predicates to ontology (fundamental or otherwise). One can pay lip service to the macro-image that still fails to square with one's favoured metasemantical principles—some kind of fudge is needed. Very well, so take a theory in which, say, 'all there is to the world is configuration space'. Then the best package, all things considered, is one that has ordinary macropredicates (like Wallace's example, 'tiger') pick out features of configuration space. Everettians, says Hawthorne, are then tempted to argue as follows:

But this shows that certain features of configuration space are *good enough* to count as tigers. And then the line of reasoning proceeds as follows: 'Even if there were extra stuff—throw Bohmian particles or whatever into the mix—we have agreed that the relevant features of configuration space are good enough to count as tigers. So whether or not that extra stuff is floating around, you should still count those features of configuration space as tigers.' (pp.151–2)

But introduce 'extra stuff', and it may be its credentials to count as things like tigers simply *swamp* those of the configuration-space features that you were stuck with before.

It may be, but does it? According to James Ladyman, in his reply to Hawthorne, the credentials of the 'empty waves' of pilot-wave theory seem no better or worse than the occupied ones (we shall have to revisit this argument when we come to Part 5 of the book). Ladyman wonders too if they are mostly about philosophical intuitions that we have no good reason to trust—which are themselves the object of empirical investigation in cognitive science. But more importantly, and the point to which he devotes most attention, he thinks Hawthorne's alternative methodologies (the conservative and liberal strategies) are not exhaustive. For example, identifications in the physical sciences are generally dynamical—'it is the dynamics of how hydrogen bonds form, disband, and reform that gives rise to the wateriness of water and not the mere aggregation of hydrogen and oxygen in the ratio of two to one' (p.158). It is one of many devices used by Halliwell and Hartle that go beyond Hawthorne's two-part distinction, according to Ladyman (he lists a number of them). When it comes to the explanatory gap between the quantum world and the macroworld, contrary to Hawthorne's claim that none of the fundamental-to-macro associations are given, Ladyman concludes, 'it must be acknowledged that Halliwell and Hartle do much to close it' (p.159).

Here is an entirely different line of attack. Might some of the devices used by Halliwell and Hartle be question-begging, in view of later discussions of probability? According to Adrian Kent in Chapter 10 and Wojciech Zurek in Chapter 13, any appeal to decoherence theory must already presuppose the idea of probability. Decoherence theory employs reduced density matrices and the trace 'and so their predictions are based on averaging' (Zurek, p.414). In the estimation of Kent, it shows that certain operators 'approximately quantifying local mass densities approximately follow classical equations of motion with probability close to one ... in other words, the ontology is *defined* by applying the Born rule' (p.338). The criticism is potentially damaging to those, like Saunders and Wallace, who seek to identify probability (or at any rate identify the quantities that a rational agent should treat as if they were probabilities) with some aspect of the ontology. Allow that the branching structure of the universal state involves objective probabilities in its definition and their arguments all but evaporate.

Zurek and Kent surely have a point: those working in decoherent histories talk freely of probabilities in their interpretation of branch structures. Witness Jonathan Halliwell in Chapter 3 in his derivation of quasiclassical hydrodynamic equations from the unitary formalism:

> The final picture we have is as follows. We can imagine an initial state for the system which contains superpositions of macroscopically very distinct states. Decoherence of histories indicates that these states may be treated separately and we thus obtain a set of trajectories which may be regarded as exclusive alternatives each occurring with some probability. Those probabilities are peaked about the average values of the local densities. We have argued that each local density eigenstate may then tend to local equilibrium, and a set of hydrodynamic equations for the average values of the local densities then follows. We thus obtain a statistical ensemble of trajectories, each of which obeys hydrodynamic equations. These equations could be very different from one trajectory to the next, having, for example, significantly different values of temperature. In the most general case they could even be in different phases, for example one a gas, one a liquid. (p.111)

But here Halliwell, like Hartle, assumes that the mod-squared amplitudes of histories can be interpreted as probabilities: he is neutral on whether all of these histories exist. Everettians at this point must speak in terms of amplitudes instead. The key question for them is whether the notion of the 'average values' of the local densities, on which the amplitudes are peaked, presupposes the notion of probability, or whether they are called the average values of the local densities because they are the values on which the amplitudes are peaked. The latter will follow if and when it is shown that the amplitudes can be interpreted in terms of probabilities—this, they say, is a task that can come *after* the delineation of the branching structure.

2.2 Probability

But is it Probability?

Chapter 6 by Simon Saunders makes the case for identifying branching and squared norms of branch amplitudes with chance processes and objective probabilities. To that end he identifies three roles played by chance. They can at best be measured by statistics, and only then with high chance; they guide rational action in the same way that objective probabilities are supposed to guide rational action, as spelled out by the principal principle; and chance processes involve uncertainty. His argument in a nutshell: all three of these roles are played by branching and branch amplitudes in Everettian quantum mechanics. Since they are more or less definitional of chance—no explanation of them is given in any conventional physical theory of probability—anything that plays all these roles should be identified with chance.

Of these the link with statistics is a straightforward dynamical question. Amplitudes cannot be measured directly in the Everett interpretation because the equation of motion is unitary. They show up at best in the statistics of repeated trials, but only on branches of comparatively high amplitude. This, says Saunders, can be uncontroversially explained.

For the argument for the link with rational action, we are referred to Wallace's chapter. The rest of Chapter 6 is on the link with uncertainty. It argues, in brief, that branching implies a form of 'self-locating uncertainty'—uncertainty as to which branch is our own. He reminds us that here there is a difficulty well known to philosophers. Suppose a large number of distinct histories are real, but that they share common parts, rather in the way that roads can overlap or, well, the way branches of a tree can overlap. Metaphysicians have considered worlds like these; they call them 'branching'. But then:

> The trouble with branching exactly is that it conflicts with our ordinary presuppositions that we have a single future. If two futures are equally mine, one with a sea fight tomorrow and one without, it is nonsense to wonder which way it will be—it will be both ways—and yet I do wonder. The theory of branching suits those who think this wondering is nonsense. Or those who think the wondering makes sense only if reconstrued: you have leave to wonder about the sea fight, provided that really you wonder not about what tomorrow will bring but about what today predetermines. But a modal realist who thinks in the ordinary way that it makes sense to wonder what the future will bring, and who distinguishes this from wondering what is already predetermined, will reject branching in favour of divergence. In divergence also there are many futures; that is, there are many later segments of worlds that begin by duplicating initial segments of our world. But in divergence, only one of these futures is truly ours. The rest belong not to us but to our otherworldly counterparts.   (Lewis [1986b p.208])

The initial segments of diverging worlds are only qualitatively, not numerically, identical.

Why not just choose divergence, then? Because things are not so simple for physicalists. Their metaphysics, if they have any, is constrained by physical theory. Indeed, Everett introduced the term 'branching' by reference to the development of a superposition of records of histories, ultimately in terms of vector-space structure, not by the philosophers' criterion of overlap. According to Saunders, this concept of branches finds a natural mathematical expression in the language of the consistent histories formalism with its attendant Heisenberg-picture vectors—an inherently tenseless four-dimensional perspective. Do worlds thus represented overlap, in the philosophers' sense, or do they diverge? The answer, he says, is *underdetermined* by the physics;[20] either metaphysical picture will do. But if either can be used, better use the one that makes sense of ordinary linguistic usage, rather than *nonsense*.

But this argument for the identification of branching and branch amplitudes with objective probability is of no use in explaining the Born rule; on the contrary it depends on it. And whilst, according to Saunders, it gives an indirect solution to the evidential problem, the remaining chapters of Part 3 favour rather the fission programme due to Hilary Greaves, in which—if only as a tactical move—talk of uncertainty is eschewed. David Papineau in Chapter 7 goes further: he questions the very desirability of an account of quantum probability in terms of uncertainty. According to him rational choice theory is better off without it. The strategy of maximizing one's expected utilities, he argues, faces a difficulty if what one really wants is the best utility—but this problem disappears in the fission picture.

Adrian Kent in Chapter 10 challenges the argument for uncertainty directly. To bring in linguistic considerations, Kent insists, is simply a *mistake*: nothing of significance to fundamental physics could turn on such questions. And the bottom line, the real reason there can be no uncertainty in the face of branching, is that there is nothing in the physics corresponding to it. Take the case of Alice, about to perform a Stern–Gerlach experiment. If she were to be unsure of what to expect, there would have to be 'a probabilistic evolution law taking brain state $|O\rangle_A$ to one of the states $|i\rangle_A$' (p.346). There is no such law; indeed, 'nothing in the mathematics corresponds to "Alice, who will see spin-up" or "Alice, who will see spin-down"' (p.347). Kent disagrees with the arguments of Part 1 that there are such laws, albeit only effective laws. He disagrees with Saunders that branch vectors are just the needed mathematical quantities.

The Born-rule Theorem

In Chapter 8 Wallace provides a formal derivation of the Born rule, making it properly speaking a theorem. Mathematically inclined readers are invited to check its validity for themselves.

But are his axioms reasonable? They are in part pragmatic constraints—constraints on the range and kind of acts that are available to an agent if the branching structure of the wavefunction is what Everettian quantum mechanics says it is. Another ('state supervenience') is an expression of physicalism: it says that an agent's preferences between acts should depend only on what state they leave his branch in. Others again are more overly rationalistic—rules that are applicable more or less whatever the physical theory.

---

[20] A point remarked on, but not taken properly to heart, in Saunders [1998 pp.399–401]. The presumption, that Everettian branching is branching in the philosophers' sense as well, is widely shared by philosophers of physics.

Of these there are only two. The first is that an agent's preferences must yield a total ordering on his available actions. The reason for this is not so much that the claim appears plausible (although Wallace thinks it is), rather it is that 'it isn't even possible, in general for an agent to formulate and act upon a coherent set of preferences violating ordering' (p.236).

The other is 'diachronic consistency': suppose an agent on performing act U has successors all of whom prefer act V to act V'; then that agent had better prefer U followed by V to U followed by V'. The rationale is roughly the same. *Local* violations of this rule may be possible, Wallace admits; thus I disingenuously tell my friend not to let me order another glass of wine after my second; but '[i]n the presence of widespread, generic violation of diachronic consistency, agency in the Everett universe is not possible at all' (p.237). Diachronic consistency is constitutive of agency, in Wallace's view.

As for the point of the axiomatization, it is that rather than pursue largely sterile arguments over the intuitive plausibility (or lack of it) to various counter-examples to the Born rule, attention can be shifted to the general principles that putatively underlie our actual epistemic practices. To that end, for each alleged counterexample to the Born rule, Wallace identifies the relevant axiom or axioms that it most obviously slights.

Chapters in Part 4 are uniformly in disagreement with Wallace's conclusions. According to Huw Price in Chapter 12, the key problem is that in moving from one world to many there is 'something new for agents to have preferences *about*' (p.370). He gives an example from political philosophy. Use of the Born rule, in that context, would amount to a form of utilitarianism (maximizing expected utility according to a certain credence function), but to that there are well-known alternatives. Why not impose some form of distributive justice instead, in which the lot of the worse off is disproportionately weighted? This is a developed and much-debated theory in political philosophy; it is simply not credible to contend that it is *irrational*. It may be that the amplitudes will have to enter into any quantitative rule, there being no a priori count of successors, but no matter: the rule thus amended will still reflect distributive rather than utilitarian goals, and hence differ from the Born rule.

It is a good question whether Price thinks this argument is independent of the notion of uncertainty. He grants that ('subjective') uncertainty may make sense in the context of branching on a certain metaphysics of personal identity, at least as first-person expectations go (citing Wallace [2006]), but he denies that it can account for uncertainty more generally. For example, he doubts whether it makes sense for events occurring long after an agent can possibly hope to survive. And, in short, he insists that metaphysical questions of personal identity be kept separate from decision theory.

But Everettians on that point can guardedly agree. Where then is the source of disagreement? His counterexample violates one or other of Wallace's axioms, obviously. That doesn't bother Price: he concludes that that is only to show that they tacitly smuggle in presuppositions appropriate to a one-world theory. As it happens, Wallace identifies the relevant axiom as more obviously a pragmatic constraint (a continuity axiom), but the real disagreement between them is closer to the surface. Price insists that, in decision theory, considerations of rationality pertain to a *single* moment in time (the time at which a decision is made). For this reason, decision theory has nothing to do with questions of personal identity even of the most deflationary kind. Price will therefore reject Wallace's axiom of diachronic consistency directly. On personal identity, he says, '[t]hese issues are essentially irrelevant to classical subjective decision theory, for whom the only "I" who matters is the "I" at the time of decision' (p.377).

Adrian Kent in Chapter 10 seeks to undermine the Born-rule theorem at several levels. There is a problem with the very idea of 'fuzzy' ontology or theory. Kent wonders how, if the branching structure is fuzzy, the mathematical precision required of the Born-rule theorem can be sustained. Mathematical precision, moreover, is not just desirable: according to Kent, one has an 'obligation to strive to express one's ideas in mathematics as far as possible' (p.346). That is the mistake of arguments from the philosophy of language: they still bring assumptions, it is just that since expressed only in words they are the more vague. Kent speaks at this point specifically of a theory of mind. Here he rejects the broadly functionalist stance of Everettians on questions of mentality. They in turn will readily welcome mathematical models of neural processes, or for that matter linguistic behaviour, but see no special role for either in quantum foundations.

Like Price, Kent offers a number of counterexamples to the Born rule. One is the 'future self elitist', who cares only about the best of his successors ('the rest are all losers'). Another is the 'rivalrous future self elitist', who cares in contrast only about the one that is the best relative to the others—someone like this will see an advantage in impoverishing all of his successors save one. And he points out the variety of (conflicting) ways in which notions like these can be quantified. They may not be particularly edifying forms of caring, true, but they are surely not *irrational*—or not when directed at a community of other people, none of them oneself.

Kent addresses Wallace's rationality axioms explicitly. In the case of inter-temporal consistency, he concludes that whilst on some occasions an agent may reasonably be required to be consistent over time, on other occasions he may not. When an agent's utilities change over time, inter-temporal consistency, Kent thinks, is impossible. His conclusion:

The best it seems to me that one might hope to say of diachronic consistency in real-world decisions is that pretty often, in the short term, it approximately holds. Clearly, that isn't a strong enough assumption to prove an interesting decision theoretic representation theorem.   (p.342).

David Albert's criticisms in Chapter 11 chime with many of Kent's. He adds the concern that an analysis of probability in the physical sciences in terms of the betting strategy of a rational agent is to simply *change the topic*—it isn't what a theory of physical chance is about. What we should be doing is *explaining the observed statistics*—in effect, our task is to solve the evidential problem. No inquiry into the nature of the pragmatic constraints on rational actors that might follow from Everettian quantum theory can ever be relevant to *that* question. The fact that an agent is required to *believe* the theory is true for Wallace's Born-rule theorem to even get going shows it is irrelevant.

But even on its own rather limited terms, Albert continues, Wallace's arguments are unsatisfactory. Counterexamples to the equivalence rule, and therefore to the Born rule, can easily be constructed. The one Albert favours is the one generalized by Kent as the 'rivalrous future self elitist': the successors that matter are those that the agent considers better in comparison to the others, specifically by being *fatter* than the others (this gives them extra gravitas). No matter if the rule is absurd (it was intended to be funny), or difficult to carry through in practice, it is not *irrational*. Albert further insists that pragmatic constraints should have nothing to do with questions of what it is right to do. In fact, Wallace's response is that Albert's 'fatness rule' violates inter-temporal consistency rather than any of the more obviously pragmatic constraints—but, of course, the latter are needed in the deduction as well. More fundamentally: for Wallace the distinction between rationality rules and pragmatic rules is anyway only a matter of degree.

The Evidential Problem

How then is Everettian quantum mechanics to be confirmed or disconfirmed by statistical evidence? The theory only says that statistics conforming to the Born rule obtain on branches of comparatively high amplitudes, whereas anomalous statistics obtain on branches of comparatively low amplitude. How is that to be empirically checked?

Recall the answer given earlier by Greaves [2007]: a general theory of statistical inference can be defined, that applies equally to branching and non-branching theories (without prejudice to either). Very well: such a confirmation theory can be defined, but why should sceptics embrace it? In Chapter 9, in collaboration with Wayne Myrvold, she argues that they must. Greaves and Myrvold show that the process of Bayesian conditionalization (updating of credences) can itself be operationalized in terms of betting preferences, where the latter are constrained by Savage's axioms. The process of statistical inference from the outcomes of an experiment, treated as 'exchangeable' in de Finetti's sense, follows in train.

This takes some unpacking. The operational definition of an agent's conditional credences $C(E|F)$ is well known from Ramsey's and de Finetti's writings:

it is the betting quotient that an agent is prepared to accept for event E, on the understanding that the bet is called off if F does not happen. It is easy to show that unless this credence satisfies the probability axiom:

$$C(E|F) = \frac{C(E \,\&\, F)}{C(F)}.$$

a Dutch book can be constructed by which an agent is bound to lose, whatever happens. Note that the credence functions on the RHS are defined prior to learning that F.

As it stands, this says nothing about how an agent's credence function should be updated in the light of new evidence. But let this be on the model of a 'pure learning experience', in Greaves and Myrvold's terminology: then, they show, $C(.|F)$ should indeed be her updated credence function (Bayesian conditionalization). For suppose:

P7. During pure learning experiences, the agent adopts the strategy of updating preferences between wagers that, on her current preferences, she ranks highest.

Then in pure learning experiences an agent's preferences among wagers, in conformity with Savage's axioms and P7, automatically induce an ordering of preferences on updating strategies. Bayesian conditionalization comes out as optimal.

Meanwhile, Greaves and Myrvold remind us, de Finetti's original representation theorem already shows how an agent who treats the order of a sequence of outcomes on repeated trials of an experimental set-up as irrelevant (as 'exchangeable', in de Finetti's terminology), and who updates her credences by conditionalization in accordance with Bayes' theorem, is *inter alia* committed to treating the outcomes of the experiment as if they were associated with definite, if unknown, probabilities.

Putting the two together, the result is an operational characterization of the entire process of Bayesian statistical inference. It is in fact a representation theorem just as much as is the Born-rule theorem—like it or not, agents who subscribe to the axioms $P1 - P7$, and who believe certain experiments are exchangeable, *have* to act as if they were updating their quasicredence functions, in the manner proposed by Greaves [2007], and accordingly updating their credences in theories. Add the requirement that one's priors not be fixed dogmatically (they can be as small as you like, but not zero), their axiom $P8$, and the resulting confirmation theory passes a variety of non-triviality tests as well. Most importantly: it *doesn't* follow that because (in some sense) everything happens, according to Everett, the theory is confirmed come what may.

The authors' challenge is now as follows. Set up the entire system of axioms in accordance with the background assumption that one has a conventional theory of chance. Now entertain the possibility that the Everett interpretation is true. How much of the framework has to be changed? The answer, according to

Greaves and Myrvold, is 'none of it' (p.284). None of their axioms make explicit mention of uncertainty, chance, or probability (and nor, so they claim, do they do so implicitly).

To all of this a variety of the objections to Wallace's methods apply. Some of them, for example Price's counterexample in terms of distributive justice, are addressed explicitly by Greaves and Wallace (see their 'answers to objections'). But the main objection, according to Albert, is that the very focus on wagers and games is misguided. Preferences of rational agents in their gambling strategies, however regimented (as in Savage's axioms), can have nothing to do with the task of *explaining* the statistics actually observed. At most they tell us how much we should bet that we will find evidence E, if we believe a scientific hypothesis H is true, not with what the probability of E would be if H were true (what we ordinarily take as an explanation, if the probability is sufficiently high, of evidence E). And betting, in the fission picture, at least once the structure of branching and amplitudes are all known, is a matter of *caring about* what goes on in some worlds, not *beliefs about* what happens in those worlds. In Albert's words:

But remember (and this is the absolutely crucial point) that deciding whether or not to bet on E, in the fission picture, has nothing whatsoever to do with guessing at whether or not E is going to occur. It is, for sure. And so is $-E$. And the business of deciding how to bet is just a matter of maximizing the payoffs on those particular branches that—for whatever reason—I happen to care most about. And if one is careful to keep all that at the centre of one's attention, and if one is careful not to be misled by the usual rhetoric of 'making a bet', then the epistemic strategy that Greaves and Myrvold recommend suddenly looks silly and sneaky and unmotivated and wrong.   (p.364).

The objection is not quite that information about self-location can have nothing to do with beliefs about whether a physical theory is true—or if it is, it is Objection 5, as considered and rejected by Greaves and Myrvold. It is that the process of confirmation in accordance with the axioms $P1 - P8$, in the case of branching worlds, is no longer *explanatory*. Indeed, the axioms themselves may no longer be reasonable. Could they be corrected? But there may be no reasonable rules *at all* by which one can statistically test for a theory of branching worlds, say Albert and Kent. One can always concoct rules by which agents in each branch will arrive at beliefs about weights of branches, on the basis of the statistics in that branch; but they would arrive at those beliefs even if a branching worlds theory were true in which there *were* no branch weights (Kent's 'weightless' case pp.325–6).

Mightn't a similar pathology arise in a one-world theory in which there is no law, deterministic or probabilistic, governing the outcomes of experiments? Again, the inhabitants of such a world will conclude, falsely, that another theory is true—one that does assign the observed outcomes weights (namely, for experimental set-ups treated as exchangeable, weights numerically equal to the observed relative frequencies). But, says Kent, there is an important difference.

In the case of many worlds, the inhabitants of each world are led to construct a spurious measure of importance that favours their own observations against the others'. '[T]his leads to an obvious absurdity. In the one-world case, observers treat what actually happened as important, and ignore what didn't happen: this doesn't lead to the same difficulty' (p.327).

A related disquiet, as made vivid by Kent's example of a 'decorative' weight multiverse (pp. 327–8), is that in the Greaves–Myrvold approach the notion of branch weight is treated as a *primitive*, with different assignment of weights counted as different theories. On one theory they may be given by the moduli squared of branch amplitudes, but on another—possibly, a theory with identical dynamics and universal state—the weights are an entirely different set of numbers altogether. So (as Albert puts it) there is either some additional physical fact about the world (giving up on the main goal of the Everett interpretation, which is to make do with the unitary theory), or else the branch weights are some non-physical facts that are supposedly confirmed or disconfirmed by the observation of relative frequencies.

We have seen this disquiet before. It is the same as Lewis's: surely branch weights cannot, any more than objective probabilities, float free of the physical facts. They should be dictated by them essentially. But on this point, say Greaves and Myrvold, their arguments are entirely neutral (pp.397–8). The objection, if pressed, anyway can be met by the Deutsch–Wallace theorem; and if it isn't pressed, then it is hardly a difficulty of their confirmation theory that this freedom is permitted. *Something* is measured, they claim, in the way that probabilities are, by an agent who obeys their axioms: any theory that predicts the value of that quantity is thus subject to empirical test.

2.3 Not (Only) Many Worlds

The remaining parts of the book bring in wider considerations. Part 5 is on realist alternatives to many worlds consistent with the unitary formalism of quantum mechanics. They go against the claim that the Everett interpretation is forced by realism alone. Part 6 is about open questions—historical, methodological, and conceptual—inspired by many worlds.

Alternatives to Many Worlds

Wojciech Zurek in Chapter 13 sketches a picture of reality in which the quantum state has a qualified ontological status consistent with a one-world reading. It is only a sketch: he cites a sizable literature (by himself and his co-workers) for the details. From an Everettian point of view, a key difference lies in his notion of 'objective existence'. This notion only applies, according to Zurek, to 'classical' states—'einselected' states—states that can be investigated in a 'pragmatic and operational' way. 'Finding out a state without prior knowledge is a necessary condition for a state to objectively exist' (p.424). This is only possible for states that 'survive decoherence'—of which multiple copies can be extracted and distributed in the environment. Survival in this sense is 'quantum Darwinism'. Meanwhile decoherence theory is not a good starting point for understanding the origins of the classical, for (in line with his complaint already mentioned) it already involves probability. Zurek substitutes ideas from information theory instead. They, and the requirement that 'evolutions are unitary', are his core principles. From them he attempts to derive those aspects of the measurement postulates that do not involve collapse.

That seems to suggest that the Schrödinger equation has unrestricted validity. But is it true in Zurek's view that the universe as a whole can be assigned a wavefunction? He says on the one hand that to whatever extent there remains a measurement problem in his framework it is solved by Everett's relative state formalism: that explains 'apparent collapse'. He notes that 'even if "everything happens", a specific observer would remember a specific sequence of past events that happened to him'. But on the other hand:

> The concept of probability does not (need not!) concern alternatives that already exist (as in classical discussions of probability, or some 'Many Worlds' discussions). Rather, it concerns future potential events one of which will become a reality upon a measurement. (p.425)

In Chapter 14 Jeff Bub and Itamar Pitowsky offer a more overtly one-world, information-theoretic account of reality. In it Everett's ideas play no role. Quantum-state realism is rejected altogether, rather than being circumscribed as in Zurek's approach. So what does exist in their picture?

Measurements, to begin with. The key idea is not only to reject the view that the quantum state is something real; it is to reject the idea that measurement cannot figure as a primitive. They are both of them 'dogmas'. The dogma about measurement (what they call 'Bell's assertion', citing Bell [1990]) is:

> [M]easurement should never be introduced as a primitive process in a fundamental mechanical theory like classical or quantum mechanics, but should always be open to a complete analysis, in principle, of how the individual outcomes come about dynamically. (p.438)

Dispense with this and quantum-state realism and the measurement problem is exposed as a pseudo-problem.

To be more specific, the measurement problem breaks down into two parts, the 'big measurement problem', namely, 'the problem of explaining how individual measurement outcomes come about dynamically', and the 'small measurement problem', which is 'the problem of accounting for our familiar experience of a classical or Boolean macroworld, given the non-Boolean character of the underlying quantum event space' (p.438). The latter they are happy to phrase as 'the problem of explaining the dynamical emergence of an effectively classical probability space of macroscopic measurement outcomes in a quantum measurement process'. Decoherence theory is the answer to the small measurement problem; but the big measurement problem should be recognized for what it is, a pseudo-problem.

Why precisely does the big problem go away if measurements are primitive and the quantum state is a matter of degrees of belief and nothing else? Because 'probability' is a primitive too: 'probabilities (objective chances) are "uniquely given from the start" by the geometry of Hilbert space' (p.444). This, and inherent information-loss, an 'irreducible and uncontrollable disturbance', follow from a deeper principle, the 'no-broadcasting' principle.

Bub and Pitowsky ask us to rethink the ways in which realism works in the physical sciences. They make a detailed parallel with the special theory of relativity: no-broadcasting (and no-cloning) and no-signalling are analogues of Einstein's relativity and light-speed principles. Minkowski spacetime is the associated 'constructive' theory—its geometry explains Einstein's phenomeno-logical principles. Analogously, the geometry of Hilbert space explains Bub and Pitowsky's information-theoretic principles. Just as Minkowski spacetime suffices, they say, to explain length contraction and time dilation, independent of any dynamical principles, Hilbert space suffices to explain the structure of quantum mechanical probabilities, independent of any dynamical analysis. In either case (in special relativity or in quantum mechanics) a dynamical analysis *can* be provided—but as a consistency proof, not as an explanation. In special relativity this involves the explicit construction of a dynamical model (it doesn't matter which, so long as it respects the spacetime symmetries). In quantum mechanics it is the 'small' measurement problem, answered by providing a construction in decoherence theory (it doesn't matter which, so long as it models the 'same' experiment) of an effectively classical probability space of macroscopic outcomes. It is because the latter is provided that their theory, in their estimation, qualifies as realist.

But is that sufficient? Omitted, according to Chris Timpson in his commentary on Bub and Pitowsky, is provision of a dynamical account of how one among these macroscopic outcomes is realized—precisely a solution to the big measurement problem. According to Timpson, 'forgo this and they forgo their realism'. In every other one-world realist interpretation–or revision—of quantum mechanics, there is an account of how one rather than another individual outcome comes about dynamically. The argument from no-broadcasting or no-cloning may show that measurement involves an irreducible, uncontrollable information loss, but that doesn't make it *indescribable*; there is nothing in the parallel with special relativity to support that contention. Bub and Pitowsky are entitled if they wish to reject the view that the measurement process—specifically, a process by which individual outcomes are obtained—be dynamically analysed, says Timpson, but the charge that it is a dogma is unargued. The claim that it can be eliminated, compatible with realism, is unsubstantiated. On the contrary, he insists, it is rather directly implied by realism.

The general advantages of an anti-realist view of the quantum state are pressed by Rüdiger Schack in Chapter 15. His perspective, like that of Bub and Pitowsky, is that of quantum information theory. In the context of Bayesian updating of beliefs on repeated measurements, Everettians have to *assume* that the same quantum state is prepared on each trial. This, says Schack, is a problem (the 'problem of repeated trials') that simply disappears if the quantum state is purely epistemic. Assumptions about the apparatus are still required, true, but they are part of an agent's priors, to be updated in the light of evidence. 'This raises the question of whether the concept of an objective quantum state has any useful role to play at all' (p.473), a question he answers in the negative.

At least in the pilot-wave theory we have a clear-cut one-world form of realism. Or do we? In Chapter 16 Antony Valentini responds to the argument that realism about the pilot wave implies many worlds.

His argument is in effect to grant that whatever the situation in equilibrium pilot-wave theory, in which the probability distribution of the Bohmian trajectories is as given by the Born rule, the charge does not apply to the non-equilibrium theory. And (his argument continues) there is every reason, if pilot-wave theory is true, to expect non-equilibrium behaviour, just as in classical statistical mechanics—it would be a conspiracy theory if the full range of dynamical behaviour in principle permitted by the theory were to be forever and in principle concealed.

But then, given a reliable source of non-equilibrium matter, one can perform 'subquantum' measurements, measurements that can be used to probe occupied and empty waves and can tell the difference between them. They will not behave as on a par. Pilot-wave theory considered in this way must in principle differ from Everettian quantum theory. Thus Valentini concludes:

At best, it can only be argued that, if approximately classical experimenters are confined to the quantum equilibrium state, so that they are unable to perform subquantum measurements, then they will encounter a phenomenological appearance of many worlds—just as they will encounter a phenomenological appearance of locality, uncertainty, and of quantum physics generally.   (pp.500–1)

In the presence of non-equilibrium phenomena, such observers will quickly discover the explanatory and predictive failings of these appearances. Therefore there is no reason to reify them—they are 'merely mathematical'. The 'basic constituents' of ordinary matter are the Bohmian particles, not wavepackets, or parts of the wavefunction indexed by the particles.

The reality of the pilot wave as a whole, however, is not in doubt. As Bell said, in a remark quoted by Valentini approvingly, 'no one can understand this theory until he is willing to think of ψ as a real objective field . . . even though it propagates not in 3-space but in $3N$-space' (Bell [1987 p.128]). For Valentini, the bottom line is its contingency: ψ simply contains too much contingent structure to be thought of as an elliptical way of stating a physical law.

But aren't worlds—patterns in the wavefunction—contingent structures too? And don't supposedly intrinsic properties of Bohmian particles like charge or mass (both gravitational and inertial mass) act, in experimental contexts, as if associated with the pilot wave rather than the particles? So asks Harvey Brown in his reply to Valentini. Most tellingly in his eyes:

[T]he reality of these patterns is *not* like locality and uncertainty, which are ultimately statistical notions and are supposed to depend on whether equilibrium holds. The patterns, on the other hand, are features of the wavefunction and are either there or they are not, regardless of the equilibrium condition.   (p.514)

It seems that we are at a stand-off: patterns in the wavefunction are epiphenomenal in a non-equilibrium theory of Bohmian trajectories, but Bohmian trajectories are epiphenomena in the Everettian theory of quantum mechanics. But not really: on this point experiment will decide. As Brown freely admits, if as Valentini hopes we were eventually to observe exotic statistics of the sort he predicts, 'Everettians would have to throw in the towel'. But he doubts that pilot-wave theory really offers grounds for that hope, even taken on its own terms.

Not Only Many Worlds

The final chapters in Part 6 of the book are by contrast friendly to Everett, but they break new ground. In Chapter 17 Peter Byrne tells the story of how Everett's ideas were initially received, and how they were encouraged and ignored—and, in certain respects, suppressed. In the 1950s and 1960s, the dead weight of Bohr's authority was clearly in evidence. But in David Deutsch's estimation, the level of debate scarcely improved in the two decades following. The reason? Because the worth of the theory should have been demonstrated at the genuinely quantum mechanical ('multiversial') level, apart from universes. Worlds, universes, are essentially the *classical* structures in quantum mechanics. Too much of the debate, according to Deutsch in Chapter 18, concerned realism in general, distorting scientific judgements in foundations. How odd, he asks, is this:

Schrödinger had the basic idea of multiple universes shortly before Everett, but he didn't publish anything. He mentioned it in a lecture in Dublin (Schrödinger [1996]), in which he predicted that his audience would think he was crazy. Isn't that a strange assertion coming from a Nobel Prize winner—that he feared being considered crazy for claiming that his own equation, the one that he won the prize for, might be *true*.   (p.544)

And how odd would it seem, Deutsch continues, if Everettian quantum theory were to be widely accepted, to talk of it as the 'interpretation' of quantum mechanics. It would be like talking of dinosaurs as the 'interpretation' of fossil records, rather than the things in the theory that explain them.

But Deutsch's main complaint is the same as Maudlin's: there has been too little progress with the really foundational questions about ontology in quantum mechanics. He goes further in demanding progress in a range of areas—probability in cosmology, quantum computers, relativistically covariant information flows—on the basis of an unfettered quantum mechanical realism. Progress on these fronts, he says, is what will settle the matter. Deutsch asks much of Everettians.

He has some takers. The links with cosmology are explored in more detail by Max Tegmark in Chapter 19. He compares and contrasts Everettian worlds with multiple universes as they arise in inflationary cosmology—or multiplicities, even, in a sufficiently large single universe—more or less independent of quantum mechanics. How do they differ? His list includes the evidential problem (under three headings), several aspects of the debates over probability, reasons for which other worlds are unseen, and more. The answer, he concludes, is surprisingly modest: decoherence, Hilbert-space structure, replaces spatiotemporal structure in explaining the invisibility of other worlds, and enters directly in the definition of probability, but in all other respects the issues are essentially unchanged. One thing he does not mention, however, is the question of whether uncertainty in a branching Everettian universe really is like uncertainty in the cosmological multiverse. He is (rightly, if the arguments of Chapter 6 are correct) insensitive to the distinction between diverging and overlapping worlds. But on this point Deutsch, who is clearly well disposed to the idea of overlap (and well disposed to the analogous manoeuvre in the case of classically diverging worlds of taking observers as sets of worlds, see p.202), may be disappointed.

Lev Vaidman in Chapter 20 takes up Deutsch's challenge more directly: what else is there in quantum mechanics apart from the universes? Vaidman considers a very specific suggestion. It is possible, in ordinary quantum mechanics, to introduce a backwards-evolving wavefunction coming from the future outcome of an experiment, as proposed by Y. Aharonov and his collaborators. The so-called 'two-vector' formalism has been put to practical use in the theory of 'weak' measurements (see Aharonov and Vaidman [2007] for a recent review): it should be available to Everettians too.

Or so Vaidman concludes. Of course in the global perspective of the Everett interpretation there is no *one* outcome—a backwards-evolving state must be introduced for every branching event—but in the case of measurement events, they have just the same uses that Aharonov advertised. All save one, perhaps the most important: it does not, according to Vaidman, define a time-symmetric theory. That is a disappointment. On the other hand, he speculates, the backwards-evolving vectors may perhaps also serve to underpin the notion of uncertainty. At the very least, it is a tool for the definition of a quantum event as part of a unique history.

Other items on Deutsch's list get little or no further mention. For better or worse, in this book we are still labouring over the question of 'interpretation'—if not, at least for the most part, the virtues of realism. Were familiarity, common sense, and intuition among them, no doubt the Everett interpretation would be rejected out of hand; but those were never the hallmarks of truth.

3 ADDENDUM: FROM RELATIVE STATES TO CONSISTENT HISTORIES

3.1 The Measurement Postulates

Measurements on a system S are formally characterized in terms of a self-adjoint operator O (an observable) on a Hilbert space $\mathcal{H}$ associated with S, and a state $|\psi\rangle$ (a unit vector in $\mathcal{H}$ up to phase). In practice there may be some uncertainty as to what the state actually prepared in an experiment is (in which case $|\psi\rangle$ is replaced by a density matrix), but we shall consider only the simplest case.

An observable in quantum mechanics is in turn associated with a range of real numbers (roughly, its possible values, or eigenvalues), the spectrum $Sp(O)$ of O. A measurement outcome is a subset $E \subseteq Sp(O)$, with associated projector P_E on $\mathcal{H}$. The most important of the measurement postulates is the rule: the outcome E will be observed on measurement of O when S is in the state $|\psi\rangle$ with probability $\Pr(E)$ given by:

Born rule $\Pr(E) = \langle\phi|P_E|\phi\rangle = \|P_E|\phi\rangle\|^2$.

If, further, the experiment is non-disturbing—on immediate repetition the same outcome E is reliably obtained—then the state must have been subject to the transition

projection postulate $|\phi\rangle \rightarrow |\phi_E\rangle = \frac{P_E|\phi\rangle}{|P_E|\phi\rangle|}$.

When E is an eigenvalue of O, the RHS is one of its eigenstates.

Thus for a non-disturbing measurement of O the overall evolution in the Schrödinger picture, in which the state (rather than operators) carries the time-dependence, is of the form:

$$|\phi\rangle \overset{\text{unitary}}{\rightarrow} |\phi'\rangle \overset{\text{collapse}}{\rightarrow} \frac{P_E|\phi'\rangle}{|P_E|\phi'\rangle|}.$$

In the case of disturbing measurements, if the measurement is probabilistic, that collapse still occurs (albeit the final state may be unknown) cannot be doubted. We may take it as a phenomenological given, independent even, of quantum-state realism.

The final stage of the measurement cannot therefore be modelled unitarily—unless, it may be, if the measurement is *not* probabilistic. Suppose it is indeed fully predictable. Then there is no obstacle, at least for certain kinds of states, for (say) reasonably massive and well-localized clusters of atoms in bound states, well localized in position and momentum space, to giving a unitary description of their motions. The spread of the wavepacket, for such massive systems, is negligible over the timescale of the experiment. Ehrenfest's theorem then takes on a strong form, showing that wavepackets like these approximately follow classical trajectories (see Hartle, Chapter 2). In terms of operators, approximate projections onto states like these form a commutative set of projectors, as shown by von Neumann [1932 pp.402–9]. They are what he called the 'elementary building blocks of the classical description of the world' (p.409). Whether in terms of wavepackets or projections of this form, the unitary equations imply approximately classical trajectories, for timescales much larger than those of the experiment, if the masses are sufficiently large.

To take the example of a Stern–Gerlach experiment for a *deterministic* measurement of electron spin with eigenstates $|\phi_\uparrow\rangle$, $|\phi_\downarrow\rangle$, the registration of the electron at the screen and subsequent amplification processes involve many-particle systems of the sort just described. If we start off with localized states for the 'ready' state of the apparatus $\mathcal{A}$ in state $|\psi^A_{\text{ready}}\rangle$, with $|\psi^A_{\text{reads spin} \uparrow}\rangle$ for the event registering 'reads spin-up', the unitary evolution is:

$$|\phi_\uparrow\rangle \otimes |\psi^A_{\text{ready}}\rangle \xrightarrow{\text{unitary}} |\phi_{\text{absorbed}}\rangle \otimes |\psi^A_{\text{reads spin} \uparrow}\rangle.$$

But then there is nothing, assuming the arbitrariness of the von Neumann cut, to including ever more aspects of the laboratory, including experimentalists and technicians. That is, as built out of the same von Neumann's projectors, one can model 'the observer' $\mathcal{O}$ well. Thus if initially in the state $|\xi^{\mathcal{O}}_{\text{ready}}\rangle$, one has by the unitary formalism:

$$|\phi_\uparrow\rangle \otimes |\psi^A_{\text{ready}}\rangle \otimes |\xi^{\mathcal{O}}_{\text{ready}}\rangle$$
$$\xrightarrow{\text{unitary}} |\phi_{\text{absorbed}}\rangle \otimes |\psi^A_{\text{reads spin} \uparrow}\rangle \otimes |\xi^{\mathcal{O}}_{\text{ready}}\rangle$$
$$\xrightarrow{\text{unitary}} |\phi_{\text{absorbed}}\rangle \otimes |\psi^A_{\text{reads spin} \uparrow}\rangle \otimes |\xi^{\mathcal{O}}_{\text{sees spin} \uparrow}\rangle.$$

If the apparatus functions properly, and reliably detects a particle in the down state of spin $\downarrow$, a similar schema will apply to that case, when the initial state of the electron is $|\phi_\downarrow\rangle$. The unitary equations, for sufficiently massive systems in states well localized in position and momentum space, appear perfectly adequate to describe such processes—highly schematic, true, but easily refined—so long as they are deterministic.

Of course the trouble with all of this if quantum mechanics is to describe the macroworld is that experiments often *aren't* deterministic, and correspondingly, however well localized the initial states of aggregates of atoms in the apparatus, the apparatus and the observer *don't* end up in states well localized in position and momentum space. For let the initial state be of the form

$$|\phi\rangle = a|\phi_\uparrow\rangle + b|\phi_\downarrow\rangle,$$

where a and b are constants. Then by the linearity of the unitary dynamics the superposition of the two final states results:

$$a|\phi_{\text{absorbed}}\rangle \otimes |\psi^A_{\text{reads spin }\uparrow}\rangle \otimes |\xi^O_{\text{reads spin }\uparrow}\rangle +$$
$$b|\phi_{\text{absorbed}}\rangle \otimes |\psi^A_{\text{reads spin }\downarrow}\rangle \otimes |\xi^O_{\text{sees spin }\downarrow}\rangle$$

and this deterministic motion doesn't seem to correspond to anything. Hence the need for the collapse postulate (with the E_k's standing for 'spin $\uparrow$' and 'spin $\downarrow$').

But note how the measurement problem, on this line of reasoning, as intimated by von Neumann [1932 ch.6], and as used by Schrödinger [1935] (in terms of the 'cat' paradox) and by Wigner [1961] (in terms of the 'friend' paradox) is changed: it is that if you allow the von Neumann chain to extend well into the macroscopic, using von Neumann's building blocks, then you find the unitary equations yield a superposition of states *each one of which tells a perfectly reasonable physical story.*

3.2 Everett's Relative States

With this background in place[21] Everett's contribution, as it appeared in Everett [1957], may seem rather modest: it was to show that on *repeated* quantum measurements (using only the unitary formalism) of the von Neumann kind one obtains a superposition of states, each of which tells a physically reasonable *statistical* story—just *as if* each sequence of states were arrived at by repeated application of the projection postulate after each trial.

Modest or not, the idea required some new notation. Everett gave a model of a quantum automaton A which combined the functions of the apparatus and the observer, but indexed, not by a single outcome, but by a string of outcomes. Its 'ready' state is $|\psi^A[\ldots\ldots]\rangle$. The measurement interaction is as before the von

---

[21] Everett had much of it: 'any general state can at any instant be analyzed into a superposition of states each of which does represent the bodies with fairly well-defined positions and momenta. Each of these states then propagates approximately according to classical laws, so that the general state can be viewed as a superposition of quasi-classical states propagating according to nearly classical trajectories' (Everett [1973 p.89]). In a footnote, Everett summarized von Neumann's construction as just discussed (but with no mention of the strong form of Ehrenfest's theorem).

Simon Saunders

Neumann model. The automaton on interacting with the system S in any of an orthogonal set of states $\{|\phi_i\rangle\}$ evolves unitarily (in the Schrödinger picture) as:

$$|\phi_i\rangle \otimes |\psi^A[.\ .\ .\ .\ .]\rangle \overset{\text{unitary}}{\rightarrow} |\phi_i\rangle \otimes |\psi^A[.\ .\ .\ .\ a_i]\rangle \qquad (1)$$

in which a_i characterizes the state $|\phi_i\rangle$ (say, the eigenvalue of an operator in the eigenstate $|\phi_i\rangle$). If the microscopic system is in the state $|\phi\rangle = \sum_i c_i |\phi_i\rangle$, it follows:

$$|\phi\rangle \otimes |\psi^A[.\ .\ .\ .\ .]\rangle \rightarrow \sum_i c_i |\phi_i\rangle \otimes |\psi^A[.\ .\ .\ .\ a_i]\rangle. \qquad (2)$$

Suppose that the system in the final state, given by the RHS of (2), is subject to the same interaction again: then there results:

$$|\phi\rangle \otimes |\psi^A[.\ .\ .\ .\ .]\rangle \rightarrow \sum_i c_i |\phi_i\rangle \otimes |\psi^A[.\ .\ .\ .\ a_i]\rangle$$
$$\rightarrow \sum_i c_i |\phi_i\rangle \otimes |\psi^A[.\ .\ .\ .\ a_i a_i]\rangle.$$

That is to say: the recorded value, on the second measurement, is precisely the same as the first, for each component of the final, total superposition—*just as if the projection postulate had been invoked at the end of the first process.*

It further follows, if there are n systems in the similarly prepared state $|\phi\rangle$, each of which is independently measured, with the results recorded by $\mathcal{A}$, that:

$$|\phi\rangle \otimes \ldots \otimes |\phi\rangle \otimes |\psi^A[.\ .\ .\ .\ .]\rangle \rightarrow$$
$$\sum_{i,j,\ldots,k} c_i c_j \ldots c_k |\phi_i\rangle \otimes |\phi_j\rangle \otimes \ldots \otimes |\phi_k\rangle \otimes |\psi^A[a_k.\ .\ .a_j a_i]\rangle,$$

whereupon a (different) sequence of results is recorded by the automaton in each state entering into the final superposition—that is, in each component, there is a record of a definite sequence of outcomes, a definite statistics.

What about the outcomes themselves, apart from the records? Everett's answer was that they have values in a 'relative' sense—that for each state $|\psi^A[a_k.\ .\ .a_j a_i]\rangle$ in the superposition there exists its *relative* state $|\phi_i\rangle \otimes |\phi_j\rangle \otimes \ldots \otimes |\phi_k\rangle$ of the n-subsystems. There is no 'true' state of a subsystem—only a state of a subsystem, relative to a state of another subsystem. This is the essential novelty of quantum mechanics in Everett's view—in fact, it already followed from the basic structure of entanglement. He summarized the matter thus quite early in his paper:

There does not, in general, exist anything like a single state for one subsystem of a composite system. Subsystems do not possess states that are independent of the states of the remainder of the system, so that the subsystem states are generally correlated with one another. One can arbitrarily choose a state for one subsystem, and be led to the relative state for the remainder. Thus we are faced with a fundamental relativity of states, which is implied by the formalism of composite systems. It is meaningless to ask the absolute state of a subsystem—one can only ask the state relative to a given state of the remainder of the subsystem. (Everett [1957 p.143], emphasis original).

That seems to invite a broadly structuralist reading of the wavefunction.

Only much later in the paper did Everett revisit the question of how, precisely, these relational structures can all coexist. But at this point, following on his analysis in terms of automata, he immediately brought the question back to the invisibility of branching—that is, to the question of what is observable. But he did make a pregnant comparison:

Arguments that the world picture presented by this theory is contradicted by experience, because we are unaware of any branching process, are like the criticism of the Copernican theory that the mobility of the earth as a real physical fact is incompatible with the common sense interpretation of nature because we feel no such motion. In both cases the argument fails when it is shown that our experience will be what it in fact is. (In the Copernican case the addition of Newtonian physics was required to be able to show that the earth's inhabitants would be unaware of any motion of the earth.) (Everett [1957] note added in proof.)

It was Galileo, of course, who supplied arguments as to why the motion of the earth would be unobservable, if the Copernican theory were true. But Everett might have elaborated the analogy. Equally, one might say that in a classical spacetime theory, only relative positions, relative velocities, are real; it is just as meaningless to ask for the absolute state of motion of a system as to ask for its absolute quantum state. But that suggests a rather different question than the one suggested, as a parallel, by Everett. Not, 'why is the motion of the earth invisible?', but 'what is motion?', and the comparison, not with Galileo, but with Descartes.

Descartes gave a purely relational account of motion just as did Everett of value-definiteness. It amounted to motion as rate of change of relative distances, and nothing else. As such it failed to explain the appearances—at best it only described them. Further dynamical principles were needed to pick out the privileged (relative) motions, the inertial motions.

It was the same with Everett's concept of relative states. He advocated the use of von Neumann's 'elementary building blocks', but equally he appealed to the Schmidt decomposition (see below), what he called the 'canonical representation' ([1973 p.47]). At times he wrote as if the superposition principle all by itself guaranteed the dynamical autonomy of components of the universal state ([1973 p.98]). What was missing were dynamical considerations to show that this was so—to pick out the *significant* motions.

Something more than the schematic and idealized dynamics of the von Neumann model or the kinematic Schmidt decomposition of the state was needed.

3.3 Quantum Histories

Equations of this kind were eventually obtained for a variety of many-particle systems—this the burgeoning field of decoherence theory. But dynamics can also be thought of in more structural terms, in terms of the possible histories of a physical system. That fits better with the philosophers' way of thinking of things.

Histories proper, retrodictions in quantum mechanics, were early on recognized as quite different from predicted courses of events. They could be fitted, sort of, into Bohr's interpretative framework, as shown by G. Hermann [1935], in a study that Heisenberg had encouraged. But the subject languished. However, dynamics as structures of histories arose in fields as diverse as optics and general relativity. Much of the impetus to develop Everett's ideas lay in hoped-for applications in quantum cosmology. The quantum histories formalism, as developed by R. Griffiths, R. Omnès, M. Gell-Mann, and J. B. Hartle in the late 1980s, had a variety of sources.

It does Everett nicely. Let $\{P_a\}$, $a = 1, 2, \ldots$ be an exhaustive, commuting set of projection operators on a Hilbert space $\mathcal{H}$, i.e.:

$$\sum_a P_a = I, \ P_a P_\beta = \delta_{a\beta} P_a.$$

They may be taken to be von Neumann's 'elementary building blocks of the classical world' (in fact, if we do this we obtain a quasiclassical domain, in Gell-Mann and Hartle's sense). Let H be the Hamiltonian—again, with no explicit time-dependency. Define the Heisenberg picture operators:

$$P_a(t) = e^{iHt/\hbar} P_a e^{-iHt/\hbar}.$$

For the simplest example of a set of histories, consider histories constructed out of sequences of projectors in $\{P_{a_1}(t_1)\}, \{P_{a_2}(t_2)\}, \ldots, \{P_{a_n}(t_n)\}$,[22] for a sequence of times $t_1 < t_2 < \ldots < t_n$ (the choice of sequence, like the choice of cells on configuration space, is for the time being arbitrary). An individual history a is now a particular sequence $(a_1, a_2, \ldots, a_n)$ and is represented by a *chain* (or in Hartle's language a *class*) operator:

$$C_a = P_{a_n}(t_n) P_{a_{n-1}}(t_{n-1}) \ldots P_{a_1}(t_1).$$

[22] Jim Hartle in Chapter 2 considers more general histories, in which different families of projectors are chosen at different times (with corresponding superscripts on the $P_{a_k}(t_k)$'s).

The operators $C_a^\dagger C_a$ are self-adjoint and positive, but they are not projectors. Acting on the state $|\Psi\rangle$ at $t = 0$ we obtain the *branch state vector* $C_a|\Psi\rangle$. It is the same as the vector (time-evolved back to $t = 0$) that *would* have been obtained in the Schrödinger picture by a sequence of non-disturbing measurements (using the measurement postulates), first of the projection $P_{a_1}^1$ at time t_1 (collapsing onto the vector $\Psi_{a_1}(t_1) = P_{a_1}e^{-iHt_1/\hbar}|\Psi(0)\rangle$), then of the projection P_{a_2} at time t_2 (collapsing onto the vector $\Psi_{a_2 a_1}(t_2) = P_{a_2}e^{-iH(t_2-t_1)/\hbar}P_{a_1}e^{-iHt_1/\hbar}|\Psi(t_1)\rangle$), and so on, with modulus square equal to the product of the probabilities for each collapse (as calculated using the measurement postulates). That is, the probability $p(a)$ for a history a is the modulus square of the branch state vector $C_a|\Psi\rangle$

$$p(a) = \||C_a|\Psi\rangle\|^2 = Tr(C_a \rho C_a^\dagger) \qquad (3)$$

where $\rho = |\Psi\rangle\langle\Psi|$ is the density matrix for the state $|\Psi\rangle$ and Tr is the trace ($Tr(O) = \sum_k \langle\phi_k|O\phi_k\rangle$, for any operator O and orthonormal basis $\{\phi_k\}$ over $\mathcal{H}$). Likewise, one can define the conditional probability of a (for $t_n < \ldots < t_{k+1}$) given β (for $t_k < \ldots < t_1$) as

$$p_\rho(a/\beta) = \frac{Tr(C_{a*\beta}\rho C_{a*\beta}^\dagger)}{Tr(C_\beta \rho C_\beta^\dagger)}, \qquad (4)$$

where $a * \beta$ is the history comprising β (up to time t_k) and a (from t_{k+1} to t_n).

But this interpretation of the quantities $p(a)$, $p(a/\beta)$ as probabilities in the context of the Everett interpretation needs justification. In general, for arbitrary choices of families of projectors $\{P_{a_k}\}$, they have nothing to do with probabilities. The use of the trace in Eqs (3) and (4) is no more than a formal device for extracting squared norms of amplitudes and transition amplitudes; they are relations in the Hilbert space norm, defined—deterministically defined, note, under the unitary equations—to facilitate the structural analysis of the state. At this stage they need mean nothing more.

But we may help ourselves to their obvious structural meaning, when these transition amplitudes are zero or one. We thus talk of anticorrelations and correlations among the associated sequences of projectors, and by extension, the configurations a and β on which they project. In the single-stage case, suppose the latter pertain to different systems, represented by projectors of the form $P_a \otimes I, I \otimes P_\beta$. Let $p_\rho(a/\beta) = 1$ and let $\rho = |\Psi\rangle\langle\Psi|$. In the special case where P_a and P_β are one-dimensional with ranges $|a\rangle$, $|\beta\rangle$, then $|a\rangle$ is the relative state of $|\beta\rangle$ in the state $|\Psi\rangle$, in Everett's sense. More generally: if $p_\rho(a/\beta) = 1$ then a is the 'relative configuration' of β in $|\Psi\rangle$.

Here is a connection with the Schmidt decomposition. It is a theorem that for any vector Ψ in the tensor product Hilbert space $\mathcal{H}^{A+B} = \mathcal{H}^A \otimes \mathcal{H}^B$ of two systems $\mathcal{A}$ and $\mathcal{B}$, there exists orthonormal basis $\{\phi_k\}$ in $\mathcal{H}^A$, and $\{\psi_k\}$ in $\mathcal{H}^B$, and complex numbers c_k such that

$$|\Psi\rangle = \sum_k c_k|\phi_k\rangle \otimes |\psi_k\rangle. \qquad (5)$$

If for $k \neq j$ $|c_k| \neq |c_j|$, then the bases $\{\phi_k\}$ in $\mathcal{H}^A$ and $\{\psi_k\}$ in $\mathcal{H}^B$ are unique. Eq. (5) is the Schmidt decomposition. If these bases diagonalize P_α and P_β respectively, then (for any dimensionality)

$$\sum_{k;\ P_\alpha|\phi_k\rangle=|\phi_k\rangle} c_k|\psi_k\rangle$$

is the relative state of $P_\alpha\sum_k c_k|\phi_k\rangle$ in the state $|\Psi\rangle$. Given this condition, relativization in Everett's sense is a symmetric relation.

3.4 Coarse-Graining and Consistency

The notion of coarse-graining of a parameter space (like configuration space) extends naturally to chain operators, as follows. Let $\{\bar{a}\}$ be a coarse-graining of $\{a\}$, so that each finer-grained cell a is contained in some coarser-grained cell $\bar{a}$ in the parameter space. We can then speak of coarser- and finer-grainings of histories too. Now consider a set of histories with chain operators $\{C_a\}$, and a coarse-graining with chain operators $\{C_{\bar{a}}\}$. Then the two are related by summation:

$$C_{\bar{a}} = \sum_{a\in\bar{a}} C_a$$

where the sum is over all finer-grained histories a contained within $\bar{a}$.

Now for a candidate fundamental ontology (Saunders [1994, 1995]): it is the system of correlations and transition amplitudes among values of self-adjoint dynamical variables and their coarse-grainings—in quantum mechanics, among values of particle variables, in quantum field theory, among values, local in space and time, of field densities. The latter mirrors, roughly, the fundamental ontology in classical general relativity theories, in terms of invariant relations among values of metric and matter fields.

As in general relativity, some order can be introduced by a formal condition. Given a Lorentzian geometry it is useful to introduce a foliation to a manifold—a collection of global three-dimensional surfaces whose tangent vectors are everywhere spacelike. It is useful, considering the structure of a quantum state, to consider families of projectors for which branch state vectors, for histories neither of which is a coarse-graining of the other, are approximately orthogonal:

$$\langle C_a \Psi | C_{a'} \Psi \rangle \approx 0, \; a \notin a' \text{ and } a' \notin a. \tag{6}$$

Such histories are called *consistent* (by Griffiths and Omnès); *(medium) decoherent* (by Hartle and Halliwell). Given consistency, Everett's relativization is a transitive relation even in time-like directions (Saunders [1995b]; it is automatically transitive in spacelike directions by virtue of microcausality).

The coarse-graining of histories exploits Hilbert-space structures, notably, the Boolean algebra of the projectors used to generate those histories. If this is used to turn a history space into a probability space (a Borel space), equipped with a σ-algebra, then the measure must be additive with respect to coarse-graining:

$$p(\bar{a}) = \sum_{a \in \bar{a}} p(a). \tag{7}$$

The analogous condition for the Schrödinger picture state (essentially, single-time histories) is automatically satisfied, given Eq. (3) (Everett turned this reasoning around: assuming additivity, he derived Eq. (3)); it is satisfied by two-time histories as well. But in the general case it fails. The consistency condition as originally defined is the necessary and sufficient condition for additivity in the sense of Eq. (7); the condition as specified, Eq. (6), is slightly stronger but somewhat simpler — this is the condition that is widely used.

It follows too that for any consistent history space there exists a fine-graining $\{P_a\}$ which is consistent and for which, for any $t_n > t_m$ and for any a_n with $P_{a_n}(t_n)|\Psi\rangle \neq 0$, there exists exactly one a_m such that

$$P_{a_n}(t_n) P_{a_m}(t_m)|\Psi\rangle \neq 0$$

(Griffiths [1993], Wallace [2010]). That is, for each a_n at time t_n, there is a *unique* history preceding it — the set of histories can be fine-grained so as to have a purely branching structure (with no recombination of branches). The connection, at this point, with the Aharonov two-vector formalism is immediate (see Vaidman's Chapter 20).

The consistency condition and the quantum histories formalism is widely advertised as providing a generalization of quantum theory as, fundamentally, a theory of probability. As such there is a continuum infinity of consistent history spaces available — new resources, for the exploration of quantum systems, indeed. But from the point of view of Everettian quantum mechanics, consistency is far too weak a condition to give substance to the notion of histories as autonomous and robust dynamical structures, and probability, as associated with branching of such structures, is too high level a concept to figure in the foundations of quantum mechanics. At any rate, consistency holds automatically given decoherence in the sense of quasiclassicality (or realms more generally), itself only an approximate condition, but still our abiding criterion for the existence of worlds.

References

Aharonov, Y., P. Bergmann, and J. Lebowitz [1964], 'Time symmetry in the quantum process of measurement', *Physical Review* **B 134**, 1410–16.

Aharonov, Y. and L. Vaidman [2007], 'The two-state vector formalism: an updated review'. Available online at arXiv.org/abs/quant-ph/0105101v2.

Albert, D. and B. Loewer [1988], 'Interpreting the many-worlds interpretation', *Synthese* 77, 195–213.

Bacciagaluppi, G. [2000], 'Delocalised properties in the modal interpretation of a continuous model of decoherence', *Foundations of Physics* 30, 1431–44.

Baker, D. [2006], 'Measurement outcomes and probability in Everettian quantum mechanics', *Studies in History and Philosophy of Modern Physics* 38, 153–69.

Ballentine, L.E. [1973], 'Can the statistical postulate of quantum theory be derived?—A critique of the many-universes interpretation', *Foundations of Physics* 3, 229.

Barrett, J. [1999], *The Quantum Mechanics of Mind and Worlds*, Oxford University Press, Oxford.

Bell, J. [1987], *Speakable and Unspeakable in Quantum Mechanics*, Cambridge University Press, Cambridge.

—— [1990], 'Against measurement'. *Physics World* 8, 33–40. Reprinted in *Sixty-Two Years of Uncertainty: Historical, Philosophical and Physical Inquiries into the Foundations of Quantum Mechanics*, Arthur Miller (ed.), Plenum, New York (1990).

Bohm, D. [1951], *Quantum Theory,* Prentice-Hall, New Jersey.

—— [1952], 'A suggested interpretation of the quantum theory in terms of "hidden" variables. 1', *Physical Review* 85, 166–79.

Brown, H.R. and Wallace, D. [2005], 'Solving the measurement problem: de Broglie–Bohm loses out to Everett', *Foundations of Physics* 35, 517–40.

Deutsch, D. [1985], 'Quantum theory as a universal physical theory', *International Journal of Theoretical Physics* 24, 1–41.

—— [1996], 'Comment on Lockwood', *British Journal for the Philosophy of Science* 47, 222–8.

—— [1997], *The Fabric of Reality*, Penguin Books.

—— [1999], 'Quantum theory of probability and decisions', *Proceedings of the Royal Society of London* **A455**, 3129–37. Available online at arXiv.org/abs/quant-ph/9906015.

DeWitt, B. [1967], 'The Everett–Wheeler interpretation of quantum mechanics', in *Battelle Rencontres, 1967 Lectures in Mathematics and Physics*, C. DeWitt, J. Wheeler, eds), W. A. Benjamin Inc., New York (1968).

—— [1970], 'Quantum mechanics and reality', *Physics Today* 23, No.9; reprinted in DeWitt and Graham [1973 pp.155–67].

—— [1993], 'How does the classical world emerge from the wavefunction?', in *Topics on Quantum Gravity and Beyond*, F. Mansouri and J.J. Scanio (eds), World Scientific, Singapore.

DeWitt, B. and N. Graham, [1973], *The Many-Worlds Interpretation of Quantum Mechanics*, Princeton University Press, Princeton.

Dowker, F. and A. Kent [1996], 'On the consistent histories approach to quantum mechanics', *Journal of Statistical Physics* **82**, 1575–646.

Everett III, H. [1957], ' "Relative state" formulation of quantum mechanics', *Reviews of Modern Physics* **29**, 454–62, reprinted in DeWitt and Graham [1973 pp.141–50].

—— [1973], 'Theory of the universal wavefunction', in DeWitt and Graham [1973 pp.3–140].

Farhi E., J. Goldstone, and S. Gutman [1989], 'How probability arises in quantum mechanics', *Annals of Physics* **192**, 368–82.

Gell-Mann, M. and J.B. Hartle [1990], 'Quantum mechanics in the light of quantum cosmology', in *Complexity, Entropy, and the Physics of Information*, W.H. Zurek (ed.), Addison-Wesley, Reading.

—— [1993], 'Classical equations for quantum systems', *Physical Review* D **47**, 3345–82. Available online at arXiv.org/abs/gr-qc/9210010.

Ghirardi, G.C., A. Rimini, and T. Weber [1986], 'Unified dynamics for microscopic and macroscopic systems', *Physical Review* D **34**, 470–91.

Ghirardi, G.C., P. Pearle, and A. Rimini [1990], 'Markov-processes in Hilbert-space and continuous spontaneous localization of systems of identical particles', *Physical Review* A**42**, 78.

Graham, N. [1970], *The Everett Interpretation of Quantum Mechanics*, PhD dissertation, Chapel Hill.

—— [1973], 'The measurement of relative frequency', in DeWitt and Graham [1973 pp.229–553].

Greaves, H. [2004], 'Understanding Deutsch's probability in a deterministic multiverse', *Studies in History and Philosophy of Modern Physics* **35**, 423–56. Available online at philsci-archive.pitt.edu/archive/00001742/.

—— [2007], 'On the Everettian epistemic problem', *Studies in History and Philosophy of Modern Physics* **38**, 120–52. Available online at philsci-archive.pitt.edu/archive/00002953.

Griffiths, R. [1984], 'Consistent histories and the interpretation of quantum mechanics', *Journal of Statistical Physics* **36**, 219–72.

—— [1993], 'Consistent interpretation of quantum mechanics using quantum trajectories', *Physical Review Letters* **70**, 2201–4.

Hartle, J. [1968], 'Quantum mechanics of individual systems', *American Journal of Physics* **36**, 704–12.

Hemmo, M. and I. Pitowsky [2007], 'Quantum probability and many worlds', *Studies in the History and Philosophy of Modern Physics* **38**, 333–50.

Hermann, G. [1935], 'Die naturphilosophischen Grundlagen der Quantenmechanik', *Abhandlungen der Fries'schen Schule* **6**, 75–152.

Kent, A. [1990], 'Against many-worlds interpretations', *International Journal of Modern Physics* A**5**, 1745–62.

Kübler, O. and H.D. Zeh [1973], 'Dynamics of quantum correlations', *Annals of Physics* (NY) **76**, 405–18.

Lewis, D. [1986a], *Philosophical Papers, Vol. 2*, Oxford University Press, Oxford.

—— [1986b], *On the Plurality of Worlds*, Blackwell.

Lewis, P. [2007], 'Uncertainty and probability for branching selves', *Studies in History and Philosophy of Modern Physics* **38**, 1–14.

Lockwood, M. [1989], *Mind, Brain, and The Quantum*, Basil Blackwell, Oxford.

Myrvold, W. [2005], 'Why I am not an Everettian', unpublished manuscript.

Ochs, W. [1977], 'On the strong law of large numbers in Quantum Probability Theory', *Journal of Philosophical Logic* **6**, 473–80.

Omnès, R. [1988], 'Logical reformulation, of quantum mechanics, *Journal of Statistical Physics* **53**, 893–975.

Papineau, D. [1996], 'Comment on Lockwood', *British Journal for the Philosophy of Science* **47**, 233–41.

Saunders, S. [1993], 'Decoherence, relative states, and evolutionary adaptation', *Foundations of Physics* **23**, 1553–85.

—— [1994], 'Remarks on decoherent histories theory and the problem of measurement', in *Stochastic Evolution of Quantum States in Open Systems and in Measurement Processes*, L. Diøsi (ed.), pp.94–105, World Scientific, Singapore.

—— [1995a], 'Time, quantum mechanics, and decoherence', *Synthese* **102**, 235–66.

—— [1995b], 'Relativism', in *Perspectives on Quantum Reality*, R. Clifton (ed.), Kluwer, Dordrecht, pp.125–42.

—— [1996a], 'Response to Lockwood', *British Journal for the Philosophy of Science* **47**, 241–8.

—— [1996b], 'Time, quantum mechanics, and tense', *Synthese* **107**, 19–53.

—— [1998], 'Time, quantum mechanics, and probability', *Synthese* **114**, pp.405–44. Available online at arXiv.org/abs/quant-ph/0112081.

—— [2001], 'Review of The Quantum Mechanics of Mind and World by J. Barrett', *Mind* **110**, 1039–43.

—— [2004], 'Derivation of the Born rule from operational assumptions', *Proceedings of the Royal Society* **A 460**, 1–18.

—— [2005], 'What is probability?', in *Quo Vadis Quantum Mechanics*, A. Elitzur, S. Dolev, and N. Kolenda (eds), Springer. Available online at arXiv.org/abs/quant-ph/0412194.

Saunders, S. and D. Wallace [2008], 'Branching and uncertainty', *British Journal for the Philosophy of Science* **59**, 293–305. Available online at philsci-archive.pitt.edu/archive/00003811/.

Schrödinger, E. [1935], 'Die gegenwärtige Situation in der Quantenmechanik', *Die Naturwissenschaften* **23**, 817–2, 823–8, 844–9. Translated as 'The present situation in quantum mechanics', in *Quantum Theory and Measurement*, J.A. Wheeler and W.H. Zurek (eds), Princeton University Press, Princeton (1983).

—— [1996], *The Interpretation of Quantum Mechanics: Dublin Seminars (1949–1955) and other unpublished essays*, M. Bitbol (ed.), OxBow Press.

Vaidman, L. [1998], 'On schizophrenic experiences of the neutron or why we should believe in the many-worlds interpretation of quantum theory', *International Studies in the Philosophy of Science* **12**, 245–61.

—— [2002], 'Many worlds interpretations of quantum mechanics', *Stanford Encyclopedia of Philosophy*. Available online at http://plato.stanford.edu/entries/qm-manyworlds/.

Von Neumann, J. [1932], *Mathematische Grundlagen Der Quantenmechanik*, translated by R.T. Beyer as *Mathematical Foundations of Quantum Mechanics*, Princeton University Press (1955).

Wallace, D. [2002], 'Quantum probability and decision theory, revisited'. Available online at arXiv.org/abs/quant-ph/0211104.

—— [2003a], 'Everett and structure', *Studies in the History and Philosophy of Physics* 34, 87–105. Available online at arXiv.org/abs/quant-ph/0107144.

—— [2003b], 'Everettian rationality: defending Deutsch's approach to probability in the Everett interpretation', *Studies in the History and Philosophy of Modern Physics* 34, 415–39. Available online at arXiv.org/abs/quant-ph/0303050.

—— [2005], 'Language use in a branching universe'. Available online at philsci-archive.pitt.edu/archive/00002554.

—— [2006], 'Epistemology quantized: circumstances in which we should come to believe in the Everett interpretation', *British Journal for the Philosophy of Science* 57, 655–89. Available online at philsci-archive.pitt.edu/archive/00002839.

—— [2007], 'Quantum probability from subjective likelihood: improving on Deutsch's proof of the probability rule', *Studies in the History and Philosophy of Modern Physics* 38, 311–32. Available online at arXiv.org/abs/quant-ph/0312157.

—— [2011], *The Emergent Multiverse*, Oxford.

Wheeler, J.A. [1957], 'Assessment of Everett's "relative state" formulation of quantum theory', *Reviews of Modern Physics* 29, 463–5, reprinted in DeWitt and Graham [1973], pp.141–50.

Wheeler, J.A., and W.H. Zurek (eds) [1983], *Quantum Theory and Measurement*, Princeton University Press, Princeton.

Wigner, E. [1961], 'Remarks on the mind–body question', in *The Scientist Speculates*, I.J. Good (ed.), Heinemann, London. Reprinted in E. Wigner, *Symmetries and Reflections*, Indiana University Press, Bloomington (1967), and in Wheeler and Zurek [1983].

Zeh, D. [1970], 'On the interpretation of measurement in quantum theory', *Foundations of Physics* 1, 69–76.

—— [1973], 'Toward a quantum theory of observation', *Foundations of Physics* 3, 109–16. Revised version available online at arXiv.quant-ph/030615v1/.

—— [1975], 'Symmetry-breaking vacuum and state vector reduction', *Foundations of Physics* 5, 371–3.

—— [2000], 'The problem of conscious observation in quantum mechanical description', *Foundations of Physics Letters* 13, 221–33. Available online at arXiv.quant-ph/9908084v3.

Zurek, W.H. [1982], 'Environment-induced superselection roles', *Physical Review* D26, 1862-80.

—— [1993], 'Negotiating the tricky border between quantum and classical', *Physics Today* 46, No.4, 13–15, 81–90.

PART I

WHY MANY WORLDS?

1

Decoherence and Ontology
(or: How I learned to stop worrying and love FAPP)

David Wallace

> The form of a philosophical theory, often enough, is: *Let's try looking over here.*
>
> (Fodor [1985 p.31])

1 INTRODUCTION: TAKING PHYSICS SERIOUSLY

NGC 1300 (shown in Fig. 1) is a spiral galaxy 65 million light years from Earth.[1] We have never been there, and (although I would love to be wrong about this) we will never go there; all we will ever know about NGC 1300 is what we can see of it from 65 million light years away, and what we can infer from our best physics.

Fortunately, 'what we can infer from our best physics' is actually quite a lot. To take a particular example: our best theory of galaxies tells us that that

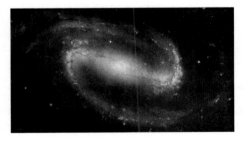

Figure 1. The spiral galaxy NGC 1300.

[1] Source: http://leda.univ-lyon1.fr/. This photo taken from http://hubblesite.org/gallery/album/galaxy/pr2005001a/, with thanks to P. Knesek (WIYN).

hazy glow is actually made up of the light of hundreds of billions of stars; our best theories of planetary formation tell us that a sizable fraction of those stars have planets circling them, and our best theories of planetology tell us that some of those planets have atmospheres with such-and-such properties. And because I think that those 'best theories' are actually pretty *good* theories, I regard those inferences as fairly *reliable*. That is: I think that there actually *are* atmospheres on the surfaces of some of the planets in NGC 1300, with pretty much the properties that our theories ascribe to them. That is: I think that those atmospheres *exist*. I think that they are *real*. I *believe* in them. And I do so despite the fact that, at 65 million light years' distance, the chance of directly observing those atmospheres is nil.

I present this example for two reasons. The first is to try to demystify—deflate, if you will—the superficially 'philosophical'—even 'metaphysical'—talk that inevitably comes up in discussions of 'the ontology of the Everett interpretation'. Talk of 'existence' and 'reality' can sound too abstract to be relevant to physics (talk of 'belief' starts to sound downright theological!) but in fact, when I say that 'I believe such-and-such is real' I intend to mean no more than that it is on a par, evidentially speaking, with the planetary atmospheres of distant galaxies.

The other reason for this example brings me to the main claim of this paper. For the form of reasoning used above goes something like this: we have good grounds to take such-and-such physical theory seriously; such-and-such physical theory, taken literally, makes such-and-such ontological claim; therefore, such-and-such ontological claim is to be taken seriously.[2]

Now, if the mark of a serious scientific theory is its breadth of application, its explanatory power, its quantitative accuracy, and its ability to make novel predictions, then it is hard to think of a theory more 'worth taking seriously' than quantum mechanics. So it seems entirely apposite to ask what ontological claims quantum mechanics makes, if taken literally, and to take those claims seriously in turn.

And quantum mechanics, taken literally, claims that we are living in a multiverse: that the world we observe around us is only one of countless quasi-classical universes ('branches') all coexisting. In general, the other branches are no more observable than the atmospheres of NGC 1300's planets, but the theory claims that they exist, and so if the theory is worth taking seriously, we should take the branches seriously too. To belabour the point:

According to our best current physics, branches are real.

Everett was the first to recognize this, but for much of the ensuing 50 years it was overlooked: Everett's claim to be 'interpreting' existing quantum mechanics,

[2] Philosophers of science will recognize that, for reasons of space, and to avoid getting bogged down, I gloss over some subtle issues in the philosophy of science; the interested reader is invited to consult, e.g., Newton-Smith [1981], Psillos [1999], or Ladyman and Ross [2007] for more on this topic.

and DeWitt's claim that 'the quantum formalism is capable of yielding its own interpretation' were regarded as too simplistic, and much discussion on the Everett interpretation (even that produced by advocates such as Deutsch [1985]) took as read that the 'preferred basis problem'—the question of how the 'branches' were to be defined—could be solved only by adding something additional to the theory. Sometimes that 'something' was additional physics, adding a multiplicity of worlds to the unitarily evolving quantum state (Deutsch [1985], Bell [1981], Barrett [1999]). Sometimes it was a purpose-built theory of consciousness: the so-called 'many-minds theories' (Lockwood [1989], Albert and Loewer [1988]). But whatever the details, the end result was a replacement of quantum mechanics by a new theory, and furthermore a new theory constructed specifically to solve the quantum measurement problem. No wonder interest in such theories was limited: if the measurement problem really does force us to change physics, hidden-variables theories like the de Broglie–Bohm theory[3] or dynamical-collapse theories like the GRW theory[4] seem to offer less extravagantly science-fictional options.

It now seems to be widely recognized that if Everett's idea really is worth taking seriously, it must be taken on Everett's own terms: as an understanding of what (unitary) quantum mechanics *already* claims, not as a proposal for how to amend it. There is precedent for this: mathematically complex and conceptually subtle theories do not always wear their ontological claims on their sleeves. In general relativity, it took decades to fully understand that the existence of gravity waves and black holes really is a claim of the theory rather than some sort of mathematical artefact.

Likewise in quantum physics, it has taken the rise of decoherence theory to illuminate the structure of quantum physics in a way which makes the reality of the branches apparent. But 20 years of decoherence theory, together with the philosophical recognition that to be a 'world' is not necessarily to be part of a theory's fundamental mathematical framework, now allow us to resolve—or, if you like, to dissolve—the preferred basis problem in a perfectly satisfactory way, as I shall attempt to show in the remainder of the paper.

2 EMERGENCE AND STRUCTURE

It is not difficult to see why Everett and DeWitt's literalism seemed unviable for so long. The axioms of unitary quantum mechanics say nothing of 'worlds' or 'branches': they speak only of a unitarily evolving quantum state, and however suggestive it may be to write that state as a superposition of (what appear to be) classically definite states, we are not justified in speaking of

---

[3] See Cushing, Fine, and Goldstein [1996] and references therein for more information.
[4] See Bassi and Ghirardi [2003] and references therein for more information.

those states as 'worlds' unless they are somehow added into the formalism of quantum mechanics. As Adrian Kent put it in his influential [1990] critique of many-worlds interpretations:

> . . . one can perhaps intuitively view the corresponding components [of the wavefunction] as describing a pair of independent worlds. But this intuitive interpretation goes beyond what the axioms justify: the axioms say nothing about the existence of multiple physical worlds corresponding to wavefunction components.

And so it appears that the Everettian has a dilemma: either the axioms of the theory must be modified to include explicit mention of 'multiple physical worlds', or the existence of these multiple worlds must be some kind of illusion. But the dilemma is false. It is simply untrue that any entity not directly represented in the basic axioms of our theory is an illusion. Rather, science is replete with perfectly respectable entities which are nowhere to be found in the underlying microphysics. Douglas Hofstadter and Daniel Dennett make this point very clearly:

> Our world is filled with things that are neither mysterious and ghostly nor simply constructed out of the building blocks of physics. Do you believe in voices? How about haircuts? Are there such things? What are they? What, in the language of the phyisicist, is a hole—not an exotic black hole, but just a hole in a piece of cheese, for instance? Is it a physical thing? What is a symphony? Where in space and time does 'The Star-Spangled Banner' exist? Is it nothing but some ink trails in the Library of Congress? Destroy that paper and the anthem would still exist. Latin still *exists* but it is no longer a living language. The language of the cavepeople of France no longer exists at all. The game of bridge is less than a hundred years old. What sort of a thing is it? It is not animal, vegetable, or mineral.
>
> These things are not physical objects with mass, or a chemical composition, but they are not purely abstract objects either—objects like the number pi, which is immutable and cannot be located in space and time. These things have birthplaces and histories. They can change, and things can happen to them. They can move about—much the way a species, a disease, or an epidemic can. We must not suppose that science teaches us that every *thing* anyone would want to take seriously is identifiable as a collection of particles moving about in space and time. Hofstadter and Dennett [1981 pp.6–7]

The generic philosophy-of-science term for entities such as these is *emergent*: they are not directly definable in the language of microphysics (try defining a haircut within the Standard Model!) but that does not mean that they are somehow independent of that underlying microphysics. To look in more detail at a particularly vivid example,[5] consider Fig. 2.[6] Tigers are (I take it!) unquestionably real, objective physical objects, but the Standard Model contains quarks, electrons and the like, but no tigers. Instead, tigers should be understood as patterns, or structures, *within* the states of that microphysical theory.

---

[5] I first presented this example in Wallace [2003].
[6] Photograph @ Philip Wallace, 2007. Reproduced with permission.

Figure 2. An object not among the basic posits of the Standard Model.

To see how this works in practice, consider how we could go about studying, say, tiger hunting patterns. In principle—and only in principle—the most reliable way to make predictions about these would be in terms of atoms and electrons, applying molecular dynamics directly to the swirl of molecules which make up, say, the Kanha National Park (one of the sadly diminishing places where Bengal tigers can be found). In practice, however (even ignoring the measurement problem itself!), this is clearly insane: no remotely imaginable computer would be able to solve the 10^{35} or so simultaneous dynamical equations which would be needed to predict what the tigers would do.

Actually, the problem is even worse than this. For in a sense, we *do* have a computer capable of telling us how the positions and momentums of all the molecules in the Kanha National Park change over time. It is called the Kanha National Park. (And it runs in real time!) Even if, *per impossibile*, we managed to build a computer simulation of the Park accurate down to the last electron, it would tell us no more than what the Park itself tells us. It would provide no explanation of any of its complexity. (It would, of course, be a superb vindication of our extant microphysics.)

If we want to understand the complex phenomena of the Park, and not just reproduce them, a more effective strategy can be found by studying the structures observable at the multi-trillion-molecule level of description of this 'swirl of molecules'. At this level, we will observe robust—though not 100% reliable—regularities, which will give us an alternative description of the tiger in a language of cell membranes, organelles, and internal fluids. The principles by which these interact will be derivable from the underlying microphysics, and will involve various assumptions and approximations; hence very occasionally they will be found to fail. Nonetheless, this slight riskiness in our description is overwhelmingly worthwhile given the enormous gain in usefulness of this new description: the language of cell biology is both explanatorily far more powerful, and practically far more useful, than the language of physics for describing tiger behaviour.

Nonetheless it is still ludicrously hard work to study tigers in this way. To reach a really practical level of description, we again look for patterns and regularities, this time in the behaviour of the cells that make up individual tigers (and other living creatures that interact with them). In doing so we will reach yet another language, that of zoology and evolutionary adaptationism, which describes the system in terms of tigers, deer, grass, camouflage, and so on. This language is, of course, the norm in studying tiger hunting patterns, and another (in practice very modest) increase in the riskiness of our description is happily accepted in exchange for another phenomenal rise in explanatory power and practical utility.

The moral of the story is: there are structural facts about many microphysical systems which, although perfectly real and objective (try telling a deer that a nearby tiger is not objectively real) simply cannot be seen if we persist in describing those systems in purely microphysical language. Talk of zoology is of course grounded in cell biology, and cell biology in molecular physics, but the entities of zoology cannot be discarded in favour of the austere ontology of molecular physics alone. Rather, those entities are structures instantiated within the molecular physics, and the task of almost all science is to study structures of this kind.

Of *which* kind? (After all, 'structure' and 'pattern' are very broad terms: almost any arrangement of atoms might be regarded as some sort of pattern.) The tiger example suggests the following answer, which I have previously (Wallace [2003 p.93]) called 'Dennett's criterion' in recognition of the very similar view proposed by Daniel Dennett (Dennett [1991]):

Dennett's criterion: A macro-object is a pattern, and the existence of a pattern as a real thing depends on the usefulness—in particular, the explanatory power and predictive reliability—of theories which admit that pattern in their ontology.

Dennett's own favourite example is worth describing briefly in order to show the ubiquity of this way of thinking: if I have a computer running a chess program, I can in principle predict its next move from analysing the electrical flow through its circuitry, but I have no chance of doing this in practice, and anyway it will give me virtually no understanding of that move. I can achieve a vastly more effective method of predictions if I know the program and am prepared to take the (very small) risk that it is being correctly implemented by the computer, but even this method will be practically very difficult to use. One more vast improvement can be gained if I don't concern myself with the details of the program, but simply assume that whatever they are, they cause the computer to play good chess. Thus I move successively from a language of electrons and silicon chips, through one of program steps, to one of intentions, beliefs, plans, and so forth—each time trading a small increase in risk for an enormous increase in predictive and explanatory power.[7]

---

[7] It is, of course, highly contentious to suppose that a chess-playing computer *really* believes, plans etc. Dennett himself would embrace such claims (see Dennett [1987] for an extensive discussion),

Nor is this account restricted to the relation between physics and the rest of science: rather, it is ubiquitous within physics itself. Statistical mechanics provides perhaps the most important example of this: the temperature of bulk matter is an emergent property, salient because of its explanatory role in the behaviour of that matter. (It is a common error in textbooks to suppose that statistical-mechanical methods are used only because in practice we cannot calculate what each atom is doing separately: even if we could do so, we would be missing important, objective properties of the system in question if we abstained from statistical-mechanical talk.) But it is somewhat unusual because (unlike the case of the tiger) the principles underlying statistical-mechanical claims are (relatively!) straightforwardly derivable from the underlying physics.

For an example from physics which is closer to the cases already discussed, consider the case of quasiparticles in solid-state physics. As is well known, vibrations in a (quantum-mechanical) crystal, although they can in principle be described entirely in terms of the individual crystal atoms and their quantum entanglement with one another, are in practice overwhelmingly simpler to describe in terms of 'phonons'—collective excitations of the crystal which behave like 'real' particles in most respects. And furthermore, this sort of thing is completely ubiquitous in solid-state physics, with different sorts of excitation described in terms of different sorts of 'quasiparticle'—crystal vibrations are described in terms of phonons; waves in the magnetization direction of a ferromagnet are described in terms of magnons, collective waves in a plasma are described in terms of plasmons, etc.

Are quasiparticles real? They can be created and annihilated; they can be scattered off one another; they can be detected (by, for instance, scattering them off 'real' particles like neutrons); sometimes we can even measure their time of flight; they play a crucial part in solid-state explanations. We have no more evidence than this that 'real' particles exist, and so it seems absurd to deny that quasiparticles exist—and yet, they consist only of a certain pattern within the constituents of the solid-state system in question.

When *exactly* are quasiparticles present? The question has no precise answer. It is essential in a quasiparticle formulation of a solid-state problem that the quasiparticles decay only slowly relative to other relevant timescales (such as their time of flight) and when this criterion (and similar ones) are met then quasiparticles are definitely present. When the decay rate is much too high, the quasiparticles decay too rapidly to behave in any 'particulate' way, and the description becomes useless explanatorily; hence, we conclude that no quasi-particles are present. It is clearly a mistake to ask *exactly* when the decay time is

---

but for the purposes of this section there is no need to resolve the issue: the computer can be taken only to 'pseudo-plan', 'pseudo-believe', and so on, without reducing the explanatory importance of a description in such terms.

short enough (2.54 × the interaction time?) for quasiparticles not to be present, but the somewhat blurred boundary between states where quasiparticles exist and states when they don't should not undermine the status of quasiparticles as real, any more than the absence of a precise boundary to a mountain undermines the existence of mountains.

One more point about emergence will be relevant in what follows. In a certain sense emergence is a bottom-up process: knowledge of all the microphysical facts about the tiger and its environment suffices to derive all the tiger-level facts (in principle, and given infinite computing power). But in another sense it is a top-down process: no *algorithmic* process, applied to a complex system, will tell us what higher-level phenomena to look for in that system. What makes it true that (say) a given lump of organic matter has intentions and desires is not something derivable algorithmically from that lump's microscopic constituents; it is the fact that, when it occurs to us to try interpreting its behaviour in terms of beliefs and desires, that strategy turns out to be highly effective.

3 DECOHERENCE AND QUASICLASSICALITY

We now return to quantum mechanics, and to the topic of decoherence. In this section I will briefly review decoherence theory, in a relatively simple context (that of non-relativistic particle mechanics) and in the environment-induced framework advocated by, e.g., Joos, Zeh, Kiefer, Giulini, Kupsch, and Stamatescu [2003] and Zurek [1991, 2003]. (An alternative formalism—the 'decoherent histories' framework advocated by, e.g., Gell-Mann and Hartle [1990] and Halliwell [1998]—is presented in the Introduction to this volume and in Halliwell's contribution, Chapter 3.)

The basic set-up is probably familiar to most readers. We assume that the Hilbert space $\mathcal{H}$ of the system we are interested in is factorized into 'system' and 'environment' subsystems, with Hilbert spaces $\mathcal{H}_S$ and $\mathcal{H}_E$ respectively—

$$\mathcal{H} = \mathcal{H}_S \otimes \mathcal{H}_E. \tag{1}$$

Here, the 'environment' might be a genuinely external environment (such as the atmosphere or the cosmic microwave background); equally, it might be an 'internal environment', such as the microscopic degrees of freedom of a fluid. For decoherence to occur, there needs to be some basis $\{|a\rangle\}$ of $\mathcal{H}_S$ such that the dynamics of the system–environment interaction give us

$$|a\rangle \otimes |\psi\rangle \longrightarrow |a\rangle \otimes |\psi; a\rangle \tag{2}$$

and

$$\langle \psi; a|\psi; \beta\rangle \simeq \delta(a - \beta) \tag{3}$$

on timescales much shorter than those on which the system itself evolves. (Here I use α as a 'schematic label'. In the case of a discrete basis $\delta(\alpha - \beta)$ is a simple Kronecker delta; in the case of a continuous basis, such as a basis of wavepacket states, then (3) should be read as requiring $\langle \alpha | \beta \rangle \simeq 0$ unless $\alpha \simeq \beta$.) In other words, the environment effectively 'measures' the state of the system and records it. (The orthogonality requirement can be glossed as 'record states are distinguishable', or as 'record states are dynamically sufficiently different', or as 'record states can themselves be measured'; all, mathematically, translate into a requirement of orthogonality.) Furthermore, we require that this measurement happens quickly: quickly, that is, relative to other relevant dynamical timescales for the system. (I use 'decoherence timescale' to refer to the characteristic timescale on which the environment measures the system.)

Decoherence has a number of well-known consequences. Probably the best known is diagonalization of the system's density operator. Of course, *any* density operator is diagonal in some basis, but decoherence guarantees that the system density operator will rapidly become diagonal in the $\{|\alpha\rangle\}$ basis, independently of its initial state: any initially non-diagonalized state will rapidly have its non-diagonal elements decay away.

Diagonalization is a synchronic result: a constraint on the system at all times (or at least, on all time intervals of the order of the decoherence timescale). But the more important consequence of decoherence is diachronic, unfolding over a period of time much longer than the decoherence timescale. Namely: because the environment is constantly measuring the system in the $\{|\alpha\rangle\}$ basis, any interference between distinct terms in this basis will be washed away. This means that, in the presence of decoherence, the system's dynamics is *quasiclassical* in an important sense. Specifically: if we want to know the expectation value of any measurement on the system at some future time, it suffices to know what it would be were the system prepared in each particular $|\alpha\rangle$ at the present time (that is, to start the system in the state $|\alpha\rangle \otimes |\psi\rangle$ — for some environment state $|\psi\rangle$ whose exact form is irrelevant within broad parameters — and evolve it forwards to the future time), and then take a weighted sum of the resultant values. Mathematically speaking, this is equivalent to treating the system as though it were in some definite but unknown $|\alpha\rangle$.

Put mathematically: suppose that the superoperator $\mathcal{R}$ governs the evolution of density operators over some given time interval, so that if the system intially has density operator ρ then it has density operator $\mathcal{R}(\rho)$ after that time interval. Then in the presence of decoherence,

$$\mathcal{R}(\rho) = \int d\alpha \langle \alpha | \rho | \alpha \rangle \mathcal{R}(|\alpha\rangle\langle\alpha|). \tag{4}$$

(Again: this integral is meant schematically, and should be read as a sum or an integral as appropriate.)

And, of course, quasiclassicality is rather special. The reason, in general, that the quantum state cannot *straightforwardly* be regarded as a probabilistic description of a determinate underlying reality is precisely that interference effects prevent the dynamics being quasiclassical. In the presence of decoherence, however, those interference effects are washed away.

4 THE SIGNIFICANCE OF DECOHERENCE

It might then be thought—perhaps, at one point, it was thought—that decoherence alone suffices to solve the measurement problem. For if decoherence picks out a certain basis for a system, and furthermore has the consequence that the dynamics of that system are quasiclassical, then—it might seem—we can with impunity treat the system not just as *quasi*classical but straightforwardly as classical. In effect, this would be to use decoherence to give a precise and observer-independent definition of the collapse of the wavefunction: the quantum state evolves unitarily as long as superpositions which are not decohered from one another do not occur; when such superpositions do occur, the quantum state collapses instantaneously into one of them. To make this completely precise would require us to discretize the dynamics so that the system evolves in discrete time steps rather than continuously. The decoherent-histories formalism mentioned earlier is a rather more natural mathematical arena to describe this than the continuous formalism that I developed in Section 3, but the result is the same in any case: decoherence allows us to extract from the unitary dynamics a space of *histories* (strings of projectors onto decoherence-preferred states) and to assign probabilities to each history in a consistent way (i.e., without interference effects causing the probability calculus to be violated).

From a conceptual point of view there is something a bit odd about this strategy. Decoherence is a dynamical process by which two components of a complex entity (the quantum state) come to evolve independently of one another, and it occurs owing to rather high-level, emergent consequences of the particular dynamics and initial state of our universe. Using this rather complex high-level process as a criterion to define a new fundamental law of physics is, at best, an exotic variation of normal scientific practice. (To take a philosophical analogy, it would be as if psychologists constructed a complex theory of the brain, complete with a physical analysis of memory, perception, reasoning, and the like, and then decreed that, as a new fundamental law of physics—and not a mere definition—a system was conscious if and only if it had those physical features.[8])

---

[8] As it happens, this is not a straw man: David Chalmers has proposed something rather similar. See Chalmers [1996] for an exposition, and Dennett [2001] for some sharp criticism.

Even aside from such conceptual worries, however, a pure-decoherence solution to the measurement problem turns out to be impossible on technical grounds: the decoherence criterion is both too strong, and too weak, to pick out an appropriate set of classical histories from the unitary quantum dynamics.

That decoherence is too *strong* a condition should be clear from the language of Section 3. Everything there was approximate, effective, for-all-practical-purposes: decoherence occurs on short timescales (not instantaneously); it causes interference effects to become negligible (not zero); it approximately diagonalizes the density operator (not exactly); it approximately selects a preferred basis (not precisely). And while approximate results are fine for calculational short cuts or for emergent phenomena, they are most unwelcome when we are trying to define new fundamental laws of physics. (Put another way, a theory cannot be 99.99804% conceptually coherent.)

That it is too *weak* is more subtle, but ultimately even more problematic. There are simply *far too many* bases picked out by decoherence—in the language of Section 3 there are far too many system–environment splits which give rise to an approximately decoherent basis for the system; in the language of decoherent histories, there are far too many choices of history that lead to consistent classical probabilities. Worse, there are good reasons (cf. Dowker and Kent [1996]) to think that many, many of these histories are wildly non-classical.

What can be done? Well, if we turn away from the abstract presentation of decoherence theory, and look at the concrete models (mathematical models and computer simulations) to which decoherence has been applied, and if, in those models, we make the sort of system–environment split that fits our natural notion of environment (so that we take the environment, as suggested previously, to be—say—the microwave background radiation, or the residual degrees of freedom of a fluid once its bulk degrees of freedom have been factored out), then we find two things.

First: the basis picked out by decoherence is approximately a coherent-state basis: that is, it is a basis of wavepackets approximately localized in both position and momentum. And second: the dynamics is quasiclassical not just in the rather abstract, bloodless sense used in Section 3, but in the sense that the behaviour of those wavepackets approximates the behaviour predicted by classical mechanics.

In more detail: let $|q, p\rangle$ denote a state of the system localized around phase-space point (q, p). Then decoherence ensures that the state of the system+environment at any time t can be written as

$$|\Psi\rangle = \int dq \, dp \, \alpha q, p; \, t|q, p\rangle \otimes |\epsilon(q, p)\rangle \qquad (5)$$

with $\langle\epsilon(q, p)|\epsilon(q', p')\rangle = 0$ unless $q \approx q'$ and $p \approx p'$. The conventional (i.e., textbook) interpretation of quantum mechanics tells us that $|\alpha(q, p)|^2$ is the probability density for finding the system in the vicinity of phase-space point

(q, p).[9] Then in the presence of decoherence, $|a|^2(q, p)$ evolves, to a good approximation, like a *classical* probability density on phase space: it evolves, approximately, under the Poisson equations

$$\frac{\mathrm{d}}{\mathrm{d}t}(|a(q,p)|^2) \simeq \frac{\partial H}{\partial q}\frac{\partial |a(q,p)|^2}{\partial p} - \frac{\partial H}{\partial p}\frac{\partial |a(q,p)|^2}{\partial q} \qquad (6)$$

where $H(q, p)$ is the Hamiltonian.

On the assumption that the system is classically non-chaotic (chaotic systems add a few subtleties), this is equivalent to the claim that each individual wavepacket follows a classical trajectory on phase space. Structurally speaking, the dynamical behaviour of each wavepacket is the same as the behaviour of a macroscopic classical system. And if there are multiple wavepackets, the system is dynamically isomorphic to a collection of independent classical systems.

(*Caveat*: this does not mean that the wavepackets are actually evolving on phase space. If phase space is understood as the position-momentum space of a collection of classical point particles, then *of course* the wavepackets are not evolving on phase space. They are evolving on a space isomorphic to phase space. Henceforth when I speak of phase space, I mean this space, not the 'real' phase space.)

So: if we pick a particular choice of system–environment split, we find a 'strong' form of quasiclassical behaviour: we find that the system is isomorphic to a collection of dynamically independent simulacra of a classical system. We did not find this isomorphism by some formal algorithm; we found it by making a fairly unprincipled choice of system–environment split and then noticing that that split led to interesting behaviour. The interesting behaviour is no less real for all that.

We can now see that all three of the objections at the start of this section point at the same—fairly obvious—fact: decoherence is an emergent process occurring *within* an already stated microphysics: unitary quantum mechanics. It is not a mechanism to define a part *of* that microphysics. If we think of quasiclassical histories as emergent in this way, then

- the 'conceptual mystery' dissolves: we are not using decoherence to define a dynamical collapse law, we are just using it as a (somewhat pragmatic) criterion for when quantum systems display quasiclassical behaviour
- there is nothing problematic about the approximateness of the decoherence process: as we saw in Section 2, this is an absolutely standard feature of emergence
- similarly, the fact that we had no algorithmic process to tell us in a bottom-up way what system–environment splits would lead to the discovery of interesting

---

[9] At a technical level, this requires the use of phase-space POVMs (i.e., positive operator-valued measures, a generalization of the standard projection-valued measures; see, e.g., Nielsen and Chuang [2000] for details): for instance, the continuous family $\{N|q,p\rangle\langle q,p|\}$ is an appropriate POVM for suitably chosen normalization constant N. Of course, this or any phase-space POVM can only be defined for measurements of accuracy $\leq \hbar$.

structure is just a special case of Section 2's observation that emergence is in general a somewhat top-down process.

Each decoherent history is an emergent structure within the underlying quantum state, on a par with tigers, tables, and the other emergent objects of Section 2—that is, on a par with practically all of the objects of science, and no less real for it.

But the price we pay for this account is that, if the fundamental dynamics are unitary, at the fundamental level there is no collapse of the quantum state. There is just a dynamical process—decoherence—whereby certain components of that state become dynamically autonomous of one another. Put another way: if each decoherent history is an emergent structure within the underlying microphysics, and if the underlying microphysics doesn't do anything to prioritize one history over another (which it doesn't) then all the histories exist. That is: a unitary quantum theory with emergent, decoherence-defined quasiclassical histories is a many-worlds theory.

5 SIMULATION OR REALITY?

At this point, a sceptic might object:

All you have shown is that certain features of the unitarily evolving quantum state are isomorphic to a classical world. If that's true, the most it shows is that the quantum state is running a simulation of the classical world. But I didn't want to recover a *simulation* of the world. I wanted to recover *the world*.

I rather hope that this objection is a straw man: as I attempted to illustrate in Section 2, this kind of structural story about higher-level ontology (the classical world is a structure instantiated in the quantum state) is totally ubiquitous in science. But it seems to be a common enough thought (at least in philosophical circles) to be worth engaging with in more detail.

Note firstly that the very assumption that a certain entity which is structurally like our world is not *our world* is manifestly question-begging. How do we know that space is three-dimensional? We look around us. How do we know that we are seeing something fundamental rather than emergent? We don't; all of our observations (*pace* Maudlin, Chapter 4) are structural observations, and only the sort of a prioristic knowledge now fundamentally discredited in philosophy could tell us more.

Furthermore, physics itself has always been totally relaxed about this sort of possibility. A few examples will suffice:

• Solid matter—described so well, and in such accord with our observations, in the language of continua—long ago turned out to be only emergently continuous, only emergently solid.

- Just as solid state physics deals with emergent quasiparticles, so—according to modern 'particle physics'—elementary particles themselves turn out to be emergent from an underlying quantum field. Indeed, the 'correct'—that is, most explanatorily and predictively useful—way of dividing up the world into particles of different types turns out to depend on the energy scales at which we are working.[10]

- The idea that particles should be emergent from some field theory is scarcely new: in the 19th century there was much exploration of the idea that particles were topological structures within some classical continuum (cf. Epple [1998]), and later, Wheeler [1962] proposed that matter was actually just a structural property of a very complex underlying spacetime. Neither proposal eventually worked out, but for technical reasons: the proposals themselves were seen as perfectly reasonable.

- The various proposals to quantize gravity have always been perfectly happy with the idea that space itself would turn out to be emergent. From Borel dust to non-commutative geometry to spin foam, programme after programme has been happy to explore the possibility that spacetime is only emergently a four-dimensional continuum.[11]

- String theory, currently the leading contender for a quantum theory of gravity, regards spacetime as fundamentally high-dimensional and only emergently four-dimensional, and the recent development of the theory makes the nature of that emergence more and more indirect (it has been suggested, for instance, that the 'extra' dimensions may be several centimetres across).[12] The criterion for emergence, here as elsewhere, is dynamical: if the functional integrals that define the cross sections have the approximate functional form of functional integrals of fields on four-dimensional space, that is regarded as sufficient to establish emergence.

Leaving aside these sorts of naturalistic[13] considerations, we might ask: *what* distinguishes a simulation of a thing from the thing itself? It seems to me that there are two relevant distinctions:

Dependency: Tigers don't interact with simulations of tigers; they interact with the computers that run those simulations. The simulations are

---

[10] The best-known example of this phenomenon occurs in quantum chromodynamics: treating the quark field in terms of approximately free quarks works well at very high energies, but at lower energies the appropriate particle states are hadrons and mesons; see, e.g., Cheng and Li [1984] and references therein for details. For a more mathematically tractable example (in which even the correct choice of whether particles are fermionic or bosonic is energy-level-dependent), see chapter 5 of Coleman [1985], esp. pp.246–53.

[11] For the concept of Borel dust, see Misner, Thorne, and Wheeler [1973 p.1205]; for references on non-commutative geometry, see http://www.alainconnes.org/en/downloads.php; for references on spin foam, see Rovelli [2004].

[12] For a brief introduction to this proposal, see Dine [2007, chapter 29].

[13] I use 'naturalism' in Quine's sense (Quine [1969]): a naturalistic philosophy is one which regards our best science as the only good guide to our best epistemology.

instantiated in 'real' things, and depend on them to remain in existence.

Parochialism: Real things have to be made of a certain sort of stuff, and/or come about in a certain sort of way. Remarkably tiger-like organisms in distant galaxies are not tigers; synthetic sparkling wine, however much it tastes like champagne, is not champagne unless its origins and make-up fit certain criteria.

Now, these considerations are themselves problematic. (Is a simulation of a person itself a person?—see (Hofstadter [1981]) for more thoughts on these matters.) But, as I hope is obvious, both considerations are question-begging in the context of the Everett interpretation: only if we begin with the assumption that our world is instantiated in a certain way can we argue that Everettian branches are instantiated in a relevantly different way.

6 HOW MANY WORLDS?

We are now in a position to answer one of the most commonly asked questions about the Everett interpretation,[14] namely: how much branching actually happens? As we have seen, branching is caused by any process which magnifies microscopic superpositions up to the level where decoherence kicks in, and there are basically three such processes:

1. Deliberate human experiments: Schrödinger's cat, the two-slit experiment, Geiger counters, and the like.
2. 'Natural quantum measurements', such as occur when radiation causes cell mutation.
3. Classically chaotic processes, which cause small variations in initial conditions to grow exponentially, and so which cause quantum states which are initially spread over small regions in phase space to spread over macroscopically large ones. (See Zurek and Paz [1994] for more details; I give a conceptually oriented introduction in Wallace [2001].)

The first is a relatively recent and rare phenomenon, but the other two are ubiquitous. Chaos, in particular, is everywhere, and where there is chaos, there is branching (the weather, for instance, is chaotic, so there will be different weather in different branches). Furthermore, there is no sense in which these phenomena lead to a naturally *discrete* branching process. Quantum chaos gives rise to macroscopic superpositions, and so to decoherence and to the emergence of a branching structure, but that structure has no natural 'grain'. To be sure, by choosing a certain discretization of (phase-)space and time, a discrete branching

---

[14] Other than 'and you believe this stuff?!', that is.

structure will emerge, but a finer or coarser choice would also give branching. And there is no 'finest' choice of branching structure: as we fine-grain our decoherent history space, we will eventually reach a point where interference between branches ceases to be negligible, but there is no precise point where this occurs. As such, the question 'how many branches are there?' does not, ultimately, make sense.

This may seem paradoxical—certainly, it is not the picture of 'parallel universes' one obtains from science fiction. But as we have seen in this chapter, it is commonplace in emergence for there to be some indeterminacy (recall: when *exactly* are quasiparticles of a certain kind present?). And nothing prevents us from making statements like:

Tomorrow, the branches in which it is sunny will have combined weight 0.7

—the combined weight of all branches having a certain macroscopic property is very (albeit not precisely) well defined. It is only if we ask: '*how many* branches are there in which it is sunny?', that we end up asking a question that has no answer.

This bears repeating, as it is central to some of the arguments about probability in the Everett interpretation:

Decoherence causes the Universe to develop an emergent branching structure. The existence of this branching is a robust (albeit emergent) feature of reality; so is the mod-squared amplitude for any *macroscopically described* history. But there is *no* non-arbitrary decomposition of macroscopically described histories into 'finest grained' histories, and *no* non-arbitrary way of counting those histories.

(Or, put another way: asking how many worlds there are is like asking how many experiences you had yesterday, or how many regrets a repentant criminal has had. It makes perfect sense to say that you had many experiences or that he had many regrets; it makes perfect sense to list the most important categories of either; but it is a non-question to ask *how many*.)

If this picture of the world seems unintuitive, a metaphor may help.

1. Imagine a world consisting of a very thin, infinitely long and wide, slab of matter, in which various complex internal processes are occurring—up to and including the presence of intelligent life, if you like. In particular one might imagine various forces acting in the plane of the slab, between one part and another.

2. Now, imagine stacking many thousands of these slabs one atop the other, but without allowing them to interact at all. If this is a 'many-worlds theory', it is a many-worlds theory only in the sense of the philosopher David Lewis (Lewis [1986]): none of the worlds are dynamically in contact, and no (putative) inhabitant of any world can gain empirical evidence about any other.

3. Now introduce a weak force normal to the plane of the slabs—a force with an effective range of 2–3 slabs, perhaps, and a force which is usually very small compared to the intra-slab force. Then other slabs will be detectable from within a slab but will not normally have much effect on events within a slab. If this is a many-worlds theory, it is a science-fiction-style many-worlds theory (or maybe a Philip Pullman or C.S. Lewis many-worlds theory[15]): there are many worlds, but each world has its own distinct identity.

4. Finally, turn up the interaction sharply: let it have an effective range of several thousand slabs, and let it be comparable in strength (over that range) with characteristic short-range interaction strengths within a slab. Now, dynamical processes will not be confined to a slab but will spread over hundreds of adjacent slabs; indeed, *evolutionary* processes will not be confined to a slab, so living creatures in this universe will exist spread over many slabs. At this point, the boundary between slabs becomes epiphenomenal. Nonetheless, this theory is *stratified* in an important sense: dynamics still occurs predominantly along the horizontal axis and events hundreds of thousands of slabs away from a given slab are dynamically irrelevant to that slab.[16] One might well, in studying such a system, divide it into layers thick relative to the range of the inter-slab force—and emergent dynamical processes in those layers would be no less real just because the exact choice of layering is arbitrary.

Ultimately, though, that a theory of the world is 'unintuitive' is no argument against it, provided it can be cleanly described in mathematical language. Our intuitions about what is "reasonable" or "imaginable" were designed to aid our ancestors on the savannahs of Africa, and the universe is not obliged to conform to them.

7 CONCLUSION

The claims of the Everett interpretation are:

- At the most fundamental level, the quantum state is all there is—quantum mechanics is about the structure and evolution of the quantum state in the same way that (e.g.) classical field theory is about the structure and evolution of the fields.

---

[15] See, for instance, Pullman's *Northern Lights* or Lewis's *The Magician's Nephew.*

[16] Obviously there would be ways of constructing the dynamics so that this was not the case: if signals could easily propagate vertically, for instance, the stratification would be lost. But it's only a thought experiment, so we can construct the dynamics how we like.

- As such, the 'Everett interpretation of quantum mechanics' is just quantum mechanics itself, taken literally (or, as a philosopher of science might put it, Realist-ically) as a description of the universe. DeWitt has been widely criticized for his claim that 'the formalism of quantum mechanics yields its own interpretation' (DeWitt [1970]), but there is nothing mysterious or Pythagorean about it: *every* scientific theory yields its own interpretation, or rather (cf. David Deutsch's contribution to this volume) the idea that one can divorce a scientific theory from its interpretation is confused.
- 'Worlds' are mutually dynamically isolated structures instantiated within the quantum state, which are structurally and dynamically 'quasiclassical'.
- The existence of these 'worlds' is established by decoherence theory.

No *postulates* about the worlds have needed to be added: the question of whether decoherence theory does indeed lead to the emergence of a quasiclassical branching structure is (at least in principle) settled a priori for any particular quantum theory once we know the initial state. It is not even a *postulate* that decoherence is the source of all 'worlds'; indeed, certain specialized experiments—notably, some algorithms on putative quantum computers—would also give rise to multiple quasiclassical worlds at least locally; cf. Deutsch [1997].[17]

I will end this discussion on a lighter note, aimed at a slightly different audience. I have frequently talked to physicists who accept Everett's interpretation, accept (at least when pressed!) that this entails a vast multiplicity of quasiclassical realities, but reject the 'many-worlds' label for the interpretation—they prefer to say that there is only one world but that it contains many non- or hardly interacting quasiclassical parts.

But, as I hope I have shown, the 'many worlds' of Everett's many-worlds interpretation are not fundamental additions to the theory. Rather, they are emergent entities which, according to the theory, are present in large numbers. In this sense, the Everett interpretation is a 'many-worlds theory' in just the same sense as African zoology is a 'many-hippos theory': that is, there are entities

---

[17] Since much hyperbole and controversy surrounds claims about Everett and quantum computation, let me add two deflationary comments:

1. There is no particular reason to assume that *all* or even *most* interesting quantum algorithms operate by any sort of 'quantum parallelism' (that is: by doing different classical calculations in a large number of terms in a superposition and then interfering them). Indeed, Grover's algorithm does not seem open to any such analysis. But Shor's algorithm, at least, does seem to operate in this way.
2. The correct claim to make about Shor's algorithm is not (*pace* Deutsch [1997]) that the calculations *could not* have been done other than by massive parallelism, but simply that the actual explanation of how they *were* done—that is, the workings of Shor's algorithm—does involve massive parallelism.

For some eloquent (albeit, in my view, mistaken) criticisms of the link between quantum computation and the Everett interpretation, see Steane [2003].

whose existence is entailed by the theory which deserve the name 'worlds'. So, to Everettians cautious about the 'many-worlds' label, I say: come on in, the water's lovely.

References

Albert, D.Z. and B. Loewer [1988], 'Interpreting the many worlds Interpretation', *Synthese* 77, 195–213.

Barrett, J.A. [1999], *The Quantum Mechanics of Minds and Worlds*, Oxford University Press, Oxford.

Bassi, A. and G. Ghirardi [2003], 'Dynamical reduction models', *Physics Reports* 379, 257. Available online at arXiv.org/abs/quant-ph/0302164.

Bell, J.S. [1981]. 'Quantum mechanics for cosmologists', in C.J. Isham, R. Penrose, and D. Sciama (eds), *Quantum Gravity 2: a second Oxford Symposium*, Clarendon Press, Oxford. Reprinted in Bell [1987], pp.117–38.

——— [1987], *Speakable and Unspeakable in Quantum Mechanics*, Cambridge University Press, Cambridge.

Chalmers, D.J. [1996], *The Conscious Mind: In Search of a Fundamental Theory*, Oxford University Press, Oxford.

Cheng, T.-P. and L.-F. Li [1984], *Gauge Theory of Elementary Particle Physics*, Oxford University Press, Oxford.

Coleman, S. [1985], *Aspects of Symmetry*, Cambridge University Press, Cambridge.

Cushing, J.T., A. Fine, and S. Goldstein (eds) [1996], *Bohmian Mechanics and Quantum Theory: An Appraisal*, Kluwer Academic Publishers, Dordrecht.

Dennett, D.C. [1987], *The Intentional Stance*, MIT Press, Cambridge, Mass.

——— [1991], 'Real patterns', *Journal of Philosophy* 87, 27–51. Reprinted in *Brainchildren*, D. Dennett, (London, Penguin 1998), pp.95–120.

——— [2001], 'The fantasy of first-person science'. Available online at ase.tufts.edu/cogstud/papers/chalmersdeb3dft.htm.

Deutsch, D. [1985], 'Quantum theory as a universal physical theory', *International Journal of Theoretical Physics* 24(1), 1–41.

——— [1997], *The Fabric of Reality*, Penguin, London.

DeWitt, B. [1970], 'Quantum mechanics and reality', *Physics Today* 23(9), 30–5. Reprinted in DeWitt and Graham [1973].

DeWitt, B. and N. Graham (eds) [1973], *The Many-Worlds Interpretation of Quantum Mechanics*, Princeton University Press, Princeton.

Dine, M. [2007], *Supersymmetry and String Theory: Beyond the Standard Model*, Cambridge University Press, Cambridge.

Dowker, F. and A. Kent [1996], 'On the consistent histories approach to quantum mechanics', *Journal of Statistical Physics* 82, 1575–646.

Epple, M. [1998], 'Topology, matter and space, i: Topological notions in 19th-century natural philosophy', *Archive for History of Exact Sciences* 52, 297–392.

Fodor, J.A. [1985], 'Fodor's guide to mental representation: the intelligent auntie's vade-mecum', *Mind* 94, 76–100. Reprinted in Jerry A. Fodor, *A Theory of Content and Other Essays* (MIT Press, 1992).

Gell-Mann, M. and J.B. Hartle [1990], 'Quantum mechanics in the light of quantum cosmology', in W.H. Zurek (ed.), *Complexity, Entropy and the Physics of Information*, pp.425–59, Addison-Wesley, Redwood City, California.

Halliwell, J.J. [1998], 'Decoherent histories and hydrodynamic equations', *Physical Review* D 35, 105015.

Hofstadter, D.R. [1981, May], 'Metamagical themas: A coffeehouse conversation on the Turing test to determine if a machine can think', *Scientific American*, 15–36. Reprinted as 'The Turing test: A coffeehouse conversation', in Hofstadter and Dennett [1981].

Hofstadter, D.R. and D.C. Dennett (eds) [1981], *The Mind's I: Fantasies and Reflections on Self and Soul*, Penguin, London.

Joos, E., H.D. Zeh, C. Kiefer, D. Giulini, J. Kubsch, and I.O. Stamatescu [2003]. *Decoherence and the Appearance of a Classical World in Quantum Theory* (2nd edn), Springer, Berlin.

Kent, A. [1990], 'Against many-worlds interpretations', *International Journal of Theoretical Physics* A5, 1764. Available at arXiv.org/abs/gr-qc/9703089.

Ladyman, J. and D. Ross [2007], *Every Thing Must Go: Metaphysics Naturalized*, Oxford University Press, Oxford.

Lewis, D. [1986], *On the Plurality of Worlds*, Basil Blackwell, Oxford.

Lockwood, M. [1989], *Mind, Brain and the Quantum: the Compound 'I'*, Blackwell Publishers, Oxford.

Misner, C.W., K.S. Thorne, and J.A. Wheeler [1973], *Gravitation*, W.H. Freeman and Company, New York.

Newton-Smith, W.S. [1981], *The Rationality of Science*, Routledge, London.

Nielsen, M.A. and I.L. Chuang [2000], *Quantum Computation and Quantum Information*. Cambridge University Press, Cambridge.

Psillos, S. [1999], *Scientific Realism: How Science Tracks Truth*, Routledge, London.

Quine, W. [1969], 'Epistemology naturalized', in *Ontological Relativity and Other Essays*, Columbia University Press, New York.

Rovelli, C. [2004], *Quantum Gravity*, Cambridge University Press, Cambridge.

Steane, A. [2003], 'A quantum computer only needs one universe', *Studies in the History and Philosophy of Modern Physics* 34, 469–78.

Wallace, D. [2001], 'Implications of quantum theory in the foundations of statistical mechanics', Available online from philsci-archive.pitt.edu.

—— [2003], 'Everett and structure', *Studies in the History and Philosophy of Modern Physics* 34, 87–105. Available online at arXiv.org/abs/quant-ph/0107144 or from philsci-archive.pitt.edu.

Wheeler, J.A. [1962], *Geometrodynamics*, Academic Press, New York.

Zurek, W.H. [1991], 'Decoherence and the transition from quantum to classical', *Physics Today* 43, 36–44. Revised version available online at arXiv.org/abs/quant-ph/0306072.

—— [2003], 'Decoherence, einselection, and the quantum origins of the classical', *Reviews of Modern Physics* 75, 715. Available online at arXiv.org/abs/quant-ph/0105127.

Zurek, W.H. and J.P. Paz [1994], 'Decoherence, chaos and the second law', *Physical Review Letters* 72(16), 2508–11.

2

Quasiclassical Realms

Jim Hartle

ABSTRACT

The most striking observable feature of our indeterministic quantum universe is the wide range of time, place, and scale on which the deterministic laws of classical physics hold to an excellent approximation. This essay describes how this domain of classical predictability of everyday experience emerges from a quantum theory of the universe's state and dynamics.

1 INTRODUCTION

The most striking observable feature of our indeterministic quantum universe is the wide range of time, place, and scale on which the deterministic laws of classical physics hold to an excellent approximation. What is the origin of this predictable quasiclassical realm in a quantum universe characterized by indeterminacy and distributed probabilities? This essay summarizes progress in answering this question both old and new.

The regularities that characterize the quasiclassical realm are described by the familiar classical equations for particles, bulk matter, and fields, together with the Einstein equation governing the regularities of classical spacetime geometry. Our observations of the universe suggest that this quasiclassical realm extends from a moment after the big bang to the far future and over the whole of the visible spatial volume. Were we to set out on a journey to arrive in the far future at a distant galaxy we would count on the regularities of classical physics holding there much as they do here. The quasiclassical realm is thus a feature of the universe independent of human cognition or decision. It is not a feature that we determine, but rather one already present that we exploit as 'information gathering and utilizing systems' (IGUSes) acting in the universe.

So manifest is the quasiclassical realm that is usually simply assumed in constructing effective physical theories that apply in the late universe. Classical spacetime for instance is the starting assumption for the standard model of the elementary particle interactions. Classical spacetime obeying the Einstein equation is assumed in cosmology to reconstruct the past history of our universe.

Even formulations of quantum mechanics assume some part of the universe's quasiclassical realm. Copenhagen quantum theory assumed a separate classical physics and quantum physics with a kind of movable boundary between them. Classical physics was the realm of observers, apparatus, measurement outcomes, and spacetime geometry. Quantum physics was the realm of the particles and quantum fields that were being measured. In the Everett formulations classical spacetime is usually assumed to define the branching histories which are their characteristic feature.

Classical behavior is not exhibited by every closed quantum-mechanical system, only a small minority of them. For example, in the simple case of a non-relativistic particle, an initial wavefunction in the form of a narrow wavepacket may predict a high probability for classical correlations in time between sufficiently coarse-grained determinations of position at a sequence of times. But a generic wavefunction will not predict high probabilities for such correlations. Classical behavior is manifested only through *certain* sets of alternative *coarse-grained* histories and then only for *particular* quantum states. In particular, we cannot expect the classical spacetime that is the central feature of our quasiclassical realm to emerge from every state in quantum gravity, although it must from the particular quantum state of our universe.

This essay summarizes progress in understanding the origin of our quasiclassical realm from a fundamental quantum theory of the universe—a quantum cosmology.[1] There are two inputs to this theory: First, there is the specification of the quantum dynamics (the Hamiltonian in the approximation of classical spacetime.) Second, there is the particular quantum state of our universe. Superstring theory is a candidate for the first input; Hawking's no-boundary wavefunction of the universe [1984] is a candidate for the second. An explanation of the quasiclassical realm from these inputs consists roughly of exhibiting sets of suitably coarse-grained alternative histories of the universe that have high probabilities for patterns of correlations in time summarized by closed systems of deterministic classical equations of motion.

The expansion of the universe together with the properties of the strong interactions mean that nuclear densities ($\sim 10^{15}$gm/cm^3) are the largest reached by ordinary matter any time past the first second after the big bang. There are nearly 80 orders of magnitude separating these densities from the Planck density (10^{93}gm/cm^3) characterizing quantum gravity. This large separation in

---

[1] This is not a review of the long history and many different approaches taken to classicality in quantum theory. Rather it is mostly a brief summary of the author's work, much of it with Murray Gell-Mann, especially (Gell-Mann and Hartle [1990, 1993, 2007]). The references should be understood in this context. For another approach to classicality in the quantum mechanics of closed systems see Zurek [2003].

scale permits the division of the explanation of the quasiclassical realm into two parts: first, the understanding of the origin of classical spacetime in quantum cosmology, and, second, the origin of the classical behavior of matter fields *assuming* classical spacetime.

This division into Planck-scale physics and below corresponds to a division in contemporary theoretical uncertainty. But, more importantly, it corresponds to a division in the nature of the explanation of the parts of the quasiclassical realm. As we shall see, the classical behavior of matter follows mostly from the conservation laws implied by the local symmetries of classical spacetime together with a few general features of the effective theory of the elementary particle interactions (e.g. locality) and the initial condition of the universe (e.g. low entropy). By contrast the emergence of classical spacetime involves the specific theory of the universe's quantum state and a further generalization of quantum theory to deal with histories of spacetime geometry.

These differences should not obscure the point that the explanation of the quasiclassical realm is a unified problem in quantum cosmology. But because of them it is convenient to explain the origin of the quasiclassical behavior of matter first and return to the classical behavior of spacetime later.

This essay is structured as follows: In Section 2 we exhibit a standard textbook derivation of classical behavior largely as a foil to the kind of treatment that we aim for. Section 3 sketches the elements of decoherent histories quantum theory. In Section 4 we consider classicality in a familiar class of oscillator models. Section 5 sketches a general approach to classicality in terms of the approximately conserved hydrodynamic variables. In Section 6 we briefly discuss the origin of the second law of thermodyamics which is necessary for the understanding of the origin of the quasiclassical realm as well as being an important feature of it. Section 7 discusses the origin of classical spacetime that is a prerequisite for a quasiclassical realm. Section 8 considers the Copenhagen approximation to decoherent histories quantum theory that is appropriate for measurement situations. Open questions are mentioned in Section 9. In Section 10 we return to the theme of the connection between fundamental physics and the quasiclassical realm.

2 CLASSICALITY FROM THE EHRENFEST THEOREM

Standard derivations of classical behavior from the laws of quantum mechanics are available in many quantum mechanics texts. One popular approach is based on Ehrenfest's theorem relating the acceleration of the expected value of a particle's position to the expected value of the force:

$$m\frac{d^2\langle x\rangle}{dt^2} = -\left\langle\frac{\partial V}{\partial x}\right\rangle \qquad (2.1)$$

(written here for one-dimensional motion). Ehrenfest's theorem is true in general. But for certain states —typically narrow wavepackets—we may approximately replace the expected value of the force with the force evaluated at the expected position, thereby obtaining a classical equation of motion for that expected value:

$$m\frac{d^2\langle x\rangle}{dt^2} = -\frac{\partial V(\langle x\rangle)}{\partial x}. \tag{2.2}$$

This equation shows that the center of a narrow wavepacket moves on an orbit obeying Newton's laws. More precisely, if we make a succession of position and momentum measurements that are crude enough not to disturb the approximation that allows (2.1) to replace (2.2), then the expected values of the results will be correlated by Newton's deterministic law.

This kind of elementary derivation is inadequate for the type of classical behavior that we hope to discuss in quantum cosmology for the following reasons:

- *Limited to expected values*: The behavior of expected values is not enough to define classical behavior. In quantum mechanics, the statement that the Moon moves on a classical orbit is properly the statement that, among a set of alternative coarse-grained histories of its position as a function of time, the probability is high for those exhibiting the correlations in time implied by Newton's law of motion and near zero for all others. To discuss classical behavior, therefore, we should be dealing with the probabilities of sets of alternative time histories, not with expected or average values.

- *Deals only with measurements*: The Ehrenfest theorem derivation deals with the results of "measurements" on an isolated system with a few degrees of freedom. However, in quantum cosmology we are interested in classical behavior over cosmological stretches of space and time, and over a wide range of subsystems, *independently* of whether these subsystems are receiving attention from observers. Certainly our observations of the Moon's orbit, or a bit of the universe's expansion, have little to do with the classical behavior of those systems. Further, we are interested not just in classical behavior as exhibited in a few variables and at a few times of our choosing, but over the bulk of the universe in as refined a description as possible, so that classical behavior becomes a feature of the universe itself and not a choice of observers.

- *Assumes that the classical equations follow from the fundamental action*: The Ehrenfest theorem derivation relies on a close connection between the equations of motion of the fundamental action and the deterministic laws that govern classical behavior. But when we speak of the classical behavior of the Moon, or of the cosmological expansion, or even of water in a pipe, we are dealing with systems with many degrees of freedom whose phenomenological classical equations of motion (e.g. the Navier–Stokes equation) may be only distantly related to the underlying fundamental theory, say superstring theory. We need a derivation which derives the *form* of the equations as well as the probabilities that they are satisfied.

- *Posits rather than derives the variables exhibiting classical behavior*: The Ehrenfest theorem derivation posits the variables—the position x—in which classical behavior is exhibited. But, as mentioned above, classical behavior is most properly defined in terms of the probabilities of histories. In a closed system we should be able to *derive* the variables that enter into the deterministic laws, especially because, for systems with many degrees of freedom, these may be only distantly related to the degrees of freedom entering the fundamental action.

- *Assumes classical spacetime*: The Ehrenfest derivation assumes classical spacetime if only to define the Schrödinger equation that underlies (2.1). But we aim at explaining the universe's quasiclassical realms from a quantum cosmology founded on a unified theory of the fundamental interactions including gravity. There, spacetime geometry is a quantum variable whose classical behavior must be explained not posited. Indeed, we do not expect to find classical spacetime geometry at the big bang where its quantum fluctuations may be large. Classical spacetime is part of a quasiclassical realm, not separate from it.

Despite these shortcomings, the elementary Ehrenfest analysis already exhibits two necessary requirements for classical behavior: some coarseness is needed in the description of the system as well as some restriction on its initial condition. Not every initial wavefunction permits the replacement of (2.1) by (2.2) and therefore leads to classical behavior; only for a certain class of wavefunctions will this be true. Even given such a suitable initial condition, if we follow the system too closely, say by measuring position exactly, thereby producing a completely delocalized state, we will invalidate the approximation that allows (2.2) to replace (2.1) and classical behavior will not be expected. Some coarseness in the description of histories is therefore also needed.

3 DECOHERENT HISTORIES QUANTUM MECHANICS

These conferences marked 50 years of Everett's formulation of quantum theory. But they were only a year away from marking 25 years of the decoherent (or consistent) histories quantum theory that can be viewed as an extension and to some extent a completion of Everett's work (e.g. Griffiths [2002], Omnès [1994], Gell-Mann [1994]). Today, decoherent histories is the only formulation of quantum theory that is logically consistent, consistent with experiment as far as is known, consistent with the rest of modern physics such as special relativity and field theory, general enough for histories, general enough for cosmology, and generalizable for quantum gravity. Quasiclassical realms are defined through the probabilities of histories of the universe. Decoherent histories quantum theory is the framework for computing them.

The basics of decoherent histories quantum mechanics in a classical background spacetime are reviewed briefly in Simon Saunders' introduction to this volume[2]. We recap the essential ingredients here: For simplicity we consider a model cosmology consisting of a closed system of particles and fields in a very large box. The basic theoretical inputs are a Hamiltonian H specifying quantum dynamics in the box and a density matrix ρ specifying the box's initial quantum state. Coarse-grained histories are represented by class operators C_α. In an operator formulation these can be chains of Heisenberg picture projections at a series of times formed with the aid of H. In a path integral formulation they can be bundles of Feynman paths $q^i(t)$ in configuration space.

Probabilities are properties of exhaustive sets of exclusive histories $\{C_\alpha\}, \alpha = 1, 2, 3, \ldots$. Decoherence is a sufficient condition for their probabilities $\{p(\alpha)\}$ to satisfy the usual rules of probability theory. The central relation defining both decoherence and probability is (see Equations (3), (6), Introduction Sec. 3)

$$D(\alpha', \alpha) \equiv Tr(C_{\alpha'} \rho C_\alpha^\dagger) \approx \delta_{\alpha' \alpha} p(\alpha). \qquad (3.1)$$

The first equality defines the *decoherence functional*. The second defines decoherence and the probabilities that are predicted from H and ρ. A decoherent set of alternative coarse-grained histories is called a *realm* for short.[3]

In a path integral formulation, sets of alternative coarse-grained histories can be defined by partitioning fine-grained configuration space paths $q^i(t)$ into exhaustive sets of exclusive classes $\{c_\alpha\}$. A useful transcription of the decoherence functional (3.1) for such coarse-grained histories on a time interval $[0, T]$ is

$$D(\alpha', \alpha) = \int_{c_{\alpha'}} \delta q' \int_{c_\alpha} \delta q \, \delta(q'_f - q_f) e^{i(S[q'(\tau)] - S[q(\tau)])/\hbar} \rho(q'_0, q_0). \qquad (3.2)$$

---

[2] Alternatively see Hartle [1993] for a tutorial in the present notation.
[3] There will generally be families of realms defined by closely related coarse-grainings that exhibit classical behavior. Realms employing slightly different intervals for defining coarse-grained position are a simple example. Thus it would be more accurate to refer to the quasiclassical realms exhibited by the universe rather than the quasiclassical realm, and we shall do so from now on.

Here, the integrals are over fine-grained paths $q^i(t)$ lying in the classes $c_{\alpha'}$ and c_α, $S[q(t)]$ is the action corresponding to the Hamiltonian H, and $\rho(q_0', q_0)$ is the configuration space representative of the initial density matrix ρ.

4 OSCILLATOR MODELS

The oscillator models pioneered in Feynman and Vernon [1963], Caldeira and Leggett [1983], Unruh and Zurek [1989], and Gell-Mann and Hartle [1993] and developed by many others provide an explicitly computable setting for understanding aspects of classicality. The following assumptions define the simplest model.

• We consider a single distinguished oscillator of mass M, frequency ω_0, and coordinate x interacting with a bath of other oscillators with coordinates Q^A, $A = 1, 2, \ldots$. The coordinates q^i in (3.2) are then $q^i = (x, Q^A)$.
• We suppose the action to be the sum of an action for the x, an action for the Q's, and an interaction that is a linear coupling between them. That is, we assume the action has the form.

$$S[q(\tau)] = S_{\text{free}}[x(\tau)] + S_0[Q(\tau)] + S_{\text{int}}[x(\tau), Q(\tau)]. \qquad (4.1)$$

More specifically, the associated Hamiltonians are

$$H_{\text{free}} = \frac{1}{2}(M\dot{x}^2 + M\omega_0^2 x^2), \qquad (4.2)$$

a similar form with different masses and frequencies for H_0, and

$$H_{\text{int}} = x \sum_A g_A Q^A \qquad (4.3)$$

for some coupling constants g_A.
• We suppose the initial density matrix ρ factors into a product of one depending on the x's and another depending on the Q's which are often called the 'bath' or the 'environment', viz:

$$\rho(q_0', q_0) = \bar{\rho}(x_0', x_0)\rho_B(Q_0', Q_0). \qquad (4.4)$$

We assume that the bath oscillators are in a thermal state ρ_B characterized by a temperature T_B.

- We restrict attention to a simple set of alternative coarse-grained histories that follows the coordinate x of the distinguished oscillator while ignoring the coordinates Q^A of the bath. The histories of the distinguished oscillator are specified by giving an exhaustive set of exclusive intervals of x at each of a series of times $t_1, t_2, \ldots t_n$. A coarse-grained history c_α is the bundle of paths $x(t)$ passing through a particular sequence of intervals $\alpha \equiv (\alpha_1, \alpha_2, \ldots \alpha_n)$ at the series of times $t_1, t_2, \ldots t_n$. For simplicity we take all the intervals to be of equal size Δ and the times to be equally separated by Δt.

Since the bath oscillators are unconstrained by the coarse-graining, the integral over the Q's in (3.2) can be carried out to give a decoherence functional just for coarse-grained histories of the x's of the form:

$$D(\alpha', \alpha) = \int_{c_{\alpha'}} \delta x' \int_{c_\alpha} \delta x \, \delta(x'_f - x_f)$$

$$\times \exp\left\{ i \left(S_{\text{free}}[x'(\tau)] - S_{\text{free}}[x(\tau)] + W[x'(\tau), x(\tau)] \right) / \hbar \right\} \overline{\rho}(x'_0, x_0) \qquad (4.5)$$

where $W[x'(\tau), x(\tau)]$, called the Feynman–Vernon influence phase, summarizes the results of integrations over the Q's.

In the especially simple case of a cut-off continuum of bath oscillators and high bath temperature, the imaginary part of the influence phase is given by Caldeira and Leggett [1983]:

$$\text{Im} \, W[x'(\tau), x(\tau)] = \frac{2M\gamma k T_B}{\hbar} \int_0^T dt (x'(t) - x(t))^2 \qquad (4.6)$$

where γ is a measure of the strength of its coupling to the bath related to the g_A in (4.3). $\text{Im} \, W$ becomes substantial when $x'(\tau)$ and $x(\tau)$ are very different and the time difference Δt is long enough. Then the off-diagonal elements of $D(\alpha', \alpha)$ are exponentially suppressed meaning that the set of alternative histories approximately decoheres [cf. (3.1)]. Roughly, the coarse-graining time required is

$$\Delta t \gtrsim t_{\text{decoh}} \equiv \frac{\hbar^2}{2M\gamma k T_B \Delta^2}. \qquad (4.7)$$

The time t_{decoh} is called the decoherence time (Zurek [1984]). This is typically very much shorter than typical dynamical timescales, for instance $1/\gamma$.

The diagonal elements of the decoherence functional (4.5) are the probabilities $p(\alpha)$ of the individual histories in the set (cf. (3.1)). With a little work, these can be expressed in the following form (Gell-Mann and Hartle [1993]):

$$p(a) = \int_{c_a} \delta x(\cdots) \exp\left[-\int dt \left(\frac{M^2}{4\hbar}\right)\left(\frac{\hbar}{2M\gamma kT_B}\right) E(x(t))^2\right] w(x_0, p_0),$$
(4.8)

the dots denoting factors irrelevant for the subsequent argument. Here $w(x_0, p_0)$ is the Wigner distribution for the density matrix of the distinguished particle $\bar{\rho}$ [cf. (4.4)] and E is

$$E(x(t)) \equiv \ddot{x} + \omega^2 x + 2\gamma\dot{x}$$
(4.9)

where ω is the frequency of the x-oscillator ω_0 renormalized by its interaction with the bath. Equation (4.8) has been organized to show that the factor in front of the imaginary part of the influence phase (4.6) appears inversely in the exponent of this relation.

$E = 0$ is the classical equation of motion for the distinguished oscillator. This includes a frictional force arising from the interaction of the particle with the bath. When the coefficient in front of E^2 in (4.8) is large, the probabilities for histories $p(a)$ will peak about histories that satisfy the classical equations of motion. Thus classical behavior of the distinguished oscillator is predicted. The width of the distribution is a measure of thermal and quantum noise causing deviations from classical predictability.

In this simple case, an analysis of the requirements for classical behavior is straightforward. Equation (4.6) shows that high values of $M\gamma kT_B/\hbar$ are needed to achieve decoherence. Put differently, a strong coupling between the distinguished oscillator and the bath is required if interference phases are to be dissipated efficiently into the bath. However, the larger this coupling is, the smaller the coefficient in the exponent of (4.8) is, decreasing the size of the exponential and *increasing* deviations from classical predictability. This is reasonable: the stronger the coupling to the bath the more noise is produced by the interactions that are carrying away the phases. To counteract that, and achieve a sharp peaking about the classical equation of motion, $M^2/4\hbar$ must be large. That is, high inertia is needed to resist the noise that arises from the interactions with the bath.

Thus, much more coarse-graining is needed to ensure classical predictability than naive arguments based on the uncertainty principle would suggest. Coarse-graining is needed to effect decoherence, and coarse-graining beyond that to achieve the inertia necessary to resist the noise that the mechanisms of decoherence produce.

This derivation of classicality deals genuinely with histories, and is not restricted to measurements. But there is still a close connection between the classical equations and the fundamental action, the variable x which behaves classically was posited, not derived, and classical spacetime was assumed. The progress in relation to the Ehrenfest derviation is summarized in the table below:

Deficiencies of the Ehrenfest derivation

√ *Limited to expected values, but classicality is defined through histories.*
√ *Deals only with measurements on isolated subsystems with a few degrees of freedom.*
× *Assumes the classical equations follow directly from the fundamental action.*
× *Posits rather than derives the variables that exhibit classical behavior.*
× *Assumes classical spacetime.*

---

5 QUASICLASSICAL COARSE-GRAININGS, LOCAL EQUILIBRIUM, AND HYDRODYNAMIC EQUATIONS

Isolated systems evolve toward equilibrium; that is a consequence of statistics. But conserved or approximately conserved quantities approach equilibrium more slowly than others. These include conserved quantities such as energy and momentum that arise from the local symmetries of classical spacetime together with conserved charges and numbers arising from the effective theory of the particle interactions. A situation of *local equilibrium* will generally be reached before complete equilibrium is established, if it ever is. This local equilibrium is characterized by the values of conserved quantities averged over small volumes. Even for systems of modest size, timescales for small volumes to relax to local equilibrium can be very, very much shorter than the timescale for reaching complete equilibrium. Once local equilibrium is established, the subsequent evolution of the approximately conserved quantities can be described by closed sets of effective classical equations of motion such as the Navier–Stokes equation. The local equilibrium determines the values of the phenomenological quantities such as pressure and viscosity that enter into these equations and the constitutive relations among them.

That in a nutshell is the explanation of the quasiclassical realms of matter given classical spacetime. It both identifies the variables in which the quasi-classical realms are defined and the mechanism by which they obey closed sets of equations of motion. To make this more concrete we will review very briefly the standard derivation (e.g. Forster [1975], Zubarev [1974]) of these equations of motion in a simple model. We follow Gell-Mann and Hartle

[2007] where more detail can be found. In his article in these proceedings and in Halliwell [1998, 1999] Jonathan Halliwell explains why sets of sufficiently coarse-grained histories of these variables decohere and lead to high probabilities for correlations in time summarized by the same equations of motion.

Consider a system of conserved particles inside a non-rotating box interacting by local short-range potentials. Let the density matrix ρ—possibly pure—describe the state of the system. Divide the box up into equal volumes of size V labeled by a discrete index $\vec{y}$. Let $T^{\alpha\beta}(\vec{x}, t)$ be the stress-energy-momentum operator in the Heisenberg picture. The energy density $\epsilon(\vec{x}, t)$ and momentum density $\pi^i(x, t)$ are $T^{tt}(\vec{x}, t)$ and $T^{ti}(\vec{x}, t)$ respectively. Let $\nu(\vec{x}, t)$ denote the number density of the conserved particles. Then define

$$\epsilon_V(\vec{y}, t) \equiv \frac{1}{V} \int_{\vec{y}} d^3x \, \epsilon(\vec{x}, t), \tag{5.1a}$$

$$\vec{\pi}_V(\vec{y}, t) \equiv \frac{1}{V} \int_{\vec{y}} d^3x \, \vec{\pi}(\vec{x}, t), \tag{5.1b}$$

$$\nu_V(\vec{y}, t) \equiv \frac{1}{V} \int_{\vec{y}} d^3x \, \nu(\vec{x}, t), \tag{5.1c}$$

where in each case the integral is over the volume labeled by $\vec{y}$. These are the quasiclassical variables for our model. We note that the densities in (5.1) are the variables for a classical hydrodynamic description of this system—for example, the variables of the Navier–Stokes equation.

Were the system in complete equilibrium, the expected values of the quasiclassical variables defined from the density matrix ρ could be accurately computed from the effective density matrix

$$\tilde{\rho}_{\text{eq}} = Z^{-1} \exp[-\beta(H - \vec{U} \cdot \vec{P} - \mu N)]. \tag{5.2}$$

Here, H, $\vec{P}$, and N are the operators for total energy, total momentum, and total conserved number inside the box—all extensive quantities. The c-number intensive quantities β, $\vec{U}$, and μ are respectively the inverse temperature (in units where Boltzmann's constant is 1), the velocity of the box, and the chemical potential. A normalizing factor Z ensures $Tr(\tilde{\rho}_{\text{eq}}) = 1$. In equilibrium the expected values are, for instance,

$$\langle \epsilon_V(\vec{y}, t) \rangle \equiv Tr(\epsilon_V(\vec{y}, t)\rho) \approx Tr(\epsilon_V(\vec{y}, t)\tilde{\rho}_{\text{eq}}). \tag{5.3}$$

Indeed, this relation, and similar ones for $\vec{\pi}_V(\vec{x}, t)$ and $\nu_V(\vec{x}, t)$, *define* equilibrium.

Local equilibrium is achieved when the decoherence functional for sets of histories of quasiclassical variables $(\epsilon, \vec{\pi}, n)$ is given approximately by the *local* version of the equilibrium density matrix (5.2)

$$\tilde{\rho}_{\text{leq}} = Z^{-1} \exp\left[-\int d^3 y \beta(\vec{y}, t)(\epsilon_V(\vec{y}, t) - \vec{u}(\vec{y}, t) \cdot \vec{\pi}_V(\vec{y}, t) - \mu(\vec{y}, t)\nu_V(\vec{y}, t)) \right].$$
(5.4)

(The sum over $\vec{y}$ has been approximated by an integral.) Expected values are given by (5.3) with $\tilde{\rho}_{\text{eq}}$ replaced by $\tilde{\rho}_{\text{leq}}$. The expected values of quasiclassical quantitites are thus functions of the intensive c-number quantities $\beta(\vec{y}, t)$, $\vec{u}(\vec{y}, t)$, and $\mu(\vec{y}, t)$. These are the local inverse temperature, velocity, and chemical potential respectively. They now vary with time and place, as the system evolves toward complete equilibrium.

A closed set of deterministic equations of motion for the expected values of $\epsilon(\vec{x}, t)$, $\vec{\pi}(x, t)$, and $\nu(\vec{x}, t)$ follows from assuming that $\tilde{\rho}_{\text{leq}}$ is an effective density matrix for computing them. To see this, begin with the Heisenberg equations for the conservation of the stress-energy-momentum operator $T^{\alpha\beta}(\vec{x}, t)$ and the number current operator $j^\alpha(\vec{x}, t)$.

$$\frac{\partial T^{\alpha\beta}}{\partial x^\beta} = 0, \quad \frac{\partial j^\alpha}{\partial x^\alpha} = 0.$$
(5.5)

Noting that $\epsilon(\vec{x}, t) = T^{tt}(\vec{x}, t)$ and $\pi^i(\vec{x}, t) = T^{ti}(\vec{x}, t)$, Eqs (5.5) can be written in a 3+1 form and their expected values taken. The result is the set of five equations:

$$\frac{\partial \langle \pi^i \rangle}{\partial t} = -\frac{\partial \langle T^{ij} \rangle}{\partial x^j},$$
(5.6a)

$$\frac{\partial \langle \epsilon \rangle}{\partial t} = -\vec{\nabla} \cdot \langle \vec{\pi} \rangle,$$
(5.6b)

$$\frac{\partial \langle \nu \rangle}{\partial t} = -\vec{\nabla} \cdot \langle \vec{j} \rangle.$$
(5.6c)

The expected values are all functions of $\vec{x}$ and t.

The set of equations (5.6) close for the following reason: Eq. (5.3) with $\tilde{\rho}_{\text{leq}}$ could in principle be inverted to express $\beta(\vec{y}, t)$, $\vec{u}(\vec{y}, t)$, $\mu(\vec{y}, t)$, and therefore $\tilde{\rho}_{\text{leq}}$ itself, in terms of the expected values (5.1). Thus the expected values on the right-hand side of (5.6) become functionals of the quasiclassical variables on the left-hand side, and the equations close.

The process of expression and inversion sketched above could be difficult to carry out in practice. The familiar classical equations of motion arise from further approximations, in particular from assuming that the gradients of all quantities are small. For example, for a non-relativistic fluid of particles of mass m, the most general Galilean-invariant form of the stress tensor that is linear in the gradients of the fluid velocity $\vec{u}(x)$ has the approximate form (e.g. Landau and Lifshitz [1959])

$$\langle T^{ij} \rangle = p\delta^{ij} + mvu^i u^j - \eta \left[\frac{\partial u^i}{\partial x^j} + \frac{\partial u^j}{\partial x^i} - \frac{2}{3} \delta_{ij} \left(\vec{\nabla} \cdot \vec{u} \right) \right]$$
$$- \zeta \delta_{ij} \left(\vec{\nabla} \cdot \vec{u} \right). \tag{5.7}$$

The pressure p and coefficients of viscosity η and ζ are themselves functions say of the expected values (5.1). This form of the stress tensor in (5.6a) leads to the Navier–Stokes equation.

What determines the volume V defining the coarse-grained variables of the quasiclassical realms? The volume V must be large enough to ensure the decoherence of histories constructed from these quasiclassical variables, and beyond that to ensure classical predictability in the face of the noise that typical mechanisms of decoherence produce. The volumes must be small enough to allow local equilibrium. Roughly speaking the volume V should be chosen as small as possible consistent with these requirements. That is, it should be chosen so the quasiclassical realms are maximally refined consistent with decoherence and predictability. Then they are a feature of our universe and not a matter of our choice.

We have now removed two more of the deficiencies of the Ehrenfest derivation as shown in the following table:

Deficiencies of the Ehrenfest derivation

√ *Limited to expected values, but classicality is defined through histories.*
√ *Deals only with measurements on isolated subsystems with a few degrees of freedom.*
√ *Assumes the classical equations follow directly from the fundamental action.*
√ *Posits rather than derives the variables which exhibit classical behavior.*
× *Assumes classical spacetime.*

There remains the origin of classical spacetime to which we turn after a brief discussion of the second law of thermodynamics.

6 THE SECOND LAW OF THERMODYNAMICS

The quasiclassical realms of our universe exhibit two important thermodynamic features that are not directly connected to classical determinism:

- the tendency of a total entropy of the universe to increase
- the tendency of this entropy for nearly isolated subsystems to increase in the same direction of time—this may be called the homogeneity of the thermodynamic arrow of time.

These two features are connected. The first follows from the second, but only in the late universe when nearly isolated subsystems are actually present. In the early universe we have only the first. Together they may be called the second law of thermodynamics.

Thermodynamics, including the second law, is an essential part of classical physics, and, indeed, a prerequisite for it. In the previous section, for example, we assumed the second law when we posited the rapid approach to local equilibrium necessary to derive a closed system of deterministic equations from the conservation relations.

Entropy is generally a measure of the information missing from a coarse-grained description of a physical system. In the case of the quasiclassical variables (5.1) we can define it at a given time as the maximum of the information measure $-Tr(\tilde{\rho}\log\tilde{\rho})$ over density matrices $\tilde{\rho}$ that preserve the expected values of the quasiclassical variables at that time. More specifically, if ρ is the state of the system, we take

$$S(t) \equiv \max_{\tilde{\rho}}[-Tr(\tilde{\rho}\log\tilde{\rho})], \tag{6.1}$$

keeping fixed for each $\vec{y}$

$$\langle \epsilon_V(\vec{y}, t) \rangle \equiv Tr(\epsilon_V(\vec{y}, t)\rho) \approx Tr(\epsilon_V(\vec{y}, t)\tilde{\rho}_{eq}), \tag{6.2}$$

together with the similar relations for $\vec{\pi}_V(\vec{y}, t)$ and $\nu_V(\vec{y}, t)$. The result is the local equilibrium density matrix (5.4).

The entropy defined this way *is* the usual entropy of chemistry, physics, and statistical mechanics. *The coarse-graining in terms of local conserved quantities that exhibits the determinism of the quasiclassical realms thus also defines the entropy for its thermodynamics.*

A special initial quantum state is needed to predict with high probability the classical spacetime whose symmetries are the origin of the conservation laws behind classical determinism. But further conditions on the state are needed for the universe to exhibit the thermodynamic features mentioned above. First, the general increase in total entropy requires that:

- the quantum state is such that the initial entropy is near the minimum it could have for the coarse-graining defining it—it then has essentially nowhere to go but up
- the relaxation time to equilibrium is long compared to the present age of the universe so that the general tendency of its entropy to increase will dominate its evolution.

In our simple model of cosmology, we have neglected gravitation for simplicity, but to understand the origin of the second law it is necessary to consider it. That is because gravity is essential to realizing the first of the conditions above. In a self-gravitating system gravitational clumping increases entropy. The matter in the early universe is not clumped and is nearly in thermal equilibrium—already at maximal entropy. But the spacetime in the early universe is approximately homogeneous, implying that the entropy has much more room to increase through the *gravitational* growth of fluctuations. In a loose sense, as far as gravity is concerned, the entropy of the early universe is low for the coarse-graining defined by quasiclassical variables. The entropy then increases. The no-boundary quantum state in particular implies that gravitational fluctuations are small in the early universe (Halliwell and Hawking [1985], Hawking et al. [1993]) giving entropy room to grow.

Coarse-graining by approximately conserved quasiclassical variables helps with the second of the two conditions above. Small volumes come to local equilibrium quickly. But the approximate conservation ensures that the whole system will approach equilibrium slowly, whether or not such equilibrium is actually attained.

The homogeneity of the thermodynamic arrow of time, which was the other aspect of the second law mentioned at the beginning of this section, cannot follow from the approximately time-reversible dynamics and statistics alone. Rather, the explanation is that the progenitors of today's nearly isolated systems were all far from equilibrium a long time ago and have been running downhill ever since. As Boltzmann put it over a century ago: "The second law of thermodynamics can be proved from the [time-reversible] mechanical theory, if one assumes that the present state of the universe . . . started to evolve from an improbable [i.e. special] state" (Boltzmann [1897]). There is thus a stronger constraint on the initial state than merely having low total entropy. It must be locally low.

The initial quantum state of our universe must be such that it leads to the decoherence of sets of quasiclassical histories that describe coarse-grained spacetime geometry and matter fields. Our observations require this now, and the successes of the classical history of the universe suggests that there was a quasiclassical realm at a very early time. In addition, the initial state must be such that the entropy of the quasiclassical coarse-graining is low in the beginning and also be such that the entropy of presently isolated systems was also low then. Then the universe can exhibit both aspects of the second law of thermodynamics.

The quasiclassical coarse-grainings are therefore distinguished from others, not only because they exhibit predictable regularities of the universe governed by approximate deterministic equations of motion, but also because they are characterized by a sufficiently low entropy in the beginning and a slow evolution towards equilibrium—two properties which make those regularities exploitable.

7 THE ORIGIN OF CLASSICAL SPACETIME

The classical behavior of matter in a given background spacetime depends only weakly on the matter's fundamental quantum physics. The forms of the dynamical equations (5.6) follow largely from conservation laws and the conditions on the interactions necessary for local equilibrium. In a sense, the quasiclassical realms shield us from quantum physics—a happy circumstance that was of great importance historically.

By contrast, the origin of classical spacetime is strongly dependent on the physics of quantum gravity and the theory of the initial quantum state of the universe. That is both the attraction of the issue and its difficulty. It is impossible to say much about this in the space made available for this paper, not least because the quantum theory sketched in Section 3 must be generalized further to deal with quantum spacetime (see, e.g. Hartle [1995, 2007]). The discussion in Sections 3–5 relied on a fixed notion of time to describe histories—a notion that is not available when spacetime itself is a quantum variable. The following heuristic discussion may however give some sense of the issues involved.

Let's first recall one way in which quantum mechanics predicts classical behavior for the motion of a non-relativistic particle. Consider a particle of mass m moving in one dimension x in a potential $V(x)$. Wavefunctions $\psi(x)$ describe its states. Consider wavefunctions that are well approximated in the semiclassical (WKB) form

$$\psi(x) \approx A(x) \exp[iS(x)/\hbar] \tag{7.1}$$

where $S(x)/\hbar$ varies rapidly with x and $A(x)$ varies slowly. Such states predict classical behavior for the particle. Specifically they imply that, in a set of alternative histories suitably coarse-grained in x at a series of times, the probabilities are high for correlations in time summarized by the classical equation of motion for the particle.

A wavefunction satisfying (7.1) also predicts probabilities for *which* classical histories satisfy the equation of motion. That is, it predicts probabilities for the initial conditions to the dynamical equations. Consider histories that pass through a position x at the time the wavefunction is specified. Non-zero probabilities are predicted only for the history with momentum p given by

$$p \equiv m\frac{dx}{dt} = -\nabla S(x), \tag{7.2}$$

and the probability (density) for this history is $|A(x)|^2$. Thus, a wavefunction of semiclassical form (7.1) predicts the probabilities of an ensemble of classical histories labeled by their initial x.

An analogous discussion of the origin of classical spacetime can be given in quantum cosmology (e.g. Hartle [1992]). In quantum gravity the metric on spacetime becomes a quantum variable that can take on any value. Consider a simple model in which the quantum metrics are restricted to be homogeneous, isotropic, and spatially closed. As a model of the matter assume a single homogeneous scalar field $\phi(t)$.

Spacetime geometry in these models is described by metrics of the form

$$ds^2 = -dt^2 + a^2(t)d\Omega_3^2 \tag{7.3}$$

where $d\Omega_3^2$ is the metric on the unit, round, three-sphere. The scale factor $a(t)$ determines how the size of the spatial geometry varies in time. Closed Friedmann–Robertson–Walker cosmological models describing the expansion of the universe from a big bang have metrics of this form with a scale factor $a(t)$ satisfying the Einstein equation. In quantum mechanics $a(t)$ could have any form. Classical behavior of these minisuperspace models means high probability for coarse-grained $a(t)$'s obeying the Einstein equation.

A wavefunction of the universe in this model is a function $\Psi(a, \phi)$ of the scale factor and homogeneous scalar field. Suppose that the wavefunction in some region of (a, ϕ) space is well approximated by the semiclassical form

$$\Psi(a, \phi) \approx A(a, \phi)\exp[iS(a, \phi)/\hbar] \tag{7.4}$$

where $S(a, \phi)/\hbar$ is rapidly varying and $A(a, \phi)$ is slowly varying. Then, from the analogy with non-relativistic quantum mechanics, we expect the wavefunction to predict an ensemble of classical spacetimes with initial data related by the analog of (7.2) and probabilities related to $|A(a, \phi)|^2$.

If our universe is a quantum mechanical system, it has a quantum state. A theory of that state is a necessary part of any 'final theory' and the goal of quantum cosmology. Hawking's no-boundary wavefunction of the universe (Hawking [1984]) is a leading candidate for this theory. In the context of the simple model the no-boundary wavefunction is specified by the following functional integral:

$$\Psi(a, \phi) = \int_C \delta a' \delta \phi' \exp(-I[a'(\tau), \phi'(\tau)]/\hbar). \tag{7.5}$$

Here, the path integration is over histories $a'(\tau)$ and $\phi'(\tau)$ of the scale factor and matter field and $I[a'(\tau), \phi'(\tau)]$ is their Euclidean action. The sum is over cosmological geometries that are regular on a manifold with only one boundary at which $a'(\tau)$ and $\phi'(\tau)$ take the values a and ϕ. The integration is carried out along a suitable complex contour C which ensures the convergence of (7.5) and the reality of the result.

Does the no-boundary quantum state predict classical spacetime for the universe, and if so what classical spacetimes does it predict? The answer to the first part of the question is 'yes'. In certain regions of (a, ϕ) space the defining path integral in (7.5) can be carried out by the method of steepest descents. The dominant contributions come from the complex extrema of the Euclidean action. The leading order approximation of one extre- mum is

$$\Psi(a, \phi) \approx \exp\{[-I_R(a, \phi) + iS(a, \phi)]/\hbar\} \tag{7.6}$$

where $I_R(a, \phi)$ and $-S(a, \phi)$ are the real and imaginary parts of the Euclidean action evaluated at the extremizing path.

When $S(a, \phi)/\hbar$ varies rapidly and $I_R(a, \phi)\hbar$ varies slowly this is a wavefunction of the universe of semiclassical form (7.4). An ensemble of classical spacetimes is predicted with different probabilities. The probabilities will be different for such things as whether the universe bounces at a minimum radius or has an intial singularity, how much matter it has, and the duration of an inflationary epoch. These are important issues for cosmology (e.g. Hartle et al. [2008]). But a quasiclassical realm of matter depends only on the local symmetries of a classical spacetime from the arguments of the preceding three sections. Each classical spacetime with any matter at all will therefore exhibit quasiclassical realms.

Our list of tasks now stands complete like this:

Deficiencies of the Ehrenfest derivation

√ *Limited to expected values, but classicality is defined through histories.*
√ *Deals only with measurements on isolated subsystems with a few degrees of freedom.*
√ *Assumes the classical equations follow directly from the fundamental action.*
√ *Posits rather than derives the variables which exhibit classical behavior.*
√ *Assumes classical spacetime.*

---

8 THE COPENHAGEN APPROXIMATION

Copenhagen quantum mechanics can be seen as an approximation to decoherent histories quantum theory that is appropriate for situations in which a series of measurements is carried out by an apparatus on an otherwise isolated subsystem.

In Copenhagen quantum mechanics, the isolated subsystem is described quantum mechanically. But the apparatus is described by the separate classical physics posited by the theory. The probabilities for the outcomes of a series of 'ideal' measurements is given by unitary evolution of the subsystem's state interrupted at the time of measurements by projections onto the values of the outcomes—the infamous reduction of the wavepacket.

In decoherent histories, apparatus and subsystem are separate parts of one closed system (most generally the universe). In a measurement, a variable of the subsystem, perhaps not quasiclassical and perhaps not otherwise decohering, becomes correlated with a quasiclassical variable of an apparatus. Histories of the measured variable decohere because of this correlation with the decohering histories of the quasiclassical realm.

The Copenhagen prescription for the probability of a series of measurement outcomes can be derived from the probabilities of decoherent histories quantum theory by modeling the measurement situations to which it applies (e.g. Hartle [1991]). Idealized measurement models have a long history in quantum theory (e.g. London and Bauer [1939]). A typical model assumes a closed system—a model universe—consisting of an apparatus, a subsystem which it measures, and perhaps other degrees of freedom. The Hilbert space is idealized as a tensor produce $\mathcal{H}_s \otimes \mathcal{H}_r$ with the factor $\mathcal{H}_s$ for the subsystem and the factor $\mathcal{H}_r$ for the rest including the apparatus. The subsystem is measured by the apparatus at a series of times t_1, t_2, . . ., t_n and is otherwise isolated from the rest of the universe. The initial state is assumed to factor into a pure state $|\psi\rangle$ in $\mathcal{H}_s$ and a density matrix for the rest.

The measurement interaction is idealized to (i) occur at definite moments of time, (ii) create a perfect correlation between the measured alternatives of the subsystem and the registrations of the apparatus—the former represented by sets of projections $\{s_\alpha(t)\}$ in $\mathcal{H}_s$ and the latter by projections $\{R_\alpha(t)\}$ in $\mathcal{H}_r$, and (iii) disturb the subsystem as little as possible (an ideal measurement). Under these assumptions the probability of the sequence of registrations can be shown (Hartle [1991]) to be given by

$$p(a_n, \cdots a_1) = ||s_{a_n}^n(t_n)\cdots s_{a_n}^1(t_1)|\psi\rangle||^2. \tag{8.1}$$

The argument of the square in (8.1) can be thought of as a state of the subsystem which evolved from the initial $|\psi\rangle$ by unitary evolution (constant state in the

Heisenberg picture) interrupted by the action of projections at the times of measurements (state reduction). This is the usual Copenhagen story.

Equation (8.1) is a huge and essential simplification when compared to the basic relation (3.1). Decoherence has been assumed rather than calculated. More importantly, (8.1) refers to a Hilbert space which may involve only a few degrees of freedom whereas (3.1) involves all the degrees of freedom in the universe.

Assumptions (i)–(ii) may hold approximately for many realistic measurement situations. But assumption (iii)—the projection postulate or second law of evolution—does not hold for most.[4] But it is in this way that Copenhagen quantum mechanics is recovered from the more general decoherent histories quantum mechanics once one has a quasiclassical realm. It is not recovered generally but only for idealized measurement situations. It is not recovered exactly but only to an approximation calculable from the more general theory—an approximation which is truly excellent for many realistic measurement situations (Hartle [1991]). The separate classical physics posited by Copenhagen quantum theory is an approximation to the quasiclassical realms. *Copenhagen quantum mechanics is thus not an alternative to decoherent histories, but rather contained within it as an approximation appropriate for idealized measurement situations.*

The founders of quantum mechanics were correct that something besides the wavefunction and Schrödinger equation were needed to understand the theory. But it is not a posited classical world to which quantum mechanics does not apply. Rather, it is the quantum state of the universe together with the theory of quantum dynamics that explains the origin of quasiclassical realms within the more general quantum mechanics of closed systems.

9 SUMMARY AND OPEN QUESTIONS

We now have a complete sketch of an explanation of the quasiclassical realms in our quantum universe in the context of today's fundamental physics. Our discussion has been top-down—proceeding from the classical world to the quantum—starting in today's universe and working backward to the beginning. To summarize, we recapitulate these developments from the bottom up.

- The particular quantum state of our universe implies the classical behavior of spacetime geometry coarse-grained on scales well above the Planck scale. Further, it predicts the homogeneity of this spacetime on cosmological scales that implies a low total entropy leading to the second law of thermodynamics.

---

[4] The idea that the two forms of evolution of the Copenhagen approximation are some kind of problem for quantum theory seems misplaced from the perspective of the quantum mechanics of closed systems which has no such division.

- Local Lorentz symmetries of classical spacetime imply conservation of energy and momentum. The effective theory of the matter interactions implies the approximate conservation of various charges and numbers at various stages in the evolution of the universe.
- Quasiclassical variables specified by ranges of values of the averages of densities of conserved or approximately conserved quantities over small volumes are definable. Sets of alternative histories of these variables decohere and define quasiclassical realms.
- When the volumes are suitably large, the approximate conservation of the quasiclassical variables ensure that they evolve predictably despite the noise that typical mechanisms of decoherence produce.
- When the volumes are suitably small, their contents approach local equilibrium on timescales short compared to those on which the quasiclassical variables are changing.
- Local equilibrium implies that the evolution of the quasiclassical variables obeys a closed, deterministic set of equations of motion incorporating constitutive relations determined by local equilibrium.

The chain above gives a broad outline of how the quasiclassical realms of our universe emerge from its fundamental quantum physics and particular quantum state. However, touch this chain where you will and there are issues that remain to make it more realistic, more general, more complete, more precise, and more quantitative. The following is a short and selective list of outstanding problems:

Decoherence of classical spacetime: Our understanding of the *emergence of* classical spacetime from particular states in quantum gravity is more primitive than our understanding of the emergence of the classical behavior of matter *given* a fixed spacetime. Even a cursory comparison of Section 7 with Section 5 reveals this. Partly this is because we lack a complete and manageable quantum theory of gravity. But even in the low energy effective theory of gravity based on general relativity we do not have precise notions of the diffeomorphism invariant coarse-grainings[5] that *define* the classical behavior of geometry in everyday situations above the Planck scale. And, perforce, we have an inadequate understanding of the mechanisms effecting their decoherence.

More realistic models: The model universe of a static box of particles interacting by short-range potentials that was discussed all too briefly in Section 5 is highly simplified. Models are needed that incorporate at least the following features of the realistic universe:

- *cosmology*: the expansion of the universe, gravitational clumping, possible eternal inflation, the decay of the proton, the formation and evaporation of black holes

---

[5] Defining diffeomorphism invariant coarse-grainings of matter fields in quantum spacetime is itself an issue, see, for example, Giddings et al. [2006].

- *degrees of freedom*: the relativistic quantum fields that are the basic variables of today's effective field theories
- *coarse-graining*: branch-dependent coarse-grainings that express narratives directly in terms of realistic hydrodynamic variables
- *maximal refinement* of coarse-grainings consistent with decoherence and classicality so that the quasiclassical realms are a feature of the universe and not a matter of human choice as discussed at the end of Section 5
- *initial states* that arise from theories of the quantum state of the universe and not from ad hoc assumptions about an environment as in (4.4).

Comparing different realms: A quantum universe can be described by many decohering sets of alternative coarse-grained histories—many realms. The quasiclassical realms are distinguished by a high level of classical predictability and a low initial entropy among other properties. Intuitively they provide the simplest description of the general regularities of the universe that are readily exploitable by IGUSes of the kinds we know about. A genuine comparison of the quasiclassical realms with others that the universe exhibits would require quantitative measures on realms of simplicity, predictability, classicality, etc. Various approaches to such measures have been explored (Gell-Mann and Hartle [1998]) but no completely satisfactory result has yet emerged.

Thus while we have gone far beyond the Ehrenfest derivation, there is still a long way to go!

10 QUASICLASSICAL REALMS AND FUNDAMENTAL PHYSICS

From the present theoretical perspective, a final theory consists of two parts: (1) a dynamical theory specifying quantum evolution (the Hamiltonian in simple models), and (2) a theory of the universe's quantum state. Without both there are no predictions of any kind. With both, probabilities for the members of every decoherent set of alternative histories of the universe are in principle predicted.

Today the search for a final theory has taken physics further and further from the determinism and unique reality that characterized classical physics. A final theory may incorporate quantum indeterminacy, mutually incompatible realms, and not have spacetime at a basic level. In that context, the seemingly prosaic quasiclassical realms of our universe appear remarkable.

On what features of the two parts of a final theory do the quasiclassical realms depend? The discussion in this essay suggests the following:

For the most part, what is required of the dynamical part of theory is an effective theory of the elementary interactions which has the properties necessary for local equilibrium at matter energies well below the Planck scale that are reached in an expanding universe. Specifically, the interactions should be approximately local and dominantly short-range.

However, the specific properties of the only unscreened long-range interaction—gravity—are crucial for the quasiclassical realms. It is the gravitationally driven expansion of the universe that ensures the separation of the energy scales of matter from those of quantum gravity. It is the attractive and universal character of gravity which allows isolated systems to form by the growth and collapse of fluctuations. And it is the relative weakness of the gravitational interaction which allows the universe to remain out of total equilibrium on the timescale of its present age.

More is required of the initial state. It must be such as to imply that histories of cosmological geometry, coarse-grained above the Planck scale, behave classically. The local symmetries of this classical spacetime imply conservation laws that determine in part the variables characterizing the quasiclassical realms. Further, the quantum state must imply an initial condition of low total entropy so that the universe can exhibit the second law of thermodynamics

Almost as important as what the quantum state is required to predict is what it is not required to predict. The beauty of quantum theory is that probabilities are basic. A simple discoverable theory of the quantum state is therefore unlikely to predict with high probability the *particular* classical history we observe with all its apparent complexity. Rather it predicts the simple dynamical regularities common to every classical history with high probability, leaving to quantum accidents the complexity of particular configuration of matter observed. Thus quantum mechanics allows the laws determining probabilities to be simple and still be consistent with present complexity.

It is possible to emphasize how specific these requirements for a quasiclassical realm are. Surely they will not be satisfied by every state in quantum gravity nor every conceivable theory of quantum dynamics. They are sufficiently specific that classicality could be important as a vacuum state selection principle (Hartle et al. [2008a, 2008b]) in theories such as string theory that permit many.

However, it is equally striking how little is required for a final theory to exhibit a quasiclassical realm. The small number and general nature of the requirements discussed above mean that there must be many states and dynamical theories that manifest a quasiclassical realm. Indeed, historically classical physics has shielded us from the nature of the final theory. Given classical spacetime, the form of the classical equations of motion was determined by conservation laws plus Maxwell's equations for the electromagnetic field and the Einstein equation for spacetime geometry. The equations of state, susceptibilities, etc. that entered into these equations could be determined phenomenologically. It was thus not necessary even to know about atoms, much less their quantum mechanics, to explore classical regularities. As far as quantum gravity is concerned, the expansion of the universe has shielded us from an immediate need to consider it by driving the characteristic scales of matter away from the Planck scale.

In these ways our particular universe has allowed a step by step, level by deeper level journey of discovery of the fundamental regularities—a journey which we have not yet completed. The quasiclassical realms of everyday experience have played a central role in this journey, both as a starting point for the exploration and as the chief observational feature of our quantum universe to be explained.

Acknowledgments

Discussions with Murray Gell-Mann and Jonathan Halliwell, over many years, on the emergence of classical behavior in the universe are gratefully acknowledged. This work was supported in part by the National Science Foundation under grant PHY05-55669.

References

Boltzmann, L. [1897], 'Zu Hrn. Zermelo's Abhandlung "Uber die mechanische Erklärung Irreversibler Vorgange" ', *Annalen der Physik* **60**, 392–8.

Caldeira, A. and A. Leggett [1983], 'Path integral approach to quantum Brownian motion', *Physica* **121A**, 587.

Feynman R.P. and J.R. Vernon [1963], 'The theory of a general quantum system interacting with a linear dissipative system', *Annals of Physics* (NY) **24**, 118.

Forster, D. [1975], *Hydrodynamic Fluctuations, Broken Symmetry, and Correlation Functions*, Addison-Wesley, Redwood City.

Gell-Mann, M. [1994], *The Quark and the Jaguar*, W.H. Freeman, New York.

Gell-Mann, M. and J.B. Hartle [1990], 'Quantum mechanics in the light of quantum cosmology', in *Complexity, Entropy, and the Physics of Information*, SFI Studies in the Sciences of Complexity, Vol. VIII, W. Zurek (ed.), Addison Wesley, Reading, MA.

—— [1993], 'Classical equations for quantum systems', *Physical Review* **D 47**. Available online at arXiv: gr-qc/9210010.

—— [1998] 'Strong decoherence', in *Proceedings of the 4th (1994) Drexel Conference on Quantum Non-Integrability: Quantum-Classical Correspondence*, D.H. Feng and B.-L. Hu (eds), International Press of Boston, Hong Kong. Available online at arXiv:gr-qc/9509054.

—— [2007], 'Quasiclassical coarse graining and thermodynamic entropy', *Physical Review* **A 76**. Available online at arXiv:quant-ph/0609190.

Giddings S.B., D. Marolf, and J.B. Hartle [2006], 'Observables in effective gravity', *Physical Review* **D 74**, 064018. Available online at arXiv: hep-th/0512200.

Griffiths, R.B. [2002], *Consistent Quantum Theory*, Cambridge University Press, Cambridge.

Halliwell, J. [1998], 'Decoherent histories and hydrodynamic equations', *Physical Review* **D 58**, 1050151–1228. Available online at arXiv:quant-ph/9805062.

—— [1999], 'Decoherent histories and the emergent classicality of local densities', *Physical Review Letters* **83**, 2481. Available online at arXiv:quant-ph/9905094.

Halliwell, J. and S.W. Hawking [1985], 'Origin of structure in the universe', *Physical Review* **D 31**, 1777.

Hartle, J.B. [1991], 'The quantum mechanics of cosmology', in *Quantum Cosmology and Baby Universes: Proceedings of the 1989 Jerusalem Winter School for Theoretical Physics*, S. Coleman, J.B. Hartle, T. Piran, and S. Weinberg (eds), World Scientific, Singapore, pp.65–157.

—— [1992], 'Spacetime quantum mechanics and the quantum mechanics of spacetime in gravitation and quantizations', *Proceedings of the 1992 Les Houches Summer School*, by B. Julia and J. Zinn-Justin, (eds), [1995], Les Houches Summer School Proceedings, Vol. LVII, North Holland, Amsterdam. Available online at gr-qc/9304006.

—— [1993] 'The quantum mechanics of closed systems', in *Directions in General Relativity, Volume 1: A Symposium and Collection of Essays in honor of Professor Charles W. Misner's 60th Birthday*, B.-L. Hu, M.P. Ryan, and C.V. Vishveshwara, (eds), Cambridge University Press Cambridge. Available online at arXiv:gr-qc/9210006.

—— [1995] 'Quantum mechanics at the Planck scale', talk given at the Workshop on Physics at the Planck Scale, Puri, India, December 1994. Available online at arXiv:gr-qc/9508023.

—— [2007], 'Generalizing quantum mechanics for quantum spacetime', in *The Quantum Structure of Space and Time: Proceedings of the 23rd Solvay Conference on Physics*, D. Gross, M. Henneaux, and A. Sevrin (eds), World Scientific Singapore. Available online at arXiv:gr-qc/0602013.

Hartle, J.B., S.W. Hawking, and T. Hertog [2008a], 'The no-boundary measure of the universe', *Physical Review Letters* **100**, 202301. Available online at arXiv:0711:463,

—— [2008b], 'The classical universes of the no-boundary quantum state' Available online at arXiv:0803.2663/.

Hawking, S.W. [1984], 'The quantum state of the universe', *Nuclear Physics* **B 239**, 257–76.

Hawking, S.W., R. Laflamme, and G.W. Lyons [1993], 'Origin of time asymmetry', *Physical Review* **D 47**, 5342–56.

Landau, L. and E. Lifshitz [1959], *Fluid Mechanics*, Pergamon, London.

London, F. and E. Bauer [1939], *La Théorie de l'Observation en Mécanique Quantique*, Hermann: Paris.

Omnès, R. [1994] *Interpretation of Quantum Mechanics*, Princeton University Press, Princeton.

Unruh W. and W. Zurek [1989], 'Reduction of a wavepacket in quantum Brownian motion', *Physical Review* **D 40**, 1071.

Zubarev, D.N. [1974], *Non-equilibrium Statistical Thermodynamics*, P. Gray and P.J. Shepherd (eds), Consultants Bureau, New York.

Zurek, W. [1984], 'The reduction of the wavepacket: how long does it take?', in *Frontiers of Non-Equilibrium Quantum Statistical Physics*, G. Moore and M. Scully (eds), Plenum Press, New York. Available online at arXiv:quant-ph/0302044.

—— [2003], 'Decoherence, einselection, and the quantum origins of the classical', *Review of Modern Physics* 75, 715.

3

Macroscopic Superpositions, Decoherent Histories, and the Emergence of Hydrodynamic Behaviour

Jonathan Halliwell

ABSTRACT

Macroscopic systems are described most completely by local densities (particle number, momentum, and energy) yet the superposition states of such physical variables, indicated by the Everett interpretation, are not observed. In order to explain this, it is argued that histories of local number, momentum, and energy density are approximately decoherent when coarse-grained over sufficiently large volumes. Decoherence arises directly from the proximity of these variables to exactly conserved quantities (which are exactly decoherent), and not from environmentally induced decoherence. We discuss the approach to local equilibrium and the subsequent emergence of hydrodynamic equations for the local densities. The results are general but we focus on a chain of oscillators as a specific example in which explicit calculations may be carried out. We discuss the relationships between environmentally induced and conservation-induced decoherence and present a unified view of these two mechanisms.

1 INTRODUCTION

If the Everett interpretation of quantum theory is to be taken seriously, there will exist superposition states for macroscopic systems, perhaps even for the entire universe. Since such states are not observed, it is therefore necessary to explain why they go away. This question is a key part of the general question of the emergence of classical behaviour from quantum theory, an issue that has received a considerable amount of attention (Hartle [1994] and Chapter 2 above).

There are a number of different approaches to emergent classicality, but common to most of them is the notion that there must be decoherence, that is, that certain types of quantum states of the system in question exhibit negligible interference, and therefore superpositions of them are effectively equivalent to statistical mixtures. Decoherence has been extensively investigated for the situation in which there is a distinguished system, such as a particle, coupled to its surrounding environment (Joos and Zeh [1985], Paz et al. [1994]). However, for many macroscopic systems, and in particular for the universe as a whole, there may be no natural split into distinguished subsystems and the rest, and another way of identifying the naturally decoherent variables is required. Most generally, decoherence comes about when the variables describing the entire system of interest naturally separate into 'slow' and 'fast', whether or not this separation corresponds to, respectively, system and environment. If the system consists of a large collection of interacting identical particles, such as a fluid for example, the natural set of slow variables are the local densities: energy, momentum, number, charge, etc. They are 'slow' because they are locally conserved. These variables, in fact, are also the variables that provide the most complete description of the classical state of a fluid at a macroscopic level. The most general demonstration of emergent classicality therefore consists of showing that, for a large collection of interacting particles described microscopically by quantum theory, the local densities become effectively classical. Although decoherence through the system—environment mechanism may play a role, since the collection of particles are coupled to each other, it is important to explore the possibility that, at least in some regimes, decoherence could come about because the local densities are almost conserved if averaged over a sufficiently large volume (Gell-Mann and Hartle [1993]). Hence, the approximate decoherence of local densities would then be due to their proximity to a set of exactly conserved quantities, and exactly conserved quantities obey superselection rules.

We will approach these questions using the decoherent histories approach to quantum theory (Gell-Mann and Hartle [1990, 1993], Griffiths [1984, 1993, 1996, 1998], Omnès [1988a, b, c, 1989, 1990, 1992], Halliwell [1994, 2002]). This approach has proved particularly useful for discussing emergent classicality in a variety of contexts. In particular, the issues outlined above are most clearly expressed in the language of decoherent histories. The central object of interest is the decoherence functional:

$$D(\underline{a}, \underline{a}') = \text{Tr}\left(P_{a_n}(t_n) \cdots P_{a_1}(t_1)\rho P_{a_1'}(t_1) \cdots P_{a_n'}(t_n)\right) \qquad (1.1)$$

The histories are characterized by the initial state ρ and by the strings of projection operators $P_a(t)$ (in the Heisenberg picture) at times t_1 to t_n (and $\underline{a}$ denotes the string of alternatives $a_1 \cdots a_n$). Intuitively, the decoherence functional is a measure of the interference between pairs of histories $\underline{a}$, $\underline{a}'$. When it is zero for $\underline{a} \neq \underline{a}'$, we say that the histories are decoherent and probabilities $p(\underline{a}) = D(\underline{a}, \underline{a})$, obeying the usual probability sum rules, may be assigned to them. One can then ask whether these probabilities are strongly peaked about trajectories obeying classical equations of motion. For the local densities, we expect that these equations will be hydrodynamic equations.

The aim of this paper is to review this programme, following primarily Halliwell [1998, 1999a, b, 2003]. We will outline the argument showing how the approximate conservation of the local densities implies negligible interference of their histories at sufficiently coarse-grained scales, and show how hydrodynamic equations of motion for them arise.

2 LOCAL DENSITIES AND HYDRODYNAMIC EQUATIONS

We are generally concerned with a system of N particles described at the microscopic level by a Hamiltonian of the form

$$H = \sum_j \frac{\mathbf{p}_j^2}{2m} + \sum_{\ell > j} V_{j\ell}(\mathbf{q}_j - \mathbf{q}_\ell). \tag{2.1}$$

We are particularly interested in the number density $n(\mathbf{x})$, the momentum density $\mathbf{g}(\mathbf{x})$, and the energy density $h(\mathbf{x})$, defined by,

$$n(\mathbf{x}) = \sum_j \delta(\mathbf{x} - \mathbf{q}_j) \tag{2.2}$$

$$\mathbf{g}(\mathbf{x}) = \sum_j \mathbf{p}_j \, \delta(\mathbf{x} - \mathbf{q}_j) \tag{2.3}$$

$$h(\mathbf{x}) = \sum_j \frac{\mathbf{p}_j^2}{2m} \delta(\mathbf{x} - \mathbf{q}_j) + \sum_{\ell > j} V_{j\ell}(\mathbf{q}_j - \mathbf{q}_\ell)\delta(\mathbf{x} - \mathbf{q}_j). \tag{2.4}$$

We are interested in the integrals of these quantities over volumes that are large compared to the microscopic scale but small compared to macroscopic physics. Integrated over an infinite volume, these become the total particle number N, total momentum P, and total energy H, which are exactly conserved. It is also often more useful to work with the Fourier transforms of the local densities:

$$n(\mathbf{k}) = \sum_j e^{i\mathbf{k} \cdot \mathbf{q}_j} \tag{2.5}$$

$$\mathbf{g}(\mathbf{k}) = \sum_j \mathbf{p}_j \, e^{i\mathbf{k} \cdot \mathbf{q}_j} \tag{2.6}$$

$$h(\mathbf{k}) = \sum_j \frac{\mathbf{p}_j^2}{2m} e^{i\mathbf{k} \cdot \mathbf{q}_j} + \sum_{\ell > j} V_{j\ell}(\mathbf{q}_j - \mathbf{q}_\ell) \, e^{i\mathbf{k} \cdot \mathbf{q}_j} \tag{2.7}$$

These quantities tend to the exactly conserved quantities in the limit $k = |\mathbf{k}| \to 0$, so we are interested in what happens for small but non-zero k.

Setting aside for the moment the issues of decoherence, there is a standard technique for deriving hydrodynamic equations for the local densities.[1,2] It starts with the continuity equations expressing local conservation, which have the form,

$$\frac{\partial \sigma}{\partial t} + \nabla \cdot \mathbf{j} = 0, \qquad (2.8)$$

where σ denotes n, $\mathbf{g}$, or h (and the current $\mathbf{j}$ is a second rank tensor in the case of $\mathbf{g}$). It is then assumed that, for a wide variety of initial states, conditions of local equilibrium are established after a short period of time. This means that, on scales small compared to the overall size of the fluid but large compared to the microscopic scale, equilibrium conditions are reached in each local region, characterized by a local temperature, pressure etc. which vary slowly in space and time. Local equilibrium is described by the density operator

$$\rho = Z^{-1} \exp\left(-\int d^3x\, \beta(\mathbf{x}) \left[h(\mathbf{x}) - \overline{\mu}(\mathbf{x}) n(\mathbf{x}) - \mathbf{v}(\mathbf{x}) \cdot \mathbf{g}(\mathbf{x}) \right] \right), \qquad (2.9)$$

where β, $\overline{\mu}$, and $\mathbf{v}$ are Lagrange multipliers and are slowly varying functions of space and time. β is the inverse temperature, $\mathbf{v}$ is the average velocity field, and $\overline{\mu}$ is related to the chemical potential which in turn is related to the average number density. (Note that the local equilibrium state is defined in relation to a particular coarse-graining, here, the anticipated calculation of average values of the local densities. Hence it embraces all possible states that are effectively equivalent to the state (2.9) for the purposes of calculating those averages.) The hydrodynamic equations follow when the continuity equations are averaged in this state. These equations form a closed set because the local equilibrium form depends (in three dimensions) only on the five Lagrange multiplier fields β, $\overline{\mu}$, $\mathbf{v}$ and there are exactly five continuity equations (2.8) for them. (More generally, it is possible to have closure up to a set of small terms which may be treated as a stochastic process. See Calzetta and Hu [1999] and Brun and Hartle [1999] for example.)

We will in this paper concentrate on the useful pedagogical example of a chain of oscillators, in which many calculations can be carried out explicitly (Halliwell [2003]). The Hamiltonian of this system is

$$H = \sum_{n=1}^{N} \left[\frac{p_n^2}{2m} + \frac{v^2}{2}(q_n - q_{n-1})^2 + \frac{K}{2}(q_n - b_n)^2 \right]. \qquad (2.10)$$

---

[1] A useful basic discussion of the derivation of hydrodynamics may be found in Huang [1987].

[2] The standard derivation of the hydrodynamic equations from their underlying microscopic origins is described in many places. An interesting discussion of the general issues is by Uhlenbeck [1957]. See also, for example, Balescu [1997], Forster [1975], Kreuzer [1975], Zubarev [1974], Zubarev et al. [1996].

There are two cases: $K = 0$ (the simple chain) and $K \neq 0$ (the harmonically bound chain). In the bound chain case, it is also useful to consider the case $b_n = 0$, which corresponds to the situation in which the whole chain moves in a harmonic potential. We consider a finite number N of particles but it is sometimes useful to approximate N as infinite.

The local densities of this system are:

$$n(x) = \sum_{n=1}^{N} \delta(q_n - x) \tag{2.11}$$

$$g(x) = \sum_{n=1}^{N} p_n \delta(q_n - x) \tag{2.12}$$

$$h(x) = \sum_{n=1}^{N} \left[\frac{p_n^2}{2m} + \frac{\nu^2}{2}(q_n - q_{n-1})^2 + \frac{1}{2}K(q_n - b_n)^2 \right] \delta(q_n - x) \tag{2.13}$$

They satisfy the local conservation laws:

$$\dot{n}(x) = -\frac{1}{m}\frac{\partial g}{\partial x} \tag{2.14}$$

$$\dot{g}(x) = -\frac{\partial \tau}{\partial x} - Kxn(x) + K\sum_j b_j \delta(q_j - x) \tag{2.15}$$

$$\dot{h}(x) = -\frac{\partial j}{\partial x} \tag{2.16}$$

The currents $\tau(x)$ and $j(x)$ are rather complicated in configuration space, except in the case where we neglect the interaction term, when they are given by:

$$\tau(x) = \sum_j \frac{p_j^2}{m} \delta(q_j - x) \tag{2.17}$$

$$j(x) = \sum_j \frac{p_j}{m} \left(\frac{p_j^2}{2m} + \frac{1}{2}K(q_j - b_j)^2 \right) \delta(q_j - x) \tag{2.18}$$

The standard derivation of the hydrodynamic equations may be carried out reasonably easily in this model. Instead of the density operator form (2.9) of the local equilibrium state, we work with the equivalent one-particle Wigner function (phase space density),

$$w_j(p_j, q_j) = f(q_j) \exp\left(-\frac{(p_j - mv(q_j))^2}{2mkT(q_j)} \right), \tag{2.19}$$

where f, v, and T are slowly varying functions of space and time (f is simply related to the chemical potential in (2.9)). This is the one-particle distribution function for particle j—it is labelled by j since the particles are distinguishable. If we now average the system (2.14)–(2.16), together with the currents $\tau(x)$, $j(x)$ in the local equilibrium state, we obtain a closed system, since we get three equations for three unknowns. In the case of negligible interactions and $b_j = 0$, we find:

$$\langle n(x) \rangle = N f(x) \tag{2.20}$$

$$\langle g(x) \rangle = m v(x) N f(x) \tag{2.21}$$

$$\langle h(x) \rangle = \left(\frac{1}{2} m v^2 + \frac{1}{2} kT + \frac{1}{2} K x^2 \right) N f(x) \tag{2.22}$$

$$\langle \tau(x) \rangle = \left(m v^2 + kT \right) N f(x) \tag{2.23}$$

$$\langle j(x) \rangle = \left(\frac{3}{2} v k T + \frac{1}{2} m v^3 \right) N f(x) + \frac{K}{2m} x^2 \langle g(x) \rangle \tag{2.24}$$

The first three equations give the explicit inversion between the averages of the local densities and the three slowly varying functions f, v, T. Inserted in (2.14)–(2.16), the above relations give a closed set of equations for the three variables f, v, and T. After some rearrangement, these equations are

$$\frac{\partial f}{\partial t} + v \frac{\partial f}{\partial x} = -f \frac{\partial v}{\partial x} \tag{2.25}$$

$$\frac{\partial v}{\partial t} + v \frac{\partial v}{\partial x} = -\frac{1}{m} \frac{\partial \theta}{\partial x} - \frac{\theta}{mf} \frac{\partial f}{\partial x} - \frac{Kx}{m} \tag{2.26}$$

$$\frac{\partial \theta}{\partial t} + v \frac{\partial \theta}{\partial x} = -2\theta \frac{\partial v}{\partial x}, \tag{2.27}$$

where $\theta = kT$. These are the equations for a one-dimensional fluid moving in a harmonic potential (Huang [1987]). Note that non-trivial equations are obtained even though we have neglected the interaction terms in deriving them. The role of interactions is to ensure the approach to local equilibrium, as we discuss below.

In these expressions, the definition of the temperature fields is essentially equivalent to,

$$\sum_j \frac{1}{2m} (\Delta p_j)^2 \, \delta(q_j - x) = \frac{1}{2} kT(x) n(x) \tag{2.28}$$

(recalling that we are working at long wavelengths, so the δ-function is coarse-grained over a scale of order k^{-1}). Hence temperature arises not from an environment, but from the momentum fluctuations averaged over a coarse-graining volume.

It is straightforward to give a decoherent histories version of the standard derivation of the hydrodynamic equations. We take the initial state to be the local equilibrium state. We take the histories to be characterized by projection operators onto broad ranges of values of the local densities. (The local densities do not commute in general, but they will approximately commute for sufficiently small k and it is not difficult to construct quasiprojectors that are well localized in all three densities). Then, it is easily shown that, for sufficiently broad projections, the histories are peaked about the average values of the local densities, averaged in the initial local equilibrium state (Halliwell [1998]). The standard derivation shows that the average values obey hydrodynamic equations, hence the probabilities are peaked about evolution according to those equations.

However, what is important here is that the decoherent histories approach to quantum theory offers the possibility of a derivation of emergent classicality much more general than that entailed in the standard derivation of hydrodynamics. The standard derivation is rather akin to the Ehrenfest theorem of elementary quantum mechanics which shows that the averages of position and momentum operators obey classical equations of motion. Yet a description of emergent classicality must involve much more than that (Hartle [1994]). Firstly, it must demonstrate decoherence of the local densities, thereby allowing us to talk about probabilities for their histories. Secondly, it should not be restricted to a special initial state. Whilst it is certainly plausible that many initial states will tend to the local equilibrium state, the standard derivation does not obviously apply to superpositions of macroscopic states, which are exactly the states that a description of emergent classicality is supposed to deal with. It is to this more general derivation that we now turn.

3 DECOHERENCE AND CONSERVATION

We begin by describing the connection between decoherence and conservation. It is well known that histories of exactly conserved quantities are exactly decoherent (Hartle et al. [1995], Giulini et al. [1995]). The simple reason for this is that the projectors onto conserved quantities commute with the Hamiltonian. The projectors P_{a_k} on one side of the decoherence functional (1.1) may therefore be brought up against the projectors $P_{a'_k}$ on the other side, hence the decoherence functional is exactly diagonal. (In the situation considered here, in which there are three conserved quantities involved, these quantities must in addition commute with each other, but this is clearly the case.)

There is another way of expressing this that is more useful for the generalization to the local densities. Suppose we take the initial state to be a pure state $|E, \mathbf{P}, N\rangle$ which is an eigenstate of the total energy, momentum and number, and consider a history of projections onto total energy, momentum and number. Clearly, unitary evolution preserves the eigenstate (except for a phase), and the projections acting on it either give back the state, or give zero. This means that

$$P_{a_n}(t_n) \cdots P_{a_1}(t_1)|E, \mathbf{P}, N\rangle \tag{3.1}$$

is equal to either $|E, \mathbf{P}, N\rangle$ or to zero. It is easy to see, by expanding an arbitrary initial state in eigenstates of the conserved quantities, that this implies exact decoherence of histories for any initial state.

Turning now to the local densities, which are most usefully discussed in the Fourier transformed form (2.5)–(2.7), the above argument shows that they define exactly decoherent sets of histories for the case $k = 0$. Now here is the key point: as k departs from zero, the decoherence functional will depart from exact diagonality, but there will still clearly be approximate decoherence if k is sufficiently small. That is, decoherence of local densities essentially follows from an expansion for small k about the exactly decoherent case, $k = 0$. The aim of much of the rest of this section is to spell out in more detail how this works out.

We generalize the above argument for exact decoherence of histories of conserved quantities to locally conserved quantities. We suppose that we have a set of histories characterized by projections onto the local densities for some value of k. We then consider states $|h, \mathbf{g}, n\rangle$ which are approximate eigenstates of the local densities. Exact eigenstates are not possible, but it is not hard to find states that are well localized in all three variables. Under time evolution, the local density eigenstates $|h, \mathbf{g}, n\rangle$ will not remain exact eigenstates, but as long as they remain approximate eigenstates (that is, well localized in the local densities), the above argument goes through and we deduce approximate decoherence.

Hence, denoting the local densities by Q, what we need to show is that, for an initial state localized in the local densities, under time evolution, Q satisfies the condition,

$$\frac{(\Delta Q(t))^2}{\langle Q(t)\rangle^2} \ll 1 \tag{3.2}$$

where

$$(\Delta Q(t))^2 = \langle Q^2(t)\rangle - \langle Q(t)\rangle^2. \tag{3.3}$$

Equation (3.2) means that the state remains strongly peaked in the variable Q under time evolution. The states are then approximate eigenstates of the projectors at each time as long as the widths of the projectors are chosen to be much greater than $(\Delta Q(t))^2$. The condition (3.2) must be true approximately for some $k \neq 0$ since it holds exactly in the limit $k \to 0$. The question is to determine the lengthscale involved.

The number and momentum density are both operators of the form,

$$A = \sum_{n=1}^{N} A_n \tag{3.4}$$

as is the local energy density, if we ignore the interaction term. For such operators it follows that

$$(\Delta A)^2 = \sum_n (\Delta A_n)^2 + \sum_{n \neq m} \sigma(A_n, A_m) \tag{3.5}$$

and

$$\langle A \rangle^2 = \sum_{n,m} \langle A_n \rangle \langle A_m \rangle \tag{3.6}$$

where the correlation function σ is defined by

$$\sigma(A, B) = \frac{1}{2} \langle AB + BA \rangle - \langle A \rangle \langle B \rangle. \tag{3.7}$$

A state will be an approximate eigenstate of the operator A if

$$\frac{(\Delta A)^2}{\langle A \rangle^2} \ll 1. \tag{3.8}$$

The expression for $\langle A \rangle^2$ potentially involves N^2 terms, as does the expression for $(\Delta A)^2$, but the latter will involve only N terms if the correlation functions $\sigma(A_n, A_m)$ are very small or zero for $n \neq m$. So simple product states will be approximate eigenstates and will have $(\Delta A)^2/\langle A \rangle^2$ of order $1/N$. (See Halliwell [1998, 1999] for more detailed examples of this argument.)

Under time evolution, the interactions cause correlations to develop. However, the states will remain approximate eigenstates as long as the correlations are sufficiently small that the second term in (3.5) is much smaller than order N^2. The interactions and the subsequent correlations are clearly necessary in order to get interesting dynamics and in particular the approach to local equilibrium. The interesting question is therefore whether there is a regime where the effects of interactions are small enough to permit decoherence but large enough to produce interesting dynamics. The fact that the variables we are interested in are locally conserved indicates that there is such a regime. The important point is that the local densities become arbitrarily close to exactly conserved quantities as $k \to 0$. This means that, at any time, $(\Delta A)^2/\langle A \rangle^2$ becomes arbitrarily close to its initial value (which is of order $1/N$) for sufficiently small k.

In specific examples, an uncorrelated initial state develops correlations with a typical lengthscale. These correlations typically then decay with time. What is found is that the second term in (3.5) will remain small as long as k^{-1} is much greater than the correlation length. Hence the key physical aspect is the locality of the interactions, meaning that only limited local correlations develop, together with the coarse-graining scale k^{-1} which may be chosen to be sufficiently large that the correlation scale is not seen. Differently put, as k increases from zero, departing from exact decoherence, it introduces a lengthscale k^{-1}. Since the decoherence functional is a dimensionless quantity, clearly nothing significant can happen until k^{-1} becomes comparable with another lengthscale in the system. The natural scale is the correlation length in the local density eigenstates.

4 CHAINS OF OSCILLATORS

Some of the claims of the physical ideas of the previous section may be seen explicitly in the oscillator model with Hamiltonian (2.10). The equations of motion are

$$m\ddot{q}_n + K(q_n - b_n) = v^2(q_{n+1} - 2q_n + q_{n-1}) \tag{4.1}$$

where we take $q_{N+1} = q_1$. This system has been discussed and solved in many places (Huerta and Robertson [1969, 1971], Feynman and Hibbs [1965], Thirring [2002], Agarwal [1971], Tegmark and Shapiro [1994]). The solution may be written,

$$q_n(t) = b_n + \sum_{r=1}^{N}\left[f_{r-n}(t)q_r(0) + \frac{g_{r-n}(t)}{m\Omega}p_r(0)\right] \tag{4.2}$$

were, $\Omega^2 = (K + 2v^2)/m$. For the bound chain, $K \neq 0$, it is most useful to work in the regime in which the interaction between particles is much weaker than the binding to their origins, so $v^2 \ll K$. In this case, the functions $f_r(t)$ and $g_r(t)$ are given by Huerta and Robertson [1969],

$$f_r(t) \approx J_r(\gamma\Omega t)\cos(\Omega t - \pi r/2) \tag{4.3}$$

and

$$g_r(t) \approx J_r(\gamma\Omega t)\sin(\Omega t - \pi r/2) \tag{4.4}$$

where $\gamma = v^2/m\Omega^2$, so $\gamma \ll 1$, and J_r is the Bessel function of order r (and we have used the convenient approximation of taking N to be infinite).

The general behaviour of the solutions is easily seen. The functions $f_{r-n}(t)$ and $g_{r-n}(t)$ loosely represent the manner in which an initial disturbance of particle r affects particle n after a time t, and is given by the properties of Bessel functions (Abramowitz and Stegun [1972]). $J_n(x)$ decays rapidly for large n at fixed x, so distant particles do not affect each other very much. Evolving in x, $J_n(x)$ starts at zero for $x = 0$ (except for $n = 0$, where $J_0(0) = 1$), grows exponentially, and then goes into a slowly decaying oscillation,

$$J_n(x) \sim \left(\frac{2}{\pi x}\right)^{1/2} \cos\left(x - \pi n/2 - \pi/4\right). \tag{4.5}$$

In this oscillatory regime, the Bessel function $J_n(x)$ has only a very limited dependence on n, namely it has the form (4.5) for some n, plus the three possible phase shifts of $\pi/2$. This means that conditions along the chain do not vary very much for reasonably large sections, which relates to the establishment of local equilibrium.

These classical solutions may be used to determine the time evolution of the correlation functions such as $\sigma(q_n, q_m)$, $\sigma(q_n, p_m)$, and $\sigma(p_n, p_m)$ which are the key quantities determining the behaviour of the local densities under time evolution. In brief, what is found is the following. An initially uncorrelated state will develop correlations, but these then decay with time, with the correlations never becoming too great. Furthermore, the quantities $(\Delta q_n)^2$ and $(\Delta p_n)^2$ become dependent only very weakly on n, indicating a situation similar to local equilibrium.

Now consider the local densities of the oscillator chain. For simplicity, we focus on the number density $n(k)$, given by the one-dimensional version (2.5). Following the general scheme outlined in the previous section, we consider initial states that are approximate eigenstates of the local densities. Gaussian states suffice, in fact, and these will be approximate eigenstates of the local densities if we choose the correlation functions $\sigma(q_n, q_m)$, $\sigma(q_n, p_m)$, and $\sigma(p_n, p_m)$ to be zero, or at least sufficiently small, for $n \neq m$.

In a general Gaussian state, we have

$$\langle n(k)\rangle = \sum_{j=1}^{N}\langle e^{ikq_j}\rangle = \sum_{j=1}^{N}\exp\left(ik\langle q_j\rangle - \frac{1}{2}k^2(\Delta q_j)^2\right) \tag{4.6}$$

and

$$(\Delta n(k))^2 = \langle n^\dagger(k)n(k)\rangle - |\langle n(k)\rangle|^2$$
$$= \sum_{j=1}^{N}\sum_{n=1}^{N}\langle e^{ikq_j}\rangle\langle e^{-ikq_n}\rangle\left(e^{k^2\sigma(q_j,q_n)} - 1\right). \tag{4.7}$$

The latter is to be compared with

$$|\langle n(k)\rangle|^2 = \sum_{j=1}^{N}\sum_{n=1}^{N}\langle e^{ikq_j}\rangle\langle e^{-ikq_n}\rangle. \qquad (4.8)$$

With an initially uncorrelated state we have $\sigma(q_j, q_n) = 0$ for $j \neq n$ and we see that

$$(\Delta n(k))^2 = \sum_{j}|\langle e^{ikq_j}\rangle|^2 \left(e^{k^2(\Delta q_j)^2} - 1\right). \qquad (4.9)$$

From this we expect that

$$(\Delta n(k))^2 \ll |\langle n(k)\rangle|^2 \qquad (4.10)$$

as long as k^{-1} does not probe on scales that are too short (compared to Δq_j), and in this case the Gaussian state is an approximate eigenstate as required.

Under time evolution, correlations develop, but we expect that the state will remain an approximate eigenstate if k^{-1} is much greater than the lengthscale of correlation. As k increases from zero we have, to leading order in small k,

$$\frac{(\Delta n(k))^2}{|\langle n(k)\rangle|^2} \sim \frac{k^2(\Delta X)^2}{N^2} \qquad (4.11)$$

where $X = \sum_j q_j$ (the centre of mass coordinate of the whole chain). This will be very small as long as k^{-1} is much larger than the typical lengthscale of a single particle. $(\Delta n(k))^2$ starts to grow very rapidly with k, and (4.10) is no longer valid, when k^{-1} becomes less than the correlation length indicated by $\sigma(q_j, q_n)$. Hence the local density eigenstate state remains strongly peaked about the mean as long as the coarse graining lengthscale k^{-1} remains much greater than the correlation length, confirming the general arguments of the previous section. Similarly, it can be argued that the local density eigenstates also remain localized in the local energy and momentum. This shows that there is approximate decoherence of histories in the oscillator chain model, confirming the general argument.

5 THE APPROACH TO LOCAL EQUILIBRIUM

Given decoherence, we may now look at the probabilities for histories and see if they are peaked around interesting evolution equations. Since we have shown that there is negligible interference between histories with an initial state consisting of a superposition of local density eigenstates, we may take the initial state in these probabilities to be a local density eigenstate. Decoherence alone is not enough to get the hydrodynamic equations. Decoherence ensures that the probabilities for histories are well defined but the probabilities may not be peaked around any particularly interesting histories and in fact will typically describe a situation which is highly stochastic. The hydrodynamic equations we seek form a *closed* set of equations. This requires at least two things in the histories description. First of all, it requires that we consider histories specified by a sufficiently large number of variables—all three of the local densities, particle number, momentum, and energy, are required. It is not enough to consider histories of just one of them. Even classically, the momentum density, for example, will generally not obey a closed evolution equation on its own. Hence, we will assume that histories of all three local densities are considered.

Secondly, the hydrodynamic equations emerge only when the initial state is a local equilibrium state. We need to show how this state, a mixed state, arises from the local density eigenstate, a pure state defined very differently. The point here is that, for sufficiently coarse-grained projections onto the local densities, the object that will determine the probabilities for histories is ρ_1, the one-particle density operator constructed by tracing the local density eigenstate. This is actually quite similar to the local equilibrium state, since they are both mixed states localized in the local densities. They differ in that ρ_1 may still contain correlations (and in particular have non-zero $\sigma(p, q)$) not contained in the local equilibrium state. However, since they are so similar, it is physically extremely plausible that ρ_1 will approach the local equilibrium form on short timescales and this has indeed been explicitly verified in the oscillator model of Halliwell [2003]. It then follows that the probabilities will be peaked about the hydrodynamic equations.

The final picture we have is as follows. We can imagine an initial state for the system which contains superpositions of macroscopically very distinct states. Decoherence of histories indicates that these states may be treated separately and we thus obtain a set of trajectories which may be regarded as exclusive alternatives each occurring with some probability. Those probabilities are peaked about the average values of the local densities. We have argued that each local density eigenstate may then tend to local equilibrium, and a set of hydrodynamic equations for the average values of the local densities then follows. We thus obtain a statistical ensemble of trajectories, each of which obeys hydrodynamic equations. These equations could be very different from one trajectory to the next, having, for example, significantly different values of temperature. In the most general case they could even be in different phases, for example one a gas, one a liquid.

Decoherence requires the coarse-graining scale k^{-1} to be much greater than the correlation length of the local density eigenstates, and the derivation of the hydrodynamic equations requires $k^{-2} \gg (\Delta q)^2$. In brief, the emergence of the classical domain occurs on lengthscales much greater than any of the scales set by the microscopic dynamics.

6 CONNECTIONS WITH ENVIRONMENTALLY INDUCED DECOHERENCE

As noted in Section 1, most studies of decoherence and emergent classicality have focused on the situation in which there is an explicit split into system and environment, and there, the decoherence comes about owing to the coarse-graining over environmental variables. What is the connection between environmentally induced decoherence (EID) and conservation-induced decoherence (CID) considered in this paper? Here we consider three different issues.

First, in EID there is the question of the split into system and environment. Here, the guiding princple is conservation. System usually means a 'large' particle, and environment a bunch of 'small' particles, but in practice the key difference between them is that large particles are slow and the small ones fast, which relates, approximately, to conservation of something, such as number or momentum density. (Although there is typically no limit of exact conservation).

Second, there is a unified way of seeing decoherence of histories in the two cases. Denote a generic variable by $A(t)$. Decoherence of histories of A follows when $A(t)$ commutes with itself at different times. Commutation and the resultant decoherent are usually not exact, so approximate decoherence follows when a condition something like this holds:

$$\| A(t_2)A(t_1) + A(t_1)A(t_2) \| \gg \| [A(t_2), A(t_1)] \| \qquad (6.1)$$

That is, the anticommutator is much bigger than the commutator in some suitably defined operator norm $\| \cdots \|$.

For CID, A is one or more of the hydrodynamic variables $n(k)$, $g(k)$, $h(k)$. These quantities are exactly conserved at $k = 0$ so commute with their values at different times. The inequality (6.1) can be satisfied because the right-hand side of this inequality may be made arbitrarily small by taking k sufficiently small.

For EID, A is typically the position x of a Brownian particle coupled to an environment, and $x(t)$ denotes evolution with the total (system plus environment) Hamiltonian. The norm includes a trace over the environment in a thermal state. The right-hand side will be proportional to $1/M$ (M is the mass of the particle) which will be 'small' due to the massiveness (slowness) of the particle and it will also be proportional to $\hbar$. However, what is more important is that, because of the thermal fluctuations, the left-hand side will be large—it is typically of order $(\Delta x)^2$ which grows with time and with the temperature of the environment. This corresponds to the known fact that EID comes about when thermal fluctuations are much larger than quantum ones.

In brief, (6.1) gives a unified picture of decoherence of histories. It is satisfied in CID because the right-hand side can be made small, and in EID because the left-hand side can be made large.

A third issue is the question of the relative roles of CID and EID in a given system, since one might generally expect that both mechanisms will operate. The point is that it is a question of lengthscales. We have demonstrated decoherence of the local densities starting with exact conservation at the largest lengthscales and then moving inwards. In this way we were able to prove decoherence without using an environment, for certain sets of histories at very coarse-grained scales whose probabilities are peaked about classical paths. However, in general, we would like to be able to assign probabilities to non-classical trajectories. For example, what is the probability that a system will follow an approximately classical trajectory at a series of times, but then at one particular time undergo a very large fluctuation away from the classical trajectory? The approach adopted here, based on conservation, would yield an approximately zero probability for this history, to the level of approximation used. Yet this is a valid question that we can test experimentally. It is at this stage that an environment becomes necessary to obtain decoherence, and indeed it is frequently seen in particular models that when there is decoherence of histories due to an environment, decoherence is obtained for a very wide variety of histories, not just histories close to classical. It is essentially a question of information. Decoherence of histories means that information about the histories of the system is stored somewhere (Gell-Mann and Hartle [1993], Halliwell [1999]). Classical histories need considerably less information to specify than non-classical ones, and indeed specification of the three local densities at any time is sufficient to specify their entire classical histories. This is not enough for non-classical histories, so an environment is required to store the information.

Related to this is the issue of timescales involved in the models considered. Decoherence through interaction with an environment involves a timescale, which is typically exceptionally short. Here, however, there is no timescale associated with decoherence by approximate conservation. The eigenstates of the local densities remain approximate eigenstates for all time. There is, however, a timescale involved in obtaining the hydrodynamic equations, namely, the time required for a local density eigenstate to approach local equilibrium. In this model, this timescale is of order $(\gamma\Omega)^{-1}$ (for the infinite chain in the $K \neq 0$ case).

Finally, and somewhat straying from the issue of EID, we briefly comment on relations to the Boltzmann equation in these models. It would also be of particular interest to look at CID models involving a gas. Many-body field theory may be the appropriate medium in which to investigate this, following the lead of Dodd and Halliwell [2003]. The decoherent histories analysis might confer some interesting advantages over conventional treatments. For example, one-particle dynamics of a gas is described by a Boltzmann equation. One of the assumptions involved in the derivation of the Boltzmann equation is that the initial state of the system contains no correlations, which is clearly very restrictive (Huang [1987]). However, in the general approach used here it is natural to break up any arbitrary initial state into a superposition of local density eigenstates, and that these may then be treated separately because of decoherence. The local density eigenstates typically have small or zero correlations. Hence, decoherence gives some justification for one of the rather restrictitve assumptions of the Boltzmann equation.

7 SUMMARY AND DISCUSSION

Physics would be impossible without conservation laws. They are respected by both classical and quantum mechanics and are the key to understanding the emergence of classical behaviour from an underlying quantum theory. We have outlined the process whereby local densities become effectively classical, using local conservation as the guiding principle. The key idea is to split the initial state into local density eigenstates and show that they are preserved in form under time evolution. The subsequent probabilities for histories are peaked about the average values of the local densities, and the equations of motion for them form a closed set of hydrodynamic form on sufficiently large scales, provided, in general, that sufficient time has elapsed for the local density eigenstates to settle down to local equilibrium.

Since this account of emergent classicality is so firmly anchored in conservation laws, and since conservation laws are so central to physics, it seems likely that this account is very general, and will apply to a wide variety of Hamiltonians and initial states: as long as there are conserved quantities there is a regime nearby of almost conserved quantities behaving quasiclassically.

An important general question is whether the quasiclassical domain derived in this way is unique. The familiar quasiclassical domain is characterized by local densities obeying closed sets of deterministic evolution equations. This domain may also be referred to as a *reduced description* of the quantum system, in which, at sufficiently coarse-grained scales, certain predictions are possible using only a limited set of variables, the local densities, without having to solve the full quantum theory. Could there be an utterly different domain, characterized by completely different variables, but still obeying deterministic evolution equations? That is, is there another completely different reduced description of the system? The derivation described here rests entirely on conservation laws, and from that point of view, the existence of another quasiclassical domain seems most unlikely, unless there are conservation laws that we have not yet discovered. So perhaps the appropriate question is to ask whether different reduced descriptions of the system are possible that do not rely on conservation laws. Little is known about this issue at present.

Acknowledgements

I am very grateful to Jim Hartle for very many discussions on emergent classicality over a long period. I would also like to thank Simon Saunders for encouraging me to write this article and for many useful discussions.

References

Abramowitz, M. and A. Stegun [1972], *Handbook of Mathematical Functions*, Dover, New York.

Agarwal, G.S. [1971], 'Entropy, the Wigner distribution function, and the approach to equilibrium of a system of coupled harmonic oscillators', *Physical Review* **A3**, 828.

Balescu, R. [1997], *Statistical Dynamics: Matter out of Equilibrium*, Imperial College Press, London.

Brun, T. and J.B. Hartle [1999], 'Classical dynamics of the quantum harmonic chain', *Physical Review* **D60**, 123503.

Calzetta, E. A. and B.L. Hu [1999], 'Influence action and decoherence of hydrodynamic modes', *Physical Review* **D59**, 065018.

Dodd, P. J. and J.J. Halliwell [2003], 'Decoherence and records for the case of a scattering environment', *Physical Review* **D67**, 105018.

Feynman, R.P. and A.R. Hibbs [1965], *Quantum Mechanics and Path Integrals*, McGraw-Hill, New York.

Forster, D. [1975], *Hydrodynamic Fluctuations, Broken Symmetry and Correlation Functions*, Benjamin, Reading, MA.

Gell-Mann, M. and J.B. Hartle, [1990], 'Quantum mechanics in the light of quantum cosmology', in *Complexity, Entropy and the Physics of Information, SFI Studies in the Sciences of Complexity*, Vol. VIII, W. H. Zurek (ed.), Addison Wesley, Reading.

—— [1993], 'Classical equations for quantum systems', *Physical Review* **D47**, 3345.

Giulini, D., C. Kiefer, and H.D.Zeh [1995], *Physics Letters* **A199**.

Griffiths, R. B. [1984], 'Consistent histories and the interpretation of quantum mechanics', *Journal of Statistical Physics* **36**, 219.

—— [1993], 'Consistent interpretation of quantum mechanics using quantum trajectories', *Physical Review Letters* **70**, 2201.

—— [1996], 'Consistent histories and quantum reasoning', *Physical Review* **A54**, 2759.

—— [1998], Choice of consistent family, and quantum incompatibility', *Physical Review* **A57**, 1604.

Halliwell, J.J. [1994], 'A review of the decoherent histories approach to quantum mechanics', in *Fundamental Problems in Quantum Theory*, D. Greenberger and A. Zeilinger (eds), *Annals of the New York Academy of Sciences* **775**.

—— [1998], 'Decoherent histories and hydrodynamic equations', *Physical Review* **D58** 105015.

—— [1999a], 'Somewhere in the universe: Where is the information stored when histories decohere?, *Physical Review* **D60**, 105031.

Halliwell, J.J. [1999b], 'Decoherent histories and the emergent classicality of local densities', *Physical Review Letters* **83**, 2481.

—— [1999c], 'Two approaches to coupling classical and quantum variables', *International Journal of Theoretical Physics* **38**, 2969.

—— [2003a], 'Some recent developments in the decoherent histories approach to quantum theory'. Available online at quant-ph/0301117. To appear in *Proceedings of the Conference, Decoherence, Information, Complexity, Entropy* (DICE), T. Elze (ed.), Piombino, Italy.

—— [2003b], 'Decoherence of histories and hydrodynamic equations for a linear oscillator chain', *Physical Review* **D68**, 025018.

Hartle, J.B. [1994], 'Quasiclassical realms in a quantum universe', *Proceedings of the Cornelius Lanczos International Centenary Conference*, J.D. Brown, M.T.Chu, D.C. Ellison and R.J.Plemmons (eds). SIAM, Philadelphia. Available online as e-print gr-qc/9404017.

Hartle, J.B., R. Laflamme, and D. Marolf [1995], 'Conservation laws in the quantum mechanics of closed systems', *Physical Review* **D51**, 7007.

Huang, K. [1987], *Statistical Mechanics*, 2nd edition, New York, Chichester, Wiley.

Huerta, M.A. and H.S. Robertson [1969], 'Entropy, information theory, and the approach to equilibrium of coupled harmonic oscillator systems', *Journal of Statistical Physics* **1**, 393.

—— [1971], 'Approach to equilibrium of coupled harmonic oscillator systems II', *Journal of Statistical Physics* **3**, 171.

Joos, E. and H.D. Zeh, [1985], 'The emergence of classical properties through interaction with the environment', *Zeitschrift für. Physik* **B59**, 223.

Kreuzer, H.J. [1981], *Non-equilibrium Thermodynamics and its Statistical Foundations*, Clarendon Press, Oxford.

Omnès, R. [1988], 'Logical reformulation of quantum mechanics, I. Foundations', *Journal of Statistical Physics* **53**, 893.

—— [1988], 'Logical reformulation of quantum mechanics, II. Interferences and the Einstein-Podolsky-Rosen experiment', *Journal of Statistical Physics* **53**, 933.

—— [1988], 'Logical reformulation of quantum mechanics. III. Classical limit and irreversibility', *Journal of Statistical Physics* **53**, 957.

—— [1989], 'Logical reformulation of quantum mechanics. IV. Projectors in semiclassical physics', *Journal of Statistical Physics* **57**, 357.

—— [1990], 'From Hilbert space to common sense: A synthesis of recent progress in the interpretation of quantum mechanics', *Annals of Physics* **201**, 354.

—— [1992], 'Consistent interpretations of quantum mechanics', *Review of Modern Physics* **64**, 339.

Paz, J.P. and W.H. Zurek [1993], 'Environment-induced decoherence, classicality, and consistency of quantum histories', *Physical Review* **D48**, 2728.

Tegmark, M. and H.S. Shapiro [1994], 'Decoherence produces coherent states: An explicit proof for harmonic chains', *Physical Review* **E50**, 2538.

Thirring, W. [2002], *Quantum Mathematical Physics: Atoms, Molecules and Large Systems*, Springer, New York.

Uhlenbeck, G.E. [1957], *Appendix, Lectures in Applied Mathematics: Probability and Related Topics in the Physics Sciences*, M.Kac, Interscience, NY.

Zubarev, D. N. [1974], *Non-equilibrium Statistical Thermodynamics*, Consultants Bureau, New York.

Zubarev, D., V. Morozov, and G. Röpke [1996], *Statistical Mechanics of Non-equilibrium Processes*, Vol.1, Akademie Verlag, Berlin.

Zurek, W.H. [1994], 'Preferred sets of states, predictability, classicality, and environment-induced decoherence', in *Physical Origins of Time Asymmetry*, J.J.Halliwell, J.Perez-Mercader, and W.Zurek (eds), Cambridge University Press, Cambridge.

——— [1998], 'Decoherence, einselection, and the existential interpretation (the rough guide)', *Philosophical Transactions of the Royal Society* of *London* **A356**, 1793–820. Available online at quant-ph/9805065.

PART II
PROBLEMS WITH ONTOLOGY

4

Can the World be Only Wavefunction?

Tim Maudlin

The argument will be organized around three citations. The first I take to express something correct but nonetheless, by conversational implication, rather misleading (at least I have been long misled by it). The second I take to express something flatly incorrect. And the third points at (but does not resolve) the fundamental issue that underlies the first two.

The first is the widely cited formulation of the measurement problem by John Bell: 'Either the wavefunction, as given by the Schrödinger equation, is not everything, or it is not right' [1987 p.201]. Bell's dilemma appears to offer us two quite distinct routes to solving the problem: either add to the ontology, so that the wavefunction is not everything, or change the dynamics, so that the linear evolution does not always obtain. Like many others, I took the first option to be the way to understand deBroglie–Bohm, and the second to be the way to understand GRW.

With respect to *one* problem, this seems perfectly adequate. That problem concerns the understanding of the probabilities employed in the quantum theory. In blank terms, the question can be put this way: what are the probabilities (i.e. numbers derived via Born's rule, numbers that in the formal sense satisfy the conditions for being a probability measure) *for*? If the wavefunction is not everything, as in Bohm's theory, then they can be probabilities for the additional variables to take one or another value. If the wavefunction is not always governed by a linear equation, as in GRW, then they can be probabilities for the wavefunction to evolve one way rather than another. And if the wavefunction always evolves deterministically and there is nothing but the wavefunction, then it is unclear how the 'probabilities' are to be understood.

Over the past decade or so, I have come to appreciate the ultimate inadequacy of this way of putting things. It does still seem to me that the Born 'probabilities' need to be probabilities for *something* physical to happen, but it has become more and more evident in addition that they have to be probabilities for *the right kind* of thing to happen. Let's start with a simple example.

I have said that, in Bohm's theory, we understand the Born rule probabilities as probabilities for the additional variables (particle positions in a common low-dimensional space) to take one or another value, that is, for the particles to evolve into one or another configuration in this space. But this alone is not *sufficient* for an adequate understanding of the theory. At least two more conditions must obtain. First, these additional variables must exactly *not* be 'hidden' variables: they had better be *manifest* variables, variables whose value is, in common experimental situations, easy to reliably determine (with the usual sorts of accuracy). Thus, suppose we want to account for an experiment that involves an archetypal pointer that can end up pointing one way or another. Then it is clearly not enough to simply have something that will end up pointing in one of two directions, such that the theory ascribes different probabilities to the different outcomes. That something ought to be the sort of thing that people can easily see: they ought to be able to reliably determine, just by looking, which way it points. If they couldn't, then the theory, despite having distinct physical alternatives that the probabilities are *for*, could not explain what we want to explain, viz., things that we take to happen in laboratories (or in everyday life).

Now in the case of Bohm's theory, this turns out not to be a problem: not only will the theory give probabilities for the pointer to point one way or another, it also has the resources to explain, at least in a schematic way, how people who look at pointers can reliably tell which way they are pointing. But when it comes to the GRW theory, things do not proceed as smoothly.

We will approach this in steps. We begin by considering GRW as an ontologically monistic theory: all it postulates to exist is a wavefunction. The wavefunction is something that evolves in a very, very, very, very high-dimensional space. (Alternatively, the wavefunction has a great number of physical degrees of freedom.) We are not yet entitled to call this space 'configuration space' since there is no low-dimensional space at all, and a fortiori there is nothing configured one way or another in it. In the sort of experimental situation we are considering (like a Schrödinger's cat case), the pure linear evolution of the wavefunction alone will result in its having two sizable lumps, in two very different regions of this high-dimensional space. And, because of the collapse dynamics, we can say what the probabilities are probabilities *for*: they are probabilities for one of the lumps to end up with a very, very, very much higher amplitude than the other. Given the right dynamics, those probabilities will be (very nearly) the squared-amplitudes of the linearly evolved lumps.

Now it should be evident that without some *further* story, this probability for the wavefunction to lump up one way or another *makes no contact at all* with our original problem. We began thinking that we were doing an experiment in a laboratory, an experiment that ended with a pointer pointing either, say, to a '1' or to a '0'. In the Bohmian picture, we have many particles arrayed in a three-dimensional space, and as the experiment is analyzed (with a fixed initial wavefunction and a somewhat uncertain initial configuration) we find that the particles will evolve either into a configuration like A or a configuration like B:

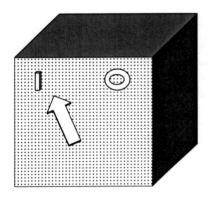   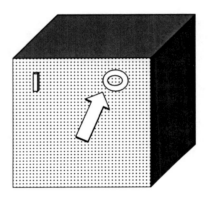

Configuration A   Configuration B

Figure 1.

Notice that I have *not* said which configuration corresponds to which outcome: I ask the reader to see if this can be determined without further interpretive machinery being added. In an obvious sense, it can.

Although I have contrived this example to be particularly simple, I insist that I have nowhere *cheated*. For example, in conveying information about the two configurations, I have used a very perspicuous diagrammatic method. If, instead, I had *listed* the locations of, say 10^{23} distinct particles in space, it would take much more effort to tell which outcome corresponds to configuration A and which to configuration B. But the perspicuous representation is just that: a perspicuous representation of physical facts that are straightforwardly postulated by the theory. I have done nothing sneaky by making it perspicuous.

I have also not claimed that merely by generating evolutions of particles into Configuration A and Configuration B I have solved the measurement problem. As noted above, if the particles in these configurations were *invisible* and *intangible*,

or somehow otherwise hard to get information about by usual human capabilities,
then merely having the particles in one of these configurations would contribute
nothing to solving the original problem. In addition, we need at least some
schematic reason to think that we could *see* these configurations. An obvious
way would be to show that, by means of the physics, the patterns of firing
of the rods and cones in the retina (under normal lighting conditions, etc.)
would correspond to the usual projection of the three-dimensional array onto
the retina by geometrical optics. We have every reason to believe that this can
be done.

I also made my life easier by having relevant facts directly encoded in the
configuration. If the numbers on the devices had been *painted* on, instead
of being raised blocks, then one might not be able to tell, simply from the
configurational information, which outcome corresponds to Configuration A
and which to B. But this simplification does not paper over any deep problem.
We have already said that merely *having* the configuration is not enough: it ought
to be *visible*. In the case of seeing configurations, this amounts to correlating the
state of the optic system to the positions of things. If the numbers are painted
on, then the locations of the numbers may well be determined more directly by
the *wavefunction* of the device than by its configuration. That's fine, since it has
both. And if we can show (at least schematically) how the different distributions
of paint interact with light to cause the same sorts of retinal firing as we got in
the case of the raised numbers, then we have done our job just as well. And this
can be done (at least schematically).

Note that this is as much as we have ever demanded of any physical theory in
history. Newtonian gravitational theory was deemed a predictive success because
it got the (relative) positions of the planets right, without bothering to give a
detailed account of how we can see the planets or know where they are. No
doubt this latter question is one to be addressed by physics, but without any
particular reason to doubt the reliability of plain eyesight or instrumentation we
start out firmly believing that we know a whole lot about the configurations of
the planets. The theory has succeeded if it can predict the very configurations
we think we already have ascertained. (There can, of course, be corrections.
Gravitational bending of light can account for where certain stars *appear* to be
without postulating that they are there. The way to make these locutions exact is
straightforward.)

So it's not merely that Bohm's theory gives us probabilities for *something
physical*, it is that it gives us probabilities for *the right kind of thing to make the
connection with experiment comprehensible*. Can the same be said for GRW?

As we have seen, if we interpret GRW as a monistic theory, then at base it is
about the evolution of a field-like object in an extremely high-dimensional space.
And corresponding to the two possible physical outcomes, assigned different
probabilities, in Bohm's theory we also have (to simplify matters) two possible

physical outcomes assigned probabilities in monistic GRW. We can represent these as below:

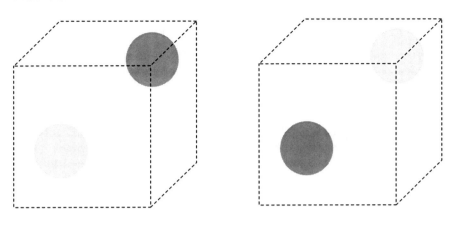

Physical state C   **Physical state D**

Figure 2.

I have evidently had to suppress the vast majority of dimensions of the space, and the shading, which represents the squared-amplitude of the wavefunction, is much too dark in the lighter section relative to the darker, but the intent should be clear. In the monistic wavefunction GRW theory, the probabilities are probabilities for, say, physical state C or physical state D to obtain at the end of a certain period of time. The difference in these states is simply a matter of what the squared-amplitude of the wavefunction is in different regions of some very-high-dimensional space. But is it C or D (or neither) that corresponds to the pointer pointing to the '1'? The situation vis-à-vis Bohm's theory could not be starker.

So although monistic GRW has unproblematic probabilities, and we even know what these are probabilities *for* (e.g. physical state C obtaining or physical state D obtaining), it is *not at all obvious* that these are the *right* sorts of things to make comprehensible how the theory connects up to laboratory operations.

This is where I now find Bell's famous dictum at least somewhat misleading. For the conversational implication of Bell's dictum is that one can solve the measurement problem either by adding more to the ontology than the wavefunction (à la Bohm) or else by simply changing the dynamics of the wavefunction (à la GRW). We have seen how adding Bohmian particles can do the job, if the particles live in a familiar low-dimensional space, and take familiar configurations, and interact with other things in the right way so as to be visible, and so on. But it is not at all obvious how the GRW collapses, by themselves, can do *any* of the work that makes the Bohmian account comprehensible. Some further story must be told, and even the skeletal outline of such a story is not evident. I will return to this question later.

Why wasn't this inadequacy of the monistic GRW evident from the beginning? I think that there are two different reasons. One involves cheating — real cheating, not just choosing easy-to-analyze cases or glossing over details that need to be filled in. The cheating consists in nothing more than using the familiar term 'configuration space' to talk about the space in which the wavefunction 'lives'. A configuration space is an abstract space, not a physical space, each of whose points represents the configuration of a multiplicity of points in a common low-dimensional space. Thus, a single point in a configuration space, by definition, represents a particular, complex, structured array of points in a common space. And, given the geometrical structure of that common space, that array can have interesting geometrical structure. Take the configuration space of four points in Euclidean space. This is a 12-dimensional space, but it is not a *uniform* or *homogeneous* space. A particular subset of points in that space — and not others — corresponds to quadruples of points that lie at the vertices of a regular tetrahedron. If one possible outcome of an experiment could be that four objects end up arranged in such a way, then we would know which points in the configuration space correspond to this outcome. In a 12-dimensional Euclidean space, in contrast, there is no such special set of points.

Furthermore (as Shelly Goldstein pointed out to me) the terminology that has the wavefunction 'live' in a space also requires care — it is unclear that it means anything more than that there are many distinct physical states that the wavefunction can be in that have a natural topology. On this understanding, the wavefunction 'lives' not in configuration space but in a Hilbert space. Perhaps a better terminology for what is meant is that the wavefunction lives *on* a certain space, in the way that a classical *field* 'lives on' spacetime. That is, the physical state of a classical field is specified by ascribing a certain number or magnitude to each point in a space, and the natural ontology holds that the existence of the field requires the existence of this underlying space. It is in *this* sense that one could hold that the wavefunction lives on a particular high-dimensional space: the state of the wavefunction is specified by associating a complex number with each point in that space.

If we *call* this 'space of the wavefunction' *configuration* space, then we are suggesting that there is something here to do with *configurations*, and that each point is the space is associated with, or that represents, a configuration. And the only question left is *which* configuration each point corresponds to. If we could secure this correspondence, then interpreting the difference between physical state C and physical state D appears to be simple: C represents the obtaining of the configuration associated with the high-amplitude region in the one diagram, and D represents the obtaining of the configuration associated with the other region. But in a monistic theory, all of this, from beginning to end, is a cheat: there *are* no configurations at all in the ontology, and so nothing *else* for a particular state of the wavefunction to 'represent' or 'be associated with'.

A clinical example of this insouciance about language can be found in Thibault Damour's book *Once Upon Einstein*. Damour, an adherent of the Everett interpretation, presents the theory this way:

> What is the essential idea of Everett's interpretation? To introduce it, let us recall the central paradox of quantum theory, such as was highlighted by the arguments of Einstein's gunpowder barrel (half-exploded, half-intact) and Schrödinger's cat (half-living, half-dead). Quantum theory describes the system consisting of the cat and its environment (the box enclosing it, the air it breathes, the lethal mechanism triggered by a radioactive atom, etc.) by a function of the configuration of the system. To each configuration q of the system is associated a (complex) number $A(q)$ that we shall simply call the amplitude of the configuration q. What is a configuration q, considered at a fixed time t, and how is it described? For example, one could describe each possible instantaneous configuration of the cat and its environment by specifying the position in space of each of the system's atoms (the atoms making up the cat, those in the air, those in the lethal mechanism, etc.) The position of each atom is specified by giving its three coordinates in space (length, width, and height). Let N be the number of atoms in the system. The number N is gigantic. Indeed, we recall that a gram of matter contains 600 thousand billion billion (6×10^{23}) atoms. A configuration of the total system is thus specified by giving a (gigantic) list of $3N$ numbers. The notation q denotes such a list. (Damour [2006 p.151])

Damour goes on to visualize the dynamics of the (uncollapsed) wavefunction as the superimposing of many movies, of different hues and intensities, which correspond to the evolution of these configurations.

I hope it is evident just how much Damour's presentation relies on the usual understanding of a *configuration*: the cat, mechanism, and environment somehow are 'made up' of a very large number of *atoms* that all are located in a *common space* and therefore can have particular *configurations*. Damour's problem is not how to associate configurations of many particles in a common space with a single point in a high-dimensional space: he simply *takes that for granted*. Damour's problem instead is how to interpret the fact that his wavefunction does nothing more than 'associate' each one of these configurations with a complex number. The question of how, in a truly *monistic* theory, to get from a high-dimensional space to a configuration space is not even asked.

This deeper problem for understanding monistic wavefunction theories has been systematically concealed in standard discussions of the theory. How, we are asked, are we to make sense of Schrödinger's cat, which is *half-dead* and *half-alive*? But this already obscures the question we are asking. The Schrödinger cat state, from the point of view of a monistic theory, is just the wavefunction with a high amplitude lumped up somewhere in the space on which the wavefunction 'lives'.[1] To assume that that particular region 'corresponds to' or 'represents' or 'is associated with' cats at all (much less 'half-alive, half-dead' cats) is to grant the association of points in the space with configurations. Thus in *posing* what is commonly taken to be the central interpretive problem, one is *masking* the more fundamental one.

The second reason that this inadequacy in the monistic theory has been missed is, I think, the general instrumentalism that permeated the birth of the theory. The microscopic world, Bohr assured us, is at least *unanschaulich* (unvisualizable) or even non-existent. Unvisualizable we can deal with—a 10-dimensional space with compactified dimensions is, I suppose, unvisualizable but still clearly describable. Non-existent is a different matter. If the subatomic world is non-existent, then there is no ontological work to be done at all, since there is nothing to describe. Bohr sometimes sounds like this: there is a classical world, a world of laboratory equipment and middle-sized dry goods, but it is not composed of atoms or electrons or *anything at all*. All of the mathematical machinery that seems to be about atoms and electrons is just part of an uninterpreted apparatus designed to predict correlations among the behaviors of the classical objects. I take it that no one pretends anymore to understand this sort of gobbledegook, but a generation of physicists raised on it might well be inclined to consider a theory adequately understood if it provides a predictive apparatus for macroscopic events, and does not require that the apparatus itself be comprehensible in any way.

If one takes this attitude, then the problem I have been trying to present will seem trivial. For there is a simple *algorithm* for associating certain clumped up wavefunctions with experimental situations: simply *pretend* that the wavefunction is defined on a configuration space, and *pretend* that there are atoms in a configuration, and read off the pretend configuration where the wavefunction is clumped up, and associate this with the state of the laboratory equipment in the obvious way. If there are no microscopic objects from which macroscopic objects are composed, then as long as the method works, there is nothing more to say. Needless to say, no one interested in the *ontology* of the world (such as a many-worlds theorist) can take this sort of instrumentalist approach.

Another quick indicator of how deeply ingrained a certain form of instrumentalism is: It is commonplace to say that the quantum formalism lends itself to two 'pictures', the Schrödinger picture and the Heisenberg picture. In the Schrödinger picture, the state of the world (or the system) changes with time, while in the Heisenberg picture the state of the system is *static* and the 'operators'

---

[1] Can we at least say this: the Schrödinger cat state, unlike a 'normal' state, is one in which the wavefunction clumps up in *two* regions of the space rather than one? Again, only if we already assume we are entitled to use 'configuration' space or 'position representation'. If one thinks that the wavefunction 'lives' in *Hilbert* space, then the Schrödinger cat state is just one of the vectors. The distinction between 'lives in' and 'lives on' is again important here.

or the 'observables' evolve. The *instrumental equivalence* of these 'pictures' is easy to see: given how predictions (given the initial state and the 'initial operators') are made, one can mathematically group the 'time evolution' part with the state or with the operators. Evidently, the predictions come out the same either way.

But if these are supposed to be alternative *ontological pictures* of what really evolves, I confess to complete incomprehension of the 'Heisenberg picture'. Go back to our simple experiment: we set up some laboratory apparatus, run a particle through it, and check at the end to see if the pointer points to '1' or to '0'. The system is the apparatus and particle. What could it possibly mean to say that *the state of the system* is no different at the end of the experiment than at the beginning: we *know*, before doing the experiment, where the pointer points *initially*, and we *don't know*, until we look, where the pointer points when the experiment is over. Intuitively, if the physical state of the pointer doesn't change, then it should point at the end where it pointed at the beginning.

The defender of the 'Heisenberg picture' can say some *words* to this effect: 'No, the physical state of the apparatus and particle is just the same at the end as it was at the beginning; it's just that *what counts as "the pointer pointing in a certain way"* has changed. If you think that the pointer changed from pointing to "0" to pointing to "1" you are wrong: The pointer it just as it was, but what counts as "pointing to '0' " and "pointing to '1' " is different.' But I simply can't make any sense out of this at all—and hence can't make sense of the 'Heisenberg picture' as a *picture* (as opposed to a purely mathematical technique for solving equations).

So the conversational implication of Bell's dictum, viz. that one way of solving the measurement problem involves *simply* changing the Schrödinger dynamics while keeping only the wavefunction in the ontology, now strikes me as wrong. There may be a way to maintain a monistic wavefunction ontology, but it is certainly not *trivial* to see what that way is, and there are quite substantial interpretive problems that stand in the way.

This brings us to the second passage I want to discuss. It comes from 'Solving the measurement problem: de Broglie–Bohm loses out to Everett' by Harvey Brown and David Wallace, and is evidently relevant to the topic of this meeting. Given what I have said so far, discussion of this should be brief.

In characterizing the function that the particles (or 'corpuscles') play in Bohm's theory, Brown and Wallace begin their discussion by posing the very question I have been discussing: whether the wavefunction *by itself* (considered as an object on 'a space', but not on a *configuration* space) is obviously adequate to account for the sorts of experimental facts from which we began our theorizing (e.g. that the pointer ended up pointing at the '0'). They start by taking the issue of probabilities or indeterminism out of the game by focusing on the deterministic case (according to Born's rule): suppose the measured system starts out in the appropriate eigenstate, so one gets a specific prediction about the outcome of the experiment:

> In this case only a single 'localised' wavepacket exists in the configuration space at the end of the measurement process—the wavepacket correlated with the initial eigenvector of the observable being measured which happens to be the initial state of the object system. The crucial question we wish to raise is this. *Does this wavepacket, in and of itself, account for the result of the measurement, or does a definite measurement outcome require, even in this case of complete predictability, the presence of the hidden variables within it?*
>
> Most discussions of the measurement problem in quantum mechanics take it for granted that no difficulties arise in this case of the predictable outcome—that the problem only rears its head in the more interesting and more general case of unpredictability, when the initial state of the object system is some linear combination of eigenvectors of the relevant observable. But if analysis of the predictable case is successful without appeal to hidden variables, then Bohm's Result Assumption in the general case is problematic. In the general case, each of the non-overlapping packets in the final joint-system configuration space wavefunction has the same credentials for representing a definite measurement outcome as the single packet does in the predictable case. The problem, if it is one, is that there is more than one of them. But the fact that only one of them carries the de Broglie–Bohm corpuscles does nothing to remove these credentials from the others. Adding the corpuscles to the picture does not interfere destructively with the empty packets. The Result Assumption appears to be inconsistent with the treatment of the predictable case, or at least to override it in some mysterious way.   (Brown and Wallace [2005 p.524])

I quite agree that the question raised is crucial, and arises even in the predictable case. I also agree that the Bohmian should insist—as all Bohmians I know of do!—that predictability has nothing to do with it: even in the predictable case, the state of the wavefunction *alone*, the wavefunction *without any particles at all* (if any sense can be made of that) is *not* sufficient to account for the result of any measurement. For if the result of a measurement consists in, say, a pointer pointing a certain way, and if the pointer is made of particles, then if there are no particles there is no pointer and hence no outcome. All of this talk of a wavepacket 'representing' an outcome is unfortunate: what the wavefunction monist has to defend is that the outcome *just is* the wavefunction taking a certain form (in some high-dimensional space). And, as we have seen, it is not at all obvious how that is to be done. But in any case, the *Bohmian* need not worry: in a theory with particles, the laboratory apparatus can be unproblematically made of particles, and the outcome can be a matter of how the particles behave. None of this implies that the wavefunction does not exist, or isn't real, just that the essential role of the wavefunction in accounting for the outcome is not to *constitute* the apparatus but to determine the dynamics of the particles that do constitute the apparatus. *Given this account of what the apparatus is*, there would evidently be no outcome if there were no particles.

The whole of Brown and Wallace's paper appears to be an attempt to simply dodge this point and to assert, without discernible argument, that the structure of the wavefunction alone is sufficient to account for macroscopic reality. According to them, since the wavefunction has the structure needed to account for macroscopic reality, the only role of the particles is as (in a different sense!) a *pointer*:

> From this viewpoint, the corpuscle's role is minimal indeed: it is in danger of being relegated to the role of a mere epiphenomenal 'pointer', irrelevantly picking out one of the many branches defined by decoherence, while the real story—dynamically and ontologically—is being told by the unfolding evolution of those branches. The 'empty wavepackets' in the configuration space which the corpuscles do not point at are none the worse for its absence: they still contain cells, dust motes, cats, people, wars and the like.   (ibid. p.527)

Note that Brown and Wallace see fit to avail themselves of the terminology 'configuration space' while simultaneously denying the very existence of the particles that might be in any configuration and, presumably, the low-dimensional space that the multiplicity of particles might commonly inhabit. So *both* of the key roles of the particles in this theory, viz. as the localized entities of which familiar objects are made and as the things that allow the 'space' of the wavefunction to be understood as a configuration space, are simply ignored by Brown and Wallace. It is hardly a surprise, then, that they find the role of the particles to be marginal.

Confronted with Bohm and Hiley saying just the things that one would expect, viz. that since cats are made of particles (according to this theory), getting a cat to come out either alive or dead demands that the particles end up in one or another configuration, Brown and Wallace respond thus:

> Consider the element in the superposition that does not contain the corpuscles—the 'empty' wavepacket in the configuration space of the joint system. Does it not describe (amongst other things) a cat that is either dead or alive? Yes, most of us would say, if the cat had actually been prepared exclusively in the cat-state that is contained in (better: is a factor state of) that element. Recalling the discussion in section 3 above of the predictable single-outcome measurement, and the arguments of the previous section, is it not more natural to say that the superposition describes both a dead and an alive cat (each correlated with distinct states of the environment) with one of these possibilities replete with corpuscles? It is hard to see how the corpuscles annul the reality of the other possibility; indeed they cannot. To argue that the 'state of being' of the cat depends 'in addition' or 'also' on the corpuscles is to admit that the wavefunction plays some role in the matter, but not one consistent with the common interpretation of the predictable, single outcome measurement scenario.   (ibid. p.531)

Note again the odd terminology: the wavefunction is said to 'describe' a cat, and then is said to be a 'possibility' of how the cat might have been 'if the cat had actually been prepared' in a certain way. But according to the wavefunction monist, wavefunctions do not 'describe' things: they *are* things. And according to the wavefunction monist, the wavefunction taking a certain form, or behaving a certain way, does not correspond to a *possibility* but to an *actuality*: that's just what cats *are* in this account. But most of all, note that Brown and Wallace's answer to Bohm and Hiley is just to note that their account is not 'consistent with the common interpretation of the predictable, single outcome measurement scenario'. So what? As they themselves have clearly pointed out, the Bohmian will *deny* the 'common interpretation' that makes the wavefunction alone ontologically sufficient, even in the predictable case. The burden of the paper ought to have been a defense of the common view. Instead, what one gets is a rejection of Bohm and Hiley's position simply because it conflicts with the common view!

The logic of Brown and Wallace's defense of Everett depends critically on not questioning the 'credentials' of the wavefunction to represent a familiar macroscopic reality in the predictable case. It is only with this in hand that the discussion of decoherence even becomes relevant. Perhaps it is worthwhile to spell this out in detail.

One way to raise trouble for Everett is this: begin by insisting that the only condition in which the wavefunction represents a familiar macroscopic reality is when it is in an appropriate eigenstate, the very eigenstate that results in the predictable case. If this is the right interpretive principle, then the non-collapse theory is in trouble in the unpredictable case, for the linear evolution does not yield such an eigenstate. End of story.

Decoherence is then invoked to get around this problem in three steps. The first step notes that when decoherence obtains, different elements or parts of the wavefunction become effectively causally decoupled from each other. The second step argues that such decoupled parts can legitimately be considered to be a *collection* of separate, non-interacting 'worlds'. And the third step argues that since each of the 'worlds' is associated with an appropriate eigenstate, each is unproblematically associated with a familiar macrostate, the macrostate that the original interpretive principle would have ascribed to the eigenstate. Hence decoherence yields a *plurality* of familiar macroscopic worlds, contrary to the original worry that it yields *none*.

Now one might object that the argument needs some supplementation: after all, the *original* interpretive principle (that there must be an appropriate eigenstate to have the familiar macroscopic reality) still implies that there is no such macroscopic reality, even if there is decoherence. But what we have been considering is a much more radical complaint: that even being in the eigenstate does not comprehensibly yield the familiar macrostate if the wavefunction is all there is. It is evident that appeal to decoherence, by itself, can do nothing whatever to solve this problem.

In sum, any theory whose physical ontology is a complete wavefunction monism automatically inherits a severe interpretational problem: if all there is the wavefunction, an extremely high-dimensional object evolving in some specified way, *how does that account for the low-dimensional world of localized objects that we start off believing in, whose apparent behavior constitutes the explanandum of physics in the first place?* A lion's share (although not all) of this burden can seem to disappear if we unjustifiably help ourselves to the phrase 'configuration space', since that phrase rather unavoidably suggests that we have to do with configurations. And it should be obvious that all of the resources of the phrase 'configuration space' are legitimately available to a non-monist who postulates a plethora of localized particles (or strings, or whatever) in a common low-dimensional space.

We should note at this juncture that the problem of accounting for the appearance of a low-dimensional spacetime inhabited by localized objects can arise even if one is *not* a wavefunction monist. In particular, David Albert's version of Bohmian mechanics faces this interpretive problem as well. Albert's theory postulates a wavefunction on a high-dimensional space and, in addition, a *single* particle (the so-called 'marvelous point') in *that same high-dimensional space*. Evidently, the addition of this single particle does nothing, by itself, to justify regarding the high-dimensional space as a configuration space of many entities in a low-dimensional space. So the issue here is not one peculiar to *monistic* theories: it is common to theories that only postulate, as onto-logically fundamental, a very, very, very high-dimensional (non-compactified) space.

But unlike Damour and unlike Brown and Wallace, Albert is *aware* of this challenge and is committed to trying to meet it. This is not the place to examine the considerations—based in dynamics—that he brings forward. And of course Albert's theory has no problem explicating what the Born probabilities are *for*: they are for the marvelous point to evolve in one way rather than another in the high-dimensional space. But what we *want* them to be for are particular experimental outcomes such as a pointer pointing some way—so we are left with the task of explaining how one sort motion of a single particle could *be* a pointer pointing one way while some other motion of the particle is a pointer pointing another way. And I do not mean to assert that the challenge cannot be met. But it does seem to me that most wavefunction monists are simply unaware of this problem, and get tremendous mileage (as we have seen Damour and Brown and Wallace do) from the use of the phrase 'configuration space', to which they have no evident claim.

Returning to Bell's dictum, then, we have seen that simply adding a non-linear part to the Schrödinger dynamics does not obviously suffice to provide an acceptable understanding of quantum ontology. We begin by thinking there are localized objects inhabiting a low-dimensional space, whose behavior we seek to explain. The obvious way for a physical theory to accomplish this task is to postulate that there *are* localized objects in a low-dimensional spacetime (maybe particles, maybe strings, maybe even flashes, with the spacetime maybe having compactified dimensions) that constitute macroscopic objects, and to provide these objects with a dynamics that yields the sort of behavior we believe occurs. The provision of the dynamics may involve the postulation of more ontology (fields, dark matter, dark energy, wavefunctions). Since the wavefunction is not a localized object in a low-dimensional spacetime, a wavefunction monist must seek some other, radically different, way to account for our initial beliefs, as must a marvelous point theorist. But leaving those projects aside, let's consider more closely what sorts of localized objects in a low-dimensional space can help solve our problem.

The original understanding of the GRW theory was as a monistic wavefunction ontology, but eventually GianCarlo Ghirardi came to accept the need for a local ontology in a low-dimensional spacetime. Ghirardi has suggested a *mass density*, mathematically related to the wavefunction in a specified way. At first glance, given the GRW dynamics (and hence the resulting dynamics of the mass density) this also seems to solve our interpretive problems. But on closer inspection, some difficulties emerge.

The first is more of a puzzle than a concrete problem. It was first pointed out to me by Nino Zanghì, and turns yet again on the particular form of the wavefunction. The wavefunction, as we have seen, is commonly explicated as being defined on a configuration space, an abstract space naturally associated with a multiplicity of localized objects or points in a low-dimensional space. But in the mass density ontology, there is no such multiplicity of objects. The natural configuration space for a *mass density* would be a space each of whose points corresponds to a complete specification of what the mass density is at every point of the low-dimensional space, and that is nothing like the space on which the wavefunction is usually defined. So there is still a somewhat peculiar mismatch between the nature of the local ontology and the nature of the wavefunction.

Let's leave that aside, though, and see where the theory takes us. In the simple case, where the Born probability is 1, we really do not have any problem. If we perform an experiment such that it is certain that the pointer will point to the '0', then the wavefunction will evolve so that the mass density takes the shape of configuration B above. And if a competent observer looks at that piece of apparatus, the mass density in the observer's brain will evolve into a configuration different from what it would take if the observer looked at configuration A instead. This is the best that any physical theory has ever been able to do to account for our initial beliefs about how macroscopic things move.

What of the case where the probabilities are non-trivial? If we arrange things so that the Born probabilities for the pointer to point to '0' and for the pointer to point to '1' are both 0.5, then without any GRW hit the mass density would evolve into a state in which half the density is in Configuration A and half in Configuration B. This does not correspond to anything that we think happens in the world, but the GRW dynamics appears to solve this difficulty. As Bell says:

'Quite generally any embarrassing macroscopic ambiguity in the usual theory is only momentary in the GRW theory. The cat is not both alive and dead for more than a split second' (Bell [1987 p.204]). A GRW hit, one way or another, will result in almost all of the mass density being in one configuration and only a tiny, tiny bit being in the other.[2] *On the assumption that this small (and ever shrinking) mass density can be safely neglected*, the post-hit state would be as satisfactory in accounting for our beliefs about the outcome as it is in the case of predictive certainty.

But on what basis, exactly, can the small mass density be neglected? After all, that mass density is *something*, and it has the same shape and (neglecting the shrinkage of scale and distortion effect mentioned in the last footnote) behavior and dispositions to behave as it would have had if the hit had left it with the lion's share of mass density. Another way of putting this is that if one adopts a fairly natural kind of *structural* or *functional* account of the apparatus, the low-density apparatus seems to have the same credentials to be a full-fledged macroscopic object as the high-density apparatus *since the density per se does not affect the structural or functional properties of the object*. Of course, the amplitude of the smaller mass density suffers continual exponential shrinking on account of subsequent hits (while the high-density piece does not), but there is no obvious sense in which these changes in amplitude relevantly affect the *structural* or *functional* organization of the low-density part.

There are, I think, several different approaches to addressing this challenge. One appeals to the treatment of extremely-low-density material in *classical* physics. For example, there may be a very-very-low-density distribution of matter in interstellar space, but we are justified in neglecting it—acting as if it weren't there at all—for most practical purposes. There is a perfectly good sense in which small enough densities of classical matter become physically irrelevant for all practical purposes (FAPP).

But as soon as it is stated explicitly in this way, the cases are seen not to be parallel at all. Low-density classical matter just does not have the same structural or functional characteristics as high-density classical matter. We can neglect the low-density matter between the planets FAPP when calculating planetary orbits because the low-density matter does not appreciably influence the orbits, *but the reverse is not true*. If we want to predict how the low-density matter will evolve, we had better take the high-density matter into consideration: if a planet sweeps by, the low-density matter will be severely affected. And, of course, classical low-density matter cannot be formed into rigid structures, and so on.

---

[2] As David Wallace has pointed out, this isn't quite accurate. Because of the way the 'tails' of a Gaussian decay, the configuration that is *not* hit will not only have its mass density reduced, but also distorted. Since the choice of the Gaussian was not made with consideration of its tails, save that they be small, I will ignore this effect in the sequel.

But in the GRW case (again, neglecting the distortion effect), the *structural* evolution of the low-density apparatus is just as autonomous from the high-density matter as the high-density matter is autonomous from the low-density. Decoherence is supposed to ensure that neither apparatus is notably influenced by the presence of the other: the high-density and low-density objects are effectively decoupled from one another. This is in stark contrast to the classical case. So the neglect of the low-density object cannot be justified by appeal to classical intuitions.

A second approach would be to rule out the low-density object by *fiat* or by *hypothesis* or by a *semantical rule* or by an *ontological analysis*. The idea here is that it is just an assertion of the physical theory itself that familiar, everyday objects are high-density objects. Probabilities for experiments to have particular outcomes are then the probabilities for the high-density objects to behave in certain ways. If this is accepted, then the problem I have been concerned with evaporates: the GRW theory with a mass density has localized high-density material objects in spacetime that behave in the way we think laboratory apparatuses, and planets, and people, behave. Of course, it is not possible, in the mass density theory, to deny the existence or the physical reality of the low-density objects: what one must rather do is find the means to deny their significance.

But one might well wonder—at least I wonder, and I'm sure partisans of many worlds will wonder—exactly how to understand the status of this rule or hypothesis or fiat connecting the physical description with the everyday description. It is not, in any obvious sense, a *physical* hypothesis (as the postulation of the mass density itself *is* a physical hypothesis). In the old logical empiricist tradition, one sometimes spoke of *bridge rules* or *correspondence rules* that were to link the theoretical vocabulary to the pre-theoretical vocabulary, and the notion seemed to be that there were no constraints whatever on what those rules could be like. They were simply part of the predictive apparatus of the theory, and were to be judged only as part of the whole package—laws stated in theoretical vocabulary together with bridge principles—whose object was making good predictions in the pre-theoretical vocabulary. But the general orientation of this sort of approach is instrumentalist: all the theory is meant to be is a device for producing predictions about observable objects, or experiences, and the whole project of postulating a physical *ontology* is left aside. Few people at this conference would endorse such an approach: we want something more from physics than an uninterpreted formal mechanism for producing predictions about macroscopic objects. We want a physical theory to tell us what macroscopic objects are.

The instrumentalist orientation is deeply embedded in many discussions of quantum theory. Arthur Fine, for example, suggested that the connection between the pure quantum description of an object (via a wavefunction) and its observable properties is achieved by a *link*: the so-called 'eigenfunction-eigenvalue' link. What is it for an object to have a particular position? Well, first one needs to associate the 'observable' position with a Hermitian operator, and then one postulates that a system only has a position if its wavefunction is in an eigenstate of that operator, with the eigenvalue giving the position. GRW theory, of course, had to reject this account from the beginning, if it wanted to have physical objects with positions, since the GRW collapses do not yield position eigenstates. (Nor, indeed, does any known experimental situation!) Note that GRW with a mass density does not attempt to recover localized physical objects by postulating a different 'link' or 'rule' or 'bridge principle' connecting the wavefunction to talk of localized objects: it rather does so by *postulating localized objects*. This is straightforward and unobjectionable. The question that remains for this form of GRW is whether the structure and behavior of these objects gives rise to the sort of physical world we think we inhabit.

'Link' talk has encouraged a somewhat cavalier approach to the issue I am concerned with here. For example, it has been suggested that the 'problem' that GRW collapses don't yield eigenstates can be solved by replacing the eigenstate-eigenvalue link with something weaker: a so-called 'fuzzy link' to the effect that, for example, in order for a system to have a position its wavefunction just has to be *sufficiently close* to an eigenstate of position. The subtext of this suggestion is that from a formal point of view the eigenstate-eigenvalue link is perfectly fine—it is, at least, an *exact* rule—and the fuzzy link is almost as good. But at a very basic level, the eigenstate-eigenvalue link is tremendously puzzling. It is typically invoked either in the service of a monistic wavefunction theory or in an instrumentalist atmosphere where there is no clear physical ontology at all. And it simply presupposes many things that cry out for explanation: what are 'observable properties'? How do they become associated with Hermitian operators? Why should (or should not) being an exact eigenstate of such an operator, as opposed to being a near eigenstate, make a significant physical difference to a system? I think we would do well to abjure all talk of 'links' or 'rules' at all: a physical theory should posit a physical ontology and a dynamics, and the rest should be a matter of *what is comprehensible in terms of that ontology*. If something is *not* easily comprehensible (as, I have argued, the existence of a low-dimensional spacetime containing localized physical objects is not easily comprehensible if a theory only postulates a wavefunction on a high-dimensional space), then what is called for is either argument or new physical postulates, not just a *rule* or a *link*. GRW with a mass density is honest about this, and makes a new postulate. But even then, as we have seen, the *adequacy* of the postulate can be a matter of dispute, and the dispute cannot be settled just by invoking a rule.

There is good reason, then, to worry that GRW with a mass density isn't the theory we all took it to be, namely a theory with a single macroscopic world that evolves in the right way. It seems that instead it too is a many-worlds theory: there is one high-density world and many, many low-density worlds, largely uninfluenced by one another, each with a right to be called 'macroscopic' and 'physically real'. One is, sure enough, of much higher density than the rest, but all are equally real, equally (structurally) stable, equally functional. What, then, is the significance of the density? Could we have any reason, right now, to feel assured that *we* are high-density rather than low-density objects?

In fact, GRW with a mass density seems to be afflicted with some of the same interpretive problems as many worlds. One problem with many worlds has always been understanding the *significance* of the square-amplitude of the wavefunction. Damour says the following:

> Let us now explain Everett's idea. It consists of taking seriously Einstein's statement: 'The theory itself defines what is observable'. Let us first take quantum theory seriously and ask it to define what is real. Each configuration *q* will have more or less 'reality' according to the value of the amplitude *A(q)*. In other words, we interpret *A* as an *existence amplitude*, and not (like in the Born-Heisenberg-Born interpretation) as a *probability amplitude*. Indeed, the notion of probability amplitude for a certain configuration *q* suggests, from the very beginning, a random process by which only one configuration, among an ensemble of possible configurations, is realized, passing from the possible to the actual. By contrast, the notion of existence amplitude suggests the simultaneous existence (within a multiply exposed frame) of all possible configurations, each actually 'existing', but with more or less intensity.   (Damour [2006 p.153])

Although Damour announces an admirable goal—to let the theory itself determine what is observable, and even more importantly what is *real*—I fail to see at all how this has been done. Quantum mechanics itself does not say that the amplitude of the wavefunction corresponds to the 'intensity of existence' of a configuration. Indeed neither quantum mechanics nor Damour give us any clue what such 'intensity of existence' could connote. The GRW with mass density, on the other hand, gives us *two* clear ways that the square-amplitude (or near enough) of the wavefunction has physical significance: one via the postulation of the mass density, which is (nearly) proportional to a suitable projection of the square-amplitude of the wavefunction (in configuration space), and another via the dynamics, which makes the *evolution* of the high-density configuration probabilistic, with probabilities (near to) the appropriate square-amplitudes. So the postulates of GRW with a mass density are all physically clear. But for all that, I am not convinced that they are adequate to the task. For, at the end of the day, we need in addition an association of the world we experience with the high-density world, and the consequent neglect of the low-density worlds, and I don't see how that is to be accomplished. In other words, why does not the mass density GRW simply boil down to many worlds? If it does, there is no advantage to be gained by the postulation of the GRW hits.

There is, though, another way to supplement the GRW wavefunction with a local ontology in spacetime, viz. Bell's 'flash' ontology. In this theory, there is a local event in spacetime associated with each GRW collapse, and the distribution of these flashes is overwhelmingly likely to correspond, in an obvious sense, with a familiar sort of macroscopic world. Furthermore, the chance of various macroscopic conditions being realized (e.g. with the pointer one way rather than the other) are the quantum mechanical probabilities. There are no 'low density' or 'low intensity' flash-sequences to be ignored or discounted or argued away. In this respect, flashy GRW is like Bohmian mechanics, where there are no 'low-density' or 'low-intensity' particles acting out the unrealized possible outcomes of experiments. (Or, to be perfectly precise, there is only a tiny and practically negligible chance that there will be *any* flashes associated with the low-amplitude part of the wavefunction through the whole life of the universe. Recall that the amplitude of this part, and hence the chance of a flash being associated with it, is drastically reduced by *every* hit on the high-density part.) So it seems to me that neither flashy GRW nor Bohmian mechanics has the sort of interpretive problems that infect many worlds, bare GRW, or even mass-density GRW.

The logical situation here is, of course, rather delicate. Bohmian mechanics has no trouble accounting for our initial opinions about a world of localized objects in a low-dimensional spacetime because it postulates a world of localized microscopic objects in a low-dimensional spacetime which behave, in gross, as we take the world to. Various simple adjustments might appear here: as noted, compactified dimensions could be added to the physical theory, or the microscopic objects could be strings, or the spatio-temporal arena could be spacetime rather than a persisting Newtonian space. In a clear sense, we do not require 'rules' or 'links' to make sense of these sorts of adjustments. And, as already remarked, the theory not only postulates these local objects but provides the means to explain physically how observers could get information about them. The situation with flashy GRW seems quite similar.

I have tried to explain in detail why the elimination of a localized object in a low-dimensional spacetime produces a much more problematic situation, and even the postulation of the GRW mass density in low-dimensional spacetime leaves interpretive loose ends. But what the philosopher would like is a clear formal analysis here, not merely examples and particular arguments. I have no such general formal analysis to give. In its place, I would like to end with a final text, a letter that Einstein wrote to Maurice Solovine in May 1952. I think that Einstein expresses the situation admirably. Here is Don Howard's translation:[3]

---

[3] The translation in the Citadel Press volume is not reliable, and disastrously translates 'uncertain' and 'certain' in this letter.

Tim Maudlin

Albert Einstein to Maurice Solovine, May 7, 1952

As regards the epistemological problem, you have fundamentally misunderstood me, probably because I expressed myself poorly. I see the situation schematically thus:

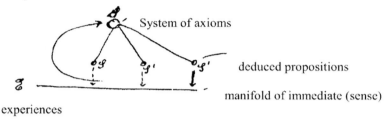

System of axioms

deduced propositions

manifold of immediate (sense) experiences

(1) The E (experiences) are given to us.

(2) The A are the axioms from which we deduce consequences. From a psychological point of view, the A rest on the E. **But there is no logical path from the E to the A, instead only an intuitive (psychological) connection that is always 'subject to cancellation.'**

(3) From the A are deduced <u>in a logical way</u> individual assertions S, which derivations can assert a claim to correctness.

(4) The S are brought into relation with the E. (Testing in experience.) **A careful examination shows that this procedure also belongs to the extra-logical (intuitive) sphere because the relation between the concepts appearing in S and the experiences E is not of a logical nature.**

But this relation of the S to the E is (pragmatically) much less uncertain than the relationship of the A to the E. (Example of the concept dog and the corresponding experiences.) If such correspondence were not achievable with great security (even though not logically comprehensible), then the logical machinery for the 'understanding of reality' would be completely worthless (example theology). –

Ultimately it comes down to the eternally problematic connection of everything conceptual with that which can be experienced (sense experiences).

Figure 3.

Einstein set himself a task more profound and difficult than any we have considered, viz. making a connection from the axioms of a physical theory all the way to *sense experience*. Such a connection would require confronting, in one form or another, the mind–body problem. Ultimately, of course, some relation between physical structure and experience is presupposed by every physical theory: if we were convinced that our experience bore *no relation at all* to the physical state of the world, then we would become skeptics about physics. But still, it has generally seemed sufficient, in physical theorizing, to connect the proprietary terminology of the physical theory to the language of everyday macroscopic reality. If *that* connection is sufficiently secure, if we *find* in the theory a world of objects that correspond to those we start out believing in, then the connection to sense experience seems to be secure enough.

We all agree, I think, that there is a *logical* gap between the derivable consequences of any physical theory and sense experience. Given that gap, the question is what, if anything, could be meant by saying that the connection between experience and the derived consequences is much less uncertain, or more secure, than the relation between the experiences and the theory that arose by reflection on them. Since the experiences are *as of* localized objects behaving in a certain way in a low-dimensional spacetime, it seems to me undeniable that the derivation from the theory of localized objects so behaving in such a spacetime is one means of achieving security. The extreme instrumentalist, in contrast, may admit no more security than is afforded by the bare posit of a 'bridge principle' connecting the theoretical description, derived from the physical axioms, and everyday talk. We also all agree, I think, that such a bare posit is inadequate. The many-worlds theory, for example, has for a long time been extolled as the natural understanding that arises *from the axioms of standard quantum theory without collapse*. No need for 'extra postulates' or 'bridge principles' is foreseen. I admire this ambition of the many-worlds approach, but do not yet see how it has been achieved.

In particular, in the absence of the direct postulation of localized objects in a low-dimensional spacetime, the sort of close correspondence between the immediate physical ontology and the world as we initially take it to be cannot be so transparently attained. And here there is an important choice point. The many-worldser, or the proponent of bare GRW, could aim to sketch a connection *directly* between the state of the wavefunction (which is all there is) and *experience*. If the experiences come out right (i.e. *as of* localized objects in a low-dimensional spacetime, behaving the way we think objects behave) then all will be well. But at this point, I don't think anyone has any idea of how such a direct connection between the physical state of anything and our experiences can be made. The other choice (this is the choice that David Albert makes in expositing the marvelous point theory) is to try to get from the fundamental high-dimensional reality to an *emergent* low-dimensional physical reality, with emergent localized objects moving in the right way (making no reference to experience at all). If this can be done, if the emergent objects and emergent low-dimensional spacetime are robust enough, then one can plausibly argue that the connection to experience is no worse off than it has been for any theory. But what we are in need of is an account of emergence and robustness that could support these claims. And as of yet, I don't think that we have such an account.

There is a certain obvious sense in which a world described by Bohmian mechanics or flashy GRW comprehensibly corresponds to the world as we experience it. If the notion of the emergence of a low-dimensional spacetime with localized objects from a high-dimensional reality can be made equally comprehensible, then bare GRW or the marvelous point or many worlds will have passed one hurdle. As we know, the understanding of probability in many worlds (and, as we have seen, perhaps even in bare GRW or GRW with a mass density) is yet another hurdle. I personally don't see the advantages that might accrue to these theories, which begin with these handicaps, but it is a better world in which they have their advocates.

There is an even deeper question that may be raised. I have been trying to articulate a sense in which various physical theories yield a comprehensible connection between the physical description and the experienced world. Yet who's to say that the world is comprehensible in this way at all? Perhaps the relation between our experience and the physical state of the world is one we are simply incapable of comprehending. Perhaps the desire for a certain sort of comprehensibility is nothing more than the desire to conceptualize the world with our primitive, and inadequate, mental resources. Perhaps the desire to understand things, after the fashion we are capable of, leads us away from the truth.

There is evidently no completely adequate answer to this sort of worry. The unparalleled predictive accuracy of physics suggests that our minds are not so badly adapted to finding out about the world, and the world not so badly adapted to being found out about, even on scales far removed from the evolutionary pressures that guided the development of our cognitive machinery. And there is also the *personal* version of this worry: maybe there is a connection between physical reality and experience that *I* cannot comprehend but others can. So if nothing else, take these remarks as a piece of autobiography: at this moment, I can feel that I can comprehend the connection between the physical description provided by Bohmian mechanics or flashy GRW and the world as I experience it. Even apart from issues of probability, I cannot similarly see such a connection to any monistic wavefunction theory or to the marvelous point theory. And when I try to comprehend GRW with a mass density (to *understand* the theory, not merely have a rule for interpreting it) I find that I cannot discount the low-density objects in the way I would have to to make me comfortable.

If the world is just not comprehensible in the way I hope, then there is nothing to be done. If it is comprehensible, but in a way I cannot yet appreciate, I can do no better than to express my puzzlement and ask for help. Consider this, then, a plea for enlightenment.

References

Bell, J.S. [1987], *Speakable and Unspeakable in Quantum Mechanics*, Cambridge University Press, Cambridge.

Brown, H. and D. Wallace [2005], 'Solving the measurement problem: de Broglie–Bohm loses out to Everett', *Foundations of Physics* **35**, 517–40.

Damour, Thibault [2006], *Once Upon Einstein*, A.K. Peters, Wellesley, Mass.

5

A Metaphysician Looks at the Everett Interpretation[1]

John Hawthorne

A metaphysician engaging with the Everett interpretation is likely to look at two kinds of questions. First, it is natural to ask what fundamental story about the structure of the world is driving the interpretation. (One imagines God writing the fundamental book of the world and asks oneself what the book would look like.) A second natural question to ask is how the familiar truths about the macroscopic world that we know and love ('the manifest image') emerge from the ground floor described by the fundamental book of the world. Assuming that we don't wish to concede that most of our ordinary beliefs about the physical world are false, we seem obliged to make the emergence of the familiar world from the ground floor intelligible to ourselves.

I'm mainly going to be looking at the second question, but let me say a little bit about how Everettians in Oxford seem to be thinking about the first question.[2] As I understand them, they are working by a few ground rules that at least constrain their vision of what the fundamental book of the world would look like. One ground rule is that there's no ineliminable vagueness: God would have no inescapable need to resort to vague predicates in writing the fundamental book of the world. The second idea is that the branching picture famously associated with Everett *will* be ineliminably vague. Owing to vagueness in the branching story, there will be ineliminable vagueness about how many branches there are, and ineliminable vagueness to the contours of a branch.

Why do they expect ineliminable vagueness here? Well, roughly, their (decoherence-theoretic) answer to the well-known basis problem concerning branching involves recourse to vague specifications of basis. They don't want to

[1] What follows is a lightly edited transcript of a talk that I gave at the Oxford Everett conference. I am grateful to Paul Tappenden for preparing the transcript from video footage of the event and to Frank Artnzenius and Hilary Greaves, Simon Saunders and David Wallace for helpful discussions. The choice of title was made by the conference organizers, not me.
    [2] See, for example, Saunders, [1997], [1998] and Wallace [2003a, b].

posit unknowable branching structure that turns on certain fine-grained facts that are inaccessible to current or future physics. In order to avoid positing such unknowable, inaccessible structure, they stick with the idea that the bottom line story about branching is vague. And, relatedly, the bottom line story about the contours of *this world* is vague. Those two ideas, the *no fundamental vagueness* thesis and the *vagueness in branching* thesis, constrain their view as to what the fundamental story is going to look like. In particular, the fundamental story can't be a fission story: the book of the world can't be a story about myriad fission of spacetimes because that would not square with the two theses that I just mentioned. If you had a fission-driven picture of the fundamental story then you'd expect a branching story in the fundamental book of the world. But then, assuming no fundamental vagueness, one would have to insist (against *vagueness in branching*) upon a non-vague branching story. In sum, Oxford Everettians assume that the branching story has to be *overlaid* on the fundamental book of the world.

Earlier in the conference James Hartle asked something like the following question: 'Could you go in for a version of the Everettian picture in which just one history is real?' Well, you could kind of tell a story like that. Suppose you posited a configuration space but maintained that a certain region of it was haloed with some special 'reality-making' property—concreteness, vim, or whatever. In this picture the one real concrete history is the region of configuration space with concreteness. Insofar as one went in for a picture of that sort one would, in effect, violate the idea that the contours of the actual world are vague. For on the picture just entertained, the contours of the actual world would be settled by what the book of the world says about where the vim is sprinkled. Insofar as the book of the world is precise, one is now going to posit unwanted determinate contours. This provides us with another nice example of how certain versions of an Everettian metaphysics are ruled out by the rules of the game that the Oxford Everettians are tacitly working with.

A number of foundational stories are compatible with the ground rules sketched so far. Here is one hypothesis that is particularly bizarre but which will be useful to us later on: The fundamental structure of the world is a Hilbert space with a world particle, like the 'marvellous point' (borrowing David Albert's lingo) in configuration space, only moving around in Hilbert space. Normally one thinks of Hilbert space as *representing* fundamental reality; on this picture this space *is* what fundamental reality comes to: the absolutely bottom line story concerns a single world particle moving around in something structurally like Hilbert space and then the rest of the world emerges from that. An alternative story is the one that Tim Maudlin talked about earlier: We posit a multi-dimensional space over which some complex scalar field is distributed. (Insofar as we call that space 'configuration space', that will merely serve to remind us that the space has an architecture that is isomorphic to the kind of architecture that is typically associated with configuration space. One shouldn't think of the book of the world as describing the space as a 'configuration space' since that would imply that it represents something else.) A third hypothesis—favored by Simon Saunders that the fundamental book described a world built out of local events—described by Heisenberg projectors—which stand in amplitude relations and spatiotemporal relations to each other. Another candidate fundamental story—one that David Wallace likes—posits a low-dimensional spacetime with various non-local properties distributed over it.

I am obviously not competent to adjudicate between these alternatives (one relevant issue—one that I am certainly not competent to deal with—is which picture will best underwrite a relativistic version of quantum theory). For the purposes of discussion let us, as a default, pretend that the Oxford Everettians are willing to embrace the 'configuration space' vision of the fundamental book of the world.

I now turn to the second issue that I gestured at initially. Let us assume that one does not simply want to deny the existence of ordinary objects like cats and stones and philosophers. There are two games that one might play when trying to generate the familiar macro-image from the fundamental image. (I'll call them 'conservative' and 'liberal'.) The conservative insists on *identifying* all those ordinary objects that he is willing to countenance with objects that are quantified-over in the fundamental book of the world. The liberal, meanwhile, posits generational principles that describe how the fundamental layer of reality necessitates the existence of additional objects that, as it were, float over the fundamental layer. Let us illustrate this contrast in a mundane way. Suppose one went in for a field-theoretic classical vision, where the fundamental story describes field values of points in Galilean spacetime. Suppose further that one wants to claim that cats and dogs exist. If one is conservative one will have to identify cats and dogs with objects mentioned in the fundamental story. The natural thing to say in that setting is that a cat or dog is identical to a region of spacetime. (I note in passing that this is a move that a number of metaphysicians, David Lewis included, have been open to.) The liberal will instead opt for the view that when there are certain patterns of field values scattered over spacetime, this necessitates the existence of concrete occupants of that spacetime—like cats and dogs and philosophers—and imagines there to be metaphysical generation principles that explain this emergence of concrete objects. The Oxford Everettians that I'm familiar with do not avail themselves of these liberal maneuvers; they are playing the conservative game, not the liberal one.

Go back to the simple conservative game as applied to the Galilean setting described earlier. It is easy to see that the conversative may struggle to accommodate our ordinary modal intuitions. Intuitively we think a particular cat could have lived longer, could have been smaller, and could have been somewhere else at a particular time; but suppose we identify that cat with a region of spacetime. It is odd to claim that a certain region of spacetime could have been more longlasting or smaller or somewhere else. Hence there seems to be a conflict.

Given all this, it is a pretty safe bet that conservatives who have reflected on modal issues are either going to do violence to our modal intuitions or else will tie themselves to the masts of certain speculative programs in the philosophy of modality—notably counterpart theory—which offer a way out in this kind of setting.[3] (Crucially, the counterpart theorist makes modal predication a context-sensitive affair: she allows that the applicability of 'could have been smaller' to be sensitive to how one is thinking about an object.) This is not the place to assess the prospects of those speculative programs. It is enough that we at least recognize that modal and counterfactual thinking poses a severe challenge to the conservative version of the Everettian program.

By way of making vivid certain other concerns, let me focus on a particular version of the conservative configuration space framework. Suppose our fundamental story posits a configuration space along with events that occur in, and features of, that space. Suppose further it identifies particular objects from the manifest image with features or events. How about the many-worlds hypothesis? Let us imagine that the predicate 'world' vaguely specifies certain regions of configuration space history: on any way of making 'world' precise, there will be many worlds, but the boundaries of and even the number of the worlds vary across precisifications. The association of the predicate 'world' with regions is guided by the (albeit vague) decoherence heuristic: on any precisification, the regions that count as worlds do not have a lot of interference with each other and hence, for the most part, each such region can be considered as a world apart for the purpose of dynamics. (As I understand it, the decoherence heuristic encodes a fundamentally dynamical conception of how to carve up the configuration space into worlds.) What this all gives us is a package that combines a vision of what the fundamental book of the world looks like with a stock of identification hypotheses that identify familiar objects with certain features or events that appear in the book. A theorist of this stripe will no doubt think that the truths about familiar objects are fixed by fundamental truths: thus he will also endorse certain supervenience hypotheses about how truths using familiar predicates supervene on truths stated in the language of the fundamental vision.

Now one thing to see at the outset is that there can be metaphysical packages that square perfectly well with the sentences that we want to utter about the macroworld but that nevertheless strike us as absurd. Let us look at a dramatic example. Suppose a philosopher came along and said that the only things that exist in reality are real numbers and sets of real numbers, crafting his theory to pay lip service to the manifest image: 'I am a real number and the property of being green is a set of real numbers. Since I am not green, the real number that is identical to me doesn't belong to that set.'

When presented with this theory, we certainly couldn't object that the theory claims that there are no people or that the theory makes false predictions about

3 See Lewis [1986].

the colors of things. That wouldn't be the right kind of objection, since the theory will have been deliberately crafted to vindicate obvious truths about the observable world. Nevertheless, considered as a vision of what the world is like, the theory still seems totally ridiculous.

A useful exercise here is to get clear about *why* that theory is ridiculous even though it pays lip service to the things we want to say about the world around us.[4] (I note in passing that, assuming that you do feel strongly that real number metaphysic is absurd, you ought to see the same kind of absurdity in a metaphysic that says that ultimate reality consists of a single world particle in Hilbert space. At any rate, I think it's a real challenge to say why one is absurd but the other is not.)

Let's try to lay down some rules of the game here for discriminating macrofriendly packages that are intolerable from ones that are tolerable, where a macrofriendly package is one which pays lip service to the claims that we ordinarily want to make about the world. The issue is one that needs to be approached with extreme care but I can suggest a few, albeit vague, guidelines.

The first guideline is, I think, more or less the one that Tim Maudlin (in his talk at this conference) had in mind. I recall that he was pretty up-front about the fact that he couldn't be very precise about what that guideline is. But to get the *feel* of the guideline let's go to a topic that many of you don't like talking about—consciousness.

Get yourself into the frame of mind of philosophers who think that the true package of the world says that there's physical stuff arranged in a certain way and then says that conscious experience supervenes on the physical stuff, so that the pattern of conscious experience is fixed by the pattern of microphysical events. Philosophers very easily get themselves into the frame of mind where they think that if the world is like that then there is going to be an uncloseable explanatory gap between the consciousness-theoretic descriptions and the physicalistic descriptions. Even supposing certain theses about the connection of microphysics to consciousness are true we seem to be in the predicament of being unable to *see* why they're true. No matter how much we look back and forth at the experiential and microphysical, we're just not going to be able to see why the true principles are the true principles and why certain false principles are false. At any rate, that's the frame of mind of those who claim to see a deep explanatory gap between the microphysical world and the world of conscious experience.

---

[4] I might mention in passing that Quine at various times entertained the theory I am calling ridiculous. After all, he took seriously the theses (i) that a spacetime point is just a quadruple of real numbers, (ii) that a property of a point—a field value for example—is a set of those quadruples, and that (iii) reality consists of spacetime points. The bottom-line story would then seem to be that all there are are real numbers and sets recursively constructed out of them. (Note further that on the hypothesis—one which Quine also took seriously—real numbers are themselves just pure sets, and the book of the world will have nothing but pure set theory in it.) A fuller discussion of Quine's work on this topic would obviously have to incorporate a treatment of anti-realist themes—notably his 'ontological relativity'. That is not my task here. See Quine [1960], [1969], [1981].

I think Tim Maudlin's heuristic is (roughly) that it's going to be a significant cost to a package if there are uncloseable explanatory gaps of the sort just described between the fundamental book of the world and the macrodescription. If you claim that certain bridge principles are true but you can't see for the life of you, no matter how much you look, why they are true while certain competing principles are false, then there's going to be an uncloseable explanatory gap. It is at least a significant cost to a theory if that is the upshot.

The explanatory gap concern has particularly forceful application to the real number metaphysic. If one hypothesizes that being green is a set of real numbers and the Eiffel Tower is a real number and the loving relation is a certain set of ordered pairs of real numbers, and so on, then you can see at the outset that there's just no way, no matter how much you look at things, that you're going to be able to see why one particular identification hypothesis is true.[5] Supposing God hands over the book of the world and you see nothing but real numbers and set theory in it, one can see that there will be no way of closing the explanatory gap between the fundamental story and familiar macro-truths. Similarly for the world particle in Hilbert space; you should be able to see that there's going to be an explanatory gap and that there will be no real hope of closing it. It is certainly not going to be closed just by getting clearer about the book of the world: even supposing you understand it fully, there's going to be nothing you can do to enable yourself as an information processor to see why the true identification claims are true and why the false ones are false.

Assuming that one wishes to hang on to a certain foundational metaphysics, there are two things you can do when faced by an apparent explanatory gap accruing to it. One kind of reaction is to claim that the gap is merely apparent. For example, there are certain physicalists about consciousness who maintain that if one thinks *enough* about consciousness and the physical ground floor then one is going to be able to see why certain bridge principles are true and why certain competitors are false. A second reaction is to live with the explanatory gap: while there are true principles about how the macrofacts are fixed by the ground floor, we are just not in a position as information processors to see why the true ones are true and the false ones are false. We can notice a few correlations here and there but there's no way we can see why the true ones are true. (Maybe one will even have a story which predicts why we, as information processors,

---

[5] I should mention that a separate concern about the real number package is that it risks positing radical disunity between the meanings of certain vocabulary in theoretical contexts and in ordinary contexts. Thus in the theoretical context we may say that no real number is part of another; meanwhile, if we identify the referents of ordinary terms with real numbers we will have to posit a very different meaning for 'part of' as it is used in ordinary contexts in so far as we wish to salvage ordinary claims. (Note that one might try to fix this problem by identifying objects with mereological sums of real numbers in a way that preserves part–whole relations. But similar problems are bound to crop up elsewhere.) I shall not pursue this theme further here.

would be unable to discern the true bridge principles from the false ones.) So there are two kinds of strategies: one is the live-with-the-gap strategy, one is the dissolve-the-gap strategy.

David Wallace earlier made it clear that he is a dissolver. His way of dissolving is to adopt functionalism. Citing Daniel Dennett as an ally, he claims to close the explanatory gap between tiger talk and the fundamental vision—configuration space, for example—via the functionalist dictum, that whatever acts like a tiger is a tiger.[6] One looks in the configuration space for something that acts like a tiger and, thanks to the functionalist dictum, we get to say that there is a tiger.

There is a very natural response to this Dennettian suggestion: 'Hold on a minute. To know that something acts like a tiger I've got to know that it growls. But the association of growling with the configuration space isn't given to us.'

The situation here is not like the one typically described by functionalists about mentality, where one earns the right to ascribe beliefs and desires to agents in a setting where various input–output facts about stimulus and behavior are already assumed to be in place. It is hard to know what the take-home message of the functionalist is in a setting where *none* of the fundamental-to-macro associations are given.

There's a version of structuralism which, if correct, would solve the problem. Take the favored macrodescription, treat all the predicates in it as variables and then say that the macrodescription is true of a chunk of reality just in case there's some assignment to those variables where, on that assignment, it comes out true.[7] But that kind of hyper-structuralism would deliver absurd results. *Inter alia*, it would commit you to the claim that actually there are tigers in real number space because, after all, there *are* some devious ways of associating predicates to sets of real numbers so that the macroclaims all come out true. (And we all agreed that that was an absurd theory.) A rampant structuralism where we just allow any realization (in the sense that I've just described) to be good enough shouldn't be tolerable to anyone. But if that's not what I'm being instructed to do then I fail to see altogether what I *am* being instructed to do when I'm told: Be like Daniel Dennett!

I wouldn't be surprised if the best version of the Everettian strategy is going to have to live with some measure of explanatory gap.[8] This may not be *so* damning. Many of us think that there *is* an explanatory gap when it comes to conscious experience and the physical world and have learned to live with a physicalism that concedes this. What makes matters very difficult—and I think Tim was sensitive

---

[6] See, for example, Dennett [1991].

[7] For a general discussion of this type of strategy, see Lewis, [1970].

[8] Some philosophers believe that this kind of concession is fundamentally unacceptable. They hold a kind of rationalism according to which supervenient truths must somehow be rationally recoverable from a book that describes the metaphysical ground floor. I am not one of those philosophers. (For a defense of this sort of rationalism, see Chalmers [1996].)

to this—is that it's a little difficult to know which explanatory gaps are liveable and which are unliveable. That's part of the confusion that we're faced with methodologically in this setting.

Let me mention another heuristic that I think is potentially important in this setting. It is a bit more theoretical, though I do think it has some purchase. Some of us are at least tentatively committed to certain metasemantical principles—principles about how words get associated with features in the world. Once one has seen that the kind of hyper-structuralism described earlier is wrong, one sees the pressing need for some fundamental metasemantical principles that select the correct interpretation from alternative—and perhaps charitable—but incorrect, interpretations, the correct model from incorrect models. Crucially it is quite possible that a package of the sort I've described will pay lip service to the macro-image but nevertheless fail to square with our favored fundamental metasemantical principles.[9]

To get the flavor of things, here are some candidate metasemantical principles: (i) The causal theory of reference: in general singular terms refer to things that they're causally connected to, albeit sometimes by a long causal chain. (ii) David Lewis's proposal: other things being equal, predicates semantically gravitate to more natural properties rather than less natural properties.[10] (iii) Timothy Williamson's principle of knowledge maximization:[11] other things being equal, if we know more according to one semantic profile then according to another, that constitutively weighs in favor of the former.

Now it is quite easy to imagine that a certain package will pay lip service to the macro-image but fail to square with one's favored metasemantical principles. The package might combine a certain worldbook with certain identificatory hypotheses and yet one's metasemantic principles might suggest that if that worldbook is actual, rather different identificatory hypotheses are true (or worse still, that if that worldbook is actual, nothing falls under the extension of certain ordinary predicates).

Admittedly, this is all very schematic. By way of illustrating the relevance of metasemantics, I wish to point to a seductive line of reasoning that Oxford Everettians appear to sometimes go in for, but which metasemantical reflection might show to be dubious. The line of reasoning I have in mind proceeds as follows: Assume that all there is to the world is configuration space; then the best package, all things considered, is one that has ordinary macropredicates pick out features of configuration space. But this shows that certain features of configuration space are *good enough* to count as tigers. And then the line of reasoning proceeds as follows: 'Even if there were extra stuff—throw Bohmian particles or whatever into the mix—we have agreed that the relevant features of configuration space are good enough to count as tigers. So whether or not that

---

[9] For more on this topic see my [2007].
[10] See Lewis [1983].   [11] See Williamson [2007, ch. 8].

extra stuff is floating around, you should *still* count those features of configuration space as tigers.'

This is at best very dicey reasoning. By way of analogy, let us suppose that, on the hypothesis that the world is a four-dimensional Galilean spacetime with field values at points, the best hypothesis is that cats are regions of spacetime (with a fudged account of modality). A parody of the line of reasoning would then claim that even if, in addition to Galilean spacetime, there are extended occupants of that spacetime that are distinct from spacetime regions and that have the modal profile we intuitively associate with cats in a more straightforward way, this could not make any difference, since it is has already been conceded that cat-shaped regions of spacetime are *good enough* to count as cats: so you should *still* say that those regions are cats.[12] It is obviously not acceptable to reason like this: the credentials of the regions to count as cats get *swamped* once you hypothesize that there are these additional things of a suitable sort. Their credentials are so bad relative to the occupants you might imagine that, on the hypothesis that there is Galilean spacetime plus, it becomes no longer plausible to say that the regions of spacetime are cats.

I do think that, at least taken at face value, some of the reasoning I see from Oxford Everettians is dicey in the way that the reasoning that I've just described is dicey. And one way to see that it is dicey is by appealing to suitable metasemantical principles. Suppose for example, that one is a fan of Lewis's naturalness principle: it might be that on the hypothesis that there's just a configuration space the predicate 'cat' gets associated with a fairly gerrymandered property, but that on certain richer metaphysics the claim of that property to be the semantic value of 'cat' is trumped by a more natural property.[13,14]

---

[12] In addition to the considerations in the text, there are yet more obvious ways that the additional occupants might make trouble. One functional feature we associate with tigers is that they get in each other's way—they can't be in the same place. But that feature would not be preserved in a setting where we regard both the region of spacetime and its occupant as tigers. This points to a very straightforward way in which something's being good enough to count as a tiger may not be independent of additional objects: the latter can make a difference to whether or not the original object has certain functional features of a relational sort.

[13] One could also easily imagine a principle of knowledge maximization being brought to bear. It would be especially disturbing, for example, if it turned out that, on some Everettians favored identification hypotheses, we know hardly anything, while on some rival set of hypotheses we know much more. I shall not try to explore whether—even assuming a knowledge maximization principle in metasemantics—that criticisms of this sort will pan out. (Note that this kind of criticism should be distinguished from another but equally pressing kind of criticism: if it turned out that, on a certain foundational metaphysics, any reasonable semantic profiles reckoned us to know hardly anything, this would generate a threat of self-defeat for one who embraced that metaphysics. In that case, the criticism would be directed not so much at the part of the package that described how the macro-image emerged from the ground floor, but instead at the theory of the ground floor itself.)

[14] There are other heuristics beyond the two that I have already mentioned. For example, it is good to be wary of exotica. If a certain package generates really bizarre objects, in addition to familiar macro-objects, that may provide some evidence that the package that you're working with is dubious. I don't know as much as I should about the technical side of things but let me say how that might start to play itself out in the current setting. We do think it would be a bit

I don't pretend by any means to have offered any decisive criticisms of the Everettian approach. For the most part I have merely gestured at two places where one might try to turn the screw. First, there is a prima facie risk that Everettians of the sort I have described will commit themselves to explanatory gaps of an intolerable sort. Second, at least certain Everettians may have overestimated the strength of their view due to an excessively naive metasemantics. I recognize that these threads cannot be pursued properly without engaging with the details of various Everettian approaches, with regard both to the competing accounts of the fundamental ground floor, and also their attempts to link that ground floor with the manifest image, but as someone who is frank about his own limitations in these areas I cannot myself offer more.

References

Chalmers, D. [1996], *The Conscious Mind*, Oxford University Press, Oxford.

Dennett, D. [1991], 'Real patterns', *Journal of Philosophy* 87, 27–51.

Hawthorne, J. [2007], 'Craziness and metasemantics', *The Philosophical Review* 116, 427–40.

Lewis, D. [1970], 'How to define theoretical terms', *Journal of Philosophy* 67, 427–46.

—— [1983], 'New work for a theory of universals', *Australasian Journal of Philosophy* 61, 343–77.

—— [1986], *The Plurality of Worlds*, Blackwell.

Quine, W.V. [1960], *Word and Object*, MIT Press, Cambridge.

—— [1969], *Ontological Relativity and Other Essays*, Columbia University Press, New York.

—— [1981], *Theories and Things*, Belknap Press.

Saunders, S. [1997], 'Naturalizing Metaphysics', *The Monist* 80, 44–69.

—— [2008], 'Time, quantum mechanics and probability', *Synthese* 114, 373–404.

Wallace, D. [2003a], 'Everett and structure', *Studies in the History and Philosophy of Physics* 33, 637–61.

—— [2003b], 'Worlds in the Everett interpretation', *Studies in the History and Philosophy of Physics* 34, 86–105.

Williamson, T. [2007], *The Philosophy of Philosophy*, Blackwell, Oxford.

weird if it turned out that as well as three-dimensional beings walking around in three-dimensional space there were universes of flatlanders, in reality, two-dimensional beings walking around in a two-dimensional space, and universes of multilanders, multidimensional beings walking around in multidimensional space. If you had a theory which said *that*, you'd start to get a little bit worried. Unless these exotic objects offered significant empirical gain, you'd think that something is probably going wrong. Supposing we were broadly Dennettian, it wouldn't be surprising that when we looked around in configuration space we could find things that looked like flatlanders and things that behaved like two-dimensional space which the flatlanders inhabited, and so on. Before long, if one wasn't careful, one might end up saying that there are all sorts of different kinds of worlds in reality; worlds like ours, flatlander worlds, multilander worlds, and that, *sub specie eternitatis*, ours is not special. The exotica generated by that package provide pretty good reason to be a bit wary of it. (Obviously the prohibition isn't absolute.)

Commentary: Reply to Hawthorne: Physics before Metaphysics[1]

James Ladyman

The metaphysical conception of the generation of the macroworld from fundamental physics that Hawthorne considers is criticized and compared with the scientific account offered by Halliwell and Hartle. It is argued that Hawthorne's critique of Everettian quantum mechanics fails.

There are three questions that we ought to distinguish in our assessment of Everettian quantum mechanics (EQM):

(1) Is EQM coherent?
(2) Is EQM the best interpretation of quantum mechanics?
(3) Are there multiple worlds?

The coherence of EQM is challenged by the basis and probability problems that the Oxford Everettians have done so much to address. One could of course grant an affirmative answer to (1), while arguing that (2) and (3) should be answered negatively because, for example, Bohm theory or dynamical collapse is superior on some grounds. One could also answer both (1) and (2) affirmatively but remain agnostic about (3), citing the fact that quantum mechanics is not a complete theory of the world, and claiming that we do not yet know how much structure the next theory to emerge in the evolution of fundamental physics will share with quantum mechanics. I shall briefly return to these issues at the end of this paper but the bulk of it addresses John Hawthorne's paper, which, as I read it, is a challenge to the Oxford Everettians' defence of an affirmative answer to (2).

Hawthorne raises a number of problems for EQM. All of them have to do with the relationship between the world of the wavefunction, and reality as it is known independently of quantum mechanics. Assuming that we have an account of the world as it fundamentally is, and as Hawthorne points out there are several candidates for such an ontology based on quantum mechanics (Hilbert space,

[1] Many thanks to Simon Saunders for comments on a previous version of this paper.

configuration space, spacetime with non-local fields, a space of operators), we can ask 'how the familiar truths about the macroscopic world that we know and love ("the manifest image") emerge from the ground floor described by the fundamental book of the world' (p. 144). This question is at the heart of Hawthorne's paper. Note that the way it is put assimilates the macroworld and the manifest image. However, they are not the same thing. The macroworld obeys classical physics to a good approximation, and classical physics conflicts with the manifest image in many ways. To start with there is Galilean relativity, the lack of absolute position, and in general the distinction between the primary and secondary qualities of things that the natural philosophers of the Scientific Revolution, including Galileo, followed the ancient Greek atomists in deploying to explain how the world of corpuscles in the void could be so unlike the manifest image. Subsequently, classical physics came to include the immaterial fields of electromagnetism, the properties of which challenge any attempt to reconstruct their workings as those of manifest matter in motion. Then of course we have relativity theory and the repudiation of absolute time and absolute length. Science is not under any obligation to recover familiar truths from the manifest image, only approximations of them, the reasonableness of asserting them even though they are false, or their persistence as illusions. The familiar 'truths' that the Earth does not move, that there is a fact about what is happening now on Alpha Centuri, that we and apes do not have common ancestors and so on, have all fallen to the progress of science. Conserving common sense is not a desideratum for physical theory or its interpretation and it may well be that 'most of our ordinary beliefs about the physical world are false' (ibid.), or at least only approximately and partially true and in other ways systematically wrong.

Thankfully most physicists did not take seriously the objections of metaphysicians to relativistic physics because it was incompatible with real frame-independent becoming. Hawthorne's problem of how the universal experience of becoming 'we know and love' is supposed to emerge from the fundamental world has not been solved, but that is no reason to reject relativity. The solution must surely be a scientific not a metaphysical one. Similarly, when it comes to explaining how the macroworld or the manifest image emerges from the quantum world we must look to science. In the case of recovering the manifest image, cognitive psychology must join forces with physics and the task is much harder. As Jonathan Halliwell and Jim Hartle (in this volume) explain, the incomplete scientific account of how the macroworld can be recovered from the fundamental quantum structure is complex and difficult enough. It is instructive to consider some of the general features of it, and how it differs radically from the kind of story that Hawthorne envisages. Before turning to that I briefly comment on a couple of other matters that arise from Hawthorne's critique.

Hawthorne points out that the Everettians suppose that there is no ineliminable vagueness in the fundamental book of the world, where vagueness is understood in terms of vague predicates. It is worth noting that there is no reason to suppose that fundamental physics be expressible in the subject–predicate form found in natural language and beloved of philosophers. The metaphor of the fundamental book of the world written by God that Hawthorne deploys brings to mind Galileo's famous claim that it is written in mathematics not natural language. The Oxford Everettians do believe that there is vagueness in matters pertaining to branches, and this is often seen as an expensive price to pay, but I think of it rather as a free gift. Branches are supposed to correspond to macroworlds, and it is quite reasonable to think that macroscopic objects are not composed of exact numbers of microparticles, and that their spatial boundaries are fuzzy. In that sense, macroscopic objects are vague objects. Most philosophers deny that there can be vague objects but their arguments for doing so depend on premises that are at least as contestable as the denial of the claim that the height of Mount Everest is not determinate to the nanometre.

When it comes to the 'generation' of the macro- or manifest images from the fundamental image, Hawthorne distinguishes between two views. The conservative involves the identification of the macro- or manifest objects with fundamental objects or aggregates of them, while the liberal posits additional objects that 'float over the fundamental layer' (p.146). However, this is a false dichotomy. A third position asserts that the macro- or manifest objects are patterns in the structure of the fundamental objects that are neither free floating nor identical with those objects. A fourth position adds that there is no fundamental layer at all, and that the allegedly fundamental objects are themselves patterns in the structure of some deeper level of reality.[2]

The two hypotheses about generation that Hawthorne considers have in common that they involve the recovery of macroscopic objects exactly from the fundamental objects, whether as aggregates of them or as necessary concomitants of them. However, it is no argument against a particular version of fundamental physics that we cannot generate from it evolution by natural selection, or the law of supply and demand. Similarly, no collection of fundamental objects is identical with or necessarily gives rise to the futures market in oil or the species *Felis silvestris catus*. Of course, philosophers who worry about what they call 'fundamental ontology' are inclined to deny the existence of species and markets, but then many of them deny the existence of tables too, or hold that tables only exist because every arbitrary sum of fundamental objects exists. On the other hand, on the two hypotheses about generation above, that Hawthorne does not consider, macroscopic objects may only appear when we coarse-grain over the structure of the more fundamental ontology, as in fact is the case in the physics of 'generation' discussed below, and so the vagueness of the macroscopic with respect to the microscopic is unavoidable.

---

[2] The 'real patterns' conception of ontology due to Daniel Dennett is elaborated and defended in Ladyman and Ross [2007], chapters 4 and 5. Chapter 3 discusses EQM and the question of whether reality has a fundamental level.

Supposing the need for an account of generation, there are two forms it could take: we could ask, given that we have classical physics and a description of the macroworld in its terms, can we recover it from quantum mechanics? Or we could ask, given that we just have quantum mechanics can we generate the resources to describe and recover the macroworld? I think that requiring the latter places an unfair burden of proof on the Everettian since most cases of successful reduction in science only work because some higher-level concepts are presupposed, as for example with the reduction of chemistry to quantum mechanics where the notion of the chemical bond and the idea of molecular shape cannot be derived *ab initio*. The physics of decoherence is a top-down reduction, but as Wallace (in this volume) also argues, none the worse for that. (Note that there is a further distinction between supposing that the Everettian account of generation must convince the non-Everettian, or merely reassure the person who already believes EQM.)

Hawthorne's paradigm of conservative generation is the identification of everyday objects with regions of Galilean spacetime. However, note that the spacetime of classical physics that respects Galilean relativity is not the product manifold $E_3 \times E_1$, but rather the fibre bundle of E_3 over E_1. In this structure there is no possibility of identifying a cat with a region of spacetime because there is not one space but many. Of course, Hawthorne is just making an analogy with this example, but my point is that the toy models of philosophers are crude with respect to the models of even classical physics and so there is no reason to expect the kind of simple generational stories that philosophers tell to be anything like the truth. Indeed, the idea that everyday objects can be identified with regions of Galilean spacetime is itself 'metaphysically ridiculous', in the sense that we have discovered that such an ontology is no more a candidate for being physically fundamental than an ontology of Aristotelian forms or the four elements. What is metaphysically ridiculous cannot be a matter of common sense but must be decided on the basis of theoretical and empirical considerations. If the idea that the world is fundamentally a Hilbert space with a 'marvellous point' turns out to be scientifically adequate then there is no reason to care whether metaphysicians think that it is ridiculous.[3]

Let us now turn to Halliwell's and Hartle's discussions of the quasiclassical realms. There are features of the account of the emergence of the classical from the quantum to which I want to draw attention:

(a) It is dynamical and hence diachronic not synchronic.

Metaphysicians expect the bridges between the ontologies of the different sciences to be synchronic but they are usually diachronic. So, for example, it

---

[3] As Wallace (in this volume) points out, our intuitions about the world are the product of our evolution which cannot be expected to have made them reliable about matters that were not relevant to the survival of our ancestors. Ladyman and Ross [2007], chapter 1, criticizes the reliance on intuition in metaphysics.

is the dynamics of how hydrogen bonds form, disband, and reform that gives rise to the wateriness of water and not the mere aggregation of hydrogen and oxygen in a ratio of two to one. Likewise it is the dynamical isolation of branches that makes them Everett worlds and not their isolation in configuration space. Hartle's account makes this absolutely clear.

(b) It is based on coarse-graining.

Histories are grouped into classes. There is an inherent vagueness about how coarse-grained the classes need to be, and the macroworld is recovered only as a coarse-grained feature of the microscopic world not as a precise and definite entity in its own right. Hence, the macroworld cannot be identified with a set of objects in the microworld as in a conservative story, but nor is it a free-floating set of objects as in the liberal story. Rather, as Wallace (in this volume) emphasizes, the decoherent histories of quasiclassical worlds are patterns in the wavefunction. As Saunders (in this volume) says, why should the task of recovering classicality from the quantum world be any different from the task of recovering the gross properties of matter from condensed matter physics, or thermodynamics from statistical physics?

(c) Modal facts are not invariant.

Hawthorne worries that there will be 'violence to our modal intuitions'. However, there is no reason why we should take it that such intuitions are a reliable guide to the modal facts. Suppose, as I believe is the case, that it is an open question whether the final fundamental physics will be deterministic. Under determinism, the only way for supposed everyday counterfactual possibilities, such as that I might have had something different for breakfast, to be genuine is if the initial conditions of the universe could have been different enough to make just that fact different, but not so different as to preclude the existence of my life this morning. However, we have no idea whether this is the case or not. Maybe the only variations in the initial conditions of the universe are ones that give rise to radically different worlds from our own, rather than ones that differ just in virtue of my breakfast. Going back to the revisions of common sense mentioned above, it is very plausible on inductive grounds that science will force us to revise our estimations of what is or is not possible (it is after all hardly intuitive that the speed of light cannot be exceeded). In particular, the modal facts are not invariant between classical physics and EQM on Hartle's account because classical determinism only holds approximately.

Note by the way, that when Hawthorne contrasts the ontology of Galilean spacetime with field values at points, in the context of which cats are regions of spacetime, with the ontology such that 'in addition to Galilean spacetime, there are extended occupants of that spacetime that are distinct from spacetime regions and which have the modal profile we intuitively associate with cats in a more straightforward way', he intends to give an analogy to the wavefunction with and without Bohmian particles. However, in the latter case the Bohmian particles don't come with a different modal profile to the empty trajectories, whereas in the former case the extra cat structures are posited to have a different modal profile, and so the analogy fails because the Bohmian particles really add nothing to the structure of the world only some of Hawthorne's 'vim'.

(d) The methodology is top-down not bottom-up.

This means that we start by helping ourselves to classical concepts and seek to recover the approximate truth of classical laws from quantum mechanics rather than expecting to find all the resources we need in quantum theory. As noted above, it is not reasonable to expect more from scientific reduction than this in many cases because we can only find the higher-level ontology in the more fundamental one when we make the right coarse-graining, and there is no reason to require that the fundamental level should tell us what that is. Indeed, if it is right that higher-level entities are always approximate and vague with respect to lower-level ones then it will be impossible for the lower level to determine the higher-level structure just because the latter is strictly incompatible with it. Consider, for example, the case of Kepler's laws of planetary motion: we cannot expect to find the exact form of them within Newtonian mechanics because the latter actually entails that perfect elliptical orbits are impossible for the planets because of the attraction they have for each other.

(e) It is based on the supposition of conservation laws.
(f) It is based on the concept of local equilibrium.
(g) It depends on the low-entropy initial state of the universe.
(h) It is valid only over relatively short timescales.[4]

These last features have no analogues in Hawthorne's idea of a generational story.

When it comes to the explanatory gap between the quantum world and the macroworld, contrary to Hawthorne's claim that 'none of the fundamental-to-macro associations are given' (p.150), it must be acknowledged that Halliwell and Hartle do much to close it. When Hawthorne says of Wallace's account of tigers as patterns in the wavefunction that 'the association of growling with the configuration space isn't given to us' (ibid.), he also overstates the explanatory gap since clearly the tiger growling is associated with the movement of its larynx and the air in its lungs and these are associated with the motions of particles in configuration space. EQM is arguably coherent and superior to Bohm theory and dynamical collapse interpretations because of the way it, unlike them, is integrated with the rest of physics. Given this advantage its alleged metaphysically absurdity is not worth worrying about, but this does not necessarily mean that we should believe in many worlds. That step need only be taken if it can be shown that Halliwell and Hartle's account requires them, and that quantum mechanics

---

[4] Ladyman and Ross [2007] defend the idea that, in general, ontology is scale-relative.

is sufficiently close to the final fundamental physics for us to be sure that the latter will also deserve an Everett interpretation.

Reference

Ladyman, J. and D. Ross [2007], *Every Thing Must Go: Metaphysics Naturalised*, Oxford University Press, Oxford.

Transcript: Ontology

Vaidman (to Saunders and Wallace)

My question is for David [Wallace] but it's also for Simon [Saunders]. You say that we need this decoherence formalism to explain the separation of worlds. In a very simple experiment of a photon, beam-splitter, and detector, the many-worlds interpretation states that there are two worlds and the standard theory says there's only one. In this example, the decoherence is obvious; there is no interference whatsoever. If you consider other experiments, like observing a diffraction pattern on a screen, then even in standard quantum mechanics one needs to calculate the time when things become different or not exactly the same. In all normal cases where decoherence is needed to explain the separation, it seems absolutely obvious; decoherence is necessary but it's absolutely obvious.

Wallace

I've got multiple responses. Firstly, of course, what's obvious to one person isn't the same for another and not all of us share your genius Lev [Vaidman]. Seriously, obviously it wasn't that nobody had thought of decoherence before Zurek came along; I mean Mott knew about decoherence. I think what decoherence brought us was a language and a sort of unified way of thinking, a way of asking and answering questions like: why do macroscopic systems end up in these branching non-interfering states—sure, situation by situation it's dead easy to model it, but what are the general features that lead to that sort of thing? And for some people that decoherence language is a perspicuous way of answering, but I think the fact of the decoherence process—what's driving it—is independent of whether one decides to call it decoherence and of what is explained situation by situation. I find the language perspicuous and also very helpful as a way of getting at questions, or sort of seeing the structuralist aspect of this and getting at questions like: what about alternate ways of decomposing the wavefunction? Look, after all, as Tim Maudlin has pointed out for instance, configuration space is not, or at least not guaranteed to be, a primary thing. Certainly it's not the configuration space of classical physics. So we need to ask something about why that particular way of representing the quantum state is special. I think there's a sort of useful, unified way of thinking about it that makes the right answers to this become a bit more obvious and a bit easier to talk about. So I don't think it's magic. It's certainly not a new theory, it's something that's always in quantum mechanics.

Saunders

What if some theorist comes along and finds a very alien domain, a non-quasiclassical domain? That is, discovers equations that seem to govern the independent autonomous behaviour of worlds, but worlds entirely different from our world. What would you say to that?

Vaidman

If in these worlds physics is not local, and interactions are not as strong as here, then there would be no consciousness—or maybe if the interactions are non-local you have non-local consciousness.

Saunders

It's nothing to do with conscious beings. The question is, is this or is this not an interesting structure in the wavefunction? Clearly the answer is 'yes'. Now, that would be an answer delivered through direct analysis in decoherence theory. You, as it happens, taking the sorts of experiences we have, can identify very obvious decohering states of affairs, but for this alien set of entities, worlds, whatever, it is not at all obvious, because you don't happen, as it were, to live in such a world and be adapted to such a world.

Wallace

Simon finds the idea of these alien worlds much more plausible than I do.

Saunders

Well, it's illustrative.

Hartle

I don't think Lev is criticizing us (Gell-Mann and Hartle) in this approach. It's just that he restricts to a certain class of histories—the ones described in quasiclassical terms. He would agree that if we calculated the interference terms between them they would vanish. But if we start from a more general perspective, in which all possible kinds of sets of histories are available to describe the world, the question becomes which ones are more useful and more reflective of the kind of top-down structure he is describing. That's a more general starting point than jumping immediately to the idea that we already know what the top-down structure is. Starting from that more general perspective, the absence of interference is necessary to get probabilities.

Hemmo (to Wallace)

I wonder about the view which would take quantum mechanics to describe a single world, using decoherence. I'm not saying it's going to be easy to do that but it wouldn't be too difficult; one can see how it would go: basically you just do whatever you guys are doing in the context of the Everett interpretation and just focus on one of these worlds as the actual one.

Wallace

I think the reasons why I don't think that's viable are, if you like, quite locked up in some quite general philosophy of science. I don't think the one-world view works because of two features of the approach I'm describing which work fine for the emerging structure story but not at all for the single-world story. One of those is the approximateness with which the decoherence basis is described, and the other is the way in which it's described in a very top-down way, and in terms of higher-level ideas that we can't describe in the microphysics. We do not think that it's appropriate in our microphysics to have fundamental axioms that make reference to all sorts of concepts which we have an irreducibly emergent, structurally defined description of. We're fine with those concepts if we can show that they're emergent from a cleanly stated starting point but we're not happy with the idea that you just put them in by hand as an axiom. Otherwise we could have solved the measurement problem like that a very long time ago. As you said, it wouldn't have been difficult: we could have used higher-level terms like 'conscious observer' and that would've resolved the measurement problem just fine. But the point is that we don't really think that that high-level way of talking is legitimate in specifying the fundamental laws of physics. And I think my reasons for thinking it's not legitimate are pretty standard and unoriginal.

Bacciagaluppi (to Saunders and Wallace)

Simon, could you explain a bit more what you were saying about the differences between relative states and many worlds and global and local projections, and, David, I didn't understand what the measurement problem is that some philosophers have.

Saunders

The localized approach uses decohering local projections and correlations (relative states) among them. I think it's clear that one can do that over small macroscopic dimensions and larger. I don't think you can do that over small microscopic dimensions. So this is a different take on how to see structure in the universal state: one builds up from localized projections, totally degenerate at large distances. One builds up a universe around us, which will be distinguished in that it is value-definite in our local region, but indefinite—with macroscopic superpositions—far from that region. But this is not in contradiction to the picture of worlds. It is in fact a superposition of worlds—all those that coincide on this local state of affairs.

Value definiteness in this relational sense is a little like simultaneity in classical spacetime. One can do something similar in spacetime terms to arrive at global times: one looks at relations between events to define three-dimensional worlds, each a world at a time. The difference for the relation of value-definiteness is that it is not an equivalence relation. You don't get a partitioning of the wavefunction into equivalence classes, you get something more complicated—a branching structure.

Wallace

The 'philosophy measurement problem' essentially is the view that there is something wrong with quantum mechanics as a physical theory, *over and above* the fact that it fails approximately to reproduce something structurally isomorphic to the pre-quantum world. So, for instance, it includes the view that in the 'problem of tails' of the GRW theory, even if the tails were unstructured there would be a problem with the theory and we'd have further interpretative work to do to the theory before it would count as a solution to the measurement problem . . . It may be a straw man but I think there is a relevant distinction to draw.

Brown (to Wallace)

I think it might be useful for some related discussion, David, to say a little bit more about a point you raised about decoherence in relation to classical systems and chaotic behaviour. Can you say a little more about that?

Wallace

OK, if you take a non-chaotic system without decoherence, a Gaussian wave-packet does a pretty good job of tracking the classical dynamics on pretty long timescales. Take a chaotic system without decoherence and it doesn't do that. The reason it doesn't do that is that what the classical dynamics wants to do is make the system become highly fibrillated and the quantum system can't cope with that when the fibrillation is thin compared to h-bar, so you start getting a violation of quasiclassicality in a certain sense. Throw the environment in and what you find is that those differences lead effectively to wavefunction collapse, and effectively what you have is a coarse-grainedness of the classical dynamics, which means that branch-by-branch, what you've got is something like a stochastic dynamics. You've still got a branch structure, the environment is still recording the full history, you haven't got a sort of merging together of branches but it's no longer the case that each branch does something classical and deterministic. And so the sort of systems that get used in chaos theory popularizations are systems whose quantum evolution leads to indeterminism and to a branching phenomenology.

Lehner

As a historian, one thing I'd like to point out about Everett himself is that the way he framed his theory was with this very simplistic definition of an observer as a recording device, right? And it seems like that's all very badly out of fashion: nobody wants to talk about that any more, and I have a suspicion people think that decoherence is the appropriate modern substitute for this model of the observer that Everett used. I would just like to point out two things. One: the two don't conflict with each other. That's Zurek's claim—that recording devices are decoherence-inducing devices—so Everett's observer is of course in itself a 'decoherence machine'. And secondly, from the philosophical point of view I do think that Everett's model is actually a very useful one and helpful to discuss a lot of philosophical points that come up, for example David Albert's metaphysical complaints about the Everett interpretation. More helpful, I would claim, than talk about approximate decoherence, which tries to absolutely avoid speaking of conscious observers, is to talk about conscious observers but be perfectly functionalistic about it. I totally subscribe to your ringing endorsement of functionalism and I think that's exactly what Everett is doing; he's being functionalistic about consciousness in this sense. It perfectly matches with decoherence but it might actually be a very useful way of talking about the philosophical questions.

Wallace

I think it has fallen out of fashion; I think it's good that it's fallen out of fashion. The reason it's good is, firstly, it lets us avoid having to get entangled in the mind–body problem and it lets us leave the job of understanding conscious observers to the people that ought to, i.e. psychologists. And secondly, I think that talking that way makes the Everett interpretation sound 'even' more weird than it actually is. To me, leaving aside issues about possibly branching into multiple people synchronically, the only weird thing about the Everett universe is that it is much, much bigger than I thought it was. But, if one talks about consciousness one starts saying things like, we don't have an actual recovery of quasiclassicality, we might have a recovery of the appearance of quasiclassicality but it wouldn't necessarily be quasiclassicality. And then one starts talking in a very unhelpful way about the fact that there isn't really any definiteness at all, just an illusion of definiteness. I think those are traps which one is walking on the edge of when one talks about observers too much.

Hartle (comment)

I didn't completely understand Tim [Maudlin]'s talk, but we (Gell-Mann and
Hartle) certainly have a view on what it means to get out the macroscopic world.
We don't think of that in terms of some large-dimensional configuration space
or phase space. Rather we think it consists of histories that describe the evolution
of fields or particles in time, say, and that are constructed in a certain way: We
look for sets of histories that, first, decohere, so we can assign probabilities, and,
second, have within them the descriptions of all the possible objects, tables, chairs
and so forth that we might conceive of (typically represented by integrals over
small volumes of densities of conserved quantities like energy) and, third, have
members that evolve in time approximately according to classical laws with high
probability. I'm not sure that would satisfy Tim but that's the way we would get
out the usual description of classical behaviour. That's how we get out classical
physics from quantum mechanics.

Wallace (to Maudlin)

My question is about going from a high-dimensional space to a low one. I'm kind
of uncertain about the rules here: I mean it seems to me that it would be very
unsatisfactory if our account of the way that physics is to be systematically related
to our intuitions says that somehow we can go a certain way away from the
manifest image, but thus far and no further. I'm a bit worried that if we do that,
philosophy of physics is in danger of becoming a debate about what our different
intuitions make plausible. And personally I don't have an intuition problem with
that sort of higher-dimensional stuff. If the higher-dimensional configuration
space on which the wavefunction lives were genuinely homogeneous, possibly I
would, but it's not homogeneous at all, it's got a very complicated structure. Do
you feel you have something critical to say about where that cut is between what
does and doesn't make sense, or is it just going to be that the cut is where my
intuitions tell me it is?

Maudlin

I think that's a very good question and that's why I said I don't have a
philosophical analysis that says that my conceptual system baulks *here*. The best
I can do is give examples—I mean, I hope everyone appreciates the smoothness
in the one case and the lack of smoothness in the other case. And I agree with
you, I can imagine a series of steps and, who knows, maybe now we're into
psychology, maybe if I were slowly acclimatized, like the frog that boils to death,
I could be slowly acclimatized so that in 50 years I became a many-worlds person
without realizing it. But all I'm doing is reporting the situation in my brain
right now.

Myrvold to Maudlin

If I just give you a mathematical structure I don't yet have a physical theory, I need what the logical positivists would call correspondence rules to tell me what bits of math map onto bits of the world. And it sounded to me like that's what you were saying when you were saying that just having the wavefunction, that's not enough, we need some kind of correspondence rules. But then I don't see why it's more of a problem than for the point particle model of classical atomism. Take the example of four classical particles and you want to know if they're in the shape of a tetrahedron. If I just give you a 12-dimensional space and say that's configuration space for particles I need to tell you which three degrees of freedom belong to one particle and which three degrees of freedom belong to the other before you can answer the question of whether or not you have a tetrahedron. Similarly when you've got a Hilbert space and you have to say which operators on that Hilbert space correspond to which observables in the world; you also have to tell me something about what happens in terms of the wavefunction. It seems like you can do that in any of these theories just by adding correspondence rules—maybe more murky than others—but it seems to me essentially the same problem in all of these theories.

Maudlin

That's where I disagree and I think this talk about operators and observables is part of the reason which makes the situation so murky. That is, there is no natural correspondence or relation between a Hermitian matrix, say, and anything that happens in the laboratory. If somebody says, 'oh, all the Hermitian matrices are observables' and I hand you this math and say 'go measure that', *of course*, you'll say, that's not anything I can interpret in and of itself. Notice in the case of the Bohmian theory I didn't talk about observables; I didn't talk about matrices. I talked about particles moving around in space; I gave you a world that you recognized without any help. It seems to me you recognized it without any help, as possibly the world you live in. So, I think you're quite right, if you help yourself to any arbitrarily definable rules you can solve any problem. What I was trying to point out is that some theories don't seem to need to do that; and in fact most of physics through history didn't seem to need to do that; it seemed to be doing something else that seemed to be more comprehensible; and there is a way to understand quantum mechanics that way, there are several ways to understand quantum mechanics that way, and to me it'd be preferable if it were like that.

Bacciagaluppi to Maudlin

There are a set of arguments which have been around in the physics literature which would *seem* to go some way at least in addressing your problem, namely, symmetries, dynamical symmetries. So this is how to get from the abstract

Hilbert space to recognizable spacetime descriptions. The Galilei group is the symmetry group of non-relativistic quantum mechanics; now, just as in the case of Newtonian mechanics, the idea that spacetime just encodes the universal symmetries of the dynamics does seem to go through also in non-relativistic quantum mechanics. Or, look at the structure of the Hamiltonian: we've got two operators p and q, it is quadratic in p and there's a potential term which is a function of q. That introduces an asymmetry which is not in the commutation relations. Again, an irreducible representation of the Galilei group allows you to make a tensor product decomposition of the Hilbert space into what you call elementary systems, and so on and so forth. Don't all these results go a long way towards giving you what you want?

Maudlin

There's no doubt that there is a lot of structure to the Hamiltonian. The natural way to understand it—and I would make exactly the same claim for the Galilean group and so on—is that the structure of the Hamiltonian is there because you already have a structure because you're dealing with a certain number of particles in a common space. And you know how to implement the Galilean transformation on that. Now, you can try and go the other way round—I think this is the sort of thing that David Albert has been trying to do—and say no, no, there is no further explanation for this, all this interesting structure in the Hamiltonian, it's just, you know, God *decreed* the Hamiltonian should have such and such a structure and then we have this discussion. Now, it's not something I've talked about here but it does seem to me: suppose I had two programmes and one programme says, look, here's a fairly simple, comprehensible, physical ontology. The postulation of that physical ontology implies that there'll be these certain symmetries in the Hamiltonian and so on—which you find. And the other one says, my theory says, you've got a Hamiltonian, it's got all these interesting symmetries—it becomes even more interesting if you've got identical particles because in a certain way the configuration space for identical particles has an interesting topological structure (I shouldn't go into that). The configuration space has this very interesting structure, the Hamiltonian has this very interesting structure, there is *no* further story about why it has that structure; it doesn't arrive from any other fundamental ontology. Interestingly enough it's *exactly* the same structure that would arise if the other ontology had been this very simple ontology. At that point I would just say I actually find one a more plausible theory than the other, as a piece of physics.

Saunders to Maudlin

I think there's a certain worry about the nature of the arguments you've presented; the target isn't represented by anybody very well, and certainly it's not represented by me. It seems to me that the issue with the Everett interpretation of quantum mechanics isn't so much the mantra 'only the wavefunction and nothing but'; the issue is whether we need to change or add to the basic equations. Physicists extract structure from these equations in a very large variety of ways; the issue is can they extract enough structure to save the appearances. Whether they do this in a way that should be called bare wavefunctional realism, or not bare, because they make use of operators and so forth—I'm not sure that that's a very interesting question.

But I want to address something else you've been saying which I think is important, which is that the present situation is somehow unprecedented. I wonder about that. The suggestion is that, historically speaking, we haven't had problems reading off a story of the world from the equations. But most of the history of modern philosophy has been quite deeply engaged with puzzles about the 'corpuscularian philosophy', as it was called in the 17th century. It seemed that its only consistent development was essentially as Boscovician atomism—in terms of point particles alone—and that was not at all transparent as a description of the world that we see. I don't think any philosopher ever found it so. This is a world without smells, without colours, without feelings, without warmth, without solidity. There was a huge project to try to extract an account of those sorts of things, 'secondary qualities' as they were mostly called. And typically the business of extracting that account of those secondary qualities was broadly functionalist, with many of the characteristics of functionalism that David was drawing attention to. Now I agree that in quantum mechanics it's harder. Absolutely. No one would suggest otherwise. But it doesn't seem to me that we are in a *dramatically* different situation. Moreover, it seems to me that one would *expect* it to get harder, the further one goes into the microworld, or to very high energy regimes, or to the early universe. So I lose any real grip on the idea that we are in an unusual situation.

Maudlin

I haven't done any systematic or even unsystematic survey of how physicists—because the first point you made is, look, physicists are able to extract this structure from the wavefunction—do this, but I think it's pretty clear, I give the example of just one guy but he is an actual physicist, an actual advocate of the theory, Thibault Damour, it's easy to see how Damour extracts the structure: he extracts it by taking the configuration space to be a configuration space; that gives you a hell of a lot of structure. He doesn't worry his head about it; and for all practical purposes, having done that, he doesn't really have much of a problem. So the fact that physicists can get along and use quantum mechanics perfectly well, of course they can. As soon as you think you've got a configuration space—all my arguments are, if it really is a configuration space, if you pretend it is, you know what to do with it and as long as they're pretending it is of course they're not going to find they have any analytical difficulties, or practical difficulties.

The second point, look, as far as secondary qualities goes, of course, I don't think physics has ever solved the mind–body problem; I don't think any physics, Newtonian mechanics, ever told us how things were warm or anything like that. So, it's not as if I thought earlier physics had solved *that* problem. What did Newtonian gravitational theory do? It gave us trajectories of planets and cannonballs and stuff like that and we came in thinking we knew a lot about the trajectories of planets and cannonballs and stuff like that. In a way, the point that I'm making is: in philosophy there was a sidetrack during the logical empiricist period where the notion was, the evidence is sense data, the evidence is described in experiential terms. And so our job, to connect the physics up to the evidence has to bridge that gap somehow. Of course nobody knew how to do it analytically, so what could you do except say, oh, there's a bridge rule, or I'm just going to tell you that when such and such occurs then red spot for me-here-now. Now, that just seems to me to mischaracterize the nature of physics. The level of contact that is made typically, at least in a case like gravitational physics, isn't at the level of anybody's experience of anything. It's at the level of motions of macroscopic objects which could on the one hand in a straightforward way be derived from the theory and on the other hand there's something we thought that, through various means, we had a pretty good evidential handle on. It's the absence of *that* that I think would be unprecedented. The absence of that meeting place between our pre-theoretical understanding of the world and the theoretical understanding.

Hartle to Hawthorne

I want to know what happens if you have to retreat. The situation in contemporary physics seems much worse than just the question of whether you have one history or all the histories in a decohering set. There are the infinite number of other possible decohering sets in the theory, which have nothing to do with everyday descriptions of classical terms, tables and chairs, cats, human beings. And if you look a little further in physics, say to string theory, we're getting even more exotic because we have perhaps ten to the thousand so-called vacua which might conceivably describe the world, most of which have nothing to do with the world which we see. Physics seems to be moving further and further away from the everyday at the fundamental level, for better or worse. So it's becoming less and less consistent with what I think you call metasemantical principles. What will you say if this trend goes on for the next 100 years? Will you still maintain these metasemantical principles which are in conflict with extant physics? Or is there room for you to modify them?

Hawthorne

I don't have anything very informative to say. The exotical thing was a bit tentative. I don't want you to think that that's the same thing as I was calling the metasemantical principles. I think this metasemantics stuff—it's a foundational part of the story about how we're located in the world. I don't think you can blow it off—and say it's just all garbage metasemantics—but on the other hand it's methodologically very hard to know how to figure out which principles are true or not. I could perfectly well imagine that our changing vision of the contours of physical reality might in some way interact with it. Just to take an example, if it turns out that causation as we normally think about it isn't a very important part of the world, then, really leaning on causation in your account of how words get hooked onto the world might start to look very naive. That's just an example. So I'm not dogmatic in the way that you suspect here.

Hemmo to Hawthorne

I want to focus on one specific point which seems to be playing a crucial role in the context of the Oxford reading of Everett—I think also James [Ladyman] said this—which is the vagueness of the branching. Maybe I don't get exactly what is the intended status of the branching in the Everett picture. I think about the fundamental book of nature; I'm not sure there's any need for vagueness in the interpretation in the context of branching. If you think about the overall picture—I mean just in elementary quantum mechanics—you think about the universal wavefunction of the whole universe. God knows it, so he knows all the complete microphysical truths about the world. He also knows exactly how we are microphysically structured. He knows the Hamiltonian, the actual Hamiltonian; then, He can figure out exactly how we interact with the environment, everything is spelled out clearly and there need be no vagueness in that picture. And so if branching is true, if the Everett theory is true then there need be no vagueness. The only kind of vagueness that would enter in the story is in the way we now use natural language and make descriptions—we use natural language predicates and we describe the world in a vague way. That's fine, but that's not fundamental—you can push the vagueness, it seems to me, all the way into the language, into natural language, and there's nothing fundamental about it.

Hawthorne

Let's just get that—it's perfectly consistent, the Oxford picture. It is at least consistent—it's perfectly consistent to say that God could write a precise world book and the branch description of reality is inevitably vague. I gave you, at least very schematically, a model of how that would go. Suppose God knows the universal state perfectly; the way He'd describe it perspicuously is as some complex scalar field over configuration space. That's the fundamental story, and when you look at the fundamental story there's not in any fundamental way fission going on, but then the Oxford Everettians notice that on the hypothesis that the world's like that, there is of course a rough-and-ready way of parcelling up configuration space via rough-and-ready decoherence heuristics into parcels.

But if that's the way that branching talk has to be understood, and the heuristics that you use are inevitably coarse and rough-and-ready, then the way that you slap that onto the fundamental book of the world will inevitably be vague. And I should just say in passing that I'm not sure that epistemicism versus non-epistemicism is too important here. I think even if you're an epistemicist you could tell a story that goes from the premise that branch talk is vague to the conclusion that branch descriptions aren't fundamental, because even for the epistemicist you, as it were, need a range of candidates, and if there's a branch structure at the absolute rock bottom then there isn't a range of candidates for branching.

Hemmo

My question is: I don't understand why you think that there must be vagueness at the fundamental level. That's all.

Hawthorne

Can I make something vivid? Suppose there's a bunch of objects and there's interference between them all but there's little bits of interference between some and lots of interference between others and basically you want to count—the rough-and-ready idea is that you want to count things as belonging to the same world if there's lots of interference between them, but in some cases that's just simplifying. If there is super-low interference then it's OK to count them as parts of different worlds, but then you realize that in the God's-eye description there are different levels of interference—He's not going to care about what you're going to count as sufficiently low interference. There's going to be a familiar Sorities kind of structure where you've got all sorts of different levels of interference, but our ways of individuating worlds requires arbitrarily selecting some level of interference.

Wallace to Hawthorne

I'm in complete agreement with you about Meir [Hemmo]'s point. We can be quite banal about this, we don't need to talk about worlds. The inflation rate of the UK economy is a perfectly real and salient feature of the world, but God could know the exact location of every electron in the universe and He wouldn't know precisely the inflation rate of the British economy. This is totally ubiquitous across the special sciences.

Let me say something about the more substantial criticisms or worries that you were raising about structuralism, and about where it fits on the conservative/liberal spectrum. I think the distinction you brought up in your Galilean spacetime field models is quite salient in showing where I think we would be on that spectrum. Suppose your Galilean model has two fields, and, as in my shadow matter example, the fields don't interact very much. Now the strategy for identifying, say, cats, as spacetime regions is no longer available to us because we have to say which field the cat is made up from. What we'd actually want to say about the cat is something rather like what you said we do in configuration space: we're actually going to identify the cat with a feature of the spacetime region or the wavefunction, or the classical field, or whatever. And a trivial point, I think, about features is that there's less danger of our doing a modal sleight of hand. The criteria for what makes a particular feature salient will, again, be various kinds of top-down things, as James points out. So, I'm not sure if you want to call our position conservative or liberal, but it's characterized by a commitment to the fact that all there is is the microphysical stuff, and the macrophysical stuff is just a set of features of the microphysical stuff.

Hawthorne

OK, you're a conservative.

Wallace

OK, fine. As far as the kind of worries that you brought up about the functionalist strategy are concerned, to some extent I would plead ignorance on the answer. I mean we *know* that unless all hell is going to break loose in science, some sort of functionalist strategy is going to have to be sustained in some way, shape or form, whatever the rules are for that strategy. And from the naturalistic perspective, and looking at what science actually does, the rules seem to be something like: functionally defined entities need to look kind of not cooked-up in terms of the microscopic theory's fundamental variables. Building all the structure into the code we describe it in isn't OK. I don't know what the right way of completing that story is and to be honest I just plead a kind of naturalistic confidence that there will be such a story because science seems to work that way.

But as a last comment on that, the one thing I think it won't be is based on what we find more intuitive. That's one concern I have about your closest-satisfier story about what makes extended bodies the closest candidates. I'm slightly worried it's based on the idea that they're more intuitive candidates. I'm not sure that's the right kind of reason to use in this kind of account.

A last very quick comment: you say you wouldn't be surprised if you found in the quantum states these exotica and all sorts of weird things. There are people in this room much more up to speed on decoherence than I am, but I at least would be very surprised. It's supposed to be a robust, substantive discovery about the Everett theory that its states contain an awful lot of this very high-level complex structure. If it turns out that by squinting in different ways we can see structure anywhere, the theory really would be in trouble. So I don't think it could be a minor worry or quibble for the theory if we found these exotica; but I also think it's a very substantive claim to say that we might. I guess that the occasional alternative structures like Simon talked about before might be a different matter.

Janssen to Hawthorne

I wanted to follow up what Jim Hartle was saying, and make a more general comment. So basically, it seems the whole talk was about the distinction between fundamental theory and everyday experience. Metaphysicians worry about it—they say let's try to make sure that those two things are not going to drift from one another because we get in all sorts of philosophical trouble when we do that. So the heuristics that you give are: beware of gaps, beware of exotica. They seem to be driving in that direction. I think as a scientist you're playing a very, very different game. I think there the situation is—you hit upon this beautiful new theory, there's a lot of empirical evidence that bears on it, and there's the painful realization that something like that is going to have to replace the old comfortable Newtonian mechanics; and so now let's get on with it. And for a scientist—I'm not sure if I'm speaking for Jim Hartle but I'd hope so—our evidence is telling us that quantum mechanics is the right theory, and so unfortunately now we're going to have to build up and recover from there all these everyday things; and it ain't going to be perfect, and that's just the usual situation. Michael Turner among cosmologists introduced what I think is a very beautiful phrase about this: scientists tend to be 'forward engaged', they constantly take advantage of stuff that still needs to be developed in order to work out the scheme.

So just to pull in a few other things: when Simon Saunders was saying in response to Maudlin and to all these problems that you're talking about—that it would be easier if it were just a sort of billiard ball ontology—Simon quite correctly pointed out look, this is a historical artefact. Go back to the 17th-century corpuscularian philosophy, and this seems to be very odd. Everybody understands this, because everybody laughed at the joke yesterday made by Wheeler saying, like, first we had Newton's incomprehensible action at a distance, and then we had Maxwell's incomprehensible denial of action at a distance. This seems to be a typical thing to me. The frustration I have listening to your talk and listening to Tim's talk is: science ain't metaphysics.

And then there's this very strong claim, for instance, well we all agree it would be completely nuts to think that we're all like numbers . . .

Hawthorne

All *like* numbers? No, no; you didn't understand what I said, I didn't say it was nuts to think we're all isomorphic to numbers—I said it's nuts to think that reality *is* just real numbers and sets of real numbers and those are the only things that there are. I don't think you understood what I said. *That* is nuts. I think if you understood that . . .

Janssen

I don't think that is nuts.

Hawthorne

Really? I don't believe you.

Janssen

You get into a situation where your best scientific theory is pointing . . .

Hawthorne

No, no; you'll never ever find a physicist saying that. Can I just make one other point, trying to be less polemical. What you find in the history of these discussions is—you get a scientific theory and philosophers cum metaphysicians cum philosophers of science know these slightly different accounts of the world that are, as it were, in the spirit of that scientific theory. So the configuration space model and the Hilbert space model and the Simon Saunders model are all fundamental theories in the spirit, in the ball park, of the Everettian picture of the world, which thinks of branches as vague. They're all in that ball park—and what you find in the history of philosophy is that philosophers, a bit more than working physicists—fair enough—care a little bit more about which of those particular versions of the theory are most plausible and elegant, natural, simple, and compelling; and they're also much more careful and sensitive and interested in the question of how our ordinary ways of talking are made true by that fundamental description of reality. And then they in fact come to know much better than physicists how metasemantics is fundamental to an understanding of our place in the physical world. And I don't see where in anything that I've just said there is some naivety or some backward thinking—and in fact the only way of getting off the boat is a way that scientifically minded people have tried and that has failed, which is by trying to say: well, the fundamental stories are just notational variants. They go super-instrumentalist. But you know, philosophers have tried that, and they have failed dismally. History seems to have proven that these are legitimate foundational questions, and compelling ones.

Maudlin (to Janssen)

I had a small comment that, since it came up, I need to respond to. It's this idea that I'm somehow deeply conservative and won't consider these things. So let me just make a quick response to this comment. The problem with what you put forward, which sounds really great, that the theory's telling us this and we're just guys who hate the theory; the problem is the story you told was simply *incoherent*. Because what you said is I have a theory that's telling me blah, and now all I have to do is figure out how to get the macroscopic world out of it. Well, if you didn't already know how to get the macroscopic world out of it how could you have any evidence that it was correct? Your evidence was stated in terms of stuff happening in laboratories, flashes on screens and so on; we all agree that there are interference patterns; we all agree, I take it, that there's a wavefunction; I have no problems with wavefunctions, but it just can't be the case—it's simply logically impossible for it to be the case—that you're in a position to say that 'we have this really strong empirical evidence that this is the right theory, the only problem with it is that with the theory so far I don't know how to make sense of the macroscopic world'. It's logically impossible, so there's just got to be a confusion somewhere.

Loewer

One thing about the functionalist programme—an obvious point but one that's troubling here—is that functionalism in philosophy of mind makes very heavy use of notions of causation. We don't find causation in the fundamental physics. At some point, it's got to be introduced and it's very, very puzzling to me how it's going to be introduced into the fundamental physics. And without it we won't be able to give an account of tigers drooling and so forth.

Hawthorne

I think things get up and running pretty well if you don't think of the functional thing initially in causal terms, just think there are, for example, tiger-shaped things, and when they're close to mouse-shaped things often the mouse-shaped things turn into mouse-archipelagos. If you could get all that going then you're at least up and running, so I think the spirit in which David [Wallace]'s talking about functionalism is not primarily insisting on, as it were, the functionalist bit in the philosophy of mind where there's a really big causal emphasis. Importantly, in the Dennett story, the 'real pattern' story, it doesn't quite have a causal emphasis either, as an account of beliefs and desires.

Loewer

You said 'close to', so you've got a guy spatially . . .

Hawthorne

Right, you do need space, I do think you need tiger-shaped things in a low-dimensional space and mouse-shaped things in a low-dimensional space, but I think Tim's worries, for example, would be appeased a great deal if you could show how there's a really salient and natural and non-cooked-up way to build a low-dimensional space with mouse-shaped and tiger-shaped things out of the ultimate configuration-theoretic story. I do agree there'd be these additional concerns, especially if you're a conservative, as to where counterfactuals and so on are coming from.

Papineau

I wanted to talk about the explanatory gaps. So, I'm personally happy when it comes to consciousness and brute identities, but I don't see why you thought that there was such a gap between the configuration space representation and something familiar.

Hawthorne

Did you see with the real number case—you did see why there's an explanatory gap there?

Papineau

I think that's a bad model. Suppose you've got two point-particles moving around in three-space and somebody says, 'oh, that's a point moving around in six-space'. Well, it's a kind of clever-tricky thing to say but I don't have an explanatory gap between that claim, that fact; between someone who says two points moving around in three-space is one point moving around in six-space. What's going on? This guy's found an economical way of representing the same fact, and once I understand how this way of talking works I can see a priori that it's the same fact.

Hawthorne

I think you've got to be very careful even there. If your fundamental description, the six-space, doesn't in any privileged way group the dimensions into threes, then you ought to start to really worry. So, I'm not saying that David's programme is going to fail; I'm saying—and I think David's very sensitive to this—you know it is very tricky in that case and you can naively think, hey, this is just a six-space with no privileging and . . .

Papineau

But in that case there is privileging and we understand it. Now there may be special reasons when we come to a multidimensional configuration space as in the Everettian theory as to why this isn't a good analogy, but you didn't give us any special reasons.

Hawthorne

I really want to bear down on this six-dimensional example because I think it is a good test case. So you start to realize there is something very artificial and gerrymandered about describing the two-particle system in this way if in the laws there's no privileging, there's no bundling of dimensions into threes. And then you realize, hey, for this to work now at the fundamental layer of description, if at that level we're not going to have two particles, and if it is not to seem cooked-up that we've got two particles, then we're going to have to have fundamental laws, or non-Lewisian laws or fundamental properties that group together the dimensions. And then you want to ask what those properties look like and you ask yourself, 'Are these the sorts of things that we'd have liked all along? Or are these awkward, artificial labels for what we wanted at the end of the day, rather than plausible posits about the fundamental structure of the world?' And I think it gets more tricky, but exactly analogous issues come up in the configuration space story. David's bet is that there'll be features that don't look artificial that make one particular, or at least one vaguely related class of interpretations, very much less cooked-up than other candidate interpretations. That's the bet of the Oxford Everettians.

PART III

PROBABILITY IN THE EVERETT INTERPRETATION

6

Chance in the Everett Interpretation

Simon Saunders

It is unanimously agreed that statistics depends somehow on probability. But, as to what probability is and how it is connected with statistics, there has seldom been such complete disagreement and breakdown of communication since the Tower of Babel.   (Savage [1954 p.2])

For the purposes of this chapter I take Everettian quantum mechanics (EQM) to be the unitary formalism of quantum mechanics divested of any probability interpretation, but including decoherence theory and the analysis of the universal state in terms of emergent, quasiclassical histories, or *branches*, along with their branch amplitudes, all entering into a vast superposition. I shall not be concerned with arguments over whether the universal state does have this structure; those arguments are explored in the pages above. (But I shall, later on, consider the mathematical expression of this idea of branches in EQM in more detail.)

My argument is that the branching structures in EQM, as quantified by branch amplitudes (and specifically as ratios of squared moduli of branch amplitudes), play the same roles that chances are supposed to play in one-world theories of physical probability. That is, in familiar theories, we know that

(i) Chance is measured by statistics, and perhaps, among observable quantities, only statistics, but only with high chance.
(ii) Chance is quantitatively linked to subjective degrees of belief, or credences: all else being equal, one who believes that the chance of E is p will set his credence in E equal to p (the so-called 'principal principle').
(iii) Chance involves uncertainty; chance events, prior to their occurrence, are uncertain.

Those seem the most important of the chance-roles.

My claim is that exactly similar statements can be shown to be true of branching in EQM. In the spirit of Saunders [2005] and Wallace [2006], we should conclude that these branching structures *just are* chances, or physical probabilities. This is the programme of 'cautious functionalism', to use Wallace's term.

The argument continues: this identification is an instance of a general procedure in the physical sciences. Probabilities turn out to be functions of branch amplitudes in much the same way that colours turn out to be functions of electromagnetic frequencies and spectral reflectances, and heat and temperature turn out to be functions of particle motions in classical statistical mechanics—and in much the same way that sensible perceptions (of the sort relevant to measurement and observation) turn out to be functions of neurobiological processes.

Just like these other examples of reduction, whether probability is thus explained, or explained away, it can no longer be viewed as fundamental. It can only have the status of the branching structure itself; it is 'emergent' (see Wallace in Chapter 1). Chance, like quasiclassicality, is then an 'effective' concept, its meaning at the microscopic level entirely derivative on the establishment of correlations, natural or man-made, with macroscopic branching. That doesn't mean that amplitudes in general (and other relations in the Hilbert space norm) have no place in the foundations of EQM—on the contrary, they are part of the fundamental ontology—but their link to probability is indirect. It is simply a mistake, if this reduction is successful, to see quantum theory as at bottom a theory of probability.[1]

1 EXPLAINING PROBABILITY

Functional reduction is not new to the sciences; functional reduction, specifically, of probability, is not new to philosophy. In any moderately serious form of physical realism a world is a system of actual events, arranged in a spatiotemporal array, defined in terms of objective, physical properties and relations alone. Where in all this are the physical probabilities? For physicalists, they can only be grounded on the actual structure of events—what Lewis has called a 'Humean tapestry' of events (Lewis [1986], Introduction, Sec.1), whereupon the same identificatory project ensues as in EQM.[2] The links (i), (ii), (iii) stand in need of explanation whatever the physics, classical or quantum, one world or many. For

---

[1] Contrary to information-theory approaches to foundations (Bub and Pitowsky in Chapter 14, Schack in Chapter 15). It is also contrary to some aspects of orthodoxy, particularly in quantum field theory (for example, the picture of the microstructure of the vacuum as a stochastic medium, even in the absence of experimental probes).

[2] Lewis explicitly contemplated extending his tapestry of events to quantum mechanics, acknowledging that novel quantum properties and relations (like amplitudes and relations among amplitudes) may have to be included; but only come the day that quantum mechanics is 'purified' (Lewis [1986a p.xi])—'of instrumental frivolity, of doublethinking deviant logic, and—most of all—of supernatural tales about the power of the observant mind to make things jump'. Quantum mechanics has certainly been purified in EQM.

the most part they have remained surprisingly hard to explain, even, and perhaps especially, in one-world theories.[3]

But *one* of the three links, in the case of one-world theories, *is* easily explained—namely (iii), the link with uncertainty. Only allow that the dynamics is indeterministic (whether in a fundamental or effective sense), so that some kinds of events in the future are (at least effectively) unpredictable, then, for the denizens of that world, the future will in some respects at least be uncertain. This is how all the conventional theories of chance explain (iii) (with varying strengths of 'unpredictability'). But according to EQM all outcomes exist, and, we may as well suppose, are known to exist (we may suppose we know the relevant branching structure of the wavefunction)—so the future is completely known. There is no uncertainty on this score.

This marks out EQM as facing a particular difficulty with (iii), the link with uncertainty, a difficulty encountered by none of the familiar theories of probability in physics. It has seemed so severe, in fact, that not only do the majority of critics of EQM believe that, as a theory of probability, it must rest on the links (i) with statistics and (ii) rationality alone (if it is to mean anything), but so do many of its friends.[4] The attempt to ground EQM on (i) and (ii) alone, disavowing all talk of probability or uncertainty, has been dubbed the *fission programme*. As a counterbalance, this chapter is skewed towards a defence of (iii), the link with uncertainty. As I shall argue in Sections 3 and 4, the status of this link is not in fact so different in EQM than in conventional theories of chance, using a possible-worlds analysis, of a sort familiar to philosophers.

I have less to say on (i) and (ii), not because they are less important to the overall argument, but because arguments for them are either well known (in the case of (i), the link with statistics) or given elsewhere in this volume (by Wallace, for (ii), the principal principle). Both links now appear to be in significantly better shape in EQM than in the other main theories of physical probability (classical statistical mechanics, deterministic hidden variable theories, and stochastic physical theories), none of which offer comparably detailed explanations of (i) and (ii). Instead they presuppose them.

First (i), the link with statistics. A first step was the demonstration that a largely formal condition of adequacy could be met: a quantum version of the Bernouilli ('law of large numbers') theorem could be derived.[5] This says that branches recording relative frequencies of outcomes of repeated measurements at variance

---

[3] According to even so committed an empiricist as B. van Fraassen, a model of a probabilistic physical theory must include elements representing alternative *possible* sequences of outcomes, and the theory can be *true* 'only if alternative possible courses of events are real.' (van Fraassen [1980 p.197]).

[4] E.g. Papineau in Chapter 7, Greaves [2004], [2007], Deutsch in Chapter 18.

[5] As sketched by Everett in the 'long' thesis (Everett [1973])—see the Introduction, Section 1, and Byrne in Chapter 17.

with the Born rule have, collectively, much smaller amplitude than those that do not (vanishingly small, given sufficiently many trials). But the Bernouilli theorem is common ground to every theory of probability. Distinctive, in EQM, is that given an account of what probabilities *actually are* (branching structures), and given that in EQM we can model any measurement process as comprehensively as we please (including 'the observer' if need be) it becomes a purely *dynamical* question as to whether and how branch amplitudes can be measured. I will give some illustrative examples in Section 2—but I take these arguments to be relatively uncontroversial.

Not so the arguments for (ii), the decision theory link. But even here, few would deny that there has been progress—and on two quite separate fronts, both involving the principal principle: the first (the sense already suggested) in showing why credence should track chance, as identified in our Everettian tapestry; the second in showing how we could have been led, by rational, empirical methods, to anything like that tapestry in the first place.

For the former argument, see Wallace in Chapter 8, deriving the principal principle. He shows that branching structures and the squared moduli of the amplitudes, insofar as they are known, *ought* to play the same decision theory role (ii) that chances play, insofar as they are known, in one-world theories. Whatever one might think of Wallace's axioms, as long as nothing underhand or sneaky is going on the result is already a milestone: nothing comparable has been achieved for any other physical theory of chance.

It is, however, limited to the context of agents who believe that the world has the branching structure EQM says it has—it is a solution to the *practical problem*, the normative problem of how agents who believe they live in a branching universe ought to achieve their ends (without any prior assumption about probability). It has no direct role in confirming or disconfirming EQM as one of a number of rival physical theories by agents uncommitted as to its truth. It leaves the *evidential problem* unsolved.

But on this front too there has been progress. Greaves in her [2007] and Greaves and Myrvold in Chapter 9 show how confirmation theory (specifically Bayesian confirmation theory) can be generalized so as to apply to branching theories and non-branching theories evenhandedly, without empirical prejudice. The latter proves a representation theorem on the basis of axioms that are entirely independent of quantum mechanics. Nor, in keeping with the fission programme, do they make explicit or even tacit appeal to the notion of probability or uncertainty.

Wallace's axioms are likewise intended to apply even in the context of the fission programme. That shows that the argument for (ii) is independent of (iii). But if the argument for (iii) (with (i) and (ii)) goes through, then the approach of this chapter *also* promises a solution to the evidential problem. As an objection to the Everett interpretation, the problem only arises if it is granted that branching and branch amplitudes *cannot* be identified with chance and probabilities (which we take to mean are *not* in fact quantities that satisfy (i), (ii), (iii)), precisely the point here in contention. Show that they can, and the evidential problem simply dissolves. Or rather, since it can be read as a problem for every chance theory, it will have the same status in the Everett interpretation as it has in any other physical theory of chance, to all of which the Greaves and Myrvold analysis applies.

I have little more to say about (ii), the decision-theory link. What follows is directed to (i), the link with statistics (Section 2), but mainly to (iii), the link with uncertainty. In Section 3 I show how branching in EQM is consistent with talk of uncertainty (following Saunders and Wallace [2008a]). That is arguably enough, with the results of Section 2 and of Wallace in Chapter 8, to draw our main conclusion: branching in EQM should be identified with chance.

But there remains a contrary reading, according to which there is no place for uncertainty after all—or not of the right kind. Given sufficient knowledge of the wavefunction, on this contrary reading uncertainty can at best concern *present* states of affairs, not future ones. In the final and more philosophical section I argue that the difference boils down to a choice of metaphysics, a choice that is under-determined by the mathematical structure of the theory. Since choices like this should be made to help in the understanding of a physical theory, rather than to frustrate it, this contrary reading should be rejected.

2 WHY CHANCE IS MEASURED BY STATISTICS

Here are a number of no-go facts about how physical probabilities are observed that we believe to be true, but that seem very difficult to explain on any conventional theory of physical probability:

(a) There is no probability meter that can measure single-case chance with chance equal to one.
(b) There is no probability meter that can measure chance on repeated trials with chance equal to one.
(c) There is no probability meter that can measure single-case chance with chance *close* to one.
(d) Absolute probability can never be measured, only relative probability.

(The list could be continued.)

On conventional thinking, someone who needs convincing of these facts about chance has not so much failed to understand a physical theory as to understand the *concept* of probability. But if chance is a definite physical magnitude, it should (at least indirectly) be measurable, like anything else. Given the dynamics, and a theory complete enough to model the experimental process itself, these facts should be *explained*. And indeed no non-trivial function of the branch amplitudes can be measured in the ways mentioned in (a)–(d), according to EQM. This is a *dynamical* claim.

On the positive side, as to how (we know) chances are in practice measured:

(e) There are probability meters that measure chances on repeated trials (perhaps simultaneous trials) with chance close to one.

This is conventionally thought of as a consequence of the axioms of probability (the law of large numbers, or Bernouilli theorem) rather than of any physical theory; in turning it around, and deriving it rather from the dynamical equations of EQM, it identifies the appropriate physical quantities that are to count as probabilities instead.

There remains one other obvious fact without which one might well think no account of chance can so much as get off the ground: chance outcomes are typically *mutually exclusive* or *incompatible*—in which case only *one* outcome obtains (which one happens by chance). That is (where I explicitly add the qualification 'observable', as we are concerned with how chances can be measured):

Presupposition: Of two incompatible, observable, chance events, only *one* event happens.

How can the presupposition be explained in EQM? In comparison with (a)–(e) it seems to have the least to do with chance—the word 'chance' could be deleted from it entirely—but we surely do talk about chance in this way. It is presupposed by any application of the concept of chance that (at least sometimes) chances apply to incompatible outcomes. The concept of incompatibility enters at the very beginning of any mathematical definition of a probability measure on a space of events (events are represented by sets, and inherit from set theory the structure of a Boolean algebra). And yet it seems to be violated straightforwardly in EQM, as Everett's crazy idea was that in a quantum measurement *all* outcomes happen.

If that was all there was to it, it would be hard to understand why EQM was ever taken seriously by anyone. The answer is that the presupposition has two different readings, the one a physical or metaphysical claim—about what chances fundamentally are—and the other a claim about what is observable (by any observer), the phenomenology of chance.

On the first reading, the presupposition *is* straightforwardly denied; it is denied if the word 'chance' is deleted too. But we already knew this: this is simply a conflict between a many-worlds and a one-world theory. Likewise for the presupposition as a metaphysical claim: we already knew EQM challenges a number of a priori claims. It is only as an epistemological claim—as to the *observed* phenomenology—that the presupposition had better still make sense. But so it does: EQM explains it very simply. No two incompatible chance outcomes can happen *in the same branch*. And since the apparatus, and the observer, and the entire observable universe, are branch-relative—they are 'in-branch' structures—no observer can ever witness incompatible outcomes simultaneously.[6] In this second sense, then, the presupposition is rather elegantly dealt with in EQM—'elegantly', because it follows directly from its account of what a chance set-up is, and from what chance is (one should *not* delete the word 'chance'!).

The other everyday ('no-go') facts about probabilities follow, not from branching alone, but from the unitarity of the equations of motion. Because I do not think this claim is particularly controversial I will simply illustrate it with a proof of (a) for relative probability (meaning a quantity that concerns the relation between chancy outcomes). Consider then a microscopic system in the state

$$|\varphi_c\rangle = c|\varphi_+\rangle + \sqrt{1 - |c|^2}|\varphi_-\rangle.$$

For simplicity, model the measurement process as a one-step history, using the Schrödinger picture, with initial state $|\omega\rangle \otimes |\varphi_c\rangle$ at $t = 0$, where $|\omega\rangle$ is the state of the rest of the universe. Let the configuration describing (among other things) the apparatus registering (a function of) the amplitude c at time t be $a(c)$. Let the projection onto this configuration be $P_{a(c)}$ and let the associated Heisenberg-picture projection be (see Section 3.3 of the Introduction)

$$P_{a(c)}(t) = e^{iHt/\hbar} P_{a(c)} e^{-iHt/\hbar}.$$

The no-go fact we need to establish is that there is no unitary dynamics $U_t = e^{-iHt/\hbar}$ by means of which:

$$|\omega\rangle \otimes |\varphi_c\rangle \rightarrow U_t|\omega\rangle \otimes |\varphi_c\rangle = U_t P_{a(c)}(t)|\omega\rangle \otimes |\varphi_c\rangle$$

for variable c. If there were it would also follow that for $c' \neq c$:

$$|\omega\rangle \otimes |\varphi_{c'}\rangle \rightarrow U_t P_{a(c')}(t)|\omega\rangle \otimes |\varphi_{c'}\rangle.$$

But the inner product of the LHS of the two initial states can be as close to one as desired, whilst that of the vectors on the RHS must be zero, as $a(c)$ and $a(c')$ must differ macroscopically, a contradiction. Thus not even relative probabilities can be measured deterministically. (The argument for absolute probabilities is even simpler, and depends only on the linearity of the Schrödinger equation.)

---

[6] If this is thought to be question-begging, there does not exist a *perception* of two incompatible outcomes, according to EQM, treating perception biochemically, or indeed as any kind of record-making process. (This point goes back to Everett [1957] and, ultimately, the von Neumann model of measurement; see the Addendum to the Introduction. See also Gell-Mann and Hartle [1990] and Saunders [1994].)

Now for (e), how chances can be measured according to EQM, with high chance. Take a number N of subsystems, each (to a good approximation) in state $\varphi_c = c\varphi_+ + \sqrt{1 - |c|^2}\varphi_-$, and arrange the linear dynamics so that

$$|\omega \otimes \varphi_\pm\rangle \rightarrow U_t P_{a(\pm)}(t)|\omega \otimes \varphi_\pm\rangle \tag{1}$$

i.e. the von Neumann model of measurement, where $a(+)$ is a configuration in which the apparatus reads 'spin-up' and $a(-)$ 'spin-down'. Applied to an initial state of the form $|\omega\rangle \otimes |\varphi_c\rangle$, it yields the superposition

$$|\omega\rangle \otimes |\varphi_c\rangle \rightarrow U_t P_{a(+)}(t)|\omega \otimes \varphi_c\rangle + U_t P_{a(-)}(t)|\omega \otimes \varphi_c\rangle$$

in which the first vector has norm $|c|$ and the second has norm $\sqrt{1 - |c|^2}$. That is, the amplitudes c, $\sqrt{1 - |c|^2}$ of components in a microscopic superposition have been promoted to macroscopic *branch* amplitudes. Consider now repeated measurements of N microscopic systems all in the same state $|\varphi_c\rangle$, whether sequentially repeated in time, or measured all at once. The latter is the simplest to model, assuming the N apparatuses are non-interacting: the result at time t will be a superposition of vectors at t of the form

$$|a_f, t\rangle = U_t P_{f(1)}(t) \otimes \ldots \otimes P_{f(N)}(t)|\omega\rangle \otimes |\varphi_c\rangle \otimes \ldots \otimes |\varphi_c\rangle$$

where $f(k)$, $k = 1, \ldots, N$ is either $+1$ or -1. Those with the same relative frequencies M/N will all have the same norm, i.e.:

$$||a_f, t\rangle| = |c|^M \sqrt{(1 - |c|^2)}^{N-M}; \quad \sum_{k=1}^{N} f(k) = 2M - N. \tag{2}$$

The unitary evolution to time t is:

$$|\omega\rangle \otimes |\varphi_c\rangle \otimes \ldots \otimes |\varphi_c\rangle \rightarrow \sum_{M=0}^{N} \left[\sum_{f: \Sigma_{k=1}^N f(k) = 2M-N} |a_f, t\rangle \right]. \tag{3}$$

The right-most summation, for fixed M, N, is over all $N!/M!(N - M)!$ distinct f's, all with the same norm Eq. (2). Since $\langle a_f, t|a_g, t\rangle = 0$ for $f \neq g$,[7] the squared norm of the RHS of Eq. (3) is

$$\sum_{M=0}^{N} \frac{N!}{M!(N - M)!}||a_f, t\rangle|^2 = \sum_{M=0}^{N} x(M)$$

---

[7] For sequences of measurements in time, the consistency condition is needed at this point. See Section 3 of the Introduction p. 45.

where $x(M)$ is:

$$x(M) = |c|^{2M}(1 - |c|^2)^{N-M}\frac{N!}{M!(N-M)!}.$$

For N large, this function is strongly peaked about $M/N = |c|^2$. The sum of norms of vectors with the 'right' relative frequency is much larger than the sum of norms of vectors with the 'wrong' relative frequency. In this sense, measured relative frequencies close to $|c|^2$ are found on relatively high amplitude branches.

It was rather crucial to this argument that we consider only relative frequency—we don't care about precisely which particles were recorded as 'spin-up' and 'spin-down' (or which of our N apparatuses recorded which result). That is, our measurement protocol requires that the measurements be treated as 'exchangeable', in de Finetti's sense (see Chapter 9). This is an assumption in de Finetti's theory of probability, that EQM as we see explains.

Might some other method be found by which functions of amplitudes may be measured, flouting (a)–(d)? Perhaps,[8] but that is unlikely to make for an objection to EQM if it is based on the unitary formalism of quantum mechanics, for that is what EQM is. If based on a rival theory, which is empirically successful, that will anyway spell the end of quantum mechanics (and with it EQM).

Now suppose, fancifully if you will, that we are entitled to identify chances with functions of branch amplitudes. Then in summary we have just shown: chances will generally be associated with incompatible observable outcomes, only one of which can happen at each time; they may not be measured in a way that is not itself chancey, and single-case chance cannot be measured at all. They can only be measured by running a chance (branching) process repeatedly, or by a single trial involving a large number of similarly prepared systems, and listing the relative frequencies (taking care to neglect the order of outcomes, or which system had which outcome). The measurement will be veridical, however, only with high chance, given that the number of systems involved is sufficiently large. None of these facts can be explained by any conventional physical theory of probability (rather, they are presupposed).

3 WHY CHANCE INVOLVES UNCERTAINTY

We cannot, however, identify chance set-ups with branch set-ups, and chances with functions of amplitudes, failing an account of (iii), the link with uncertainty. In the absence of this, to arrive, by the Deutsch–Wallace representation theorem,

---

[8] For example, 'protective measurements' (Aharonov and Vaidman [1993]), although see the criticism of Uffink [2000]; or 'weak measurements' (Aharonov et al. [1988]); see Vaidman in Chapter 20.

at an agent's credence function (in conformity with the Born rule), raises a puzzle on its own. For what does a rational agent believe to some degree when she uses that credence function? Given the theorem, certainly we can explain credence by reference to behaviour; but in this case, better, perhaps, to simply explain it away. A rational agent must order her priorities somehow, whether or not there is anything of which she is uncertain; but it is hard to see how is it possible to have degrees of belief about physical reality different from one and zero if you know everything about physical facts there is to know.

The functionalists' response is to more or less recapitulate the representation theorem: credences derive their meaning from their function in rational action, in determining expected utilities, and hence agents' preferences. They will also note that meaning is determined by use—that if anything has been agreed by philosophers of language in recent decades it is that:

> Words have no function save as they play a role in sentences: their semantic features are abstracted from the semantic features of sentences, just as the semantic features of sentences are abstracted from their part in helping people achieve goals or realize intentions. (Davidson [2001 p.220])

So there is every reason to talk of expected utilities and credences. By the same broadly functionalist philosophy, if these credences play the same role that credences about chance events play, they *are* credences about chance events. Talk of uncertainty and unpredictability then falls into place along with the rest of our ordinary use of words.

I think this argument is essentially correct, but it leaves unanswered—brief explanations please!—just *how* talk of uncertainty is to fall into place, and just *what* to say in answer to the question of what these degrees of belief are about.

First a warm-up. In the previous section I argued that functions of branch amplitudes are measured—they are manifested in-branch—in just the way that chances are thought to be measured in one-world theories. A part of that argument was that the presupposition about incompatible outcomes follows (rather straightforwardly) from EQM: on each trial, since macroscopically distinct, only one outcome can be obtained in each branch. So from the point of view of accumulating information that can actually be used—'in-branch' information—events are registered sequentially, just as they are in a non-branching theory. This was what Everett took pains to show (see the Introduction, Section 3.2). But isn't sequential increase in information increase in knowledge? So isn't there something that is being learned in each branch?

Consider a concrete example. Alice, we suppose, is about to perform a Stern–Gerlach experiment. She understands the structure of the apparatus and the state preparation device, and she is convinced EQM is true. In what sense does she learn, post-branching, something new? The answer is that *each* Alice, post-branching, learns something new (or is in a position to learn something new)—each will say something (namely, 'I see the outcome is spin-up (respectively, spin-down), and not spin-down (respectively, spin-up)') that Alice prior to branching cannot say. It is true that Alice, prior to branching knows *that* this is what each successor will say—but still she herself cannot speak in this way.

To make this vivid, imagine that each of her successors simply *closes her eyes*, when the result is obtained; in that state, each is genuinely ignorant of the result (so-called 'Vaidman ignorance'; Vaidman [1998]). This has usually been considered (as by Vaidman himself) as ignorance that Alice, prior to branching, does not have, but it can be turned around the other way: each Alice, closing her eyes, perpetuates a state of ignorance that she already had.

The implication of this line of thought[9] is that, appearances notwithstanding, prior to branching Alice does *not* know everything there is to know. What is it she does not know? I say 'appearances notwithstanding', for of course in one sense (we may suppose) Alice does know everything there is to know; she knows (we might as well assume) the entire corpus of impersonal, scientific knowledge. But what that does not tell her is *just which person she is*—or *where she is located*—in the wavefunction of the universe.

The point is a familiar one to philosophers. One can know that there are such-and-such people, but not which one of them is me. Or that such-and-such events occur at various places, but not which of those places is here; or times, but not which of those times is now.

Such knowledge that is omitted is sometimes called 'indexical' knowledge, by philosophers, also 'knowledge de se', and 'self-locating knowledge' (we shall use the last). But why should knowledge like this be lacking? Vaidman has Alice hide her eyes. Another example, due to J.R. Perry, likewise suggests some kind of impairment is needed. Perry asks us to consider an amnesiac who has lost his way in the library at Stanford. He does not know *who* he is; he does not know *where* he is—not even were he to read every book in the library, not even were he to read his own *biography*, would he be any the wiser. But Alice is not so impaired. She does not hide her eyes, and no more is she an amnesiac. She sees where she is in the wavefunction of the universe and self-locates accordingly—and differently—from where her successors locate themselves, for, obviously, she is at a different place from them. She is not ignorant of anything her successors know; she simply reports different self-locating facts from them.

And so she would, *but only given certain assumptions as to what, exactly, she is—of how she is represented in the physics*. For example, if she at t_j is represented by the configuration β_j, or by the Heisenberg-picture vector $|\beta_j\rangle = P_{\beta_j}(t_j)|\Omega\rangle$, where $|\Omega\rangle$ is the universal state, with her successors at $t_k > t_j$ similarly represented by configurations β_k, β'_k, or vectors $|\beta_k\rangle, |\beta'_k\rangle$, the result surely follows. But this treats her as strictly a momentary thing, independent of the history in which she

---

[9] Picked up also by Ismael [2003], although she develops it in a rather different way.

is located. There is an alternative: if she at t_j is represented instead by an entire history α, (or vector $|\alpha\rangle$) containing t_j, then of course, at t_j, she does not know which history or branch vector is hers, and quite the opposite result follows.

Which of the two is correct? But questions like these (or rather their classical counterparts) are well known in metaphysics (and specifically personal identity). Most philosophers are agreed they cannot be settled on the basis of the physics alone. Here then is a metaphysics friendly to the Everett interpretation: let persons be spacetime worldtubes from cradle to grave (in the jargon, 'maximal continuants'). If Alice is a person (of course she is) then we must say, even prior to branching at t_j, that there are *many* Alices present, atom-for-atom duplicates up to t_j, each behaving in exactly the same way and saying just the same words. If that is the right metaphysical picture for EQM, then Alice *should* be uncertain after all—*each* Alice should be uncertain—for each as of t_j does not (and as a matter of principle, cannot) know which of these branching persons she is.

That, I take it, establishes that the question of whether or not there is uncertainty in the Everett interpretation can be settled either way, depending on a metaphysics of personal identity. But rather than invoke a metaphysical assumption, we can make do with a different kind of claim—a proposal, not about the ultimate natures of persons, but about the reference of the word 'person' in EQM terms. The proposal is that persons (and things) be relativized as to branch:[10]

S1 By a 'person' or 'thing' is meant a branch-part or ordered pair $(\beta,|\alpha\rangle)$, where $\alpha \in \beta$.

Here $\alpha \in \beta$ means that the sequence of configurations β is obtained by a coarse-graining of α, which we suppose is temporally and spatially much more finely grained. Likewise, a person at time t_j' is represented by an ordered triple $(\beta_j, \beta, |\alpha\rangle)$, where $\alpha \in \beta \in \beta_j$, or by an ordered pair $(\beta_j,|\alpha\rangle)$. More is needed for tensed sentences (involving 'was', 'is', 'will be' etc.), but for this, and on the question of how *S1* can be justified, see Section 4.

For now note *S1*'s virtues. First, it is clearly permissible; it makes use of nothing but the available mathematics of EQM in terms of branch vectors $|\alpha\rangle$ and sequences of configurations α, β, etc., as ordered pairs $\{(\beta,|\alpha\rangle); \alpha \in \beta\}$.[11] True, this requires the consistent histories formalism, in which branch vectors are Heisenberg-picture vectors (so that a branch vector $|\alpha\rangle$ describes the entire history α, not any particular instant of it—see the Introduction, Section 3.3). The perspective is atemporal. But branching itself is a process defined by the dynamics over time. The idea of a decoherence basis, defined at an instant of time independent of what comes before and what comes after, is a fiction.

---

[10] This semantics was discussed briefly in Wallace [2005], [2006], and in more detail in Saunders and Wallace [2008a], on which the argument that follows is based.

[11] Contrary to the criticism of Kent in Chapter 10 (pp. 346–7).

Second, using *S1*, not only is Alice entitled—bound—to be unsure of the outcome of the experiment, but she has a genuine gain in knowledge when it is learned. For after branching, using *S1*, on observing the outcome, each Alice self-locates *better* than she did before—*knows more* than she did before—and so has learned something that, prior to branching, she could not have known. Vaidman's ignorance is ignorance that each Alice already had. Likewise, there can be no algorithm, whose input is data about a branch up to one time, and whose output uniquely specifies that branch at a later time; for the same algorithm must operate in every other branch which is exactly the same up to that time but which differs thereafter. Branching events are algorithmically uncertain too—they are *indeterministic*.

Third, *S1* is neutral on some (but not all) metaphysical questions about personal identity. Specifically, those that arise given four-dimensionalism, assuming a single world, are likely to play out the same under *S1*, in the context of branching worlds, if only it is permitted to relativize one's favoured candidate for β to worlds (represented by Heisenberg-picture branch vectors $|a\rangle$), where $a \in \beta$).

And the bottom line: under *S1*, prior to branching, uncertainty is assured: Alice doesn't know if she will see spin-down or spin-up, as she doesn't know which branch she is in. Unless some hidden contradiction is involved, *S1* is the right rule for making sense of quantum mechanics in realist terms.

Conclusion. Branching, the development of superpositions of the universal state with respect to the decoherence basis, plays all the chance roles (i), (ii), (iii): it produces the same phenomenology as chance (Section 2), the branch amplitudes are of practical relevance to decision theory in the way that chances normally are (Wallace in Chapter 8), and branching involves uncertainty (using *S1* above), just as chances do (Section 3). In EQM, branching and squared norms of branch amplitudes are demonstrably functionally equivalent to chance, in these three central respects; therefore, they *are* chance processes, and chances *are* these physical magnitudes.

4 OVERLAP AND DIVERGENCE

There are alternatives to the rule *S1*, however. Invoking it seems to compromise a chief selling point of the Everett interpretation, which is that many-worlds follows from the unitary dynamics, with no added principles or special assumptions. That is what puts the Everett interpretation in a class of its own when it comes to the quantum realism problem: there are plenty of avenues for obtaining (at least non-relativistic) one-world theories if we are prepared to violate this precept.

On the other hand *S1* is on the face of it *just* a semantic rule—it is *merely* a linguistic matter. The referents of terms are constrained by their contexts of use, granted; but over and above those constraints, not even in the God's-eye view is their meaning determined. It is up to us to say what, precisely, among the structures in the universal state, our words really mean. And if that is all there is to uncertainty, nothing much should hang on the matter either way. The challenge remains: to justify the probability interpretation of EQM on the grounds of (i) and (ii) alone.

I think there is something right about this argument. Whether or not there is genuine uncertainty in the Everett interpretation appears rather less substantive than might have been thought. But in that case, let satisfaction of the chance roles (i) and (ii) be enough for branching structures to count as chance; our conclusion stands.[12]

But on three counts this would be too quick. The first is that, for many, *S1 really is* a metaphysical claim, for all my talk of semantics—and that there *is* a substantive question of whether it is true. Certainly there are alternative metaphysical claims that could be made that cause problems to *S1*. I shall have something to say about this sort of argument, but not much, for I do not believe there are metaphysical truths of this sort, independent of ordinary language and natural science. Or let me put the point more constructively: metaphysics should primarily be answerable to language and to science; it should be 'naturalized' (Saunders [1997], Ladyman and Ross [2007]). If this is right, *S1* may still be a metaphysical claim, a claim of substance, but in the naturalized sense.

The second count is that we should take rather more seriously the task of accounting for language use, for that too is emergent structure, a part of the physical world to be studied as such by scientific means. As with any form of functional reduction, we know what we are looking for in advance: the problem is to identify the right sort of structure at the more fundamental physical level to account for phenomena that we are already familiar with at the less. Devising semantic rules for ordinary sayings, whose truth conditions are fixed by reference to the underlying theory, is only a variation on the same procedure—it is reductionism as it is appropriate to linguistics. It is no different in kind from fixing on certain variables, for example hydrodynamic variables, in decoherence theory, to derive quasiclassicality. Sure, there remain the *wrong* variables, ones that give no hint of classicality; and wrong semantic rules too, which are inadequate for explaining our linguistic behaviour—and which give no hint of uncertainty. They should be eschewed.

This argument is a variant of Wallace's [2005], which puts the matter in terms of 'the principle of charity'. This principle says that the most important criterion of 'good' translation, in the radical case, where no prior standard of translation has been established between two languages, is that it maximizes truth. Working out what to say about branching, if EQM is true, is like radical translation—of how to translate into our own tongue the expressions of some alien language on the basis

---

[12] This is not the fission programme, which would have us renounce talk of objective chance and genuine probability (along with uncertainty).

of observable linguistic behaviour alone. If EQM is true, it has always been true, and we have always used words like 'uncertainty' and 'chance' in the context of branching so—like it or not!—that is what those words have always been about. Whatever we might say about novel scenarios like teleportation machines or brain transplants (the ways in which philosophers have tended to imagine persons being divided), our use of words in ordinary contexts is not in doubt. The referents of ordinary words and truth conditions of everyday sentences (as specified in a physical theory) may wait to be correctly identified, but the criteria for 'correctness' is decided, not by a priori metaphysics, but by their adequacy, as follows from the theory, to make sense of what we ordinarily say.

Greaves herself accepts this argument (Greaves [2007 p.124]), but not all Everettians do, and surely not all sceptics of EQM do. But this dispute is independent of EQM per se. It should be settled if possible by some other example of the general method—say, in the arena of a semantics for temporal affairs, involving tense and becoming, in classical spacetime theories (as argued by Saunders [1996], Wallace [2005]), or in the arena of a semantics for agency, moral responsibility, and free-will, in deterministic theories.

The third count on which the case for uncertainty can be too quickly deflated is a strengthened version of the first. I have already hinted at it: maybe *S1 does* harbour some hidden contradiction, or some failure of integrity more broadly construed. The semantics should not *misrepresent* our situation, were EQM true.

This, it seems to me, is the only serious concern, and the only one I shall consider in the rest of this chapter. But it needs explaining. The very worry seems strange. How can a sentence misrepresent a theory, if it is true by that very theory?

Some more detail on the semantics will be helpful.[13] Let $\alpha = \langle a_+, \ldots, a_- \rangle$, where t_+ is much later than any time we are interested in, and t_- is much earlier. Call branches like this *maximal*. By 'world', I mean maximal branches, represented by maximal branch vectors. Let F be true of some branches, false of others. Then since, by *S1*, speakers and things are parts of branches, so are utterances:

S2 An utterance of 'F' in branch $|\alpha\rangle$ is true if and only if F is true in α.

Now for tensed statements. Let $\alpha^+(t) \underset{def}{=} \langle a_+, .., a_k \rangle \ni a, t_k > t \geq t_{k-1}$ be *the future of α at t*; let $\alpha^-(t) \underset{def}{=} \langle a_k, .., a_- \rangle \ni a, t_{k+1} > t \geq t_k$ be the *past of α at t*. Suppose that F refers to some temporally local state of affairs (it is an 'occasion sentence' in Quinean terms). The rule for future physical contingencies is:

S3 An utterance of 'F will be the case' in branch $|\alpha\rangle$ at t is true if and only if F is true in the future of α at t.

---

[13] Other rules may be possible as well: see Wallace [2006] for some alternatives. The rules that follow are illustrative.

A first stab at a rule for future possibilities is:

S4 An utterance of '*F* might happen' in branch $|a\rangle$ at *t* is true if and only if for
some branch $|a'\rangle$, *F* is true in the future of a' at *t*, where the past of a' and
a at *t* is the same.

But the latter will obviously need to be restricted to branch vectors $|a'\rangle$ whose
norm, conditional on $a^-(t)$, is non-negligible (or else it will turn out that pretty
well *any*thing might happen). And similarly for counterfactuals.

S2 and *S3*, fairly obviously, satisfy the principle of charity. *S4* promises to, at
least if 'might happen' is taken to mean 'might happen by chance'—meaning,
for present purposes, in accordance with the laws of quantum mechanics. Apart
from this proviso, it fits well enough with standard ideas in modal metaphysics.
One has only to replace 'branch $|a\rangle$' by 'possible world' and the 'sameness
of the past' relation entering in *S4* by a 'nearest counterpart' relation (Lewis
[1973])—although philosophers typically are interested in a notion of possibility
of much broader scope.

Now notice that the 'sameness of the past' relation is neutral on the question
of whether, in the past, (maximal) branches (or rather spatiotemporal parts of
such branches) are *numerically* the same, or only *qualitatively* the same. They are
neutral on the question of whether (to use a technical term in philosophy) they
overlap or merely *diverge*. But this difference is crucial, say metaphysicians (see
the Introduction, Section 1). Rules like *S1*–*S4* have been offered by philosophers
(almost always) as a semantics for the *latter* sort, for 'diverging' worlds, not for
the former—or, as philosophers also call overlapping worlds, 'branching' worlds.

Which is it, in EQM? This question is in danger of being settled on the basis of
an accident of terminology. Let us agree that 'branching' means the development
of superpositions with respect to the decoherence basis. For the sense intended
by philosophers—where there are numerical identities among spatiotemporal
parts—we will speak of 'overlap'. Mundane examples of overlap are everywhere.
Thus roads overlap if they share a same stretch of asphalt; houses and roofs, cars
and steering wheels, hands and fingers, all overlap. Overlapping, like Everettian
branching, supposedly has a formal definition too, but not in terms of Hilbert
space structure. Rather, it is defined in terms of 'mereology', the general theory
of parts and wholes. (I shall come back to this shortly.)

As for 'divergence', its definition is more equivocal. It is sometimes used
to mean that there exist no physical relations between worlds, contrary to the
situation for worlds in EQM,[14] but we shall take it as simply the opposite of
overlap: 'diverging' means 'non-overlapping'.

---

[14] This can cause some confusion, evident in Saunders [1998, section 5], where I erroneously said
that 'fatalism' (a near-neighbour to the position I am currently defending) involved the replacement
of the superposition of histories (the universal state) by an incoherent mixture of histories, and must
thus be rejected.

Now to the point: if worlds overlap in EQM then (definitionally) there are genuine transworld identities. The very same thing exists in different worlds. In particular, since branching is massive in EQM, if worlds overlap then the very same person is part of vast numbers of worlds, differing, maybe, only in respects remote in space and time—differing, maybe, only after centuries. All this is contrary to *S1*. In that case any uncertainty to be had can have nothing to do with self-locating belief. But is there any uncertainty to be had?

The worry is not that overlapping worlds are unintelligible or inconsistent; it is that they make nonsense of ordinary beliefs. As Lewis put it:[15]

Respect for common sense gives us reason to reject any theory that says that we ourselves are involved in branching [overlapping]. . . . But we needn't reject the very possibility that a world branches [overlaps]. The unfortunate inhabitants of such a world, if they think of 'the future' as we do, are of course sorely deceived, and their peculiar circumstances do make nonsense of how they ordinarily think. But that is their problem; not ours, as it would be if the worlds generally branched [overlapped] rather than diverged. (Lewis [1986b p.209])

Diverging worlds, composed of objects and events that do not overlap (that are qualitatively but not numerically identical) do not suffer from this problem.

Of course, 'common sense' does not cut it much in the physical sciences; and Lewis's final sentence could not more comprehensively beg the question. But let us grant this much: he who believes he is a part of each of a number of worlds cannot also wonder which of them he's in.

We are at the nub of the matter: do worlds—maximal branches, sequences of relative configurations of particles and fields, as described by EQM—overlap in the philosophers' sense, or do they diverge?

We should discount two considerations. First, the coincidence in the terminology 'branching' (because, as introduced by Everett, it referred to the mathematical formalism of quantum mechanics, not to the philosophers' criterion of overlap). Second, the fact that worlds in EQM do not diverge in the sense of being physically disconnected (they are not physically disconnected, because they superpose, but the issue is whether or not they overlap).

There remains another consideration, however. There is a clear parallel with simultaneous rather than temporal overlap. For (ignoring entanglement) let the Schrödinger-picture state of a composite system of observer and environment at time t be

$$|\psi(t)\rangle \otimes (c|\chi(t)\rangle + c'|\chi'(t)\rangle) \tag{4}$$

---

[15] Not everyone agrees with Lewis on this point. Thus Johnston [1989] suggested the rule *S3* in the explicit context of overlapping persons, arguing that semantic rules like this were underdetermined by the metaphysics.

where $|\psi(t)\rangle$ is the state of the observer, and $c|\chi(t)\rangle + c'|\chi'(t)\rangle$ is the state of the environment, a superposition of macroscopically distinct states $|\chi(t)\rangle$, $|\chi'(t)\rangle$. Perhaps the latter only differ with respect to macroscopic objects at enormously large spacelike distances—say, a radioactive decay that triggers a macroscopically significant event on a planetary system in the far side of Andromeda. In such a case, it is not at all obvious that the observer should be uncertain which branch, $|\psi(t)\rangle \otimes |\chi(t)\rangle$ or $|\psi(t)\rangle \otimes |\chi'(t)\rangle$, she belongs to. Her relative state in Everett's sense is the superposition $c|\chi(t)\rangle + c'|\chi'(t)\rangle$; in this there is no hint of uncertainty.

True enough, but for the mathematical identity:

$$|\psi(t)\rangle \otimes (c|\chi(t)\rangle + c'|\chi'(t)\rangle) = c|\psi(t)\rangle \otimes |\chi(t)\rangle + c'|\psi(t)\rangle \otimes |\chi'(t)\rangle.$$

We can read the relativization the other way: of the observer $|\psi(t)\rangle$ relative to the environment $|\chi(t)\rangle$ with amplitude c, and of the observer $|\psi(t)\rangle$ relative to the environment $|\chi'(t)\rangle$ with amplitude c'. It cannot be necessary to count the two observers as numerically the same, without further assumptions. Suppose, for example, they have their amplitudes as properties; if $c \neq c'$ they are not even qualitatively the same.

Pursuit of the question of quantum non-locality leads on to relativity, where it connects with ordinary beliefs about probability in much the same way that the relativity of simultaneity connects with ordinary beliefs about tense (Saunders [1995, 1996]). Here we shall stick to probability as it applies to events related by unambiguously timelike relations, as it figures in our practical lives.

What is needed is an atemporal perspective. That takes us to the quantum histories formalism: how does the distinction between overlap and divergence play out in the Heisenberg picture?

Schrödinger picture states at time t are in $1 - 1$ relation to histories terminating at t, i.e. of the form $a^-(t)$. We used a special case of this earlier (for single-time histories). More generally, up to normalization:

$$|a^-(t), t\rangle = \exp^{-iHt/\hbar} C_{a^-(t)}|\Omega\rangle = \exp^{-iHt/\hbar} \sum_{a \in a^-(t)} C_a|\Omega\rangle \qquad (5)$$

That is, the Schrödinger picture state at time t is (the forward evolution to t) of a superposition of Heisenberg picture states $|a\rangle = C_a|\Omega\rangle$, all with the same sequence of configurations $a^-(t)$ up to time t. Let 'worlds', as before, be represented by maximal branch vectors $|a\rangle$. Do worlds overlap or diverge?

Here is the point made graphically (Fig. 1): the orbit of a Schrödinger picture wavefunction Eq.(5) is schematically depicted in Fig. 1a (which explains how branches came by their name). But the branches themselves can be depicted in two ways, by either Fig. 1b or Fig. 1c. In Fig. 1c it seems that branches do not overlap, whereas in Fig. 1b it seems that they do. Which of them is correct?

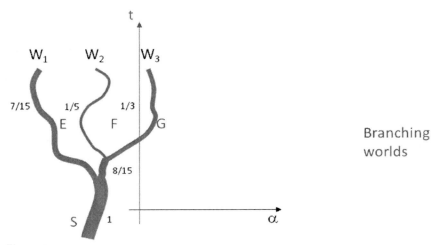

Figure 1a.

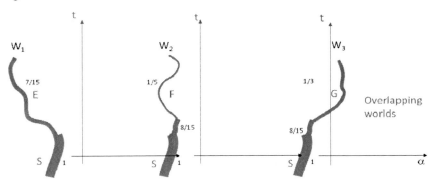

Figure 1b.

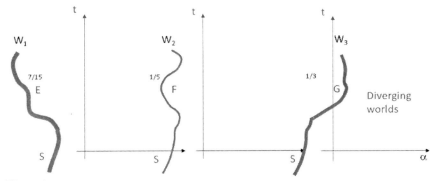

Figure 1c.

Figure 1.

Evidently the difference concerns only their amplitudes. Of the two, Fig. 1c is the obvious representation for Heisenberg picture vectors, which each have a unique amplitude. Fig. 1b fits better with a graph of the orbit of a Schrödinger picture state subject to recurrent collapse (whose amplitude is constantly changing). Which of them is the 'correct' picture? But once stated in this way, the suspicion is that whether worlds in EQM diverge or overlap is *underdetermined* by the mathematics. One can use either picture; they are better or worse adapted to different purposes.

If so, it is pretty clear which is the right one for making sense of uncertainty. But, of course, Fig. 1c is only a visual aid: can we be sure that the Heisenberg-picture branch vectors do not overlap? This, perhaps, is a *technical* philosophical question, to be settled by a theory of parts and wholes for vectors—by a vector mereology.

Alas, there is no such theory, or none on which there is any agreement. Mereology, if it is close to any branch of mathematics, is close to set theory (Lewis [1991]), but even there its links are controversial. We are going to have to make this up as we go along. And it is a good question, at this point, as to where the burden of proof really lies. Mereology has paid little or no dividends in pure mathematics, let alone in physics. It is at bottom an a priori metaphysical theory. A very good desideratum on any reasonable metaphysics is that it makes sense of our best physical theory, rather than nonsense.

Metaphysicians can, however, reasonably insist on an existence proof—a demonstration that at least one vector space mereology can be defined in terms of which maximal branches in EQM do not overlap. Here is a simple construction that depends essentially on the branching structure of a consistent history space. Let maximal branch vectors $|a\rangle$ be defined as before. Any non-maximal branch vector $C_\beta|\Omega\rangle = |\beta\rangle$ can then be written:

$$|\beta\rangle = \sum_{a \in \beta} |a\rangle.$$

By the consistency condition, $\langle a|a'\rangle = 0$ for $a \neq a'$; therefore for any two branch vectors $|\beta\rangle$, $|\gamma\rangle$ if $\langle \beta|\gamma\rangle \neq 0$ there exists a unique branch vector $|\delta\rangle$ and orthogonal branch vectors $|\beta'\rangle$, $|\gamma'\rangle$ such that

$$|\beta\rangle = |\beta'\rangle + |\delta\rangle, |\gamma\rangle = |\gamma'\rangle + |\delta\rangle$$

namely the vector:

$$|\delta\rangle = \sum_{a \in \beta \cap \gamma} |a\rangle.$$

Our candidate mereology is then: a branch vector $|\epsilon\rangle$ is part of $|\epsilon'\rangle)$ (denote $P(|\epsilon\rangle, |\epsilon'\rangle)$) if and only if either, (a) $\epsilon = \epsilon'$ or, (b) there exists a non-zero branch vector $|\delta\rangle$ such that $\langle\epsilon|\delta\rangle = 0$, $|\epsilon\rangle + |\delta\rangle = |\epsilon'\rangle$.

Standard axioms of mereology are listed in the Appendix below; it is easy to check that they are satisfied by the parthood relation just defined. From the overlap relation derived from P, we see that orthogonal vectors do not overlap.

It follows that every branch vector is part of the universal state $|\Omega\rangle$, as seems appropriate, but that orthogonal branch vectors have no parts in common. Since, by consistency, branching always produces orthogonal branch vectors, the branch vector representing Alice seeing spin-down has no part in common with that representing Alice seeing spin-up. On this basis branches of this kind diverge, in the philosophical sense, as do maximal branches.

But in what sense, it might be asked, is Alice a part of a maximal branch vector $|a\rangle$? For the latter (on our candidate mereology) has no vectors as proper parts. The answer is that Alice is represented not by a vector at all, but by a subsequence β of configurations of $a = \langle a_+, \ldots, a_-\rangle$ (see again *S1*, and our use of ordered pairs $(\beta, |a\rangle)$.[16] The point is not that there must exist a uniform theory of mereology that applies across the board to emergent ontology. Far from it: that is a piece of metaphysics that has had little or no success in the special sciences. We concede only that one might be constructible in each domain.

With that, it should be clear that our vector mereology on its own does not settle the question of uncertainty. Alice's successors downstream of branching diverge from each other, but they may still overlap with Alice prior to branching, even given our vector mereology. A branch can be coarse-grained from the time t of branching in such a way that it is degenerate with respect to Alice seeing spin-up and Alice seeing spin-down. Thus a branch vector that represents only Alice's past $a^-(t)$ at t (of the form $|a^-(t)\rangle = P_{a_k}(t_k) \ldots P_{a_-}(t_-)|\Omega\rangle$, $t_k \leq t$, the Heisenberg-picture analogue of Eq. (5)) represents Alice as containing all her possible future selves, that is, as overlapping completely, according to our mereology, with all her possible future selves, even if they they do not overlap with each other.

And there, in a nutshell, is the contrary view: by all means take the parthood relation as specified, but let 'Alice' be indexed to time, and be represented by a Schrödinger picture vector accordingly (as given by Eq. (5)), or by the Heisenberg-picture branch vector $|a^-(t)\rangle$, rather than by relativization to a maximal branch. Alice in this sense overlaps with all of her successors, and the grounds for self-locating uncertainty evaporate.

We always knew we had this option, however. This is to reject *S1* not on the grounds that *worlds* in EQM really overlap (on our proposed mereology they

---

[16] There may well be other devices possible, for example the two-vector formalism. See Vaidman in Chapter 20.

do not), but because *superpositions of worlds* overlap, and because (so goes the objection) persons and things *should* be represented by superpositions of worlds, superpositions of vectors.

But why should they? We can do the same in the case of diverging worlds of the sort usually considered by metaphysicians (having nothing to do with quantum mechanics), or as they arise in cosmic inflation (see Tegmark in Chapter 19). In either case it is uncontroversial that worlds do not overlap, but still, *sets of* worlds do. Let a person or a thing up to time t be a *set* of worlds, namely of those that contain that person or thing or one qualitatively identical up to time t (which could be far in the future): then there is no uncertainty of where a person or thing is located among worlds that diverge after that time, either. But metaphysicians are unlikely to take this proposal very seriously, not least because it makes nonsense of much of what we think. Nor should we in the analogous case in EQM.[17]

To conclude: there is no good reason to think EQM is really a theory of overlapping worlds. If questions of overlap of branches are to be settled by appeal to the underlying mathematics, in terms of vector space structure, then there is at least one natural mereology in terms of which worlds that differ in some feature, since orthogonal, are non-overlapping. The semantics *S1–S4* does *not* misrepresent the underlying mathematical structure of the theory; in terms of this mereology, it correctly describes worlds as non-overlapping.

But it does follow that the word 'branching' is something of a misnomer, in the context of EQM, in the philosophers' sense of the word. But then the word 'diverging' is also infelicitous. 'Branching' (inappropriately in EQM) suggests overlap, 'divergence' (inappropriately in EQM) suggests the absence of physical relations. I see nothing wrong with continuing to call EQM a theory of branching worlds, but only because the expression is well established among physicists[18] and because it is fundamental to EQM that worlds, by superposition, do make up a dynamical unity—that they are all parts of the universal state. At any rate, it will mark a new phase in the status of the Everett interpretation if the debate is over what the theory should be called.

Acknowledgements

My debt to David Wallace is obvious, but additional thanks are due to Harvey Brown, Hilary Greaves, and especially Alastair Wilson, for helpful comments and suggestions.

[17] Parallels but also differences in the two cases have been argued by Paul Tappenden. See, in particular, his [2008] and Wallace's and my reply ([2008b]).

[18] Although some talk of 'parallel' worlds instead (see e.g. Tegmark in Chapter 19). There is clearly a case for this terminology.

Appendix: Axioms of Mereology

We write 'Pxy' for 'x is part of y'; this relation is reflexive, transitive, and antisymmetric. Defined notions are:

Overlap $Oxy \underset{def}{=} \exists z(Pzx \& Pzy)$.

Underlap $Uxy \underset{def}{=} \exists z(Pxz \& Pyz)$.

Fusion $x \sqcup y \underset{def}{=} \iota z \forall w(Pxw \& Pyw \rightarrow Pzw)$.

(In words, x and y overlap if a part of x is a part of y; they underlap if there is something of which they are both parts; and their fusion is the unique thing that is a part of anything of which x and y are both parts.) The elementary axioms of mereology are:

M1 $\forall x(x \sqcup x) = x$
M2 $\forall x \forall y(Uxy \rightarrow \exists z(z = x \sqcup y))$
M3 $\forall z(Pxz \& Pyz \rightarrow Px \sqcup yz)$

(the fusion of anything with itself itself; if two things underlap then their fusion exists; anything which underlaps two things is part of the fusion of those things). Define the parthood relation as above:

Vector Part $P(|\beta\rangle, |\gamma\rangle)$ if and only if either (a) $\beta = \gamma$, or (b) there exists a non-zero branch $|\delta\rangle$ such that $\langle\delta|\beta\rangle = 0$, $|\beta\rangle + |\delta\rangle = |\gamma\rangle$.

It is simple to check that P is reflexive, transitive, and antisymmetric. The fusion operation, as defined by P, is:

- If $\langle\beta|\gamma\rangle = 0$, $|\beta\rangle \sqcup |\gamma\rangle = |\beta\rangle + |\gamma\rangle$
- If $\langle\beta|\gamma\rangle \neq 0$, $\beta \in \gamma$, $|\beta\rangle \sqcup |\gamma\rangle = |\gamma\rangle$; $\gamma \in \beta$, $|\beta\rangle \sqcup |\gamma\rangle = |\beta\rangle$.

It clearly satisfies $M1-M3$.

References

Aharonov, Y., D. Albert, and L. Vaidman [1988], 'How the result of a measurement of a component of the spin of a spin half particle can turn out to be 100' *Physics Review Letters* **60**, 1351–4.

Aharonov, Y. and L. Vaidman [1993], 'Protective measurements', *Physics Letters* **A 178**, 38.

Davidson, D. [2001], *Inquiries into Truth and Interpretation* (2nd edn), Oxford University Press, Oxford.

Everett III, H. [1957], ' "Relative state" formulation of quantum mechanics', *Reviews of Modern Physics* **29**, 454–62, reprinted in DeWitt and Graham [1973 pp.141–50].

Gell-Mann, M. and J.B. Hartle [1990], 'Quantum mechanics in the light of quantum cosmology', in *Complexity, Entropy, and the Physics of Information*, W.H. Zurek, (ed.) Addison-Wesley, Reading.

Greaves, H. [2004], 'Understanding Deutsch's probability in a deterministic multiverse', *Studies in History and Philosophy of Modern Physics* **35**, 423–56. Available online at philsci-archive.pitt.edu/archive/00001742/.

—— [2007], 'On the Everettian epistemic problem', *Studies in History and Philosophy of Modern Physics* **38**, 120–52. Available online at philsci-archive.pitt.edu/archive/00002953.

Ismael, J. [2003], 'How to combine chance and determinism: Thinking about the future in an Everett universe', *Philosophy of Science* **70**, 776–90.

Johnston, M. [1989], 'Fission and the facts', *Philosophical Perspectives* **3**, 369–97.

Ladyman, J. and D. Ross [2007], *Every Thing Must Go: Metaphysics Naturalized*, Oxford.

Lewis, D. [1973], *Counterfactuals*, Harvard.

—— [1980], 'A subjectivist's guide to objective chance', in R.C. Jeffries (ed.), *Studies in Inductive Logic and Probability*, Vol. 2, University of California Press, (1980); reprinted in *Philosophical Papers*, Vol. 2, Oxford University Press, Oxford (1986).

—— [1986a], *Philosophical Papers*, Vol. 2, Oxford University Press, Oxford.

—— [1986b], *On the Plurality of Worlds*, Blackwell.

—— [1991], *Parts of Classes*, Blackwell.

Saunders, S. [1994], 'Decoherence and evolutionary adaptation', *Physics Letters* **A 184**, 1–5.

—— [1996], 'Time, quantum mechanics, and tense', *Synthese* **107**, 19–53.

—— [1997], 'Naturalizing metaphysics', *The Monist* **80**, 44–69.

—— [1998], 'Time, quantum mechanics, and probability', *Synthese* **114**, 405–44. Available online at arXiv.org/abs/quant-ph/0112081.

—— [2005], 'What is probability?', in *Quo Vadis Quantum Mechanics*, A. Elitzur, S. Dolev, and N. Kolenda, (eds), Springer. Available online at arXiv.org/abs/quant-ph/0412194.

Saunders, S. and D. Wallace [2008a], 'Branching and uncertainty', *British Journal for the Philosophy of Science* **59**, 293–305. Available online at philsci-archive.pitt.edu/archive/00003811/.

—— [2008b], 'Saunders and Wallace reply', *British Journal for the Philosophy of Science* **59**, 315–17.

Savage, L. [1954], *The Foundations of Statistics*, Wiley.

Tappenden, P. [2008], 'Comment on Saunders and Wallace', *British Journal for the Philosophy of Science* **59**, 306–14.

Uffink, J. [2000], 'How to protect the interpretation of the wavefunction against protective measurements'. Available online at arXiv.org/abs/quant-ph/9903007v1.

Vaidman, L. [1998], 'On schizophrenic experiences of the neutron or why we should believe in the many-worlds interpretation of quantum theory', *International Studies in the Philosophy of Science* **12**, 245–61.

Van Fraassen, B. [1980], *The Scientific Image*, Clarendon Press, Oxford.

Wallace, D. [2002], 'Quantum probability and decision theory, revisited'. Available online at arXiv.org/abs/quant-ph/0211104.

—— [2003], 'Everettian rationality: defending Deutsch's approach to probability in the Everett interpretation', *Studies in the History and Philosophy of Modern Physics* **34**, 415–39. Available online at arXiv.org/abs/quant-ph/0303050.

—— [2005], 'Language use in a branching universe', available online at philsci-archive.pitt.edu/archive/00002554.

—— [2006], 'Epistemology quantized: circumstances in which we should come to believe in the Everett interpretation', *British Journal for the Philosophy of Science* 57, 655–89. Available online at philsci-archive.pitt.edu/archive/00002839.

—— [2007], 'Quantum probability from subjective likelihood: improving on Deutsch's proof of the probability rule', *Studies in the History and Philosophy of Modern Physics* 38, 311–32. Available online at arXiv.org/abs/quant-ph/0312157.

7

A Fair Deal for Everettians

David Papineau

1 INTRODUCTION

It is widely supposed that the Everettian account of quantum mechanics has difficulties with probability. In this paper I shall argue that those who argue against the Everettian interpretation on this basis are employing a double standard. It is certainly true that there are philosophical puzzles about probability within the Everettian theory. But I shall show that orthodox metaphysics has even worse problems with probability than Everettianism. From this perspective, orthodox metaphysicians who criticize Everettians about probability are a classic case of a pot calling a kettle black.

Why might anybody think that Everettians have trouble with probability? The basic reason is that Everettian quantum mechanics is a deterministic theory. It implies that the conditions obtaining at a given time together with the relevant laws completely fix the future course of nature. True, this future will normally involve a proliferation of branching alternatives, and the Everettian theory will attach numbers to these branches—the squared moduli of their wavefunction amplitudes—that behave numerically like probabilities. But critics argue that these numbers cannot be understood as probabilities, since they lack any essential connection with ignorance about the future. On the Everettian theory, a knowledgeable subject faced with a chancy situation, like an x-spin measurement on an electron in an eigenstate of z-spin, will know exactly what is going to happen—reality will split into as many branches as there are possible macroscopic outcomes. Given this, the critics argue that the squared moduli of the amplitudes of the branches cannot possibly be probabilities. Probabilities are measures of ignorance about what is going to happen. If the Everettian interpretation holds that there is nothing to be ignorant about, then surely this leaves no place for probability.

I think that Everettians should be unmoved by this line of thought. They should simply insist that the relevant squared moduli are probabilities, despite the lack of any connection with ignorance. In support of this, they can point out that the squared moduli behave remarkably like orthodox probabilities even for an Everettian with complete knowledge of the future. These numbers will respond to evidence and guide decisions in just the same way that probabilities are supposed to do within orthodoxy. Moreover, if we want to know *why* these numbers should so respond to evidence and guide decisions, it turns out that Everettians are at least as well placed to answer these questions as orthodoxy.

So I think Everettians should simply reject the idea that probabilities require ignorance. They should say that this is simply another of the mistakes imposed on us by orthodox metaphysics. Just as orthodoxy is wrong to suppose that only one of the possible outcomes will occur in a chancy situation, so it is wrong to suppose that the ascription of a non-unitary probability to some outcome is incompatible with knowing for sure that it will occur.

A number of writers sympathetic to Everettianism have recently sought to show that, despite first appearances, some element of ignorance about the future can still be found within the Everettian scheme of things (Albert and Loewer [1988], Vaidman [1998], Saunders and Wallace [2008]). One motivation for this move is to show that Everettians can still avail themselves of ordinary ignorance-based thinking about probability. From my point of view, this motivation is quite misplaced. It presupposes that orthodoxy has a coherent story to tell about probability. But in truth there is no such story. Orthodox ignorance-based thinking about probability is pretty close to incoherent. Everettians are thus doing themselves a disservice when they seek to ally themselves with such orthodox thinking. They will do much better to cut the tie between probability and ignorance and forget about trying to mimic orthodoxy.

Some will feel that it is simply a contradiction in terms to talk about non-unitary probabilities in the absence of ignorance. (What can it *mean* to say that a measurement of x-spin 'up' is 50% probable, if we know for sure that this result is determined to occur?) Indeed some writers who agree with me that Everettians can combine lack of ignorance with quantitative future expectations maintain that even so it is better to drop the term 'probability' for squared wavefunction amplitudes. Thus Hilary Greaves [2004] talks about a 'caring measure' and David Lewis [2004] about 'intensity'. However, I myself see no reason for this squeamishness. I think it is perfectly appropriate to talk about 'probability' even within the context of no-ignorance Everettianism.

Still, if somebody wants to insist that it is a matter of definition that non-unitary 'probabilities' imply ignorance, I am happy to let this pass. There is no substantial issue here, just a matter of terminology (cf. Greaves [2004]). To see this, suppose that we can show that a no-ignorance Everettianism has room for quantities that play just the same role within the Everettian scheme of things as probabilities play within orthodoxy, save that Everettians sometimes ascribe non-unitary values to circumstances that they know are certain to occur. And suppose that the resulting metaphysical theory gives a better overall account of the working of the world and our place in it than the orthodox picture that makes play with 'probability'. Then what does it matter if definitional niceties require us to withhold the term 'probability' from these quantities, and call them 'schprobabilties' (or 'caring measures' or 'intensities') instead? We've got just as a good theory, with just the same virtues, even so. The quantities involved might be less familiar than orthodox probabilities, in that futures that are sure to occur can still have non-unitary values. But I trust that nobody who is serious about the interpretation of quantum mechanics will take unfamiliarity per se as a reason for rejecting an otherwise cogent view.

As I said, even on the definitional level I see no reason why no-ignorance Everettians should not hang on to the term 'probability' if they like. I think that there is quite enough overlap between the role that squared amplitudes play in Everettianism and orthodoxy respectively to justify Everettians retaining this term even after the tie with ignorance is cut. But, as I said, there is no substantial issue here. If you don't like this usage, feel free to substitute 'schprobability' whenever I talk about Everettian probabilities.

In this paper I shall proceed as follows. The next three sections will rehearse a number of background claims about probability that I hope will be uncontroversial. The following four sections will then argue that Everettianism is no worse off than orthodoxy in accounting for the central properties of probability. The final section will then argue that Everettianism is actually better off than orthodoxy when it comes to explaining how probability guides decisions.

2 SUBJECTIVE AND OBJECTIVE PROBABILITIES

Any satisfactory account of probability needs to distinguish subjective and objective probabilities and explain the relation between them.

Subjective probabilities, or 'credences', are measures of psychological degrees of belief. You may not be fully convinced that it is going to rain, but you leave home with an umbrella anyway. In such a case you have a partial belief that it will rain—some degree of belief lower than one but higher than zero. In general, degrees of belief manifest themselves in behaviour, as here your degree of belief in rain is manifested by your carrying the umbrella.

A tradition of analysis going back to Ramsey [1926] and continued by Savage [1954] and Jeffrey [1965] equates degrees of belief with dispositions to behaviour, including betting behaviour, and thereby shows that we can ideally quantify agents' degrees of belief by numbers in the range zero to one. There is plenty of controversy about what exactly these analyses establish, but we can bypass this here. For present purposes I shall simply assume that subjective probabilities correspond to agents' subjective expectations of future outcomes and that these expectations guide agents when they assess the relative worth of alternative actions, via familiar calculations of expected utility.

In addition to subjective probabilities, we need also to recognize objective probabilities, or 'chances'. These are quantities that represent the objective tendency for chancy situations to issue in different future circumstances. There is such an objective tendency for an electron in an eigenstate of z-spin to display 'up' on an x-spin measurement. Objective probabilities are prior to any subjective matters. There would still have been objective probabilities in the world even if no decision-making agents had ever evolved.

In a moment, I shall say a bit more about objective probabilities, and in particular about their availability within Everettianism. But first it will be helpful to observe that there is a basic connection between objective and subjective probabilities. You will do well to match your subjective probabilities to the objective probabilities. When making choices, you want your subjective assessment of the probabilities to match the objective probabilities.

This is (roughly) the principle that David Lewis [1980] dubbed the "Principal Principle". I shall formulate it as follows:

The Principal Principle: It is rational to set your subjective probabilities equal to the objective probabilities.[1]

In effect, this principle says that you ought to allow your actions to be guided by the objective probabilities—you ought to bet with the objective odds, not against them. (This follows because subjective degrees of belief are constitutively tied to behavioural choices—having such-and-such subjective degrees of belief simply consists in choosing those actions that maximize expectations of utility weighted by those degrees of belief. So someone who sets their *subjective* degrees of belief equal to the objective probabilities will per se maximize *objective* expected utility.)

3 FREQUENCIES AND PROPENSITIES

Now, some will feel that these brief remarks about objective and subjective probabilities are already enough to show why Everettianism is in trouble with probability. I introduced objective probabilities by talking about the 'tendency for chancy situations to issue in different future circumstances'. But aren't such 'objective tendencies' essentially a matter of long-run *frequencies*? And isn't the problem that Everettianism has no room for long-term frequencies—or rather that it has room for too many of them?

Take some case where we are interested in the frequency of result R on repeated trials of kind T (for example, the frequency of 'up' when we measure

---

[1] Lewis's own formulation of the principle specifies that we should match our credences about outcomes to our credences about the objective probabilities, rather than to the objective probabilities themselves. For my reasons for stating it more objectively see Beebee and Papineau [1997]. This difference will not matter to the arguments of this paper.

x-spin on electrons in an eigenstate of z-spin). Where orthodoxy sees one actual future, in which there is a definite sequence of results if this trial is repeated, Everettianism must see a multiplicity of futures, given that a plurality of branches will be generated every time the trial is repeated. Orthodoxy can thus happily speak of *the* relative frequency of Rs displayed in *the* sequence of all trials T. But Everettianism cannot do this, for its reality will contain many different actual sequences with different relative frequencies. (Thus, where orthodoxy has, say, the actual sequence of results 'U', 'D', 'D', 'U', 'D', . . . , Everettianism will have this sequence *plus* 'D', 'D', 'D', 'U', 'D', . . . , *plus* 'U', 'U', 'D', 'U', 'D', . . . and so on, for all the different possible strings of outcomes, including ones on which the relative frequency of 'up' is quite different from the objective probability of 'up'.)

This may look like an immediate knock-down reason why Everettians can't have probabilities. If objective probabilities are *the* relative frequencies of Rs in Ts, then Everettians won't have any objective probabilities, for they won't have any unique such relative frequencies.[2] Moreover, if we are interested in the subjective probabilities imposed on agents by the principal principle, then Everettians won't have any such subjective probabilities either, for lack of any objective probabilities to set them equal to.

However, it is important to realize that this whole line of objection hinges on equating objective probabilities with long-run frequencies, and that this equation is by no means uncontentious. Over the past 50 years, many theorists of probability have come to favour *propensity* theories of probability over frequency theories. Where frequency theories aim to *reduce* objective probabilities to long-term frequencies, the propensity theory takes objective probabilities to be primitive quantities that are fully present in each chancy single case. On this view, a particular electron in a z-spin eigenstate will have a 0.5 propensity to display x-spin 'up', quite independently of what any other similar electrons may do. No doubt the frequency of spin 'up' in repeated measurement of similar electrons will tend to be close to 0.5. But this is a consequence of the particular electrons having the relevant propensities, and not the essence of objective probability itself.[3]

---

[2] A tradition of writers from Everett himself onwards [1957 p. 461] has observed that in the infinite limit the squared amplitude will be zero for the union of all sequences in which the frequency differs from the underlying squared amplitude of R. But this is not to the point in the present context. We are looking for a unique actual sequence so that we can identify the probability of R with its frequency in that sequence. Everett's observation does nothing to ensure a unique such sequence (especially given that any specific infinite sequence in which the frequency *is* equal to the underlying squared amplitude of R will also have a zero squared amplitude).

[3] I am not entirely happy about the terminology of 'propensity theory'. This term is often understood more specifically than I intend it here (cf. Gillies [2000]); indeed, Karl Popper's original 'propensity' theory is really a version of the frequency theory. I have in mind any theory that takes single-case probabilities to be basic, and not reducible to anything frequency-like. I was brought up to use 'chance' in this sense, following Mellor [1971] and Lewis [1980]. However, this term could

Everettians will do well to adopt the propensity approach to objective probability rather than the frequency theory. While the frequency theory is clearly inconsistent with Everettian metaphysics, this is not true of the propensity theory. According to the propensity theory, objective probabilities measure the brute tendencies of particular chancy situations to produce various results. There is no obvious conflict between this and the idea that all those results actually occur, each weighted by the relevant probability. Indeed the two ideas seem to fit very well together.

There is nothing ad hoc about this appeal to the propensity theory on the part of Everettians. As I said, many probability theorists have come to favour the propensity interpretation over the frequency theory for reasons quite independent of Everettian metaphysics. It will be useful at this point briefly to rehearse some of the reasons that have led them to turn against the frequency theory.

The most obvious difficulty facing the frequency theory is to ensure that there is always a suitable sequence of trials for any event to which we want to ascribe an objective probability.[4] Aren't there some genuinely chancy events that only occur once in the history of the universe? But we don't want to ascribe them an objective probability of one just on that account. This worry led classical frequency theorists to equate objective probabilities not with the frequency of Rs among the *actual* Ts, but rather with the hypothetical limiting frequency that *would* be displayed if Ts were repeated indefinitely (von Mises [1957]).

This suggestion, however, is open to its own objections. Why suppose that there is a definite fact of the matter about what *would* be observed if electron spins *were* measured indefinitely? From a metaphysical point of view this looks highly dubious. Note that it isn't enough to suppose there will be a certain 'number' of x-spin 'up's and 'down's in all the hypothetical future tosses. If there is a denumerable infinity of tosses, then presumably there will be

now be misleading, given the extent to which Lewis's later work takes 'chances' to be reductively constituted by actual frequencies [1986, 1994].

    [4] Are not frequency theories of probability also in immediate danger of contradicting themselves? The axioms of probability imply that there is always room for long-run sequences in which the frequency differs from the probability (even in infinite sequences such divergence isn't absolutely ruled out, even if it 'almost never' happens). Yet the reductive identification of probability with the long-run frequency would seem to mean that any such divergence is quite impossible. However, this worry isn't as bad as it looks. Frequency theorists can say that this supposed contradiction equivocates on the phrase 'the frequency of Rs'. Suppose that in this world there is a definite relative frequency of Rs among Ts, equal to p say. Then a frequency theory of probability does indeed entail that p is the objective probability of Rs in Ts. But this doesn't deny that things *might* have been different, in line with the axioms of probability, and that in different possible worlds, so to speak, there are different limiting frequencies. There is no contradiction between the thoughts that the frequency is (a) p in this world but that (b) it might have been different. Of course, a frequency theory of probability does imply that, if the long-term frequency had been different, the objective probability would have been different too. That is, the objective probability will not be p in those worlds where the long-run frequency differs from p. And this means that frequency theories rule out any possible worlds that display limiting frequencies different from their own objective probabilities. This in itself may strike some readers as odd. But even if it is odd, it is not obviously contradictory.

a denumerable infinity of both 'up's and 'down's. And this fact by itself is compatible with any limiting relative frequency whatsoever, given that different arrangements of the order in which these 'up's and 'down's occur will yield different limiting frequencies. So, if we want the hypothetical future tosses to determine a definite limiting relative frequency, the hypothetical results must come in a definite order. It is very hard indeed to believe that there is a fact of the matter about exactly which ordered sequence of 'up's and 'down's we would observe, if we counterfactually carried on measuring electron spins indefinitely.

These points are not necessarily fatal to any frequency theory. Recent theorists have tended to revert to the equation of objective probabilities with *actual* frequencies, and have suggested that we can avoid the difficulty of single or rarely repeated trials by lumping together superficially distinguishable types of trials into sufficiently large similarity classes.[5] Still, I trust that I have done enough to indicate why the propensity theory strikes many philosophers of probability as an attractive alternative to the frequency account.

4 STATISTICAL INFERENCE

Some readers might be wondering whether the metaphysical attractions of the propensity theory are not undermined by its epistemological shortcomings. By equating objective probabilities with long-run frequencies, the frequency theory suggests an obvious recipe for measuring objective probabilities—namely, by observing frequencies. By contrast, the propensity theory cuts any constitutive link between objective probabilities and frequencies, and so seems to leave us in the dark about how to measure objective probabilities.

However, this contrast is an illusion. The propensity theory can deal with the epistemology of objective probability in just the same way as the frequency theory. To see this, consider in more detail how the epistemology will work on the frequency theory. This theory equates the real probability with the eventual long-run frequency. However, this eventual long-run frequency is not itself directly observable. Even if we have an actual frequentist theory, we will have to wait until the end of time to observe the proportion of Rs among *all* the Ts. And on a hypothetical infinite frequency theory the relevant frequency will be even less directly observable. So frequency theorists need to recognize that in practice we must estimate the real probability from the observed sample frequency in some *finite partial* sequence of Ts.

---

[5] David Lewis's mature 'best theory' Humean theory of chances [1986, 1994] contains many elements of an actualist frequency theory. His 'best theory' aims to be sensitive to symmetries as well as frequencies, but he also asks of the 'best theory' that it optimize the probability of the actual course of history—and this will tend to fix the probability of Rs in Ts as their actual frequency. See also Loewer [2004].

Now, there is no agreed view about the logic of inferring real probabilities from finite sample statistics. It is easy enough to work out the probability of the observed statistics *given* various alternative hypotheses about the real probabilities. But what exactly is the logic of getting from this to the acceptability of the alternative hypotheses themselves *given* the observed statistics? There are a number of quite different approaches to the logic of statistical inference, including the Fisherian, Neyman–Pearson, and Bayesian accounts. Still, they do at least all agree that such inferences hinge crucially on the extent to which alternative hypotheses about the real probability imply that the observed statistic was itself probable. (For example, how probable are 40 'up's in 100 measurements if the true probability of 'up' is 0.5?) In one way or another, the different theories of statistical inference all take hypotheses about the real probability to be more favoured the more they imply that the observed statistic was likely.

Now, there is nothing in this last thought that is not available to the propensity theorist. On the propensity theory, real probabilities are primitive quantities, and not to be equated with long-term frequencies. But this does not prevent real probabilities from having implications about the probability of observing such-and-such finite statistics in finite samples. This simply falls out of the fact that propensities obey the axioms of probability. (For example, if the propensity of 'up' in one measurement is 0.5, the propensity of 40 'up's in 100 measurements is $^{100}C_{40}\ 0.5^{100}$.) So the propensity theorist can say exactly the same thing about the epistemology of probability as the frequentist—hypotheses about the real probability are to be favoured the more that they imply that the observed statistics are likely.

It is important to realize that the propensity theorist does not deny that there is an important connection between objective probabilities and frequencies. But this is not the frequentist's proposed *reduction* of objective probabilities to long-run frequencies, an equation that is of course rejected by the propensity theorist. Rather it is simply the connection implied by the axioms of probability, namely, that in repetitions of identical and independent trials in which the real probability is such-and-such, it is so-and-so probable that the observed frequency will be thus-and-so. And this connection is of course as much available to the propensity theorist as to the frequency theorist.

It is perhaps also worth observing that, from the perspective of the logic of statistical inference, frequencies in large numbers of repeated trials have no special status. They are just one example of an observed statistic that can yield information relevant to the assessment of hypotheses about true probabilities. To see this, suppose that for some reason you suspect that a given coin is strongly biased towards heads, as opposed to being fair. You toss it just once and see tails. This single toss is already significant evidence that the coin is fair rather than biased, since tails would be less likely if the coin were biased than if it were fair.

The only difference between this case and frequencies in large numbers of repeated trials is that the latter will normally carry far more discriminatory information. Suppose that we are trying to decide between a range of alternative hypotheses about the true probabilities. A single observation will often be reasonably likely on various different hypotheses from this range, and so be relatively indecisive in choosing between them. But a specific frequency in a large sample will normally be highly unlikely on all but one of the hypotheses being evaluated, and so will favour that hypothesis very strongly over the others. But this is a quantitative difference, not a qualitative one. Single observations also carry just the same *kind* of statistical information as large-sample frequencies.

5 STATISTICAL INFERENCE AND EVERETTIANISM

The last section pointed out that propensity theorists can make just the same statistical inferences as frequency theorists. Let us now apply the moral to the interpretation of quantum mechanics. Given that Everettians will think of objective probabilities as propensities, they too can make just the same statistical inferences.

In particular, Everettians can follow orthodoxy in using observed results to estimate the probabilities of different values for the measured spins, positions, momenta, energies and so on of quantum systems. Just like orthodoxy, they will favour those hypotheses about these probabilities that imply that the observed results were probable, and disfavour those hypotheses that imply that they were improbable.

Now, we can take it that Everettians and orthodoxy alike will conclude from such investigations that the probabilities of quantum events correspond to the squared moduli of their wavefunction amplitudes. Equating the relevant probabilities with these numbers implies a far higher probability for the observed results than any other assumptions.

Of course, Everettians will have to recognize that there are some branches of reality on which the observed statistics will be misleading as to the true probabilities. (For example, there is a branch of reality in which x-spin measurements on electrons in z-spin eigenstates always show 'up'.) But it is not clear that this undermines the symmetry between Everettian and orthodox statistical thinking. For orthodoxy must also recognize that it is always epistemologically *possible* that the so-far observed statistics do not reflect the underlying probabilities. In the face of this possibility, orthodox statistical thinking simply proceeds on the assumption that we are *not* the victims of misleading samples, by favouring those hypotheses that make what we observe probable, rather than hypotheses that make the observations *im*probable. Everettians can once more do the same, proceeding similarly on the assumption that our observations are not atypical.

Of course, the rationale for proceeding in this way remains to be explained. This brings us back once more to the logic of statistical inference. Now, as I said above, there are different views on this issue—Fisherian, Neyman–Pearson, and

Bayesian—and there is no question of going into details here. But we can at least ask whether there is any reason to suppose that Everettianism will be worse off than orthodoxy in accounting for statistical inference. I cannot see why there should be. Everettians take the objective probability of a statistic to be a measure over the branches of reality on which it occurs, where orthodoxy takes it to be a measure of the tendency of the corresponding possible branches of reality to be actualized. Both Everettians and orthodoxy then favour those hypotheses that imply that the statistics we observe were objectively probable. The exact rationale for this move is then the subject of dispute between the different theories of statistical inference. But there is nothing in any of these theories, as far as I can see, that requires an orthodox understanding of objective probability rather than an Everettian one.

Perhaps it is worth making one last comment about statistical inference. Most philosophers concerned with statistical inference have now turned away from the contortions of Fisherian and Neyman–Pearson accounts of statistical inference and embraced the relative comforts of Bayesianism (the reasons are given in Howson and Urbach [1989]). In the Bayesian context, the rationale for statistical inference reduces to the rationale for Bayesian updating. Why should we increase our credence in hypothesis H in proportion to the extent that H makes the observed statistics more probable than they would otherwise be? Now of course Bayesian updating is itself the subject of a huge literature, which again lies beyond the scope of this paper. But one recent and highly persuasive suggestion, due to Hilary Greaves and David Wallace [2006], is that Bayesian updating can be seen as a special case of the decision-theoretic maximization of expected utility. If this is right, then the question of whether Everettianism can account for statistical inference collapses into the question of whether it can account for decision-theoretic expected utility maximization. It is to this question that I now turn.

6 EVERETT AND THE PRINCIPAL PRINCIPLE

Maybe Everettians can mimic orthodoxy in using standard statistical reasoning to infer that the 'objective probabilities' of quantum results are given by their squared amplitudes. But I am sure that many readers will as yet remain unpersuaded that the numbers so inferred warrant being called 'probabilities'. Maybe Everettians can attach numbers to future branches that behave numerically like probabilities. But Section 2 told us that something more is required of objective probabilities. They also need to figure in the Principal Principle. Objective probabilities are quantities to which subjective probabilities and hence decisions ought rationally to conform.

As I said in my introduction, we don't want to get bogged down in the terminological issue of exactly what justifies calling some quantity an objective 'probability', and in particular whether it needs to be associated with ignorance about the future. But I am happy to concede that at least some association with subjective probability and rational decision is a minimal requirement. There would be little point to calling something objective 'probability' if it had nothing to do with rational decision. It is here that many will feel that Everettianism falls down. Their numbers have no appropriate connection with rational decision. Where orthodoxy can maintain that it is rational to orientate decisions to the squared amplitudes, Everettians are unable to do this.

At this point I think that Everettians should simply *assert* that it is rational to match subjective probabilities to the squared amplitudes. In making this move, Everettians will thereby commit themselves to favouring actions that bring rewards in future circumstances with large squared amplitudes, and attach relatively little significance to how things will turn out in futures with low squared amplitudes. Of course, Everettians won't think of risky choices in quite the same way that orthodoxy does. They won't think that only one future will become real, and that their actions are designed to do well, if things pan out as expected, in *that* future. Rather, they will think that all futures with positive amplitudes will become real, and that their action is designed to maximize utility across all these real futures, weighted by their objective probabilities. But, even so, they will treat the squared moduli just as orthodoxy does in their expected utility calculations.

Still, many will want to ask, are Everettians *entitled* to enter the squared amplitudes into expected utility calculations? I just said that Everettians should 'simply assert that it is rational to match subjective probabilities to the squared amplitudes'. But many will feel that this is not enough. Maybe Everettians can simply assert that it is rational to match subjective probabilities to the squared amplitudes. But cannot orthodoxy go further, and *show why* it is rational to have these subjective probabilities? Those suspicious of Everett will feel that their 'simple assertion' of the principal principle has all the advantages of theft over honest toil. Aren't they just helping themselves to an assumption that can only be properly accounted for within orthodoxy?

The underlying thought here was forcefully stated by David Lewis in 'Humean Supevenience Debugged' [1994]. (He wasn't thinking of Everettianism, but he may as well have been.) 'Be my guest—posit all the . . . whatnots you like . . . But play fair in naming your whatnots. Don't call any alleged feature of reality "chance" unless you've already shown that you have something, knowledge of which could constrain rational credence . . . Again, I can only agree that the whatnots deserve the name of chancemakers if I can already see, disregarding the names they allegedly deserve, how knowledge of them constrains rational credence in accordance with the Principal Principle' [pp.484–5].[6]

---

[6] We can see Lewis here as raising the standards for something to qualify as 'objective probability'. Where his earlier work required only that 'objective probability' should be that quantity to which it *is* rational to match your credence, here he seems to be requiring in addition that we be able to '*see how*' knowledge of this quantity constrains rational credence.

Lewis is here suggesting that within an orthodox metaphysical framework certain quantities can be *shown* to deserve the name of objective probability—by showing how knowledge of them constrains rational credence. If this cannot be done within Everettianism—if Everettianism is simply reduced to *asserting* that Everettian squared amplitudes should constrain rational credence, where orthodoxy can explain why—then this would certainly count strongly against Everettianism. However, it will be a central contention of this paper that there is no real contrast here. Orthodoxy cannot deliver on Lewis's promise. It does no better than Everettianism when it comes to explaining why we should match our subjective probabilities to the squared amplitudes. In the end, orthodoxy too can do no better than simply asserting without further justification that it is a basic requirement of rationality that we should set our degrees of belief equal to the squared amplitudes.

The next two sections will be devoted to this point. That is, I shall show that orthodoxy has no good way of explaining why we should bet in line with the squared amplitudes, as opposed to simply asserting this. In this respect orthodoxy is no better off than Everettianism. The following section will then seek to show that orthodoxy is rather *worse* placed than Everettianism when it comes to accounting for the connection between the squared amplitudes and subjective probabilities.

7 ORTHODOXY AND THE JUSTIFICATION OF BETTING WITH THE OBJECTIVE PROPENSITIES

So—can orthodoxy *justify* setting degrees of belief equal to the squared amplitudes of quantum results? We are taking it that, for both orthodoxy and Everett, standard methods of statistical inference will indicate that these squared amplitudes measure the objective probabilities of those results. And we can also take it, following the last section, that Everettians will follow orthodoxy in asserting *that* it is rational to set degrees of belief equal to these objective probabilities. The crucial question is then whether orthodoxy can do better than Everettianism in showing *why* it is rational to allow these objective probabilities to guide us in making decisions.

As we saw earlier, orthodoxy can choose between two different theories about the nature of objective probabilities. On the frequency theory, objective probabilities are identified with long-run frequencies. On the propensity theory, by contrast, objective probabilities are primitive and fully present in each single case—while of course still having (probable) implications about what sample frequencies will be observed.

In the rest of this section, I shall show that an orthodox propensity theorist can give no good account of why it is rational to bet with the objective probabilities.

In the next section, I shall consider whether orthodoxy can do any better by bringing in the frequency theory of probability.

At first sight it might seem as if orthodoxy has an obvious advantage over Everettianism in explaining why we should bet with the objective probabilities. Cannot orthodoxy just say that betting with the probabilities will bring desired results in *the* actual future? By contrast, Everettians cannot say this, since they don't think there is *the* single actual future, but a branching future which includes every outcome with a non-zero squared amplitude—which means that they have to admit that betting with the objective propensities will bring bad results on some low-amplitude branches, as well as good results on high-amplitude ones.

But the contrast here is spurious. (Don't confuse familiarity with cogency.) After all, orthodoxy *cannot* really say that betting with the propensities *will* bring desired results in the actual future. Good bets don't always pay off. Even odds-on favourites can fail to win. At best, orthodoxy can say that betting with the propensities makes it objectively *probable* that you will win.

But of course Everettians can say this too (or at least that it's objectively 'schprobable' that you will win). And for neither does this amount to any kind of independent justification of betting with the squared amplitudes. If you want a further reason for choosing actions that give a high probability to desired results, it's scarcely any answer to be told that those choices make desired results highly probable.

What about the thought that you will win in the *long run*, even if not in the single-case, if you keep making bets in line with the objective propensities? Here again it may look as if orthodoxy does have a real advantage. There is no Everettian counterpart to the orthodox thought that you will win in the long run if you bet with the objective odds. For, as we saw earlier, Everettians must recognize a *plurality* of long-run futures, encompassing all long-run sequences with non-zero amplitudes, and on many of these sequences, betting with the objective odds will lead to long-run losses.

But this line of thought doesn't help orthodox propensity theorists either. For they cannot really say that you *will* win in the long run, any more than they could say this about the single case. All that they can say is once more that you will *probably* win. After all, if you can be unlucky once, you can be unlucky repeatedly. Of course, if you are betting with the odds every time, then the probability of your losing overall will become smaller the more bets you make. But familiar calculations show that there will remain a finite probability of losing overall for any finite sequence of bets. So once more the putative justification for betting with the objective odds collapses into the empty thought that this strategy makes winning objectively probable.

What about the *infinite* long run? Isn't it guaranteed that you will win if you go on making advantageous bets for ever? Even this isn't obvious, on the current propensity understanding of probability. Let us put to one side for the moment worries about the idea of betting for ever and about the relevance of distantly deferred guarantees of success. Even so, there remains the point that the axioms of probability imply that infinite sequences where you end up losing remain *possible* (even if they 'almost never' occur in the technical sense that the probability measure of such sequences taken collectively is zero). After all, to take a special case, all specific infinite sequences of tosses with a fair coin are equally likely. So it would be ad hoc to say that some of them—the ones where the limiting frequency of heads differs from 0.5—are absolutely impossible, solely on the grounds that there are far fewer of them. After all, any specific such 'rogue' sequence is just as likely as any specific 'normal' sequence.

So here too it turns out that orthodoxy is no better off than Everettianism. We still can't say that you *will* win if you bet for ever, only that the probability of doing so will be very high indeed, which once more begs the question.

8 ORTHODOXY AND THE JUSTIFICATION OF BETTING WITH THE FREQUENCIES

By this stage the friends of orthodoxy might be thinking that it was a mistake to turn away from the frequency theory in favour of propensities. We have just seen that, if orthodoxy thinks of probabilities as primitive propensities, then it is unable to guarantee a definite gain for somebody who bets with the objective odds, even in the long run. At best it can say that such a gain is highly probable. However, a frequency theory promises to remedy this problem. For on the frequency theory, orthodoxy can indeed assert that betting with the objective odds will guarantee success in the *eventual* long run.

To see how this might work, suppose that the objective probability of R on any trial T can be reductively identified with the eventual frequency of Rs among some all-encompassing class of Ts. Then we can be sure of a positive result if we consistently bet a constant amount at objectively advantageous odds on *all* the Ts in that class. (If the probability equals the frequency, then the objective advantageousness of each bet will ensure that, when you have been through all the Ts, the total amount you won on the Rs will outweigh the total amount you lost on the not-Rs.)

Perhaps this is the kind of thing that David Lewis had in mind when he suggests that his frequency-style analysis of objective probability ('chance') allows us to understand *why* it is rational to match degrees of belief to objective probability. Thus he claims 'And we can well understand how frequencies, if known, could constrain rational credence' [p.476]. And later he says in similar vein 'I think I see, dimly but well enough, how knowledge of frequencies and symmetries and best systems could constrain rational credences' [p.484].

If Lewis were right here, then it would certainly follow that Everettians are less able to make sense of probabilities than orthodoxy. For, as we saw earlier, there is no possibility of Everettians embracing the frequency theory of probability. Since they recognize many possible futures displaying many eventual frequencies, it makes no sense for them to talk about probability being the same as *the* eventual long-run frequency. So they have no route to the claim that betting with the objective odds is *guaranteed* to bring eventual long-run success. They remain stuck with the non-explanatory claim that betting with the odds is highly *likely* to bring success.

Still, it remains to be shown that Lewis is justified in claiming that the frequency theory really does afford orthodoxy a good justification for matching credences to the objective probabilities. Lewis says he can see this 'dimly, but well enough'. But he does not elaborate, and it is by no means clear how the story is supposed to go.

Let us put to one side the general worries about the frequency approach to probability rehearsed in Section 3 above. Even if we take the frequency theory as given, there is an obvious worry about the proposed long-run justification for betting with the objective odds. As J.M. Keynes put it, 'In the long run we're all dead'.

We are trying to explain why any particular agent in any particular situation is well advised to bet in line with the objective odds. By way of explanation we are given the thought that, if we were to continue betting like this in every similar situation, then we would be sure to win in the end. It is not at all clear that this is to the point. Why is it an argument for my betting with the odds *now* that this would pay off in the eventual long run?

The problem is particularly clear on the hypothetical infinite version of the frequency theory. On this version, the supposed justification is that you *would* be guaranteed to win, if you were to bet an infinite number of times. It is not at all clear what bearing this counterfactual truth has on agents who will in fact only make a limited number of decisions in their lifetimes.

Nor does the problem go away if we switch to an actual frequency theory. Now the probability is equated with the relative frequency of Rs among all Ts in the actual world. The guarantee is correspondingly that you will win for sure if you bet at advantageous odds on all these actual Ts. This is still of no obvious relevance to actual agents. They aren't going to bet until the end of time, but at most until the ends of their lives. There is no obvious way of showing why such mortal beings are rationally required to adopt a strategy that isn't guaranteed to succeed until the end of time.

To bring out the point, consider some feckless agent with no thought for the future. ('I want some drugs *now*—who cares what happens tomorrow?') Suppose that this agent is offered some wager, and is allowed to decide whether to bet with the objective odds or against. I take it that the rational choice for this agent is to bet with the objective odds. Yet it is clearly of no relevance to this agent that repeating this strategy will pay off at the end of time. This agent just doesn't care about the future. Whatever makes the agent's choice rational, it looks as if it is independent of what will happen in the long run.

I conclude that frequency-based long-run justifications for betting with the odds are looking in the wrong place. The basic fact is that the right choice in any situation is to bet with the objective odds. A special case of this is that you should also bet with the objective odds in a complex trial consisting of repeated single trials. But we can't justify the basic fact in terms of the special case, even given the frequency theory of probability.

Perhaps David Lewis was thinking of something different from a long-run justification when he said he could 'see dimly, but well enough' how frequencies could constrain rational credence. But if so he gives us no clue of what the alternative story might be. And it is hard to see what else, apart from a long-run story, would give frequencies as opposed to 'whatnots' (like squared moduli of amplitudes) an advantage in constructing the requisite justification.

9 ORTHODOXY IS WORSE OFF

The last two sections were devoted to showing that orthodoxy has no good way of explaining why the squared moduli of wavefunction amplitudes should play the decision-theoretic role that the principal principle requires of objective probabilities. In the end, orthodoxy can do no better than simply assert without further explanation that it is rational to conform credences to the squared amplitudes. Since Everettians can also simply assert this, they are certainly no worse off than orthodoxy in forging a link between credences and squared amplitudes.

In this section, I want to show that Everettians are actually better off with respect to this connection. One possible reason why they might be better off has been explored by David Deutsch and David Wallace. They have sought to show that some very basic assumptions about rationality dictate that Everettian agents should match their degrees of belief to the squared amplitudes (Deutsch [1999], Wallace [2003]). However, I shall not rest anything on this somewhat controversial line of argument (for doubts, see Greaves [2004], Price [2006]). Rather I shall seek to show that, even if the Deutsch–Wallace argument fails, there is another respect in which Everettians can make better sense than orthodoxy of the connection between credences and squared amplitudes.

Go back to the thought that neither orthodoxy nor Everett can do better than simply *assert* that credences should match the squared amplitudes—equivalently, that rational agents should bet in line with the squared amplitudes. There is a sense in which the inability to further justify this claim is far more embarrassing for orthodoxy than for Everett. This is because orthodoxy thinks of the aim of rational action differently from Everettianism. For orthodoxy, an action is successful if it produces good results in the presumed one actual future. For Everettianism, by contrast, an action is successful if it maximizes utility over all future worlds weighted by their squared amplitudes. Given this difference, it seems as if orthodoxy, but not Everettianism, still faces a further question even *after* it has asserted that rational agents should bet in line with the squared amplitudes. What is the connection between this strategy and the ultimate aim of action, namely, good results in the actual world? But, of course, this is a question that orthodoxy cannot answer, as the last two sections have made clear.

The trouble is that orthodoxy seems to be making its primitive commitment in the wrong place. It commits itself to the maximization of objective expected utility. But that's not what orthodox actions are aimed at. Rather they aim at gain in the one actual future. And this then creates a demand for some further explanation of why maximizing objective expected utility is a good means to that further end. But of course, to repeat, there's no good way for orthodoxy to answer the awkward question it poses. There's no way of showing that betting with the objective odds will bring an actual pay-off.

True, we are assuming that Everettians are also unable to give any further justification for betting with the squared amplitudes. But Everettians have far less need of such a justification. For they regard the maximization of objective expected utility as itself the ultimate aim of action. From an Everettian perspective, there is no sense in which an action that maximizes objective expected utility can nevertheless turn out to be unsuccessful. There is no danger that such an action might prove to have been the wrong choice if some unlikely actual future comes to pass. Since all futures with any probability are sure to happen, an action that maximizes expected utility over all of them cannot be bettered. So Everettians, unlike orthodoxy, face no awkward and unanswerable further question about the connection between their primitive commitment and some further ultimate aim.

Let me bring out the point by an analogy. Consider a figure that I shall call 'the committed gambler'. The committed gambler doesn't care about money. Over the years she has already made a fortune. What she prides herself on is making the right bet. She wants to be in the pot if it offers better than 4–1 odds on a flush draw, and to fold if it doesn't. But she doesn't care whether or not she actually makes her flush—that's nothing to do with good poker. For the committed gambler, it is vulgar to be concerned with actual pay-offs. The important thing is to identify bets that offer 'good value' and to get in on the action. That's the true mark of success—only amateurs care about the actual fall of the cards.

Now compare the committed gambler with the 'ordinary gambler' who just wants to make money. For the ordinary gambler, there's no virtue in being in the pot just because it offers 'good value'—such a bet will only be a good thing if the flush is actually made and the pot won. Now, faced with the ordinary gambler, we can ask the awkward question I pressed above—why do you think it's such a good idea to maximize objective expected money, when what you want is actual money? And, as we have repeatedly seen, there is no good answer to this question.

But this question gets no grip on the committed gambler. After all, the committed gambler doesn't want the actual money—all she wants is to maximize objectively expected money. So the committed gambler, unlike the ordinary gambler, owes us no further explanation of why her decision strategy is a good idea. From her perspective, maximizing objective expected money isn't a means to some further end, but itself the ultimate aim of action, and so there is no need for any further explanation of its connection to some supposed further end.

The analogy with the knowledgeable Everettian agent should be clear. Like the committed gambler, the Everettian agent doesn't regard the maximization of objective expected utility as some means to a further end, success in the one actual future. Rather, it is itself the essential characteristic of a fully successful choice. So Everettian agents, unlike ordinary agents understood within orthodoxy, face no nagging question of why maximizing expected utility is a good means to actual success—a question to which, as we have seen, orthodoxy has no good answer.

Of course, the analogy between the knowledgeable Everettian agent and the committed gambler is not perfect. To see where it breaks down, we can usefully invoke another analogy. Hilary Greaves [2004] has suggested that a knowledgeable Everettian agent is like someone who adopts a differential 'caring measure' over branching futures—the agent will be concerned about these futures to different degrees, in proportion to their squared amplitudes, just as a mother might be concerned about the fates of different offspring to different degrees.

Now, the virtue of Greaves' analogy is that it emphasizes how the choices of a knowledgeable Everettian agent are not risky: their success is not hostage to future fortune. In this respect they are like the choice of a mother who gives more attention to one child than another just because she wants to: there's no sense in which this mother can then fail to achieve what she is aiming at.

But in another way, talk of a 'caring' measure can be misleading. It is not as if the general run of Everettian agents will have unusual desires compared with ordinary agents. Just like ordinary agents, they will wish only for food, comfort, fame, the welfare of their offspring, and other normal kinds of success. While their decisions can be modelled by supposing that, in addition, they 'desire' to do well more in some futures than others, these extra decision-informing attitudes are not really desires. The rationale for these attitudes is quite unlike the rationale for normal desires. The Everettian agent doesn't favour some worlds more than others because results in those worlds have some extra qualitative virtue, but simply because those worlds have higher squared amplitudes.

Given this, we will do well to distinguish these differential Everettian attitudes to future worlds from normal desires. The obvious way is to mark the difference by styling them 'probabilities'. But even those who are uneasy about this terminology will still need to distinguish these attitudes from ordinary desires, and to recognize that Everettain choices are designed to maximize the satisfaction of ordinary desires weighted by these other attitudes. That is, we still need something like the distinction between utilities and probabilities, even after we recognize that the success of a rational Everettian choice offers no hostage to future fortune.

This now enables us better to appreciate the limits of the analogy between the committed gambler and an ordinary knowledgeable Everettian. Both avoid the nasty question that can be put to the ordinary gambler—what's so good about maximizing *expected* money, when what you want is *actual* money? And moreover they both avoid it because they *don't* want actual money. But the reasons are rather different—where the committed gambler doesn't want actual *money*, the knowledgeable Everettian doesn't want *actual* money.

The point is that the committed gambler has non-standard desires—her peculiar history as a gambler means that she has no longer any interest in winning money per se, but rather takes pride in finding 'good value' bets. So the reason that she cannot be pressed to further justify her money-maximizing choices is simply that she doesn't care about money. An ordinary knowledgeable Everettian, by contrast, will have normal desires, including a desire for money. The reason the Everettian agent doesn't face the nasty question—why maximize expected money, when what you want is actual money?—is rather that this notion of 'actual' money gets no grip on the Everettian. The Everettian doesn't think there is just one actual future, which may or may not contain money, but rather a range of different futures with different squared amplitudes, some containing money and some without.

The similarity between the committed gambler and the knowledgeable Everettian is that both regard the maximization of objective expected utility as the ultimate criterion of a successful action. But for the committed gambler this is because of non-standard desires, whereas for the Everettian it is because of non-orthodox metaphysics. When Everettians weight alternative futures differentially in making choices, this isn't because they are so peculiar as to prize utility-maximizing choices in themselves, but simply because they are committed to weighing all the different futures by their squared amplitudes when making choices. For Everettians, this is a basic and *sui generis* commitment—the commitment to match subjective probabilities to squared amplitudes—and not to be viewed as akin to the committed gambler's peculiar pride in seeking out 'good value' bets wherever they can be found. It's not that Everettians desire to make good bets independently of their consequences, like the committed gambler. On the contrary, it is precisely the consequences that they are interested in, and they pursue them in the way dictated by their basic commitment to bet with the objective probabilities.

In making a basic commitment to match subjective probabilities to squared amplitudes, Everettians are akin to ordinary orthodox agents, rather than to the committed gambler. There are no unusual desires in play, just the basic commitment to betting with the objective odds. Moreover, I am allowing that

Everettians are also like orthodox agents in being able to offer no further justification for this basic commitment. Still, the point remains that in another respect Everettians line up with the committed gambler rather than with orthodox agents. For even after the basic commitment has been made, orthodox agents have something further to explain, in a way that Everettians and the committed gambler do not.

According to orthodoxy, the action that maximizes objective expected utility might or might not turn out to have been the right choice: the pot may offer very good odds for your flush draw, and so you bet, but even so you may fail to make the flush, in which case you would have been better off not betting. Given this, we cannot help asking what is so good about maximizing expected utility, given that it may well turn out to be the wrong choice—even though we know that there is no answer to this inescapable question. It counts strongly in favour of Everettian metaphysics that it renders this awkward question unaskable.

References

Albert, D. and B. Loewer [1988], 'Interpreting the many worlds interpretation', *Synthese* 77, 195–213.

Beebee, H. and D. Papineau [1997], 'Probability as a guide to life', *Journal of Philosophy* 94.

Deutsch, D. [1999], 'Quantum theory of probability and decisions', *Proceedings of the Royal Society of London* A455, 3129–37.

Everett, H. [1957], '"Relative state" formulation of quantum mechanics', *Reviews of Modern Physics* 29, 454–62.

Gillies, D. [2000], *Philosophical Theories of Probability*, Routledge, London.

Greaves, H. [2004], 'Understanding Deutsch's probability in a deterministic multiverse', *Studies in History and Philosophy of Modern Physics* 35.

Greaves, H. and D. Wallace [2006], 'Justifying conditionalization: Conditionalization maximizes expected epistemic utility', *Mind* 115, 607–32.

Howson, C. and P. Urbach [1989], *Scientific Reasoning: The Bayesian Approach*, Open Court, New York.

Jeffrey, R. [1965], *The Logic of Decision*, McGraw Hill, New York.

Lewis, D. [1980], 'A subjectivist's guide to objective chance', in R. Jeffery (ed.), *Studies in Inductive Logic and Probability, Vol II*. University of California Press, Berkeley, reprinted in D. Lewis [1986].

—— [1986], *Philosophical Papers Volume II*, Oxford University Press, Oxford.

—— [1994], 'Humean supervenience debugged', *Mind* 103, 473–90.

—— [2004], 'How many lives has Schrödinger's cat?', *Australasian Journal of Philosophy* 82.

Loewer, B. [2004], 'David Lewis's Humean theory of objective chance', *Philosophy of Science* 71, 1115–25.

Mellor, D.H. [1971], *The Matter of Chance*, Cambridge University Press, Cambridge.

Price, H. [2006, 'Probability in the Everett world; Comments on Wallace and Greaves'. Available online at philsci-archive.pitt.edu/archive/00002719/01/WallaceGreavesComments.pdf.

Ramsey, F.P. [1926], 'Truth and probability', in *The Foundations of Mathematics* [1931], Routledge & Kegan Paul, London.

Saunders, S. and D. Wallace [2008] 'Branching and uncertainty', *British Journal of the Philosophy of Science* **59**, 293–305. Available online at philsci-archive.pitt.edu/archive/00003383/.

Savage, L.J. [1954], *The Foundations of Statistics*, Wiley, New York.

Vaidman, L. [1998], 'On schizophrenic experiences of the neutron or why we should believe in the many-worlds interpretation of quantum theory', *International Studies in the Philosophy of Science* **12**, 245–61.

von Mises R. [1957], *Probability, Statistics and Truth*, Macmillan, New York.

Wallace, D. [2003] 'Everettian rationality: Defending Deutsch's approach to probability in the Everett interpretation', *Studies in the History and Philosophy of Modern Physics*, **34**, 415–39.

8

How to Prove the Born Rule

David Wallace

Thus we see that quantum theory permits what philosophy would hitherto have regarded as a formal impossibility, akin to 'deriving an ought from an is', namely deriving a probability statement from a factual statement. This could be called deriving a 'tends to' from a 'does'. (Deutsch [1999])

1 INTRODUCTION

The 'Everett interpretation of quantum mechanics' is just unitary quantum mechanics, taken literally as a description of the world; it is a 'many-worlds' theory because it instantiates multiple, emergent, branching quasiclassical realities. That much is commonplace amongst contemporary Everettians; I argued for it *in extenso* in Chapter 1, and for the purposes of *this* chapter I will take it as read.

It is widely held, however, that a problem remains: namely, how does *probability* fit into this story? It is not in dispute what physical magnitude is *supposed* to be (or to stand in for) probability: the probability of a branch is supposed to be its weight (i.e., its mod-squared amplitude). More formally, the probability of a history α represented by a history operator $\widehat{C}_\alpha$ (in the consistent-histories formalism) is supposed to be

$$\Pr(\alpha) = \langle \psi | \widehat{C}_\alpha | \psi \rangle \tag{1}$$

where $|\psi\rangle$ is the universal (Heisenberg-picture) state. In more parochial language, if an observer's branch has weight w_0, if it is going to split into multiple branches, and if those branches in which X happens have weight w_X, then the probability for that observer of X happening is supposed to be w_X/w_0.

What is in dispute is why, and how, this physical magnitude can be probability. One might ask: how can it even *make sense* for anything to 'be probability' in a theory where all possible outcomes occur; less confrontationally, one might ask what kind of *argument* can be given to justify the claim that mod-squared amplitude is probability. (I have previously referred to these as the *Incoherence*

Problem and the *Quantitative Problem* respectively, though I shall not make much of the distinction in this article.)

One quite legitimate response to this question, I think, is bemusement. After all, formally speaking the measure defined by mod-squared amplitude on any given space of consistent histories satisfies the axioms for a probability. Indeed, mathematically the set-up is identical to any stochastic physical theory, which ultimately is specified by a measure on a space of kinematically possible histories (albeit that measure is usually given indirectly, via a stochastic differential equation). In Everett-interpreted quantum mechanics, the correct space of consistent histories is the space of quasiclassical histories; as my earlier chapter argued, this space may be imprecisely defined but this is no reason not to take it seriously as emergent structure. So (goes the response) all the formal requirements to take mod-squared amplitude as probability are in place; to ask for more is no more justified than to ask why the physical quantity represented by the metric in Newtonian space "is" length, or why other mathematical features of classical physics represent mass or charge. (Saunders [1998] develops this response in more depth.)

I have a good deal of sympathy for this response, but in this chapter I wish to discuss a more positive answer to the question, which might be called the *decision-theoretic strategy*.

Decision-theoretic strategy: Probability gets its meaning in quantum mechanics through the rational preferences of agents. In particular, a rational agent who knows that the Born-rule weight of an outcome is p is rationally compelled to act as if that outcome had probability p.

The decision-theoretic strategy was advocated by Deutsch [1999]. In the same paper he presented an informal proof, from principles of decision theory that (he argued) did not themselves invoke probabilistic notions, that the only rational strategy for an agent in an Everettian universe is to follow the Born rule. I developed Deutsch's proof further, and presented alternative versions of it, in Wallace [2003] and Wallace [2007].

The argument has met with its share of criticism. Some of the criticism (e.g. Barnum et al. [2000], Lewis [2005]) has been directed at the proof itself; some (e.g. Albert and Price's contributions to this volume, Chapters 11 and 12) has sought to undermine the possibility of a proof by proposing other, (allegedly) equally rationally justifiable alternatives to the Born rule. (These, I should make clear, are not proposed as positive *suggestions* as to how we should act if the Everett interpretation is correct; they are intended rather as *reductios*.)

My purpose in this chapter is to give a self-contained defence of the decision-theoretic strategy, culminating in a formal proof of the rational necessity of the Born rule from axioms of decision theory which I will defend. Formalization is not often an aid to understanding, but when a result is controversial it can be helpful to see *exactly* what is and is not required. Along the way I will showcase some of the various proposed alternative strategies for Everettian rationality, and show exactly how they conflict with the assumptions in the argument; in doing so, I hope, the argument for making those assumptions will become clearer.

My focus in the chapter is deliberately narrow. A proof that rational agents in an Everett universe must act in accordance with the Born-rule probabilities falls short of a full solution of the probability problem: we might also ask how this decision-theoretic notion of probability connects with our use of probability in assessing the evidence for quantum mechanics, or with our ordinary, pre-theoretic notions of probability as a guide to action in cases of uncertainty. I shall address neither question here, though. (For discussions of the former question, see Greaves and Myrvold's contribution in Chapter 9, Wallace [2006], and part II of Wallace [2011]; for the latter, see Wallace [2005], part III of Wallace [2011], Saunders [1998], and Saunders' contribution in Chapter 6.)

I shall begin in Section 2 with a brief discussion of decision theory *in general*, and go on in Section 3 to see how decision theory works in the Everett interpretation. The main part of the paper (Sections 4–8) states and proves, first informally and then in full mathematical rigour, that the Born rule is the unique rational strategy available in the Everett interpretation. I illustrate this result (Section 9) by considering a number of other strategies, proposed at various times and places as counterexamples to the necessity of the Born rule, and show why those strategies are not in fact valid alternatives. I conclude with a few more general observations about why the Born rule can indeed be proven, and why the Everett interpretation is essential in such a proof.

2 PREAMBLE: THE DECISION-THEORETIC APPROACH

Suppose a coin is to be tossed in five minutes time, and suppose that an agent bets five dollars (at even odds) that it will land heads. There are two interestingly different possible results: (i) the measurement gives result 'up' and the agent gets $5; (ii) the measurement gives result 'down' and the agent loses $5. If the result is heads, the agent will be pleased about the bet; if it is tails he will be less delighted, though (if he is of an appropriate character) he may well still regard the bet as having been the right choice given his information before the coin toss. (There are of course vastly more than two microscopically distinct possible results; the division into two sets is based on the pragmatic interests of the agent.)

In deciding whether to accept this bet as opposed to any number of other bets, the agent has to weigh the cost to himself if the result is 'down' against the benefit if it is 'up'. Decision theory gives a precise answer to the question of how he should carry out this weighting: he should assign some utility (some real number) $\mathcal{V}(+\$5)$ to receiving five dollars, and some other utility $\mathcal{V}(-\$5)$ to losing five dollars, and some third utility $\mathcal{V}(\$0)$ to neither getting nor losing it, and he should assign a probability $\Pr(H)$ to heads; and he should take the bet only if

$$\Pr(H) \times \mathcal{V}(+\$5) + (1 - \Pr(H)) \times \mathcal{V}(-\$5) > \mathcal{V}(\$0). \qquad (2)$$

More generally, decision theory mandates that an agent should assign a utility to each payoff, and a probability to each outcome, and that faced with *any* decision, the agent should choose that option which maximizes expected utility with respect to those assignments.

(In elementary discussions, it is common just to assume $\mathcal{V}(+\$N) = N$, but this is too simplistic: it is not irrational to refuse to trade your house for a one-in-a-thousand chance of winning Microsoft. In fact, in a decision-theoretic framework, what it *means* to say that one reward is twice as valuable as another is that a 50% chance of getting the first is as valuable as getting the second with certainty. (See Savage [1972 pp.91–104] for more on this point.))

Why should an agent behave this way? Prima facie, it isn't obvious at all that he should try to maximize expected utility rather than, say, maximizing utility with respect to the square of the probability function; or maximizing the logarithm of utility; or just maximizing the utility of the least good possible outcome.

Decision theory has a standard answer to this question: if an agent has a definite preference (which might be indifference) between any two bets, and if that preference order obeys certain constraints which are purported to be necessary conditions of rationality, and if the set of available bets has a sufficiently rich structure, then it is possible to prove a *representation theorem*: a theorem that for any such preference order there is a unique probability function, and an essentially unique[1] utility function, such that one bet is preferred to another iff it has a higher expected utility. It follows that any agent whose preferences cannot so be represented must be acting irrationally: that is, must somewhere be violating a principle which (again, purportedly) is a necessary constraint on rational action.

It will be instructive to present two such principles (both drawn from Savage [1972]). The first is *transitivity*: if an agent prefers a to b and b to c, he should prefer a to c. The second might be called *dominance*: if an agent will do better through a than b *whatever happens*, he should choose a over b. (For instance, a bet that pays ten dollars if a coin-toss lands heads and nothing if it lands tails is to be preferred over a bet on the same toss that pays five dollars on heads and minus five on tails.)

There is an important weakness in this decision-theoretic argument which needs to be stressed. It is a proof that rational agents must bet according to *some*

---

[1] By 'essentially unique' I mean unique up to positive affine transforms $x \rightarrow ax + b$, with a positive. Fairly obviously, such transformations serve only to scale the expected utility of all bets in the same way.

probability function, but it is silent on the connection between that function and the 'real' probabilities. No decision-theoretic principle is contradicted by an agent who assigns probability 99/100 to an apparently fair coin landing 'heads', for instance. A minority of advocates of the decision-theoretic approach simply deny that there is any such thing as objective or 'real' probability; the majority just take it as a bare postulate that an agent should conform his subjective probabilities to the objective probabilities when he knows the latter.[2]

This weakness actually rather undermines the use of probability as a criticism of the Everett interpretation, even without the arguments of this paper: if classical probability can give no justification of its probability rule, why ask the Everettian for such a justification? But in fact, we will see that in the Everett interpretation, not only can we use rationality considerations to make sense of probabilities as well as in conventional decision theory, we can prove and not merely postulate the link between those probabilities and the quantum-mechanical weights.

3 EVERETTIAN RATIONALITY

So, consider the Everettian version of the coin-toss. Instead of a coin, we have a particle in a superposition of spin-up and spin-down (in some fixed direction); instead of a coin-toss, we have a spin measurement. And instead of there being two interestingly different *possibilities*, there are two interestingly different sets of branches: the spin-up branches, where (if the agent took the bet) he gets five dollars, and the spin-down ones, where he loses five dollars. In deciding whether to accept this bet as opposed to any number of other bets, the agent has to weigh the benefit to himself in the branches where the result is up, against the cost in the branches where the result is down. So the notion of a bet is at least meaningful in the Everettian context.

Sceptic: The benefit isn't to the agent: it's to copies of that agent in the future.

Author: Sure. But that's true in the non-Everettian just as in the Everettian case. In either case, the reason the agent makes one choice rather than another is because of his concern about his future interests—that is, about the interests of his future self or selves.

And an agent's future self *is* his future self just by virtue of the causal, structural, dynamical relations between it and the agent's past self. There is (I assume!) no indivisible, immaterial soul which passes through my life and magically makes me a single being: what makes the stages of me at different times all *me* is that they are appropriately

---

[2] This is a simplification: a more general statement is that an agent's subjective probability in X conditional on the objective probability of X being p should in turn be p (this is known as the principal principle, following Lewis [1980]). My main point stands, however: the principal principle is postulated, not derived.

related. And it seems, at least, that an Everettian agent's future selves
stand in all the same relations to him as a non-Everettian agent's future
selves stand to *him*.

Sceptic: There's a pretty obvious disanalogy, though. In the Everettian case,
there's more than one future self!

Author: Fair enough. (Though there are some subtleties here: see Saunders and
Wallace [2008], and also Saunders in Chapter 6, for a construal of
personal identity in which this is not the case.) But it's hard to see
why, in the Everett case, I should regard my future self as any the
less *me*—why I should not treat his goals and desires, his hopes and
dreams, as my own—just because I actually have multiple such selves.
And so it's hard to see why those future selves should be less relevant to
my considerations now—to my decision-theoretic preferences—than
would be the case in the absence of Everettian branching.

So an Everettian agent can be in a decision problem—can be faced with a
choice of bets—just as can a non-Everettian. And certainly *one* strategy available
to him is what might be called the 'Born-rule strategy': choose that bet which
maximizes expected utility with respect to the Born-rule weights (that is, the mod-
squared amplitudes). An Everettian agent who adopted the Born-rule strategy
would make exactly the same choices between bets as would a non-Everettian
who adopted the principal principle with respect to the Born-rule weights. The
two would be indistinguishable in terms of their behavioural dispositions.

Is it the *only* strategy? Advocates of Deutsch's decision-theoretic strategy say
that it is. More precisely, they argue that given certain principles of rationality,
and given knowledge of quantum mechanics, it can be proved that any strategy
other than the Born-rule strategy violates some rational constraint on action. By
analogy with the Representation Theorems of classical decision theory, we might
call such a result a (purported) *Quantum Representation Theorem*.

Such a theorem can in fact be proved formally, and in Sections 7–8 I will
give a formal proof of such a theorem. But since a fully formalized proof of a
result does not make for accessible reading, firstly I will give an *informal* version.
In Sections 4–5 I will state informally, and motivate, the axioms I wish to use;
in Section 7 I will argue informally why these axioms jointly entail a quantum
representation theorem.

4 THE QUANTUM DECISION PROBLEM

The situation I wish to consider is the following. A quantum state is to be
prepared in some superposition; the system is measured in some basis; a bet is
made by the agent on the outcome of that measurement. Our agent knows (we
assume) that the Everett interpretation is correct; he is also assumed to know

the universal quantum state, or at least the state of his branch. (The latter is an unrealistic but convenient assumption; in practice, however, it suffices for the agent to know the mod-squared amplitudes for each outcome of a measurement.) His preferences can be represented by an ordering relation on these bets.

Since (in Everettian quantum mechanics, at any rate) preparations, measurements, and payments made to agents are all physical processes, there is a certain simplification available: any preparation-followed-by-measurement-followed-by-payments can be represented by a single unitary transformation. So our agent's rational preference is actually representable by an ordering on unitary transformations.

We should acknowledge that not all unitary transformations represent something physically possible.[3] In particular, transformations which lead to *recoherence*—that is, to Everett branches merging—are certainly not performable by any agent localized to a specific branch. But nonetheless we will consider a fairly wide set of transformations to be available—exactly how wide is something that the axioms will spell out. (It might be worth recalling at this stage that decision theory is concerned with the preferences an agent would have when confronted with a particular decision—his dispositional preferences, in philosophers' language—and not just with what actually happens. It is most unlikely that I will be offered a choice between the presidency of the World Bank and the deputy leadership of the al-Qaeda terror group, but I have a definite preference between the two. As such, the assumption of a reasonably wide set of transformations seems reasonable enough.)

We should also acknowledge in our decision-theoretic set-up that decoherence imposes a certain structure on the Hilbert space. We can represent this by a resolution of the identity on the Hilbert space: that is, by a decomposition of the space into subspaces, with each subspace π corresponding to a possible macrostate. The choice of macrostates is largely fixed by decoherence, although the precise fineness of the grain of the decomposition is underspecified. (In the model, of course, it will be precisely specified, but this just illustrates that the model is artificially precise.) We call a macrostate *available* to an agent if there is an available act which, when performed, leaves some of his future selves in that macrostate.

Part of the point of the decomposition into macrostates is that an agent can be assumed not to care exactly what the microstate is within a given macrostate (if he does care, we have defined the macrostates too coarsely). But in fact, usually an agent will also be indifferent between a great many macrostates: for instance,

---

[3] Doesn't *only one* unitary transformation represent something physically possible? Doesn't the Hamiltonian of the universe uniquely determine which transformation is performed? If this is a problem, it is not specific to Everett: it is the ancient debate of free will vs. determinism. Rather than get into this morass (though I recommend Dennett [1984] for reassurance that the two are compatible), let me just note that we can talk about *rational strategies* even if an individual agent is not free to choose whether or not his strategy is rational.

if offered a million dollars, I am indifferent as to the colour of the cheque.[4] It will be useful to consider a coarse-graining of the macrostate subspaces into *reward subspaces*, such that an agent's only preference is to which reward subspace he is in. Formally speaking, 'reward subspace' is a derived concept within the decision theory.

In fact, for mathematical reasons it will be convenient to work both with the set of macrostates and with the Boolean algebra $\mathcal{E}$ of arbitrary disjunctions of macrostates,[5] which we call the *event space*. The formal development of the theory will not actually require the assumption that the event space can be constructed from a set of macrostates (though it does not rule out that assumption). Indeed, since the fineness of grain of branches is indeed underspecified, the branch structure might be best idealized in some particular situation by a model in which the algebra is not constructed this way. For instance, if the Hilbert space is $L^2(R^N) \otimes \mathcal{H}_E$, where $\mathcal{H}_E$ represents some subsystem of environmental degrees of freedom, then we might wish to take the elements of $\mathcal{E}$ to be the subspaces

$$\Sigma_E = \{ f \otimes v : E \text{ is an open subspace of } R^N \text{ and } f \text{ has support in E} \} \quad (3)$$

which cannot be generated from macrostates (unless we are willing to relax rigour and consider eigenstates of position).

For simplicity, we will refer to the set of unitary transformations over which an agent's preference order is defined as *acts*. A different set may be relevant for different physical states of the universe, so we will have cause to speak of the *acts available at* a macrostate π. (In view of the previous paragraph's comment, we might do better to talk of the acts that are *contemplatable* at π; I avoid this terminology mostly because it's cumbersome.)

In fact, it will be simpler to talk of which acts are available at a given event (not just a macrostate)—informally an act available at an event $E = \pi_1 \vee \pi_2 \vee \cdots \vee \pi_N$ is the conditional act 'if the macrostate is actually π_i, perform U_i'. This makes it much more straightforward to talk about the composition of acts: if U is available at an event E, and V is available at the smallest event containing the range of V, for instance, then VU ought to be the act of performing U and V sequentially and so also should be available at E. In the formal development we will state explicit rules to ensure that these and similar compositions are available; for now we take it as tacit that they are.

We now need to represent the agent's preferences between acts. Since those preferences may well depend on the state, we write it as follows: if the agent prefers (at ψ) act U to act U', we write

$$\widehat{U} \succ^\psi \widehat{U}'. \quad (4)$$

---

[4] The reader who doubts this claim is encouraged to test it empirically.
[5] Recall that the disjunction $E \vee F$ of two subspaces of a Hilbert space is the closure of the span of their union.

To be meaningful, of course, this requires that U and U' are both available at ψ's macrostate. So $\succ^\psi$ is to be a two-place relation on the set of acts available at that macrostate. In the event formalism we use later, we will require $\succ^\psi$ to be a two-place relation on the acts available at each event which contains ψ.

So much for the set-up; now for the axioms. They come in two categories: axioms of *richness*, which concern which acts are available to the agent (how rich the structure of the set of acts is) and which are not connected to a particular agent's preference order; and axioms of *rationality*, which constrain that preference order.

The richness axioms, then, are:

Reward availability: All rewards are available to the agent at any macrostate: that is, the set of available acts always includes ones which give all of the agent's future selves the reward.

Branching availability: Given any set of positive real numbers $p_1, \ldots, p_n$ summing to unity, an agent can always choose some act which has n different macrostates as possible outcomes, and gives weight p_i to the ith outcome.

Erasure: Given a pair of states $\psi \in E$ and $\varphi \in F$ in the same reward, there is an act $\widehat{U}$ available at E and an act $\widehat{V}$ available at F such that $\widehat{U}\psi = \widehat{V}\varphi$.

Problem continuity: For each event E, the set of acts available at E is an open subset of the set of unitary transformations from E to $\mathcal{H}$.

These should mostly be uncontroversial. Branching availability and reward availability are consequences of the relatively stylized decision problem we are considering, where measurements are being made and payments are being provided; they reflect the facts (respectively) that quantum systems can be prepared in arbitrary states and that envelopes of cash can always be given to people.

Erasure is slightly more complicated. It effectively guarantees that an agent can just forget any facts about his situation that don't concern things he cares about (that, is, by definition: that don't concern where in the reward space he is). In thinking about it, it helps to assume that any reward space has an 'erasure subspace' available (whose states correspond to the agent throwing the preparation system away after receiving the payoff but without recording the actual result of the measurement, say). An 'erasure act' is then an act which takes the quantum state of the agent's branch into the erasure subspace; the agent is (by construction) indifferent to performing any erasure act, and since he lacks the fine control to know which act he is performing, all erasures should be counted as available if any are. It follows that, since for any two such agents all erasures are available, in particular there will be two erasures available satisfying the axiom.

I postpone a discussion of problem continuity until the axioms of rationality have been introduced.

5 THE DICTATES OF RATIONALITY

Moving on to the rationality axioms, they come in two groups. The first two axioms are very general principles of rationality, as relevant in the classical as in the quantum context.

Ordering: The relation $\succeq^\psi$ is a total ordering for each ψ on the set of acts available at ψ, for each ψ (that is: it is transitive, irreflexive, and asymmetric, and if we define $U \sim^\psi V$ as holding whenever $U \succ^\psi V$ and $V \succ^\psi U$ fail to hold, then $\sim^\psi$ is an equivalence relation).

Diachronic consistency: If U is available at ψ, and (for each i) if in the ith branch after U is performed there are acts V_i, V_i' available, and (again for each i) if the agent's future self in the ith branch will prefer V_i to V_i', then the agent prefers performing U followed by the V_is to performing U followed by the V_i's.

Ordering is utterly familiar (indeed, built in to our use of the $\succ^\psi$ symbol) and hopefully uncontroversial. But it is worth stressing that the *reason* it is uncontroversial is not (just!) that it would be unintuitive for an agent's preferences to violate ordering, but because it isn't even possible, in general, for an agent to formulate and act upon a coherent set of preferences violating ordering.

Of course, in stylized and artificial special cases, it might be. If an agent knows that he will be offered three acts chosen from a set of ten, he can arbitrarily pick one element from each three-element subset, and elect to choose that one. But of course, real decision problems aren't that cleanly specified: the precise number of acts available is vague or just indeterminate and the cognitive cost of trying to pin down the size of that set is prohibitive (even when the very act of trying to pin it down does not change the problem out of recognition). Excluding stylized and occasional exceptions, then, ordering is *constitutive* of rationality, not just intuitively necessary for it.

I have stressed this because, in fact, very much the same defence can be offered of the less-familiar diachronic consistency principle, which in effect rules out the possibility of a conflict of interest between an agent and his future selves. In philosophy examples one often speaks of a (classical) agent as if he were a continuum of independent entities, one for each time, each having his own preference ordering. But of course actual decision-making takes place over time. An agent's actions take time to carry out; his desires and goals take time to be realized. If his preferences do not remain consistent over this timescale, deliberative action is not possible at all.

Of course, there are plenty of *localized* violations of diachronic consistency even outside the Everettian context. If I tell my friend not to let me order another glass of wine after my second, I acknowledge that my desires at that point will conflict with my desires now. But notice that such situations

(a) are generally not taken to be rational;
(b) are indeed analysed as situations of conflict, where my present self acts to prevent my future self having access to his preferred choice;
(c) are localized, taking place against a general assumption of diachronic consistency in myself and others (as when I assume that my friend's future self will indeed act on her agreement not to let me order the wine, or that the morning after the night before, I'll be glad that she did).[6]

Similarly, in a branching universe, to accept a conflict of interest between my pre-branch and post-branch selves is to cease to see them as the same person. If branching were an isolated occurrence, this might be possible: it is arguably callous to make a copy of myself and send him off to do a dangerous or disagreeable task—and, crucially for the point, to take actions designed to prevent him shirking that task, but it is not *irrational*.[7] But *Everettian* branching is ubiquitous: agents branch all the time (trillions of times per second at least, though really any count is arbitrary). In the presence of *widespread, generic* violation of diachronic consistency, agency in the Everett universe is not possible at all.

Sceptic: Stop there. You're trying to argue that rationality (agency, if you like) even makes sense in an Everett universe. You can't do that by saying that rationality is impossible unless such-and-such. Maybe there just isn't any coherent notion of rationality in the Everett interpretation?

Author: You misunderstand. I'm just saying that rationality requires diachronic consistency: that any rational strategy is a diachronically consistent strategy. So I'm constraining the space of rationally possible behaviours. If it turns out to be empty, of course, we're in trouble. But it won't: the Born-rule strategy is diachronically consistent and satisfies all the other axioms. All I'm doing is restricting (eventually to zero) the set of non-Born strategies.

Sceptic: What if the Born rule is also irrational?

Author: Which is to say: what if it violates some rationally required constraint on action? Then we're sunk. But it doesn't.

Sceptic: What about—

Author: Yes, yes, 'it's rationally required to weight each branch equally'. We'll come to that.

---

[6] For arguments that ascriptions of irrationality *only* make sense against a presumed backdrop of rationality, see Davidson [1973, 2004], Dennett [1987 pp.83–116], and Lewis [1974].

[7] See the first part of Greg Egan's novel *Permutation City* for a science-fictional exploration of the idea—but notice that its plausibility relies on the copy's actions being causally relevant to the original, something not possible in the Everettian universe.

Incidentally, the very idea of composing acts to make further acts, also presupposes diachronic consistency: only if an agent can think of future decisions he will make as *his decisions*, so that he can meaningfully make those decisions (for all that there is always some possibility that he will change his mind), does it make sense to consider composite acts.

The remaining rationality axioms are more specific to the Everettian context. Their precise statements get a bit more technical, so I phrase them fairly loosely here; as always, see Section 7 for details and for reassurance that there isn't sleight of hand going on.

Microstate indifference: An agent doesn't care what the microstate is provided it's within a particular macrostate.

Branching indifference: An agent doesn't care about branching per se: if a certain measurement leaves his future selves in N different macrostates but doesn't change any of their rewards, he is indifferent as to whether or not the measurement is performed.

State supervenience: An agent's preferences between acts depend only on what physical state they actually leave his branch in: that is, if $U\psi = U'\psi'$ and $V\psi = V'\psi'$, then an agent who prefers U to V given that the initial state is ψ should also prefer U' to V' given that the initial state is ψ'—$U \succ^\psi V$ iff $U' \succ^{\psi'} V'$.

Solution continuity: If for some state ψ $\widehat{U} \succ^\psi \widehat{U}'$, then sufficiently small permutations of $\widehat{U}$ and $\widehat{U}'$ will not change this.

Macrostate indifference is hopefully uncontroversial: it's built into the definition of macrostates, in fact (the point being that an agent can have no practical control as to what state he gets, within a particular macrostate, on familiar statistical-mechanics and decoherence grounds).

Solution continuity and branching indifference—and indeed problem continuity—can be understood in the same way, in terms of the limitations of any physically realizable agent. Any discontinuous preference order would require an agent to make arbitrarily precise distinctions between different acts, something which is not physically possible. Any preference order which could not be extended to allow for arbitrarily small changes in the acts being considered would have the same requirement. And a preference order which is not indifferent to branching per se would in practice be impossible to act on: branching is uncontrollable and ever present in an Everettian universe.[8]

Sceptic: Why assume a priori that the rational strategy must be physically possible? Even if there is some strategy in an Everettian universe which counts as rational, maybe it's not physically possible to carry out that strategy.

---

[8] The main source of branching is probably classically chaotic systems; see Zurek and Paz [1994] for technical details, and Wallace [2001] for discussion.

Author: That's confused. Firstly, we already know there's at least one possible rational strategy: the Born rule. Secondly, what would it even be for a strategy to be rational, but physically impossible? By that token, the rational strategy for a trader is 'always buy shares that are going to increase in value'.

To be fair, a strategy might be literally impossible but be an idealization of a possible strategy—after all, perfect rationality itself is an idealization. One might *possibly* relax the assumption of continuity on these grounds (and I'll make some comments on that later), though I don't really think it's justified. But no strategy can approximate caring about branch number, as we'll see.

The other way to understand these assumptions is as prohibitions on strategies that just exploit artefacts of our model. The branching structure—including the well-defined number of branches associated with any act—is derived from the set of macrostates, which is in turn derived from decoherence. But as I argued in Chapter 1 that this structure has a significant degree of arbitrariness associated with it, primarily in terms of the coarseness of the grain of the macrostates (see also James Hartle in Chapter 2). Put simply, in the actual physics there is no such thing as a well-defined branch number. Similarly, in the actual physics there is no division of the dynamics into discrete branching events followed by evolution of individual branches: branching, rather, is continuous. But if branching is always going on, and cannot be quantified in a non-arbitrary manner, then no strategy can be formulated which is other than indifferent to the presence of branching.

A quick defence of state supervenience would be: the agent's preferences supervene on the actual state of the branch; transformations which differ only in how they would affect non-actual quantum states do not differ in any relevant respect.

Sceptic: Hang on. This brings out a tacit assumption in the formalism you've adopted: the idea that acts can be represented by *single* unitary transformations rather than by *sequences* of unitary transformations. Why regard a sequence of measurements as decision-theoretically equivalent to a single measurement just because the same unitary transformation is enacted by both?

Author: Here's one possible defence. The agent is playing a sequence of games which result in rewards that he spends only after the sequence is done. In this case, what does he care about what happens during the brief period in which the games are being played (when having or not having rewards makes no difference to his status)—should he not care only about the state of the universe after the payouts are all made?

Sceptic: Well, that sounds intuitive, but so what? We're discussing *the Everett interpretation*—appeals to intuition are going to ring a little hollow here.

Author: Fair enough. A far better defence is to observe that caring about the final state only is the diachronic equivalent of branch indifference, and can be defended in the same way. There is no 'real' branching structure beyond a certain fineness of grain, so the details of that structure can only be included in terms of their coarse-grained consequences.

Put another way: we could have defined our decision theory in terms of preferences, not over final states, but over consistent history spaces. But if we had done so, we would have needed both synchronic and diachronic indifference assumptions: indifference both to the fineness of grain of the history projectors at each time, and to the size of the temporal gaps between history projectors. Translated back into our setting, where we consider sequences of decisions made only over very short periods of time, the former assumption entails branch indifference and the latter entails that acts can be represented by single unitary transformations.

6 A QUANTUM REPRESENTATION THEOREM

We can now prove, in succession, three results, the first three of which are (trivially) entailed by the fourth.

Equivalence lemma: If two acts assign the same weight to each reward, the agent must be indifferent between them.

Nullity lemma: An agent is indifferent to a possible outcome of an act iff that act has weight zero.

Dominance lemma: Suppose that two acts each only have two possible rewards r_1, r_2 as outcomes, with $r_1 \succ r_2$[9] and that the first act assigns a higher weight to r_1 than the second act does. Then the first act must be preferred to the second.

Born-rule theorem: There is a utility function on the set of rewards, unique up to affine transformations, such that one act is preferred to another iff its expected utility, calculated with respect to this utility function and to the quantum-mechanical weights of each reward, is higher.

Since all these results are proved *formally* in Section 8, my purpose in this section is explanation and not persuasion: I wish simply to show the general shape of the proof.

The equivalence lemma is best illustrated by examples (here I basically follow the argument of Wallace [2007]). For a simple case, suppose we have two acts

[9] That is, with an act which returns some microstate in r_1 with certainty preferred to one which returns some microstate in r_2 with certainty; that this determines a well-defined ordering over rewards follows from microstate indifference.

(A and B, say): in each, a system is prepared in a linear superposition $\alpha|+\rangle +$ $\beta|-\rangle$ and then measured in the $\{|+\rangle,|-\rangle\}$. On act A, a reward is then given if the result is '$+$'; on B, the same reward is given on '$-$' instead. The resultant states are

$$\text{A:} \quad \alpha|+\rangle \otimes |\text{reward}\rangle + \beta|-\rangle \otimes |\text{no reward}\rangle; \tag{5}$$

$$\text{B:} \quad \alpha|+\rangle \otimes |\text{no reward}\rangle + \beta|-\rangle \otimes |\text{reward}\rangle. \tag{6}$$

By erasure, there will exist acts available to the agent's future self in the reward branch (for both A and B) which erase the result of what was measured, leaving only the reward. Performing these transformations, and the equivalent erasures in the no-reward branch, leaves

$$\text{A-plus-erasure:} \quad \alpha|0\rangle \otimes |\text{reward}\rangle + \beta|0'\rangle \otimes |\text{no reward}\rangle; \tag{7}$$

$$\text{B-plus-erasure:} \quad \beta|0\rangle \otimes |\text{reward}\rangle + \alpha|0'\rangle \otimes |\text{no reward}\rangle. \tag{8}$$

Now, by branch indifference, the agent's future selves are indifferent to whether this erasure is or is not performed. (Branch indifference is needed because we have no guarantee that erasures are non-branching; if we did, microstate indifference would suffice.) So by diachronic consistency, the original agent is indifferent between A and A-plus-erasure, and between B and B-plus-erasure.

But now: if $\alpha = \beta$, then A-plus-erasure and B-plus-erasure leave the system in the same quantum state. So by state supervenience, the agent is indifferent between them. Since we know from ordering that preferences are transitive, the agent must also be indifferent between A and B. Indeed, we actually require only that $|\alpha| = |\beta|$, for phase differences too can be erased.

For a slightly more complicated case, suppose game C involves a two-state system being prepared in state

$$\sqrt{2/3}|+\rangle + \sqrt{1/3}|-\rangle$$

and a reward being given on '$+$', and game D involves a three-state system being prepared in state

$$\sqrt{1/3}(|+\rangle + |0\rangle + |-\rangle)$$

and a reward being given on '$+$' and on '0'. The resultant states are then

$$C : \sqrt{2/3}|+\rangle \otimes |\text{reward}\rangle + \sqrt{1/3}|-\rangle \otimes |\text{no reward}\rangle; \tag{9}$$

$$D : \sqrt{1/3}|+\rangle \otimes |\text{reward}\rangle + \sqrt{1/3}|0\rangle \otimes |\text{reward}\rangle + \sqrt{1/3}|-\rangle \otimes |\text{no reward}\rangle. \tag{10}$$

But by erasure, there is an act available for the future self of the agent in the 'reward' branch of game C which creates two equally weighted branches:

$$|+\rangle \otimes |\text{reward}\rangle \longrightarrow \sqrt{1/2}|X\rangle \otimes |\text{reward}\rangle + \sqrt{1/2}|Y\rangle \otimes |\text{reward}\rangle \qquad (11)$$

Since by branch indifference the agent's future self is indifferent to performing this act or not, by diachronic consistency the original agent is indifferent between C and C-plus-branching. But the state produced by C-plus-branching is

$$\text{C-plus-branching} : \sqrt{1/3}|X\rangle \otimes |\text{reward}\rangle + \sqrt{1/3}|Y\rangle \otimes |\text{reward}\rangle +$$
$$\sqrt{1/3}|-\rangle \otimes |\text{no reward}\rangle. \qquad (12)$$

By a generalization of our earlier argument, the agent is indifferent between C-plus-branching and D, and so between C and D.

By arguments of this kind, the equivalence lemma can be proved for any act with finitely many outcomes. The null and dominance lemmas are easy further steps, using the second clause of diachronic consistency.

We are now nearly done: the remainder of the proof is actually a standard decision-theoretic method for constructing utilities. Pick two rewards R and S with $R \succ S$, and assign R utility 1 and S utility 0. For any reward T satisfying $R \succeq T \succeq S$, there is a unique number $U(T)$ such that the agent is indifferent between getting T with certainty, and getting R on a branch of weight $U(T)$ and S otherwise. (We need continuity to establish this and rule out the possibility of rewards whose utilities differ only infinitesimally.)

Now consider an act which leads to rewards R, S, T with weights w(R), w(S), and w(T) respectively. The agent's future selves in the T branch are indifferent between doing nothing and performing an act that delivers R with weight $U(T)$ and S otherwise. Applying diachronic consistency once more, the original agent is indifferent between the original act and an act which delivers R with weight $w(R) + w(T)U(T)$ and an act which delivers S with weight $w(S) + (1 - U(T))w(T)$. Note that the utilities of these acts are the same: in this particular case, the agent is indifferent between two acts iff they have the same utility. Generalizing the argument, and applying the dominance lemma, tells us that one act is preferred to another iff its utility is higher.

The continuity axioms play only a limited role in these arguments. They serve to rule out situations where two rewards are infinitesimally, or infinitely, different in value; they are also required to handle the generalization to acts which have infinitely many rewards as possible outcomes.

7 FORMAL STATEMENT OF THE AXIOMS

As promised, in this section and the next I lay out the formal version of my decision theory and its associated proofs. The reader who is happy to take on trust my mathematics—and my reassurances that there has been no sleight of hand—is welcome to skip to Section 9.

A *quantum decision problem* is specified by:

- A separable Hilbert space $\mathcal{H}$. Given a set $\mathcal{S}$ of subspaces of $\mathcal{H}$, I write $\vee\mathcal{S}$ (the *disjunction* of $\mathcal{S}$) for the closure of the span of $\cup\mathcal{S}$, and $\wedge\mathcal{S}$ (the *conjunction* of $\mathcal{S}$) for the closure of $\cap\mathcal{S}$; Given subspaces E and F, I define $E \vee F = \vee\{E, F\}$ and likewise for $\wedge$, and I write Π_E for the projector onto E.
- A complete Boolean algebra $\mathcal{E}$ of subspaces of $\mathcal{H}$, the *event space*. (So $\mathcal{E}$ contains $\mathcal{H}$ and is closed under $\vee$, $\wedge$, and taking the complement.) I define a *partition* of an event E to be a set of mutually orthogonal events whose conjunction is E.
- A subset $\mathcal{M}$ of $\mathcal{E}$, the *macrostates*, such that for any event E, there is a partition of E by macrostates.
- For each $E \in \mathcal{E}$, a set $\mathcal{U}_E$ of unitary operators from E into $\mathcal{H}$, which we call the set of *acts available at* E. We write $\mathcal{O}_U$ for the smallest event containing the range of the act U[10] and require that the choice of available acts satisfies:
  1. *Restriction*: If $E, F \in \mathcal{E}$ and $F \subset E$, then if U is available at E then the unitary map $U|_F$, defined by $U\psi = U|_F\psi$ whenever $\psi \in F$, is available at F.
  2. *Composition*: If U is available at E, and V is available at $\mathcal{O}_U$, then VU is available at E.
  3. *Indolence*: For any event E, if there are any acts available at E then the identity $\widehat{1}_E$ is available at E. (More precisely, the embedding map of E into $\mathcal{H}$ is available at E.)
  4. *Continuation*: If U is available at some E, then there is some act available at $\mathcal{O}_U$.
  5. *Irreversibility*: If U is available at $E \vee F$, $\mathcal{O}_{U|_E} \wedge \mathcal{O}_{U|_F} = \emptyset$.
- A partition $\mathcal{R}$ of $\mathcal{E}$ (that is, a set of mutually orthogonal elements of $\mathcal{E}$ whose disjunction is $\mathcal{H}$), the set of *rewards*. These represent payoffs an agent could get.

The simplest choice of macrostates and event space is to pick some particular set of orthogonal subspaces of $\mathcal{H}$ whose disjunction is $\mathcal{H}$, take this as $\mathcal{M}$, and take $\mathcal{E}$ to be the set of all disjunctions of subsets of $\mathcal{M}$; this is the sense of 'macrostate' and 'event' used in the informal version of the proof. However, we could equally well take $\mathcal{E}$ to be an arbitrary Boolean algebra of subspaces and define $\mathcal{E} = \mathcal{M}$. (As was noted previously, this sort of formalization might be more appropriate for decision problems with a less natural discrete structure.)

Rays within $\mathcal{H}$, as usual, are called states. I adopt the usual convention of representing a ray by any vector within it and of blurring the distinction between

---

[10] We can define $\mathcal{O}_U$ explicitly as the conjunction of all events containing the range of U; this suffices to show that $\mathcal{O}_U$ is well defined.

the two; I do not require that vectors representing states be normalized. (This is just for notational convenience.) If $\mathcal{B}(E, \mathcal{H})$ is the set of unitary maps from E into $\mathcal{H}$, it can naturally be regarded as a subset of $\mathcal{B}(\mathcal{H}, \mathcal{H})$ by identifying U with $U\Pi_E$; as such, $\mathcal{B}(E, \mathcal{H})$ inherits the norm topology.

I introduce a few derived concepts. The *weight* $\mathcal{W}_\psi(E|U)$ of an event E with respect to a state ψ and an act U is defined by

$$\mathcal{W}_\psi(E|U) = \|\Pi_E U|\psi\rangle\|^2 = \langle\psi|U^\dagger\Pi_E U|\psi\rangle. \tag{13}$$

A *reward function* is any function from $\mathcal{R}$ to $[0, 1]$ such that $\sum_{r\in\mathcal{R}} w(r) = 1$. Any pair of a state $\psi \in E$ and an act U available at E determines a reward function

$$R_{\psi,U}(r) = \mathcal{W}_\psi(r|U) \tag{14}$$

which I call the *characteristic reward function* of U and ψ.

A set $\mathcal{F}$ of events is *available* if they are mutually orthogonal and there is at least one act available at $\vee\mathcal{F}$. (An event is available iff its singleton set is available.)

Finally, if $\mathcal{S}$ is any set of rewards, I say that an act A *has rewards in* $\mathcal{S}$ iff its range is a subset of $\vee\mathcal{S}$. If u is a real function of $\mathcal{S}$, and U is an act whose rewards are in $\mathcal{S}$, the *expected utility* of U with respect to a state ψ (and, tacitly, with respect to u) is

$$\mathrm{EU}_\psi(U) = \sum_{r\in\mathcal{S}} \mathcal{W}_\psi(r|U)u(r) \equiv \sum_{r\in\mathcal{S}} R_{\psi,U}(r)u(r). \tag{15}$$

Stating the richness axioms is a little fiddly, because of the need to make sure not only that certain acts (erasures, branchings etc.) are available everywhere, but to make sure that they are available on multiple branches concurrently. To state them in a concise way, I make the following definitions. First, if $\mathcal{P} = \{p_1, p_2, \ldots\}$ is a (countable or finite) set of positive real numbers whose sum is unity, and $\psi \in M \subset r$ for some state ψ, macrostate M, and reward r, then a $\mathcal{P}$-*branching* of ψ is some act U available at M such that $\mathcal{O}_U \subset r$ and such that there is a partition $\mathcal{M} = \{M_1, M_2, \ldots\}$ of $\mathcal{O}_U$ by macrostates with $\mathcal{W}_\psi(M_i|U) = p_i$. (Informally, a $\mathcal{P}$-branching is an act which splits the agents branch into many branches, each having the same weight as an element of $\mathcal{P}$, but without changing the rewards that the agent gets.)

Second, if M and M' are macrostates with $M \subset r$ and $M' \subset r$ for some reward r, and ψ, ψ' are states in M, M' respectively, then an *erasure* of ψ and ψ' is a pair of acts U, U' available at M and M' respectively, such that $\mathcal{O}_U$ and $\mathcal{O}_{U'}$ are both subsets of r and $U\psi = U'\psi'$.

And third, if $\mathcal{F}$ is an available set of events, an *act function* $\mathcal{U}$ for that set is a function which assigns to each $F \in \mathcal{F}$ an act $\mathcal{U}(F)$ available at F. An act function is *compatible* if

$$\sum_{F \in \mathcal{F}} \mathcal{U}(F) \Pi_F \tag{16}$$

is available at $\vee \mathcal{F}$.

The richness axioms are now stateable:

Reward availability: Suppose that $\mathcal{F}$ is an available set of macrostates and f is a function from $\mathcal{F}$ into rewards.
Then there is a compatible act function $\mathcal{U}$ for $\mathcal{F}$ with $\mathcal{U}(F) \subset f(F)$ for all $F \in \mathcal{F}$.

Branching availability: Suppose that $\mathcal{F}$ is an available set of macrostates and for each $F \in \mathcal{F}$, ψ_F is a non-zero state in F, and $\mathcal{P}_F$ is a (finite or countable) set of positive real numbers summing to unity.
Then there is a compatible act function $\mathcal{U}$ for $\mathcal{F}$ such that, for each $F \in \mathcal{F}$, $\mathcal{U}(F)$ is a $\mathcal{P}_F$-branching of ψ_F.

Erasure: Suppose that $\{r_1, r_2, \ldots\}$ is a (finite or countable) set of rewards, that $\mathcal{M} = \{M_1, M_2, \ldots\}$ and $\mathcal{N} = \{N_1, N_2, \ldots\}$ are two available sets of macrostates with $M_i \subset r_i$ and $N_i \subset r_i$, and that for each i, $\psi_i \in M_i$ and $\varphi_i \in N_i$ are non-zero states.
Then there are compatible act functions $\mathcal{U}$ for $\mathcal{M}$, and $\mathcal{V}$ for $\mathcal{N}$ such that, for each i, $(\mathcal{U}(M_i), \mathcal{V}(N_i))$ is an erasure of ψ_i and φ_i.

Problem continuity: For every available E, the set of acts available at E is an open subset (in operator norm topology) of the set of unitary maps from E to $\mathcal{H}$.[11]

Notice that reward availability and preparation together entail that, for any reward function and any $\psi \in E$, there is an act U available at E such that ψ and U have that reward function as their characteristic reward function.

We now define a *state-dependent solution* to a decision problem as specified by an assigment to every available macrostate E, and every state $\psi \in E$, of a two-place relation $\succ^\psi$ on the acts available at E. (Strictly our notation should include E but for simplicity, its value will always be tacit.)

We call an event N *null* for a given state ψ and act U iff, whenever acts V_1 and V_2 are identical on the complement of N, $V_1 U \sim^\psi V_2 U$. (So an event

---

[11] The operator norm topology on the set of linear maps between normed spaces V and W is defined by the norm $\|U\| = \sup\{\|Ux\| : \|x\| = 1\}$. The set of unitary maps from E to $\mathcal{H}$ is a subset of the set of all maps between those two spaces, and inherits the latter's topology.

is null if the agent doesn't care what happens to his future selves, if any, in the branch defined by that event. We will shortly see that, as expected, an event is null iff there are in fact no such future selves.) It is easy to see that any finite union of null sets is null, as is any subset of a null set.

We can now state the rationality axioms:

Ordering: For every ψ for which it is defined, $\succ^\psi$ is a total ordering. That is: it is transitive, asymmetric, and the relation $\sim^\psi$, defined by $E \sim^\psi F$ iff neither $E \succ^\psi F$ nor $F \succ^\psi E$, is an equivalence relation. (As usual, we write '$E \succeq^\psi F$' as an abbreviation for 'either $E \succ^\psi F$ or $E \sim^\psi F$'.)

Diachronic Consistency: Suppose U is available at E, and V_1 and V_2 are available at $\mathcal{O}_U$. Then:

   (i) If there is some partition $\mathcal{P}$ of $\mathcal{O}_U$ into macrostates such that $V_1|_E \succeq^{\Pi_E U\psi} V_2|_E$ for every element E of the partition not null with respect to ψ and U, then $V_1 U \succeq^\psi V_2 U$.

  (ii) If, in addition, $V_1|_E \succ^{\Pi_E U\psi} V_2|_E$ for at least one such E, then $V_1 U \succ^\psi V_2 U$.

Macrostate indifference: If:

- U, V are acts available at M;
- U', V' are acts available at M';
- $\mathcal{O}_U \subset M_1 \wedge r_1$ and $\mathcal{O}_{U'} \subset M_1 \wedge r_2$ for some macrostate M_1 and reward r_1;
- $\mathcal{O}_V \subset M_2 \wedge r_2$ and $\mathcal{O}_{V'} \subset M_2 \wedge r_2$ for some macrostate M_2 and reward r_2

then for any ψ, ψ' with $\psi \in M$ and $\psi' \in M'$, $U \succeq^\psi V$ iff $U' \succeq^{\psi'} V'$.

Branching indifference: If:

- r is a reward;
- M is a macrostate with $M \subset r$;
- U is available at M;
- $\psi \in M$ and $U\psi \in r$

then $U \sim^\psi \widehat{1}_M$.

State supervenience: If:

- $\psi \in E$ and $\psi' \in E'$ for macrostates E, E';
- U and V are available at E, and U' and V' are available at E';
- $U\psi = U'\psi'$ and $V\psi = V'\psi'$

then $U \succ^\psi V$ iff $U' \succ^{\psi'} V'$.

Solution continuity: If E is a macrostate and $\psi \in E$, if U, U' are available at E, and if $U \succ^\psi U'$, then in the space of unitary maps from E into $\mathcal{H}$ there are neighbourhoods (in norm topology) $\mathcal{N}, \mathcal{N}'$ of U, U' respectively such that any act in $\mathcal{N}$ available at E is preferred (at ψ) to any act in $\mathcal{N}'$ available at E.

Given a solution to a quantum decision problem, we can use it to define a preference ordering on rewards: for any two rewards, $r_1 \succ r_2$ iff there is some macrostate E, some state $\psi \in E$, and acts U_1, U_2 available at E such that $\mathcal{O}_{U_i} \subset r_i$ and $U_1 \succ^\psi U_2$. Provided that the problem is reward-available and the solution is macrostate-indifferent and branching-indifferent, this preference order is a total ordering on $\mathcal{R}$. If r and s are rewards with $r \preceq s$, I will say that a reward t is between r and s iff $s \succeq t \succeq r$; I write $[r, s]$ for the set of rewards between r and s.

If $\mathcal{M}$ consists of some set of orthonormal subspaces (as in the informal proof), then this observation more or less exhausts the usefulness of macrostate indifference. At the other extreme, if $\mathcal{M} = \mathcal{E}$ then macrostate indifference actually entails branch indifference. The distinction between the axioms, then, is a matter of how we mathematically represent the branching structure—which is appropriate, since the motivation for branching indifference itself is that the details of that structure are an unphysical artefact of the mathematics.

(The mathematically inclined reader may be wondering at this point if the axioms are consistent. To show that they are, consider the following model. Let $\mathcal{H}_R$ be a two-dimensional Hilbert space with an orthogonal basis $\{|+\rangle, |-\rangle\}$; for each $N > 0$ let $\{\mathcal{H}_N\}$ be an N-dimensional Hilbert space with an orthonormal basis $\{|N, 1\rangle, |N, 2\rangle, \ldots |N, N\rangle\}$.

Now: take the Hilbert space of our decision problem to be

$$\mathcal{H} = \mathcal{H}_R \otimes \left(\oplus_{I=1}^\infty \mathcal{H}_I\right), \tag{17}$$

so that a complete basis of states is

$$|\pm\rangle \otimes |N, M\rangle \quad (M \leq N), \tag{18}$$

and take the macrostates to consist of all the one-dimensional subspaces spanned by each of these states, and the events to be all disjunctions of macrostates. The available events are all those which are contained in some fixed $\mathcal{H}_R \otimes \mathcal{H}_N$, and the acts available at an available event contained in $\mathcal{H}_R \otimes \mathcal{H}_N$ are all unitary maps from $\mathcal{H}_R \otimes \mathcal{H}_N$ to $\mathcal{H}_R \otimes \mathcal{H}_{N'}$, with $N' > N$. The reward subspaces are $\mathcal{H}^\pm = \{\text{Span} |\pm\rangle\} \otimes \mathcal{H}$. Finally, an act U is preferred to an act U' at $|\psi\rangle$ iff

$$\left\| (|+\rangle\langle+| \otimes \widehat{1}) U |\psi\rangle \right\| > \left\| (|+\rangle\langle+| \otimes \widehat{1}) U' |\psi\rangle \right\|. \tag{19}$$

I leave readers to satisfy themselves that this system does indeed obey the axioms; the preference order is, of course, the Born rule.)

8 FORMAL STATEMENT AND PROOF OF THE REPRESENTATION THEOREM

Equivalence lemma: Suppose that:

(i) $\mathcal{P}$ is a quantum decision problem satisfying erasure, branch availability, and reward availability;

(ii) $\succ^{\psi}$ is a state-dependent solution to $\mathcal{P}$ satisfying ordering, diachronic consistency, macrostate indifference, branching indifference, and state supervenience;

(iii) U and V are available at E, and U' and V' are available at E';

(iv) $\psi \in E$ and $\psi' \in E'$;

(v) $R_{\psi,U} = R_{\psi',U'}$ and $R_{\psi,V} = R_{\psi',V'}$.

(vi) The reward functions of the acts are each non-zero for only finitely many rewards

then $U \succ^{\psi} V$ iff $U' \succ^{\psi'} V'$.

Proof: For each reward r for which $R_{\psi,U}(r) \neq 0$, let $\mathcal{M}_r$ and $\mathcal{N}_r$ be partitions of $\mathcal{O}_U \wedge r$ and $\mathcal{O}_{U'} \wedge r$ respectively, and let $\#M_r$ and $\#N_r$ be the number of elements (finite or infinite) in $\mathcal{M}_r$ and $\mathcal{N}_r$ respectively.

Define the sets $\mathcal{P}_r$ (for each r)

$$\mathcal{P}_r = \{\mathcal{W}_{\psi'}(N|U')/\mathcal{W}_{\psi'}(r|U') : N \in \mathcal{N}_r\}. \qquad (20)$$

These are sets of positive real numbers summing to unity, so by branching availability there is an act W available at $\mathcal{O}_U$ such that, for each r and each $M \in \mathcal{M}_r$, $W|_M$ is a $\mathcal{P}_r$-branching of $\Pi_M U\psi$: it splits $\Pi_M U\psi$, which has weight $\mathcal{W}_{\psi}(M|U)$, into $\#N_r$ states, one for each $N \in \mathcal{N}_r$, with weights $\mathcal{W}_{\psi}(M|U) \times \mathcal{W}_{\psi'}(N|U')/\mathcal{W}_{\psi'}(r|U')$. There is therefore[12] a partition $\mathcal{W}$ of $\mathcal{O}_W$ into macrostates, such that:

- For each reward r there are $\#M_r \times \#N_r$ elements of $\mathcal{W}$ in r.
- Each such element can be labelled by pairs of elements from $\mathcal{M}_r$ and $\mathcal{N}_r$: let us write it as $K^r_{M,N}$.
- $\mathcal{W}_{\psi}(K^r_{M,N}|WU) = \mathcal{W}_{\psi}(M|U) \times \mathcal{W}_{\psi'}(N|U')/\mathcal{W}_{\psi'}(r|U')$.

Furthermore, by branching indifference, $W|_M \sim^{\Pi_M U\psi} \widehat{1}_M$ for any macrostate M, and hence by diachronic consistency, $WU \sim^{\psi} U$.

---

[12] We appeal here to the irreversibility requirement on decision problems.

Applying the same procedure with U and U' reversed, yields an act W' such that $W'U \sim^\psi U$, and a partition $\mathcal{W}$ of $\mathcal{O}_{W'}$ by macrostates, such that:

- For each reward r there are $\#\mathcal{M}_r \times \#\mathcal{N}_r$ elements of $\mathcal{W}$ in r.
- Each such element can be labelled by pairs of elements from $\mathcal{M}_r$ and $\mathcal{N}_r$: we write it as $K'^r_{M,N}$.
- $\mathcal{W}_{\psi'}(K'^r_{M,N}|WU) = \mathcal{W}_\psi(M|U) \times \mathcal{W}_{\psi'}(N|U')/\mathcal{W}_\psi(r|U)$.

But since

$$\mathcal{W}_\psi(r|U) \equiv \mathcal{R}_{\psi,U}(r) = \mathcal{R}_{\psi',U'}(r) \equiv \mathcal{W}_{\psi'}(r|U'), \qquad (21)$$

it follows that $\mathcal{W}_\psi(K'^r_{M,N}|WU) = \mathcal{W}_{\psi'}(K'^r_{M,N}|W'U')$.

So we have constructed acts W, W', and partitions $\mathcal{W} = \{W_1, \dots\}$, $\mathcal{W} = \{W'_1, \dots\}$ of $\mathcal{O}_W$, $\mathcal{O}_{W'}$ by macrostates such that:

1. For any i, W_i exists iff W'_i does (i.e., the two partitions have the same number of elements) and there is some reward r such that W_i and W'_i are elements of r.
2. $\mathcal{W}_\psi(W_i|WU) = \mathcal{W}_\psi(W'_1|W'U)$ for all W_i.

Now define

$$\chi_i = \Pi_{W_i} WU\psi/\|\Pi_{W_i} WU\psi\| \qquad (22)$$

and

$$\chi'_i = \Pi_{W'_i} W'U'\psi'/\|\Pi_{W'_i} W'U'\psi'\|. \qquad (23)$$

By erasure, there exist acts X, X' available at $\mathcal{O}_W$, $\mathcal{O}_{W'}$ such that $(X|_{W_i})\chi_i = (X'|_{W'_i})\chi'_i$. By branching indifference, $X|_{W_i} \sim^{\chi_i} \widehat{1}_{W_i}$, so by diachronic consistency, $XWU \sim^\psi WU \sim^\psi U$; similarly, $X'W'U' \sim^{\psi'} U'$.

Since

$$XWU\psi = \sum_i \mathcal{W}_\psi(W_i|WU)(X|_{W_i})\chi_i, \qquad (24)$$

it follows that $XWU\psi = X'W'U'\psi'$.

So: for U and U', we have found acts $Y = XWU$ and $Y' = X'W'U'$ such that $U \sim^\psi Y$, $U' \sim^{\psi'} Y'$, and $Y\psi = Y'\psi'$. Repeating this process for V and V', we can find acts Z, Z' such that $Z \sim^\psi V$, $Z' \sim^{\psi'} V'$, and $Z\psi = Z'\psi'$. The conclusion now follows immediately from state supervenience. $\square$

Because of the equivalence lemma, there is a unique total ordering defined on the set of all reward functions, which we once again write as $\succ$ (note that it is state-independent).

Nullity Lemma: Suppose that:

 (i) $\mathcal{P}$ is a quantum decision problem satisfying erasure, branch availability, and reward availability;
 (ii) $\succ^{\psi}$ is a state-dependent solution to $\mathcal{P}$ satisfying ordering, diachronic consistency, macrostate indifference, branching indifference, and state supervenience;
 (iii) There exist rewards r, s with $r \succ s$.

Then an event E is null with respect to a state ψ and an act U iff $\langle\psi|U^{\dagger}\Pi_{E}U|\psi\rangle = 0$.

Proof: Let $\langle\psi|U^{\dagger}\Pi_{E}U|\psi\rangle = a$. An event is null if and only if, given acts V and W available at $\mathcal{O}_{U}$ which are identical except on E, $VU \sim^{\psi} WU$. Given the equivalence lemma, any two such acts are equivalent whenever they have the same weight function, so if E is null for ψ and U, any event E' is null with respect to some U' and ψ' whenever $\langle\psi'|U'^{\dagger}\Pi'_{E}U'|\psi'\rangle = a$. If $a > 0$, then $a > 1/N$ for some N. By combining branch availability with reward availability, we can construct some act V and state φ with weight function

$$\mathcal{W}_{\varphi}(E_1|V) = 1/N$$
$$\mathcal{W}_{\varphi}(E_2|V) = a - 1/N$$
$$\mathcal{W}_{\varphi}(E_3|V) = 1 - a.$$

$E_1 \vee E_2$ is null (wrt φ and V), hence E_1 is, hence any event with weight $1/N$ is. Applying branch availability and reward availability again, we can find φ', W and $F_1, \ldots F_N$ such that $\mathcal{W}_{\varphi'}(F_i|W) = 1/N$. Each F_i is null wrt φ' and W, hence so is $\mathcal{E}$. This contradicts premise (iii), since if all events are null then all rewards are equivalent.

Conversely, suppose that some event has weight zero. Its nullity now follows from state supervenience, since no change to the physical state is enacted by any transformation restricted to that event. $\square$

Dominance lemma: Suppose that:

 (i) $\mathcal{P}$ is a quantum decision problem satisfying erasure, branch availability, and reward availability;
 (ii) $\succ^{\psi}$ is a state-dependent solution to $\mathcal{P}$ satisfying ordering, diachronic consistency, macrostate indifference, branching indifference, and state supervenience;
 (iii) s, t are rewards with $s \succ t$;

(iv) $f[a]$ is the reward function defined by $f[a](s) = a, f[a](t) = 1 - a, f[a](r) = 0$ for all other r.

Then $f[a] \succ f[\beta]$ iff $a > \beta$.

Proof: This is an easy corollary of the nullity lemma. Suppose $a > \beta$, then by branch availability and reward availability, there will be some act A and state φ with weight function

$$W_\varphi(E_1|A) = \beta$$
$$W_\varphi(E_2|A) = a - \beta$$
$$W_\varphi(E_3|A) = 1 - a.$$

By reward availability there exist sets of compatible acts $\{U_1, U_2, U_3\}$ and $\{V_1, V_2, V_3\}$ such that U_i and V_i are available at E_i, and such that U_1, V_1, and U_2 have outcomes all lying in s and V_2, U_3, and V_3 have outcomes all lying in t. By macrostate indifference and branching indifference $U_i \simeq^\chi V_i$ for any $\chi \in E_i$ and in particular $U_2 \succ^\chi V_2$ for any $\chi \in E_2$.

If we define

$$W_a = U_1 \Pi_{E_1} + U_2 \Pi_{E_2} + U_3 \Pi_{E_3} \tag{25}$$

and

$$W_\beta = V_1 \Pi_{E_1} + V_2 \Pi_{E_2} + V_3 \Pi_{E_3} \tag{26}$$

then by diachronic consistency, since E_2 is not null then $W_a \cdot A \succ^\psi W_\beta \cdot A$. But the reward functions of $W_a \cdot A$ and $W_\beta \cdot A$ are $f[a]$ and $f[\beta]$ respectively, and the conclusion follows.

Utility lemma: Suppose that:

(i) $\mathcal{P}$ is a quantum decision problem satisfying erasure, branch availability, and reward availability;

(ii) $\succ^\psi$ is a state-dependent solution to $\mathcal{P}$ satisfying ordering, diachronic consistency, macrostate indifference, branching indifference, and state supervenience;

(iii) s, t are rewards with $s \succ t$;

(iv) u_s, u_t are real numbers with $u_s > u_t$.

Then there is a unique real function u on the set $[t, s]$ of rewards between t and s such that for any macrostate E, any state $\psi \in E$, and any two acts U, V available at E whose rewards lie a finite subset of $\mathcal{S}$,

$$U \succ_\psi V \text{ whenever } EU_\psi(U) > EU_\psi(V) \tag{27}$$

(where the expected utilities are defined with respect to u, of course) and such that $u(s) = u_s$ and $u(t) = u_t$.

Proof: For simplicity we assume $u_s = 1$ and $u_t = 0$ (other values lead to a simple affine transformation of the utility function). We define the following reward functions: $f[a]$ is defined as in the dominance lemma, and $g[r]$ is defined by $g[r](r') = \delta_{r,r'}$.

We now define $u(r)$ by

$$u(r) = \text{lub}\{a : g[r] \succ f[a]\}. \tag{28}$$

Let $\{u_n(r)\}$ be a sequence of functions such that $u_m(r) \leq u(r)$ and $\lim_{n\to\infty} u_n(r) = u(r)$, and let U be any act available at E whose rewards lie in S. We write E_r for $\mathcal{O}_U \wedge r$ and χ_r for the normalized projection of ψ onto E_r.

From branching availability and reward availability, for each n we can find a compatible set of states $\{A_n(r) : R_{\psi,U}(r) \neq 0\}$ such that $A_n(r)$ is available at E_r and A_n has reward function $f[u_n(r)]$; we define $\mathcal{A}_n = \sum_{r \in S} A_n(r) \Pi_{E_r}$. By construction, $\widehat{1}_{E_r} \succeq^{\chi_r} A_n(r)$ for all r and n, so by diachronic consistency $U \succeq^{\psi} \mathcal{A}_n \cdot U$.

By definition, the reward function of $\mathcal{A}_n \cdot U$ (with respect to ψ) is $f[\lambda_n]$, where

$$\lambda_n = \sum_{r \in S} \mathcal{W}_\psi(r|U) u_n(r). \tag{29}$$

So if $f[U]$ is the reward function of U (with respect to ψ), we have established that $f[U] \succeq f[\lambda_n]$, and hence by the dominance lemma, $f[U] \succeq f[\lambda]$ whenever $\lambda < \lambda_n$ for some n. Since $u_n(r) \to u(r)$ for each n and r, $\lambda_n \to \text{EU}_\psi(U)$, and hence $f[U] \succ f[\lambda]$ whenever $\lambda < \text{EU}_\psi(U)$. Applying the same argument with a decreasing sequence, $f[U] \prec f[\lambda]$ whenever $\lambda > \text{EU}_\psi(U)$.

Now suppose that U and V are two such acts with $\text{EU}_\psi(U) > \text{EU}_\psi(V)$. Then for any a lying between the two expected utilities, there will exist an act W with reward function (wrt ψ) $f[a]$. We have proved that $U \succ^\psi W$, and $W \succ^\psi V$, so it follows that $U \succ^\psi V$.

To see that this utility function is unique, note that if there were another utility function u' we could construct acts whose utilities were the same as calculated by this second utility, but not as calculated by the first; this contradicts the requirements on u'. $\square$

Born-rule theorem: Suppose that:

(i) $\mathcal{P}$ is a quantum decision problem satisfying erasure, branch availability, reward availability, and problem continuity;

(ii) $\succ^{\psi}$ is a state-dependent solution to $\mathcal{P}$ satisfying ordering, diachronic consistency, macrostate indifference, branching indifference, state supervenience, and solution continuity.

Then there is a function u on the rewards of $\mathcal{P}$, unique up to positive affine transformations, such that if EU denotes the expected utility with respect to this function,

$$U \succ^{\psi} V \text{ iff } \mathrm{EU}_{\psi}(U) > \mathrm{EU}_{\psi}(V). \tag{30}$$

Proof: Note that problem continuity and solution continuity jointly entail that if $U \succ^{\psi} U'$, there are neighbourhoods $\mathcal{N}, \mathcal{N}'$ of U and U' respectively such that all acts in $\mathcal{N}$ and $\mathcal{N}'$ are available and all acts in $\mathcal{N}$ are preferred (given ψ) to all acts in $\mathcal{N}'$. For simplicity I shall refer to this simply as continuity.

We begin by proving that if $s \succ r_1 \succeq r_2 \succ t$, then if the utilities determined by the utility lemma (via this choice of s and t) for r_1 and r_2 coincide, then $r_1 \sim r_2$. Let this utility function be u and again, for convenience take $u(s) = 1$ and $u(t) = 0$. Fix E and $\psi \in E$, and let U_1 and U_2 be acts available at E whose ranges lie in r_1 and r_2 respectively (by reward availability, some such acts exist). If $r_1 \succ r_2$, then $U_1 \succ^{\psi} U_2$. By continuity, there must exist neighbourhoods $\mathcal{N}_1, \mathcal{N}_2$ of U_1 and U_2 such that any available act in $\mathcal{N}_1$ is preferred (given ψ) to any available act in $\mathcal{N}_2$.

Now let $f_1[a]$ and $f_2[a]$ be reward functions with $f_1[a](r_1) = 1 - a, f_1[a](t) = a$ and $f_2[a](r_2) = 1 - a, f_2[a](s) = a$. By branch availability and reward availability, there must exist some a, and some acts $U_{i,a}$, such that $U_{i,a} \in \mathcal{N}_i$ and the reward function of $U_{i,a}$ (with respect to ψ) is $f_i[a]$.

So we have that $U_{1,a} \succ U_{2,a}$. But $\mathrm{EU}_{\psi}(U_{1,a}) < \mathrm{EU}(U_1) \equiv u(r_1)$, and $\mathrm{EU}_{\psi}(U_{2,a}) > \mathrm{EU}(U_2) \equiv u(r_2)$. So by the utility lemma we must have that $u(r_1) > u(r_2)$.

We can now define a utility function for the whole of $\mathcal{R}$. For any rewards r_1, r_2 with $r_1 \succ r_2$, and any real numbers x_1, x_2 with $x_1 > x_2$, I will write $u[r_1, r_2, x_1, x_2]$ for the unique utility function determined on $[r_2, r_1]$ by setting the utility of r_i to x_i.

Now, let s, t be any two rewards with $s \succ t$ (if there are no such rewards, the theorem is true trivially). I define the utility of any reward r by:

- If $s \succeq r \succeq t$, $u(r) = u[s, t, 1, 0](r)$.
- If $r \succ s$, $u(r)$ is the unique value fixed by requiring that $u[r, t, u(r), 0](s) = 1$.
- If $t \succ r$, $u(r)$ is the unique value fixed by requiring that $u[s, r, 1, u(r)](s) = 0$.

(Notice that this definition relies on the assumption that the utilities of s and t are guaranteed to be distinct.)

I now prove that for acts with finitely many rewards, if $U_1 \succ^\psi U_2$ then $EU_\psi(U_1) > EU_\psi(U_2)$. For suppose that $U_1 \succ^\psi U_2$. By continuity, if f is the reward function of $\widehat{U}_1$ (with respect to ψ) then it will be possible to find some act V with reward function g such that, for some rewards r_1 and r_2 with $r_1 \succ r_2$:

- $V \succ^\psi U$;
- If $r \neq r_1$ and $r \neq r_2$, $g(r) = f(r)$;
- $g(r_1) < f(r_1)$; $g(r_2) > f(r_2)$.

This means that we must have $\mathrm{EU}_\psi(V) \geq \mathrm{EU}_\psi(U_2)$; since $\mathrm{EU}_\psi(V) < \mathrm{EU}_\psi(U_1)$, it follows that $EU_\psi(U_1) > EU_\psi(U_2)$.

This suffices to prove the Born-rule theorem under the assumption that any act has only finitely many non-null rewards. To extend to the infinite case, let U_1 and U_2 be arbitrary acts, and suppose for some ψ that $U_1 \succ^\psi U_2$. By continuity, if f_1 and f_2 are the reward functions (given ψ) of U_1 and U_2, it will be possible to find a finite subset $\mathcal{R}_0$ of $\mathcal{R}$, and acts V_1, V_2 with reward functions g_1, g_2, such that:

- $V_1 \succ^\psi V_2$;
- $g_i(r) = f_i(r)$ for $r \in \mathcal{R}_0$;
- If $r \notin \mathcal{R}_0$, then $g_1(r) = s$, and $g_2(r) = t$, where $s \succ t$.

Since V_1 and V_2 have only finitely many non-null rewards, $\mathrm{EU}_\psi(V_1) > \mathrm{EU}_\psi(V_2)$. But by construction $\mathrm{EU}_\psi(U_1) > \mathrm{EU}_\psi(V_1)$ and $\mathrm{EU}_\psi(U_2) < \mathrm{EU}_\psi(V_2)$, so $\mathrm{EU}_\psi(U_1) > \mathrm{EU}_\psi(U_2)$. $\square$

9 OTHER PROPOSED STRATEGIES FOR ACTION

In the nine years since Deutsch's original paper on decision-theoretic probability, a bewildering variety of alternative strategies for rational action have been proposed in the literature and in discussion. Some of these strategies have independent motivations; some are purely meant as counterexamples; all contradict the Born rule, and so all violate the decision-theoretic axioms of this paper.

This being the case, perhaps there is little need to discuss the alternative strategies: a proof is a proof. On the other hand, it may be instructive to show exactly how some of these alternative proposals violate my axiom scheme: apart from casting light on the motivation for the axioms, this may show how what appear to be coherent and even plausible strategies come apart on close inspection.

The proposed counterexamples, as will become apparent, break into four categories. There are the 'wrong-probability' rules, which also require an agent to maximize expected utility but with respect to some probability measure other than the Born rule. There are the 'no-probability' rules, which (purportedly) cannot be represented in terms of expected utilities at all. There are what might be called the 'I-don't-want-to-play' rules, which are not so much positive strategies as arguments against the existence of any strategy. And one special group, the contextual strategies, deserve a category of their own.

Branch Counting

Description: Each branch is given an equal probability, so that if there are N branches following a particular experiment, each branch is given probability $1/N$. Utility is then maximized with respect to this probability.

Origin: Has been reinvented innumerable times, but the first proponent may have been Graham, in DeWitt and Graham [1973].

Rationale: Each branch contains a copy of me; none of them can detect, nor care about, their quantum-mechanical weight; so I should not care about that weight either, and so I have no reason to prefer one over another.

Why it is irrational: The first thing to note about branch counting is that it can't actually be motivated or even defined given the structure of quantum mechanics. There is no such thing as 'branch count': as I noted earlier, the branching structure emergent from unitary quantum mechanics does not provide us with a well-defined notion of how many branches there are. All quantum mechanics really allows us to say is that there are *some* versions of me for each outcome.

But within the stylized context of my decision theory, the branch count is defined, so of course (given the representation theorem) the branch counting rule must violate some of my axioms. In fact, it violates the combination of branching indifference and diachronic consistency. For consider two acts $A1$ and $A2$: $A1$ consists of a two-outcome measurement (a spin measurement, say) followed by a reward of utility r in the spin-up branch. $A2$ consists of $A1$ followed by another two-outcome measurement in the spin-up branch. By branching indifference, the agent who gets the reward is indifferent about whether or not he makes a further measurement; by diachronic consistency, then, the original agent is indifferent between $A1$ and $A2$. But the utility of $A1$ (in which there are two branches, one of which provides a reward) is $r/2$; the utility of $A2$ is $2r/3$.

The Fatness Rule

Description: each branch is given a probability proportional to its quantum-mechanical weight multiplied by the mass of the agent in kilograms (such that the total probability is equal to one). Utility is maximized with respect to this probability.

Origin: David Albert (in conversation, and in his contribution in Chapter 11).

Rationale: Albert says, tongue-in-cheek, that an agent should care about branches where he is fatter because 'there is more of him' on that branch. He isn't serious, though: the rule is purely presented as a counterexample.

Why it is irrational: It violates diachronic consistency. Albert's agent is (*ex hypothesi*) indifferent to dieting. But he is not indifferent to whether his future selves diet: he wants the ones on branches with good outcomes to gain weight, and the ones on branches with bad outcomes to lose weight.

This is perhaps a good point to recall the rationale for diachronic consistency: rational action takes place over time and is incompatible with widespread conflict between stages of an agent's life. In the case of the fatness rule, agents have motivation to coerce their future selves—by hiring 'minders', say—into dietary programmes that they will resist. Multiply this conflict indefinitely many times (for branching is ubiquitous) and rational action becomes impossible.

(To object 'maybe rational action is impossible in the Everett interpretation' would, as noted before, be facile. It's perfectly possible for an agent following the Born rule.)

The Fake-State Rule

Description: The agent maximizes expected utilities as for the Born rule, but using a quantum state other than the physically real one.

Origin: Suggested many times in conversation.

Rationale: None in particular, though it is often intended to undermine the connection between the 'real' state and the physics.

Why it is irrational: It violates state supervenience. There will be cases where two acts produce the same physical state but where one produces a different fake state than the other. (This is inevitable: any two distinct quantum states are invariant under different sets of transformations.) The fake-state rule will then give the acts different utilities; state supervenience rules this out. Or, put another way: the fake state rule assigns different values to the same physical state under two different descriptions.

Note that it is crucial here—as elsewhere in decision theory—that the agent has a choice between different actions, and therefore between different sets of histories and weights. Of course, in a deterministic universe it is fixed which action will actually occur, but this does not remove the necessity of defining preferences, and hence indirectly probabilities, over a wide range of actions.

The Distributive-Justice Rule

Description: The agent does not maximize expected utilities at all. He treats his various successors in rather the way that a just ruler would treat his various subjects: in particular, he will not allow the suffering of one even if it brings great advantage to others.

Origin: Huw Price (in Chapter 12).

Rationale: Any action we choose generates a multitude of individuals; we have a duty to treat them all ethically, and in particular we would not be morally justified in letting one suffer unduly for the others' benefit.

Why it is irrational: The rule is very underspecified, so it isn't easy to answer this, but on natural precisifications it either violates continuity or is not actually a counterexample to the Born rule.

To expand: a large part of what Price wants can be achieved by an appropriate utility function. An agent moved by Price's concerns can drastically increase the disutility of bad consequences and scale down the utility of good consequences, with the effect that trade-offs of the sort he considers get a much lower utility and so will tend to be rejected in favour of more equitable options. There is nothing in Everettian decision theory that prevents an agent from making such modifications to their utility function on recognizing the ethical consequences of the Everett interpretation.[13] If Price wants to hold that *no* amount of suffering, however low-weight the branch on which it occurs, is acceptable, then this strategy will not work, but there is a clash with continuity. Suppose there are three rewards r_1 and r_2 with $r_1 > r_2$, and a (dire) punishment p. Price will prefer r_1 to r_2 but will prefer r_2 to $(1 - w)r_1 + wp$, whatever the value of w; clearly this violates continuity.

Now, I think the physical arguments for continuity are pretty unassailable, but it is worth noting that the principle is only really used in my proof precisely to rule out infinite or infinitesimal utilities. (The only other use is for the mathematically convenient but physically tangential purpose of extending the Born rule to the case of infinitely many rewards.) If such utilities are allowed, there is no problem with extending the Born rule to cover even Price's case (though the utility function will have to be modelled in non-standard analysis and the maths will start getting fiddly). And in fact, precisely the same situation has arisen in *classical* decision theory, and the structure axioms of classical decision theory are selected precisely to rule out the case of infinite (dis)utility.

The Variety Rule

Description: An agent prefers A to B, but prefers receiving A in half the branches and B in the other half to either A or B.

Origin: Suggested in a seminar by Adam Elga in 2004; has not appeared in print as far as I am aware.

---

[13] Personally, though, I don't feel inclined to. Call me callous.

Rationale: An agent may regret having to make one choice or another, and may rather like the idea that one version of himself makes one choice, one another. (In Elga's example, a student prefers physics to history but likes both; that student might prefer to do history in one branch, physics in the other.)

Why it is irrational: It either violates diachronic consistency, or it isn't a counterexample to the Born rule.

To expand: suppose you are the agent who chose history. What prevents you changing your mind and switching to physics? It doesn't, after all, hurt your counterpart in the physics branch. This would clearly violate diachronic consistency.

But perhaps you wouldn't choose to switch back. That's to say that although you prefer doing physics to doing history, you prefer doing history *as a result of a situation in which a certain process chose history for you* rather than doing physics *against the result of that process*. In that case, the utility you are assigning to (history-after-process) is higher than the utility you assign to (physics-against-process), and indeed higher than (physics-without-process). The different situations in which you end up doing history count as different rewards.

Exactly analogous situations can arise in classical decision theory. A student might decide that on balance he'd rather do physics than history, but nonetheless resolves to decide by the toss of a coin (because, say, he finds it comforting to have the decision taken from his hands; the reader can probably supply other motivations). That student, again, will place a higher utility on (history-after-coin-toss) than on (physics).

Of course, if every outcome's utility depended sensitively on the circumstances in which that reward arose, decision theory couldn't get off the ground: there would be no way to define probability without being able to have the same reward available in different acts. But again, this is not specific to quantum decision theory.

The Anything-goes Rule

Description: Not so much a 'rule' as a rejection of the need to have one: according to this position, any transitive preference ordering over acts is rationally acceptable.

Origin: Suggested by Tim Maudlin in seminars on multiple occasions; frequently suggested in conversations.

Rationale: Everettian quantum theory is deterministic, and we already have a perfectly acceptable deterministic decision theory: its only axiom is transitivity. So any transitive ordering should be fine.

Why it is irrational: Even in deterministic decision theory, transitivity is not the only constraint. Rational agency is not possible without diachronic consistency; in addition, preference orders have to be defined on actual physical acts, so mathematical modelling of those orders should require an agent to be indifferent between the same state of affairs differently defined. Furthermore, the only interesting decision-theoretic strategies are those which are physically performable in at least an idealized sense. All of the rationality axioms of this paper fit into one of these categories; even in deterministic decision theory, then, they are rationally required.

The Curl-up-and-die Rule

Description: The converse of the anything-goes rule, this is not so much a 'rule' for rational action as the claim that *no* rational strategy is possible in Everettian quantum theory.
Origin: Frequently suggested in conversation.
Rationale: Various; see below.
Why it is irrational: Unless there is something concretely wrong with the Born rule, there is no case to be made that no rational strategy is available: the Born rule is available.

I am aware of two general objections to the rationality of the Born rule, though. The first is that it is rationally compulsory for an agent to weight each branch equally; since the Born rule violates this requirement, it cannot be rational (and if only the Born rule is rational, rationality is impossible in an Everettian universe). Arguments are seldom given for the suggestion that this is a rational requirement (I can see that at best it might be a rational *desideratum*, but it's not at all clear to me why, in a universe where it isn't physically possible to obey the requirement, we should be unable to settle for some second-best option). In any case, though (at the risk of repetitiveness) there is no coherent notion of branch count available in quantum mechanics, so it's not even meaningful to talk of 'weighting each branch equally'.

The other objection (frequently made in discussions, and made in print by Hemmo and Pitowsky [2007]) is that no strategy can be rational if it can be known in advance by those adopting it that some of them (or some of their successors) will make wrong decisions. So in particular, it is a corollary of the Born rule that an agent measuring a long succession of identical quantum systems should regard the observed frequencies as a guide to what state each system is in; but since all sequences of results occur somewhere, some of the agent's successors will get the wrong outcome.

Now, it is true that some agents will indeed be misled in this way. But there is nothing particularly quantum-mechanical about this. If the universe is spatially infinite (as current observations support), we can guarantee that somewhere in the universe are people as similar to us as you like but whose observed statistics have systematically misled them. Even on Earth, one can fairly easily come up with similar examples. Suppose that the British government declared that it puts some people under (non-covert) surveillance at random, but that there are very few such people: only one in ten million. And suppose it is claimed that the government is lying, and actually puts many more people than that (tens of thousands, say) under surveillance. Then each person in Britain is rational to adopt the strategy: if I am under surveillance, the government is (almost certainly) lying—even though they know that if the government is not lying, five or six people in Britain will be misled into thinking it was.

Ultimately, some people get unlucky. There is no contradiction between this and the rationality of a decision-theoretic strategy, provided that strategy tells us not to care about the unlucky cases. The Born rule tells us exactly that.

Contextual Rules

Description: An agent's preferences conform to a probability rule that violates the principle of non-contextuality: that is, it assigns different probabilities to the outcomes of a measurement of operator $\hat{X}$ according to whether or not a compatible operator $\hat{Y}$ is measured at the same time.

Origin: Various, but a particularly forceful advocacy can be found in Hemmo and Pitowsky [2007].

Rationale: As is well known, any non-contextual quantum probability rule (and hence, any strategy for rational action expressible in terms of such a rule) can be proved to be the Born rule applied to some (possibly mixed) state.[14] The suspicion, then, is that the decision-theoretic arguments are just a combination of Gleason's theorem (or a relative of it) with an unjustified assumption of non-contextuality.

Why it is irrational: Probably the easiest way to explain what is wrong with contextual rules is that they violate state supervenience. If we regard measurements as physical processes rather than as primitive, which operator(s) are being measured in a given process is dependent on the interests of the experimenter, and cannot simply be read off from the physics. (Consider the Stern–Gerlach experiment, for instance: is it a measurement of spin or of position?) For a decision rule to be contextual, then, is for a rational agent to prefer a given act to the same act (knowably the same act, in fact) under a different description, which obviously violates state supervenience (and, I hope, is obviously irrational).

---

[14] This is usually explained in terms of Gleason's theorem, but this is a rather outdated approach now that POVMs, not PVMs, are widely—and in my view correctly—seen as the best way to represent measurements in quantum theory. Most of the mathematical complexity of Gleason's theorem can be dispensed with if we require our probability function to be defined on POVMs and not just PVMs. See Caves, Fuchs, Manne, and Renes [2004] for further discussion.

It is fair to note, though, that just as a non-primitive approach to measurement allows one and the same physical process to count as multiple abstractly construed measurements, it also allows one and the same abstractly construed measurement to be performed by multiple physical processes. It is then a non-trivial fact, and in a sense a physical analogue of non-contextuality, that rational agents are indifferent to which particular process realizes a given measurement.

In earlier work (Wallace [2003]; Wallace [2007]) I called this fact *measurement neutrality*. It is indeed a tacit premise in Deutsch's original [1999] proof of the Born rule, as I argued in Wallace [2003]. In this paper, it is a theorem (a trivial corollary of the main representation theorem, in fact) that measurement neutrality is rationally required. The short answer as to why, is that two acts which correspond to the same abstractly construed measurement can be transformed into the same act via processes to which rational agents are indifference. To see the long answer, reread Sections 4–8.

Incidentally, Gleason's theorem (or more accurately its POVM generalization) is much more directly needed if we wish to generalize the results of this paper to situations where the quantum state is unknown to the agent. The details are somewhat involved; see Wallace [2011] for an account.

10 CONCLUSION

A rational agent, believing that the Everett interpretation is true and that the quantum state of a given system is $|\psi\rangle$, knows that measurements on that state will generally split his part of the multiverse into multiple branches, with different measurement outcomes, and different versions of the agent, on different branches; he also knows that the relative weights of these branches are given by the Born rule, applied to the post-measurement state of the system and measurement device. Rationality considerations not different in kind from those which apply in single-universe decision-making then compel the agent to act as if a set of branches of relative weight w has probability w. In other words, he is rationally required to act as if the Born rule were true.

As I noted in the introduction, my focus here is deliberately narrow and I leave it to other chapters in this volume (and to my own work elsewhere) to make the case that such a result suffices to justify the general role of probability in the Everett interpretation. Yet even on its own terms it is a rather remarkable result, as Deutsch's opening quotation notes, and one which to the best of my knowledge has no analogue outside the branching-universe context.

And how does this result actually come about? The decision-theoretic language in which this paper is written is no doubt necessary to make a properly rigorous case and to respond to those who doubt the very coherence of Everettian probability, but in a way the central core of the argument is not decision-theoretic at all. What is really going on is that the quantum state has certain symmetries, and the probabilities are being constrained by those symmetries.

This is actually a throwback to an older idea of probability. Quantitative probability has been concerned with symmetry ever since it was applied to the throw of dice in the 17th century: what makes it reasonable to regard each side of a die as equiprobable is that we have no reason to regard one as more probable than another, and what prevents us having reason is the rotational symmetry of the die that maps one side to another. But real dice—real classical dice, at any rate—must break the symmetry by their initial conditions, or else how in a deterministic universe could the die land one way rather than another. We then have to impose a certain probability distribution on the die's initial conditions, and any prospect of a reductive analysis of probability is lost. In Everettian quantum mechanics, there is no one actual outcome, no requirement for the symmetry to be broken by the actual state of the system, and so a programme of deriving the probabilities from the symmetries remains viable. The language of decision theory makes rigorous sense of what such a derivation would look like, and shows—I claim—that the programme can indeed be carried out.

Acknowledgements

This work has drawn heavily on conversations and correspondences over a number of years with Harvey Brown, Jeremy Butterfield, David Deutsch, Hilary Greaves, Chris Timpson and Wayne Myrvold, and above all, Simon Saunders.

References

Barnum, H., C.M. Caves, J. Finkelstein, C.A. Fuchs, and R. Schack [2000], 'Quantum probability from decision theory?', *Proceedings of the Royal Society of London* **A456**, 1175–82. Available online at arXiv.org/abs/quant-ph/9907024.

Caves, C.M., C.A. Fuchs, K. Manne, and J.M. Renes [2004], 'Gleason-type derivations of the quantum probability rule for generalized measurements', *Foundations of Physics* **34**, 193.

Davidson, D. [1973], 'Radical interpretation', *Dialectica* **27**, 313–28.

—— [2004], 'Paradoxes of irrationality', in *Problems of Rationality*, Oxford University Press, Oxford.

Dennett, D.C. [1984], *Elbow Room: the Varieties of Free Will Worth Wanting*, Oxford University Press, Oxford.

—— [1987], *The Intentional Stance*, MIT Press, Cambridge, Mass.

Deutsch, D. [1999], 'Quantum theory of probability and decisions', *Proceedings of the Royal Society of London* **A455**, 3129–37. Available online at arXiv.org/abs/quant-ph/9906015.

DeWitt, B. and N. Graham (eds) [1973], *The Many-Worlds Interpretation of Quantum Mechanics*, Princeton University Press, Princeton.

Hemmo, M. and I. Pitowsky [2007], 'Quantum probability and many worlds', *Studies in the History and Philosophy of Modern Physics* **38**, 333–50.

Lewis, D. [1974], 'Radical interpretation', *Synthese* **23**, 331–44. Reprinted in David Lewis, *Philosophical Papers*, Volume I, Oxford University Press, Oxford [1983].

—— [1980], 'A subjectivist's guide to objective chance', in R.C. Jeffrey (ed.), *Studies in Inductive Logic and Probability*, Volume II, University of California Press, Berkeley. Reprinted in David Lewis, *Philosophical Papers*, Volume II, Oxford University Press, Oxford [1986].

Lewis, P.J. [2005], 'Probability in Everettian quantum mechanics'. Available online at phil-sci.pitt.edu.

Saunders, S. [1998], 'Time, quantum mechanics, and probability', *Synthese* **114**, 373–404.

Saunders, S. and D. Wallace [2008], 'Branching and uncertainty', *British Journal for the Philosophy of Science* **59**, 293–305.

Savage, L.J. [1972], *The Foundations of Statistics* (2nd edn), Dover, New York.

Wallace, D. [2001], 'Implications of quantum theory in the foundations of statistical mechanics', Available online at philsci-archive.pitt.edu.

—— [2003], 'Everettian rationality: defending Deutsch's approach to probability in the Everett interpretation', *Studies in the History and Philosophy of Modern Physics* **34**, 415–39. Available online at arXiv.org/abs/quant-ph/0303050 or from philsci-archive.pitt.edu.

—— [2005], 'Language use in a branching universe'. Available online from philsci-archive.pitt.edu.

—— [2006], 'Epistemology quantized: circumstances in which we should come to believe in the Everett interpretation', *British Journal for the Philosophy of Science* **57**, 655–89. Available online from philsci-archive.pitt.edu.

—— [2007], 'Quantum probability from subjective likelihood: Improving on Deutsch's proof of the probability rule', *Studies in the History and Philosophy of Modern Physics* **38**, 311–32.

—— [2011]. *The Everett Interpretation of Quantum Mechanics*, Oxford University Press, Oxford.

Zurek, W.H. and J.P. Paz [1994], 'Decoherence, chaos and the second law', *Physical Review Letters* **72**(16), 2508–11.

9

Everett and Evidence

Hilary Greaves and Wayne Myrvold

In the midst of this perplexity, I received from Oxford the manuscript you have examined. I lingered, naturally, on the sentence: *I leave to the various futures (not to all) my garden of forking paths*. Almost instantly, I understood: 'the garden of forking paths' was the chaotic novel; the phrase 'the various futures (not to all)' suggested to me the forking in time, not in space.

Jorge Luis Borges, 'The Garden of Forking Paths'

ABSTRACT

Much of the evidence for quantum mechanics is statistical in nature. Relative frequency data summarizing the results of repeated experiments is compared to probabilities calculated from the theory; close agreement between the observed relative frequencies and calculated probabilities is taken as evidence in favour of the theory. The Everett interpretation, if it is to be a candidate for serious consideration, must be capable of doing justice to this sort of reasoning. Since, on the Everett interpretation, all outcomes with non-zero amplitude are actualized on different branches, it is not obvious that sense can be made of ascribing probabilities to outcomes of experiments, and this poses a prima facie problem for statistical inference. It is incumbent on the Everettian either to make sense of ascribing probabilities to outcomes of experiments in the Everett interpretation, or to find a substitute on which the usual statistical analysis of experimental results continues to count as evidence for quantum mechanics, and, since it is the very evidence for quantum mechanics that is at stake, this must be done in a way that does not presuppose the correctness of Everettian quantum mechanics. This requires an account of theory confirmation that applies to branching-universe theories but does not presuppose the correctness of any such theory. In this paper, we supply and defend such an account. The account has the consequence that statistical evidence can confirm a branching-universe theory such as Everettian quantum mechanics in the same way in which it can confirm a non-branching probabilistic theory.

1 INTRODUCTION

Quantum mechanics, standardly interpreted, yields, via the Born rule, statements about the probabilities of outcomes of experiments. These probabilities are, at least in many interesting cases, different from what would be expected on the basis of classical mechanics. Moreover, we can subject the claims made by standard quantum mechanics about the probabilities of outcomes of experiments to empirical test, and the results of such tests favour quantum mechanics over classical. This sort of empirical testing of probabilistic claims forms a substantial part of the evidence we have for accepting quantum mechanics as a theory that is empirically superior to classical mechanics.

Consider, for example, Bell-inequality experiments. Here we compare the probabilistic correlations yielded by a quantum-mechanical calculation to those that could be yielded by some local hidden-variables theory. Relative frequencies of outcomes in repeated trials are compared with probabilities calculated from quantum mechanics, and with probabilities that could be yielded by a local hidden-variables theory. The fact that the observed relative frequencies closely match the quantum probabilities, and exhibit statistically significant violations of Bell inequalities, is correctly taken to favour quantum mechanics over local hidden-variable theories. Although it is possible to lose sight of the fact in discussing the bearing of such experiments on theory, the reasoning is essentially probabilistic. *Any* sequence of outcomes of such an experiment is compatible both with quantum mechanics and with local hidden-variables theories. In particular, even if some local hidden-variables theory is correct, a sequence of outcomes is *possible* (though highly improbable) in which the relative frequencies violate the Bell inequalities. We take the observed results to rule out the latter because the results actually obtained are astronomically less probable on the assumption of a local hidden-variables theory than they are on the assumption of quantum mechanics. Similar considerations apply to the double-slit experiment. The quantum-mechanical calculation yields a probability distribution for absorption of particles by the screen. From this can be calculated a probability for any possible pattern of absorption events. The probability will be high that the observed pattern of detection events shows bands of intensity corresponding to a diffraction pattern, but we should not lose sight of the fact that *any* pattern is *consistent* with quantum mechanics, including one that matches classical expectations. The occurrence of a pattern that is much more probable on the assumption that quantum mechanics is correct than on the assumption of classical mechanics is taken to provide empirical evidence that quantum mechanics is getting the probabilities right, or approximately so.

Any interpretation of quantum mechanics that is worthy of serious consideration is going to have to make sense of this sort of reasoning. If it cannot, it runs the risk of undermining the very reasons we have for accepting quantum mechanics in the first place.

On the Everett interpretation, the quantum state vector after a typical measurement interaction is a superposition of terms on which the measurement apparatus records different outcomes. Moreover, the quantum state is taken as a complete description of physical reality, so that there is nothing that distinguishes one of these branches as uniquely real. As has often been pointed out (see, e.g. Albert and Loewer [1988]), this poses a problem for interpreting probabilistic statements in an Everettian context. There is no obvious sense in which one can ask what the probability is that a certain result will be *the* result of the experiment, since all possible results occur in the post-experiment state, on different branches of the superposition.

There is a danger, in discussing the Everett interpretation, of talking as if the goal is to provide a coherent interpretation that is consistent with our experience. But if that were the goal, the Everettian would have no need of probabilities; it would suffice merely to note that, for every outcome normally regarded as possible, the theory *entailed* that that outcome would occur on some branch. The goal is actually much higher: it is incumbent upon the Everettian to provide an interpretation in which the statistical analysis of the outcomes of repeated experiments provides empirical support for the theory. This is why the apparent lack of room for probability statements in the Everett interpretation threatens to create a problem for that interpretation. The problem is not one of *deriving* the correct probabilities within the theory; it is one of either making sense of ascribing probabilities to outcomes of experiments in the Everett interpretation, or of finding a substitute on which the usual statistical analysis of experimental results continues to count as evidence for quantum mechanics.

Call this the *Everettian evidential problem*. In our opinion the best hope for meeting this challenge lies in a decision-theoretic approach. The use of decision-theoretic ideas in connection with Everettian quantum mechanics was pioneered by Deutsch [1999], and elaborated, in different ways, by Wallace [2003, 2007], Saunders [2005], and Greaves [2004, 2007a]; see Greaves [2007b] for a recent survey of the approach. Deutsch's argument and the variants on it presuppose an agent who accepts Everettian quantum mechanics. In order to meet the evidential problem, we need a framework for appraising theories, including branching-universe theories, that does not presuppose the acceptance of Everettian quantum mechanics or any other theory.[1] This, after some preliminary discussion in Section 2, will be laid out in Section 3, and applied to branching-universe theories, such as Everett's, in Section 4. Section 5 discusses and replies

[1] This point has also been made by Wallace [2006].

to objections. Our conclusion (Section 6) is that the framework presented here suffices to solve the evidential problem.

2 TESTING PROBABILISTIC THEORIES

We will not be in any position to address the question of whether or not statistical data can be evidence for Everettian quantum mechanics unless we are crystal clear about how *exactly* such data can be evidence for uncontroversially probabilistic theories. We therefore start by stepping back from quantum mechanics and the Everett interpretation, and reviewing some general considerations about probability statements in physics and their evaluation in the light of experimental data.

Consider the questions:

1. A pair of fair dice is about to be tossed 24 times. Which is preferable: an offer of $1000 if a pair of sixes comes up at least once, or an offer of $1000 if a pair of sixes never comes up?
2. *This* pair of dice is about to be tossed 24 times on *this* table, using *this* cup, by me. Which is preferable: an offer of $1000 if a pair of sixes comes up at least once, or an offer of $1000 if a pair of sixes never comes up?

The first question is a purely mathematical one, or close to it. Provided that you prefer receiving $1000 to not receiving anything, then the question is one that can be answered by calculation, and is in fact the question that was posed by the Chevalier de Méré and answered by Pascal.[2]

The second question is not purely mathematical; it is, at least in part, a question about the physical world. To answer it, we need to know whether there is something about the physical set-up—the dice, or the way they are tossed, or the make-up of the table on which they land—that biases the results towards a pair of sixes. Questions of this type can be answered in two sorts of ways:

1. By direct empirical test. Typically this involves repeated throws of the dice, and statistical analysis of the results.
2. Theoretically. This involves a theoretical model of the set-up, plus a physical theory that says something about which factors are, and which factors are not, relevant to the outcome of the tosses.

A splendid example of the latter is found in Diaconis et al.'s model of a coin toss (Diaconis et al., [2007]). They construct a simple coin-tossing machine (crucially, the coin is not permitted to bounce upon landing), model the dynamics of the tossed coin, come to conclusions regarding the probability of landing with heads or tails up, given an initial orientation, and conclude that

---

[2] See Ore [1960] for a lucid account of this incident.

the coin toss is biased towards landing with the same side facing up that it started with. These conclusions are corroborated by data from repeated trials with the machine.

Even when a theoretical model is available, empirical testing is not superfluous, as we will want to satisfy ourselves of the appropriateness of the model to the case at hand. We will not, therefore, be able to do without the first way of answering the question. It is possible to overlook this point, because such calculations are usually made on the basis of symmetry considerations, and these can create the illusion that the results are truths known a priori. But judgements of symmetry are judgements that certain factors are irrelevant to the outcome, and this is a matter of physics. An account of probability based *exclusively* on a principle of indifference will not do.

Nor can probability concepts be replaced by relative frequencies, in either actual or hypothetical sequences of experiments, though relative frequency data will often be our most important sources of information about chances, or physical probabilities. Consider, for example, a case in which balls are drawn, with replacement, from an urn containing N balls in total, of which M are black, in such a way that each ball has an equal chance of being drawn. The chance, on each draw, that the drawn ball is black, is, in this case, equal to M/N, which is also the proportion of black balls in the urn. Suppose that we perform n drawings, with replacement, and let m be the number of times in these n trials that a black ball is drawn. Then, for large n, the chance is high that the sample relative frequency m/n will be close to the proportion of black balls in the population, M/N. Moreover, if the sequence of drawings be extended without end, then, with chance 1, the sample relative frequency will converge to the single-case chance M/N. Therefore, if we are unable to examine the contents of the urn, information about its contents can be gained by successive drawings. Similar ideas are behind statistical sampling techniques; one wishes to gain information about a population by a sampling of the population, and one attempts to construct one's sampling procedure such that the chance of any individual being chosen for the sample is independent of whether or not that person has the property whose proportion in the population is to be estimated. This intimate relation between chances and relative frequencies has suggested to some that chances can be *defined* in terms of relative frequencies. In spite of their intimate relation between chance and relative frequency, the former is not eliminable in favour of the latter. Notice that in the urn model, it is necessary to stipulate that each ball has an equal chance of being drawn; it is only this stipulation that makes the proportion of black balls in the urn, M/N, equal to the chance of drawing a black ball. Nor can chances be eliminated in terms of limiting relative frequencies in infinite sequences. That the relative frequency converges to the single-case chance is not the only *logically* possible outcome of the sequences of trials; it is rather the only outcome that has non-zero *chance*—and note that one cannot identify 'zero chance' with 'impossible', since even an outcome according to which relative frequency *does* match chance is an (infinite) disjunction of zero-chance outcomes. Thus, the conclusion that the limiting relative frequency will exist, and be equal to the single-case chance, requires the use of a notion of chance distinct from the notion of limiting relative frequency.

How, then, *does* the process of confirming or disconfirming statements of probability in physics work? On our view, the best way to make sense of such confirmation involves a role for two sorts of quantities that have sometimes been called 'probability'. The first is degree of belief, or credence, which is subjective in the sense of being attached to an (idealized) epistemic agent. Accepting this does not entail eliminating any notion of physical probability. Among the things our epistemic agent can have degrees of belief about are the chances of experimental outcomes, which are characteristic of the experimental set-up, and hence the sort of things that a physical theory can have something to say about. We test such claims by performing repeated experiments—a sequence of experiments that we regard as equivalent, or near enough, with respect to the chances of outcomes—and comparing the calculated chances with the observed relative frequencies. Conditionalization on these observations raises degrees of belief in theories whose calculated chances are near the observed relative frequency and lowers degrees of belief in theories whose calculated chances are far from the observed relative frequencies. That, in short, is the story of statistical confirmation of theories with experiments construed in the usual way. Its core can be summed up by the following confirmation-theoretic principle:

CC (**confirmation-theoretic role of chances**). If *S* observes something to which theory *T* assigned a chance higher (lower) than the average chance assigned to that same event by rival theories, then theory *T* is confirmed (disconfirmed) for *S*, relative to those theories.

Note that all three concepts—credence or degree of belief, physical chance, and relative frequency—have important roles to play in this story. The story will be elaborated upon in Section 3, below, in which we provide a set of conditions, based on Savage's axioms for decision theory, and on de Finetti's concept of *exchangeability*, that are sufficient to ensure that the agent will act as if she thinks of an experiment as having chances associated with its possible outcomes, and repeated experiments as informative about the values of those chances. This permits her to experimentally test the claims a physical theory makes about chances of outcomes.

We wish to argue that a precisely analogous story can be told if the agent thinks of experiments, not in the usual way, but as involving a branching of the world, with all possible outcomes occurring on some branch or another. We claim that the conditions we introduce remain reasonable under this supposition, and that the agent will act as if she regards branches as associated with quantities, which we will call *weights*, that play in this context a role analogous to that played by chances on the usual way of viewing things. The short version of *this* story is summed up by the principle

CW (confirmation-theoretic role of branch weights). If S observes something to which theory T assigned a branch weight higher (lower) than the average chance-or-branch-weight assigned to that same event by rival theories, then theory T is confirmed (disconfirmed) for S, relative to those theories.

In particular, according to our account, the agent will regard relative frequency data from repeated experiments as informative about values of branch weights in exactly the same way that, on the usual view, they are informative about chances of outcomes. If, therefore, Everettian quantum mechanics is taken as a physical theory that makes claims about branch weights, these claims can be tested by experiment.

In the general case, the agent will have non-neglible credence in some theories in which experiments are construed, in the usual way, as chance set-ups, and in some in which they are construed as branch set-ups. The account permits both to be handled simultaneously; what are estimated via repeated experiments are quantities that are to be interpreted as being *either* physical chances or physical branch weights.

The framework will take as its starting point the notion of preferences between wagers on outcomes of experiments.[3] We will first lay out a set of conditions on preferences between wagers, based on Savage's axioms, which suffice for a representation theorem, Theorem 1, according to which an agent's preferences can be represented as maximizing expected utility. On the usual interpretation, this expected utility is a weighted average of utilities across alternative epistemic possibilities, with the weighting function representing the agent's degrees of belief in these alternative possibilities. We will argue that the constraints on the agent's preferences are reasonable, also, if the agent thinks of experiments as branching events; in this case the weighting function becomes what Greaves [2007a] has called a 'quasicredence' function. We then argue that, upon learning the results of experiments, the agent ought to update this credence-or-quasicredence function in a manner equivalent to Bayesian conditionalization (Theorem 2). We can then take on board the de Finetti representation theorem (Theorem 3), which shows that, for an exchangeable sequence of experiments, the agent's credence-or-quasicredence function is a weighted average of certain extremal

---

[3] This may seem an odd place to start. But it is perhaps worth noting that this is where modern probability theory started, too. What we now call the mathematical theory of probability has its origins in the Fermat–Pascal correspondence (reprinted in Smith [1959]), and in the treatise of Huygens [1660]. Modern readers may be surprised that these authors never calculate what we would call a probability. They are concerned, instead, with the values of wagers (expectation values, in modern parlance). It was Jacob Bernoulli's *Ars Conjectandi* that, 50 years later, introduced probabilities into the theory. 'Before Bernoulli, the mathematics of games of chance had been developed by Pascal, Fermat, Huygens, and others largely without using the word (or concept of) "probability"' (E.D. Sylla, 'Preface' to Bernoulli [2006]).

functions that, as we will argue, can, under certain circumstances, be thought of as objective chances-or-branch-weights associated with outcomes of experiments. The weighting function (called μ in Theorem 3), under these circumstances, represents the agent's degrees of belief about which set of chances-or-branch-weights is correct. This opens the way for repeated experiments to be informative about the values of these chances-or-branch-weights: updating on observed outcomes of experiments updates the μ-function.

Some remarks on the relationship of the present paper to the existing Everettian literature are in order; these occupy the remainder of this section.

The account of decision-making and empirical confirmation of branching theories that is defended in this paper is the same as that proposed in Greaves [2007a]. The main difference between the two papers is that the present paper offers arguments (in the form of representation theorems) for two key claims that were taken as basic assumptions in Greaves [2007a]. Firstly, in Greaves [2007a] it was *assumed* that, in the branching case, decisions are to be made via maximizing a weighted mean of utilities of rewards-on-branches. The present paper, in contrast, lays out a set of (Savage-style) axioms constraining rational preferences between wagers in a branching context, and spells out how the claim concerning maximization of expected utility (MEU) follows via Savage's representation theorem. While these theorems themselves are not new, their applicability to the branching case has not previously been discussed in any detail. Secondly, in Greaves [2007a] it was also *assumed* that the agent's quasicredence function satisfies the two principles PC and PW: the principal principle for chances, and for branch weights, respectively. In the present paper, we use the de Finetti representation theorem as a vehicle for attributing to an agent degrees of belief about chances and/or branch weights, in such a way that these principles are satisfied. In both cases, our aim, in highlighting the applicability of these representation theorems to the branching case, is to shift the locus of discussion from the MEU and PW claims themselves to the axioms: if the account of rational decision-making and/or confirmation advocated here and in Greaves [2007a] is not correct then it must be that one or more of the axioms is not correct, and we urge objectors to identify which axiom they think this is.

We remark also on the relationship of the present paper to the representation theorems proved by Deutsch [1999] and Wallace [2003, 2007]. Two points are worthy of note. (i) The Deutsch–Wallace approach aims to derive the Born rule from the 'non-probabilistic' part of Everettian quantum mechanics: that is, it seeks to prove that, conditional on the truth of Everettian quantum mechanics and the given initial state for a given measurement, the rational agent's betting quotients *must* equal the corresponding amplitude mod-squares. This is not something we claim to do in the present paper. In this respect, Deutsch and Wallace's claims are stronger than (but consistent with) ours. (ii) The decision theories developed by Deutsch and Wallace assume the truth of quantum mechanics (specifically, Everettian quantum mechanics). This means that they are not general enough to address the evidential problem. The axioms we adopt in the present paper, in contrast, are much more theory-general.

3 THE FRAMEWORK

In this section, we present a simplified framework, not meant to be a model for all decisions, but rather, applicable to a limited class of decisions, involving payoffs contingent on outcomes of experiments. We apply Savage's axioms to this restricted setting (Savage [1972]). These are intended to be thought of as rationality constraints on an agent's preferences between wagers.

Suppose, therefore, that we have a set of possible experiments. Associated with an experiment A there is a set S^A of possible outcomes. We do not assume that the outcome space is finite or even countable. We assume a set of payoffs, which are in the first instance the objects of our agent's preferences, and that there is a set $\mathcal{F}^A$ of subsets of S^A (the *wagerable* subsets of S^A) with which we can associate payoffs. $\mathcal{F}^A$ will be assumed to be closed under intersections, complements, and unions. An association of payoffs with the elements of a finite partition $\{F_i | i = 1, \ldots, n\}$ composed of elements of $\mathcal{F}^A$ will be called a *wager*. It will sometimes be helpful to imagine these payoffs as sums of money paid by a bookie to an agent who accepts the wager. But the framework is not limited to such cases. In particular, we allow for preferences between states of affairs that do not differ with respect to any effect on the agent (e.g. a sum of money paid to someone else), including states of affairs in which the agent is not present. It is not irrational to accept a wager that pays a large sum of money to your heirs in the event of your death!

To make things simpler, we will assume that, for any experiment A and any finite partition Π^A of S^A, any assignment of payoffs to elements of Π^A is a possible wager. This means that the agent is indifferent about the outcomes of experiments for their own sake, and has preferences only in so far as these outcomes lead to further consequences. One can, of course, imagine situations in which this condition does not obtain, but what matters, for our purposes, is that there is a sufficiently rich set of experiments and outcomes that are such that this condition is, for all intents and purposes, realized.

A wager $\mathbf{f}$ on a partition Π^A of the outcome space of an experiment A can be represented by the function that associates payoffs with the outcomes of A. If A is an experiment, $\{F_i | i = 1, \ldots, n\}$ a partition of S^A, and $\{a_i\}$ a set of payoffs, we will write $[F_i \to a_i]$ for the wager on A that pays a_i on outcomes in F_i. We will also write $[F \to a, \neg F \to b]$ for $[F \to a, (S^A - F) \to b]$.

Performing one experiment may preclude performance of another. We assume that there is a relation of compatibility on the set of experiments. Bets on compatible experiments can be combined. For any two compatible experiments

A, B, there is a third experiment C, with outcome space $S^A \times S^B$, such that outcome $(s, t) \in S^C$ occurs iff $s \in S^A$ and $t \in S^B$ occur. For any subset $F \subseteq S^A$, there will be a corresponding subset $F \times S^B$ consisting of all $(s, t) \in S^A \times S^B$ such that $s \in F$. For notational convenience, we will occasionally ignore the distinction between F and $F \times S^B$, and will write $F \cap G$ for the set of $(s, t) \in S^A \times S^B$ such that $s \in F$ and $t \in G$: that is, $(F \times S^B) \cap (S^A \times G)$.

We assume that our agent has a preference ordering $\preceq$ on the set of wagers. The following axioms, based on those of Savage (1972), are to be taken as rationality constraints on this preference ordering.

P1. a) $\preceq$ is transitive. That is, for all wagers $\mathbf{f}$, $\mathbf{g}$, $\mathbf{h}$, if $\mathbf{f} \preceq \mathbf{g}$ and $\mathbf{g} \preceq \mathbf{h}$, then $\mathbf{f} \preceq \mathbf{h}$.

   b) $\preceq$ is a total ordering. That is, for all wagers $\mathbf{f}$, $\mathbf{g}$, $\mathbf{f} \preceq \mathbf{g}$ or $\mathbf{g} \preceq \mathbf{f}$.

(Note that reflexivity of $\preceq$ follows from (b)).

We define an equivalence relation $\approx$ by,

$$\mathbf{f} \approx \mathbf{g} \text{ iff } \mathbf{f} \preceq \mathbf{g} \text{ and } \mathbf{g} \preceq \mathbf{f}.$$

We define strict preference $\prec$ by,

$$\mathbf{f} \prec \mathbf{g} \text{ iff } \mathbf{f} \preceq \mathbf{g} \text{ and } \mathbf{g} \not\preceq \mathbf{f}.$$

We introduce the concept of a *null* outcome set as one that is disregarded in all considerations of desirability of wagers. Obviously, the empty set is a null set; we leave open the possibility that there might be others, regarded as by the agent as negligible in all deliberations regarding preferences between wagers. (Heuristically: in the probabilistic case, null outcomes are those to which the agent ascribes zero probability.)

Definition. Let A be an experiment, $F \in \mathcal{F}^A$. F is *null* iff, for all wagers $\mathbf{f}$, $\mathbf{g}$ that differ only on F, $\mathbf{f} \approx \mathbf{g}$.

The next axiom says that preferences between wagers depend only on their payoffs on the class of outcomes on which the wagers disagree. If I have wagers $\mathbf{f}$, $\mathbf{g}$ on an experiment A, that differ only on an outcome set F and agree (yield the same payoffs) on $S^A - F$, then I can replace them by wagers $\mathbf{f}'$, $\mathbf{g}'$ that agree with $\mathbf{f}$, $\mathbf{g}$, respectively, on F, and agree with each other on $S^A - F$, without changing the preference ordering.

P2. Let A be an experiment, $F \in \mathcal{F}^A$ a set of outcomes of A, and let $\mathbf{f}$, $\mathbf{f}'$, $\mathbf{g}$, $\mathbf{g}'$ be wagers on A such that, on F, $\mathbf{f}$ agrees with $\mathbf{f}'$ and $\mathbf{g}$ agrees with $\mathbf{g}'$, and on $S^A - F$, $\mathbf{f}$ agrees with $\mathbf{g}$ and $\mathbf{f}'$ agrees with $\mathbf{g}'$. If $\mathbf{f} \preceq \mathbf{g}$, then $\mathbf{f}' \preceq \mathbf{g}'$.

The next axiom is context-independence of preferences between payoffs. A preference for receiving b to a as a result of one wager carries over to other wagers.

P3. Let A, B be experiments, let $\mathbf{f}$, $\mathbf{f}'$ be wagers on A that pay a, b, respectively, on $F \in \mathcal{F}^A$, and coincide otherwise, and let $\mathbf{g}$, $\mathbf{g}'$ be wagers on B that pay a, b, respectively, on $G \in \mathcal{F}^B$, and coincide otherwise. If $\mathbf{f} \prec \mathbf{f}'$, then $\mathbf{g} \preceq \mathbf{g}'$, and $\mathbf{g} \approx \mathbf{g}'$ only if G is null.

For any payoff a and any experiment A, there will be a trivial wager $I^A(a)$ that pays a no matter what happens. P3 ensures that preferences between such trivial wagers are independent of the experiment performed. With this axiom in place, the preference order on wagers induces a preference order on payoffs: $a \preceq b$ iff $I^A(a) \preceq I^A(b)$ for some experiment A (hence, by P3, for all experiments).

Suppose I am given a choice between wagers:

$\mathbf{f}$: Receive \$1000 on F, nothing otherwise.
$\mathbf{g}$: Receive \$1000 on G, nothing otherwise.

Suppose I prefer $\mathbf{g}$ to $\mathbf{f}$. Then it is reasonable to expect that this preference would not change if some other payoff that I prefer to receiving nothing were substituted for the \$1000. Then (assuming I like chocolate cupcakes) I should therefore also prefer prefer $\mathbf{g}'$ to $\mathbf{f}'$, where these are defined by

$\mathbf{f}'$: Receive a chocolate cupcake on F, nothing otherwise.
$\mathbf{g}'$: Receive a chocolate cupcake on G, nothing otherwise.

The next axiom is meant to capture this intuition.

P4. Let A, B be experiments, $F \in \mathcal{F}^A$, $G \in \mathcal{F}^B$. If a, b, a', b' are payoffs such that $b \prec a$ and $b' \prec a'$, and $[F \to a, \neg F \to b] \preceq [G \to a, \neg G \to b]$, then $[F \to a', \neg F \to b'] \preceq [G \to a', \neg G \to b']$.

With this axiom in place, we can define an ordering $\preccurlyeq$ on wagerable outcome sets.

Definition. For $F \in \mathcal{F}^A$, $G \in \mathcal{F}^B$, $F \preccurlyeq G$ iff there exist payoffs a, b such that $a \prec b$ and $[F \to a, \neg F \to b] \preceq [G \to a, \neg G \to b]$.

It is easy to check that $\preccurlyeq$, so defined, is a reflexive, transitive, total ordering. We define an equivalence relation $F \sim G$ as $F \preccurlyeq G$ and $G \preccurlyeq F$, and a strict order $F \prec G$ as: $F \preccurlyeq G$ and not $F \sim G$. (Heuristically: in the probabilistic case, if $F \sim G$ then F and G are regarded as equally likely by the agent.) If $F \prec G$, then G counts for more in our agent's deliberations than F. Differences in payoffs attached to G have more effect on desirability of the overall wager than differences in payoffs attached to F. If $F \sim G$, then F and G hold the same weight in our agent's deliberations.

So far, everything that has been said is compatible with the preference ordering being a trivial one: $\mathbf{f} \approx \mathbf{g}$ for all wagers $\mathbf{f}$, $\mathbf{g}$. This is the preference ordering of an agent who has achieved a state of sublime detachment. To exclude such a state of nirvana, we add a non-triviality axiom.

P5. There exist payoffs a, b such that b is strictly preferred to a, that is, $a \prec b$.

We want to be able to turn the qualitative relation $\preccurlyeq$ into a quantitative one. That is, we want to associate with each outcome set F a number $a(F)$ such that $F \preccurlyeq G$ iff $a(F) \leq a(G)$. We can do this if, for every n, there is an experiment A and an n-element partition $\{F_i\}$ of S^A such that $F_i \sim F_j$ for all i, j. Assigning $a(\emptyset) = 0$ and $a(S^A) = 1$ then gives us $a(F) = m/n$ for any union of m distinct elements of this partition. Armed with sets of outcomes on which a takes on all rational values, the fact that $\preccurlyeq$ is a total ordering gives us for any outcome-set G a real number value $a(G)$.

It turns out that we can assume something a bit weaker. If we can always find experiments such that all outcomes are arbitrarily low in the $\preccurlyeq$-ordering, then we can construct n-partitions that are arbitrarily close to being equivalent, and so get a real-valued ordering function in that way. This is Savage's procedure. Thus we add one last axiom,

P6. Let $\mathbf{f}$, $\mathbf{g}$ be wagers on experiments A, B, respectively, such that $\mathbf{f} \prec \mathbf{g}$. Then, for any payoff a, there is an experiment C, compatible with both A and B, and a partition Π^C of S^C, such that, for each element $F \in \Pi^C$, if we consider the modified wager $\mathbf{f}'$ on the combination of A and C that pays a on F, and coincides with $\mathbf{f}$ otherwise, we have $\mathbf{f}' \prec \mathbf{g}$. Similarly, if we form $\mathbf{g}'$ by paying a on F and retaining $\mathbf{g}$'s payoff otherwise, then we have $\mathbf{f} \prec \mathbf{g}'$.

We now have all the conditions we need for a representation theorem.

Theorem 1 *(Savage). If the preference ordering $\preceq$ satisfies P1–P6, then there exists a utility function u on the set of payoffs (unique up to positive linear transformations), a function a (unique up to a scale factor), which takes as arguments wagerable subsets of experimental outcome-spaces, and a function U on the set of possible wagers, such that, for any experiment A, wagerable partition $\{F_i | i = 1, \ldots n\}$ of S^A, and wager $\mathbf{f} = [F_i \rightarrow a_i]$,*

$$U(\mathbf{f}) = \sum_{i=1}^{n} a(F_i)\, u(a_i)$$

and, for all wagers $\mathbf{f}$, $\mathbf{g}$, $U(\mathbf{f}) \leq U(\mathbf{g})$ iff $\mathbf{f} \preceq \mathbf{g}$.

Theorem 1 says that our agent's judgements, if they satisfy P1–P6, are as if the agent is maximizing expected utility with u giving the utilities attached to payoffs, and the a function acting as if it represents degrees of belief in the outcomes of experiments. See Savage [1972] for proof.

3.1 Learning

Our agent may revise her judgements about wagers on future experiments upon learning the results of past experiments: she may learn from experience. Suppose an experiment A is to be performed, and that our agent is to learn which member of an n-element partition $\{D_i^A\}$ the outcome of A falls into, after which she will be given a choice between wagers $\mathbf{f}$ and $\mathbf{g}$, defined on a partition $\{E_j^B \mid j = 1, \dots, m\}$ of S^B. Her choice of wager on B may, in general, depend on the outcome of A. There are 2^n strategies that she can adopt, specifying, for each D_i^A, whether her choice would be $\mathbf{f}$ or $\mathbf{g}$ were she to learn that outcome of A was in D_i^A. Her choice of which strategy to adopt is equivalent to a choice among a set of 2^n wagers on the combined outcome of A and B. Each such wager consists of specifying, for each i, whether the payoff on $D_i^A \cap E_j^B$ will be $\mathbf{f}$'s payoff on E_j^B for every j, or $\mathbf{g}$'s payoff on E_j^B. Our agent's preference ordering on wagers therefore induces a preference ordering on updating strategies.

We wish to consider changes of preference that can be regarded as pure learning experiences. This means: changes that do not involve a re-evaluation of the agent's prior judgements, and come about solely as a result of acquiring a new piece of information. We do not claim that no other change of preference is rational; the agent may reassess her judgements and revise them as a result of mere cogitation. For changes that are *not* of this sort, the following axiom is a reasonable constraint (and may even be taken as part of what one *means* by a "pure learning experience").

P7. During pure learning experiences, the agent adopts the strategy of updating preferences between wagers that, on her current preferences, she ranks highest.

This preferred updating strategy is easy to characterize.

Theorem 2. *Define the updated utility that assigns the value*

$$U_i^A(\mathbf{f}) = \sum_{j=1}^m a_i^A(E_j^B)\, u(f_j),$$

to the wager $[E_j^B \to f_j]$ *on B, where* a_i^A *is defined by*

$$a_i^A(E_j^B) = \frac{a(D_i^A \cap E_j^B)}{a(D_i^A)}.$$

for non-null D_i^A. *The strategy that recommends, upon learning that the outcome of experiment A is in* D_i^A, *that subsequent choices of wagers be made on the basis of* U_i^A, *is preferred to any other updating strategy.*

Theorem 2 says that the strategy that ranks highest in our agent's preference ordering is the strategy equivalent to updating by conditionalization. See Appendix for proof.

3.2 Repeatable Experiments

We are interested in repeatable experiments. Now, no two experiments are exactly alike (for one thing, they occur at different places or different times, which in practice means that the physical environment is different in *some* respect). But our agent might regard two experiments as essentially the same, at least with respect to preferences between wagers on outcomes. Suppose we have a sequence $\mathcal{A}$ of mutually compatible experiments $\{A_1, A_2, \ldots\}$, with isomorphic outcome spaces. For ease of locution, we will simply identify the outcome spaces, and speak as if two elements of the sequence can yield the same outcome. If, for every composite wager formed from independent wagers on each of a finite subsequence $\mathcal{E}$ of $\mathcal{A}$, an agent's assessment of the value of the wager is unchanged if the payoffs attached to any two elements of $\mathcal{E}$ are switched, we will say, following de Finetti, that the sequence $\mathcal{A}$ is an *exchangeable* sequence for that agent. This condition on a preference ordering is intended to reflect a judgement that the sequence consists of repeated instances of essentially the same type of experiment.

Note that exchangeability is a characteristic, not of the sequence of experiments, but of an agent's preference ordering over wagers on the outcomes of the experiments, and reflects judgements that the agent makes about which factors are irrelevant to the value of a wager. Note also that our agent's judgements about wagers on experiments in a sequence of repeatable experiments need not be independent of each other. Knowing the outcome of one experiment might be relevant to judgements about the value of wagers on other members of the sequence.

We can use de Finetti's representation theorem to characterize the α-functions and, hence, the utility functions U, on which a sequence will be exchangeable. Among utility functions that make $\mathcal{A}$ an exchangeable sequence there are some that make wagers independent of each other, in the sense that knowing the outcome of some subset of experiments in the sequence makes no difference to the evaluation of wagers on the other elements of the sequence. The de Finetti theorem specifies the form of these utility functions, and says that any utility function that makes $\mathcal{A}$ exchangeable is a mixture of such utilities.

First, some definitions that will facilitate stating the theorem. Let $\mathcal{A}$ be a sequence of mutually compatible experiments, and let $\Pi^A = \{F_i | i = 1, \ldots, n\}$ be a partition of their common outcome space S^A. For any finite subsequence $\mathcal{E}$ of $\mathcal{A}$, let $\mathfrak{A}_{\mathcal{E}}$ be the composite experiment consisting of elements of $\mathcal{E}$. If $\mathcal{E}$ is an m-element subset, the outcome space of $\mathfrak{A}_{\mathcal{E}}$ is $S^A \times \ldots \times S^A$ (m times). Form the partition Σ of this outcome space whose elements (n^m of them) correspond to specifying, for each experiment $A_i \in \mathcal{E}$, which member of the partition Π^A the outcome of A_i falls into. For each $s \in \Sigma$, let $\mathbf{k}(s)$ be the vector $(k_1, k_2, \ldots, k_n)$, where k_i specifies how many times an outcome in F_i occurs in s. For example, if $m = 10$, and Π^A is a two-element partition $\{F_1, F_2\}$, one element of Σ would be

$$s = (1, 2, 2, 2, 1, 1, 2, 2, 2, 1),$$

and we would have $k_1(s) = 4$, $k_2(s) = 6$. Note that we must have

$$\sum_{i=1}^{n} k_i = m.$$

We will be interested in wagers on which payoffs are paid independently on elements of $\mathcal{E}$; that is, wagers $\mathbf{f}$ composed of wagers $\mathbf{f}_j = [F_i \to a_{ji}]$ on $A_j \in \mathcal{E}$. The sequence $\mathcal{A}$ is exchangeable if, for every finite subsequence $\mathcal{E}$, and any such composite wager $\mathbf{f}$ on $\mathfrak{A}_{\mathcal{E}}$, the value of $\mathbf{f}$ is unchanged by permutations of the component wagers $\mathbf{f}_j$.

Let Λ_n be the $(n-1)$-dimensional simplex consisting of vectors $\lambda = (\lambda_1, \ldots, \lambda_n)$ satisfying the constraint:

$$\sum_{i=1}^{n} \lambda_i = 1.$$

For any $\lambda \in \Lambda_n$, we can define an a-function,

$$a_\lambda(s) = \lambda_1^{k_1(s)} \lambda_2^{k_2(s)} \ldots \lambda_n^{k_n(s)}.$$

It is easy to check that the utility functions that assign values to wagers on m-member subsets of $\mathcal{A}$ by

$$U_\lambda(\mathbf{f}) = \sum_{s \in \Sigma} a_\lambda(s) \, u(f_s)$$

are ones on which $\mathcal{A}$ is exchangeable, and, moreover, are ones on which elements of the sequence are independent of each other. What de Finetti showed is that *any* utility function on which $\mathcal{A}$ is exchangeable can be written as a mixture of such functions.

Theorem 3 *(de Finetti, 1937). Let $\mathcal{A}$ be an exchangeable sequence of experiments, $\{F_i | i = 1, \ldots, n\}$ a partition of their common outcome space S^A. Then there is a measure μ on Λ_n such that, for any wager f on the outcomes of a finite subsequence $\mathcal{E}$ of experiments in $\mathcal{A}$,*

$$U(\mathbf{f}) = \int_{\Lambda_n} d\mu(\lambda) \, U_\lambda(\mathbf{f}).$$

We are now close to having all the conditions required for our agent to take relative frequency of results of past experiments in an exchangeable sequence as a guide to future preferences between wagers. Close, but not quite there. Consider an agent who initially bets at even odds on a coin toss. Suppose, now, that the coin is tossed 100 times, with heads coming up each time. We would regard it as reasonable for the agent to favour heads on the next toss: she should prefer a reward on heads to the same reward on tails. It is, however, compatible with all the conditions above, including that she treat successive coin tosses as exchangeable, that our agent should resist learning from past experience, and continue to bet at even odds. We therefore add a condition that her preferences be non-dogmatic.

P8. For any exchangeable sequence $\mathcal{A}$, the measure μ appearing in the de Finetti representation should not assign measure zero to any open subset of Λ_n.

We now have learning from experience within an exchangeable sequence. As an example, consider a repeated coin flip. Since we have only two possible outcomes for each flip, Λ_n is just the unit interval $[0, 1]$, and the extremal α functions can be characterized by a single parameter λ. These extremal α-functions are those that assign, to a sequence s of N flips containing m heads and $n = N - m$ tails, the value

$$a_\lambda(s) = \lambda^m (1 - \lambda)^n.$$

Suppose that the measure μ is represented by a density function $\mu(\lambda)$.

$$a = \int_0^1 d\lambda \, \mu(\lambda) \, a_\lambda$$

After observing a sequence s of N tosses containing m heads and n tails, our agent updates the α-function that she uses to evaluate subsequent wagers by conditionalization,

$$a \to a_s,$$

which is equivalent to updating the density function μ via

$$\mu \to \mu_s,$$

where

$$\mu_s(\lambda) \propto \lambda^m (1 - \lambda)^n \, \mu(\lambda).$$

The function

$$l(\lambda) = \lambda^m (1 - \lambda)^n$$

is peaked at $\lambda = m/N$, which is the relative frequency of heads in the observed sequence s. Moreover, it is more sharply peaked (with a width that goes as $1/\sqrt{N}$), the larger the observed sequence. Thus, if our agent's initial α-function is non-dogmatic, for sufficiently large N the density μ will end up concentrated on an interval around the observed relative frequency, with width of order $1/\sqrt{N}$.

Strict exchangeability is a condition that will rarely be satisfied for agents with realistic judgments about wagers. The agent might not be *completely* certain that differences between elements of the sequence $\mathcal{A}$ ought to be regarded as irrelevant. If they are successive throws of a die, for example, our agent might not completely disregard the possibility that some observable feature of the environment is relevant to the outcomes of the die. Her α-function, accordingly, will be a mixture of one on which the sequence is exchangeable, and others containing correlations between the elements of $\mathcal{A}$ and the results of other possible experiments. There are generalizations of the de Finetti representation theorem that encompass such situations. Not surprisingly, they have the result that the agent can learn which experiments she ought to take as correlated and which she ought to take as independent, and may converge towards a judgement of exchangeability regarding a sequence of possible experiments. See Diaconis and Freedman [1980], Skyrms [1984, ch. 3], and Skyrms [1994] for discussions of such generalizations.

3.3 On the Notion of Physical Chance

The de Finetti representation theorem shows that an agent whose degrees of belief make a sequence of coin tosses an exchangeable sequence will bet in exactly the same way as someone who believes that there is an objective chance, perhaps imperfectly known, for each toss to come up heads, and who has degrees of belief concerning the value of this chance, which mesh with her degrees of belief concerning outcomes of the tosses in the way prescribed by Lewis's principal principle (Lewis [1980]). Furthermore, if our agent's degrees of belief are non-dogmatic, she will, upon learning the results of an initial finite sequence of tosses, update her betting preferences in exactly the same way as someone who takes these tosses to be informative about the chance of heads on the next toss. This has been taken by some—and was so taken by de Finetti—to indicate that the notion of objective chance is eliminable. There is another way to look at it, however: the agent's degrees of belief are, implicitly, degrees of belief about objective chances. An extremal α-function α_λ represents a chance distribution on which the chance of obtaining a result in F_i is λ_i, and the mixture represents the agent's degrees of belief about which of these functions give the actual chances.[4]

Should we, then, in some circumstances at least, ascribe beliefs about objective chances to agents? Note that, even if we start with the idea that probabilities are subjective, we are not thereby committed to denying that some probability assignments are better adapted to the world than others. De Finetti famously declared that the only criterion of admissibility of probability assignments is that of *coherence*; all probability assignments 'are admissible assignments: each of these evaluations corresponds to a coherent opinion, to an opinion legitimate in itself, and every individual is free to adopt that one of these opinions which he prefers, or, to put it more plainly, that which he *feels*' (de Finetti [1980 p.64]). Such language suggests that all probability assignments are equally valuable, but note that de Finetti is careful not to say that. Once an agent has adopted a probability assignment, she will not freely exchange it for any other. Nor will an agent always regard her own judgements to be the best. Suppose that Alice and Bob both have degrees of belief on which a certain sequence of experiments is exchangeable, and that their priors are the same, or close enough that differences are negligible. Suppose Alice learns the result of the first 100 elements of the sequence, and Bob does not, and that they are both offered wagers on the result of the 101st. Unless Bob has zero degree of belief that learning what Alice knows would affect his judgement about the wager he is about to undertake, coherence requires that he strictly prefer betting according to Alice's judgements to betting according to his own current judgements, if offered the choice. He does *not* regard all assignments of probability as equally valuable, and does not even rank his own highest.

Suppose, now, that there is a sequence A of experiments that Bob judges to be exchangeable, and that there are no other experiments except those in A that he takes to be relevant to elements of the sequence, and suppose his preferences are non-dogmatic. Then, if offered the opportunity to accept or reject wagers on an element A of the sequence, he would certainly prefer to have knowledge of outcomes of other elements of the sequence that have already been performed. Furthermore, if there are elements of the sequence that have not been performed, but could have been, he would prefer that they had been performed, because knowledge of the outcomes of these would improve his betting situation. There will, however, typically be no experiments that either have or could have been performed that would lead him to certainty regarding the outcome of A.

---

[4] There is an analogy here with the relationship of the principle of maximizing expected utility (MEU) to the Savage representation theorem. The Savage theorem shows that, given certain constraints on preferences, there will be a credence and utility functions according to which her preferences satisfy MEU. We are, in effect, using MEU to ascribe credences and utilities to the agent. Similarly, the de Finetti theorem shows that, given certain constraints, the agent acts as if she has credences about chances, credences that satisfy the principal principle. One could say: it is via this principle that we ascribe beliefs about chances to the agent.

However, he is certain that there is some probability function over the potential outcomes of A to which his degrees of belief, and those of any other agent who judged the sequence exchangeable and was non-dogmatic, would converge, were they to learn the results of sufficiently many other members of the sequence.

Suppose that on Bob's credences, the results of experiments not in the sequence $\mathcal{A}$ are irrelevant to experiments in $\mathcal{A}$. Then, the extremal α-functions are invariant under conditionalization on the results of any experiment that has been or could have been performed prior to betting on a given element of $\mathcal{A}$. They are, in this sense, regarded by Bob as candidates for being the maximally well-informed, or optimal betting strategy. He does not currently know which one of them is in fact optimal, but his current betting preferences are epistemically weighted averages reflecting his current degrees of belief about what the optimal strategy is. The optimal strategy is not subjective, in the sense of being the betting strategy of any agent. It is something that Bob regards as optimal for bets on a certain class of experimental set-ups. Furthermore, when he conditionalizes on the results of elements of the sequence, he learns about what the optimal strategy is, and he is certain that any agent with non-dogmatic priors on which the sequence of experiments is exchangeable will converge to the same optimal strategy. If this is not the same as believing that there are objective chances, then it is something that serves the same purpose. Rather than eliminate the notion of objective chance, we have uncovered, in Bob's belief state, implicit beliefs about chances—or, at least, about something that plays the same role in his epistemic life.

To generalize beyond the case of exchangeability: suppose that Bob has degrees of belief regarding the outcome of an experiment A, which can be represented as mixtures of probability functions that he regards as states of maximal accessible knowledge, in the sense of being invariant under conditionalization on results of all experiments that either actually have or could have been performed prior to A, and suppose that we can show that, with probability one, Bob's beliefs would converge to one of these, given a sufficient body of information of the sort that could be accessible to an agent about to bet on the outcome of A. Then Bob's preferences between wagers are as if he thinks that one of these extremal, maximally informed probability distributions is the correct chance distribution, and his preferences reflect degrees of belief about what the chance distribution is.[5]

Presumably, the physics of an experimental set-up is relevant to which betting strategy on outcomes of the experiment is optimal. Bob may formulate theories about what the optimal strategy is for a given experimental set-up. Experiments that he regards as informative about these optimal strategies will accordingly raise or lower his degrees of belief in such theories. One sort of theory would be one in which the dynamical laws are stochastic, invoking an irreducible chance element.

---

[5] This discussion is heavily indebted to that found in chapter 3 of Skyrms [1984]. See also Skyrms [1994].

The theory could also have deterministic dynamics. Though such a theory will map initial conditions into outcomes of experiments, it might nevertheless be the case that the maximal *accessible* information (confined to learning the results of all experiments that have or could have been performed, prior to the experiment on which the wager is placed) falls short of information sufficient to decide with certainty between experimental outcomes. This is the case with the Bohm theory. Though the theory is deterministic, it is a consequence of the theory that no agent can have knowledge of particle positions that would permit an improvement over betting according to Born-rule probabilities. In this context, these maximally informed degrees of belief play the role of objective chances.

Lewis remarked, of the notion of objective chance, 'Like it or not, we have this concept' (Lewis [1980 p.269]). To which we might add: like it or not, an agent with suitable preferences acts as if she believes that there are objective chances associated with outcomes of the experiments, about which she can learn, provided she is non-dogmatic. This, together with the assumption that physical theories may have something to say about these chances, is all we require for our account of theory confirmation. There may be more to be said about the nature and ontological status of such chances, but, whatever more is said, it should not affect the basic picture of confirmation we have sketched.

Though the notion of physical chance is not reducible either to epistemic probability or frequency, the three are intimately related. An agent who updates her epistemic probabilities by conditionalizing on the results of repeated experiments will take the relative frequencies of outcomes in these experiments as evidence about the values of physical chances. In this way theories that say something about physical chances are confirmed or disconfirmed by experiment. Note that we have not needed to pass to an infinite limit to achieve such confirmation. Nor is there any need for a substantive additional assumption such as 'Assume that your data are typical.' It is a consequence of conditionalizing on the data that degree of belief is raised in theories that posit chances that are close to the observed relative frequencies and lowered in theories that posit chances that are far from the observed relative frequencies.

4 THE GARDEN OF FORKING PATHS

Suppose now that our agent, having read Borges' 'The Garden of Forking Paths', (Borges [1941, 1962]) thinks of an experiment as an event in which the world divides into branches, with each outcome occurring on some branch. On each of the branches is a copy of herself, along with copies of everyone else in the world, and each payoff is actually paid on those branches on which the outcome associated with that payoff occurs. How much of the foregoing analysis would have to be revised?

We claim: none of it. The Savage axioms are requirements on the preferences of a rational agent, whether the agent conceives of an experiment in the usual way, with only one outcome, or as a branching occurrence, with all of the payoffs actually paid on some branch or another. The reader is invited to go back and reconsider the axioms in this light. (We will discuss some possible objections to this claim in Section 5.)

Reinterpreting experiments in this way, however, does force a reinterpretation of the a-functions that appear in the representation of the agent's preferences. The reason is that on a branching interpretation of experiments, $a(F)$ cannot in general be interpreted as degree of belief that the outcome of the experiment will lie in the set F: our agent may have degree of belief 1 that each outcome associated with a non-null subset of S^A will occur (on some branch), but still in general $a(F) < 1$. What we *can* say, on the basis of the way it (still) feeds into the maximization of expected utility formula, is that the function $a(F)$ is a measure of the weight that the agent attaches in her deliberations to branches having outcomes in F.

What the de Finetti representation shows (now) is that, for an exchangeable sequence, the agent's a-function will have the form of degrees of belief concerning optimal branch weights, where these 'branch weights' play the role of physical chances in her deliberations. When our agent updates her preferences by conditionalization on experimental results, she will take the results of previous experiments in an exchangeable sequence as informative about branch weights (rather than about chances).

Ordinary quantum mechanics consists of the Hilbert-space framework, plus interpretive rules that tell us how to associate operators with experimental set-ups and state vectors (or density operators) with preparation procedures, plus the Born rule, which tells us to interpret the squares of amplitudes as chances of outcomes of experiments. It is this latter rule that gives the theory much of its empirical content; theories that make claims about physical chances are confirmed or disconfirmed in the manner described in the previous section.

Now consider Everettian quantum mechanics as a theory that retains the Hilbert-space framework, the same associations of operators with experimental set-ups and state vectors or density operators with preparation procedures, but replaces the Born rule with the rule: the squares of amplitudes are to be interpreted, not as chances of outcomes, but as branch weights. The calculated values can be compared with the results of experiments, and Everettian quantum mechanics is confirmed in much the same way as quantum mechanics with Born-rule chances.

On this view, we (as agents who are agnostic about whether or not our world is a branching one) should be taking relative frequency data as informative about quantities that are *either* physical chances *or* physical branch weights. A hypothesis that makes claims about physical branch weights is confirmed by the data to precisely the same extent as a hypothesis that attributes the same numerical values to chances. As with chance, there may be more to be said about the nature and ontological status of these branch weights, but such a further account is not expected to affect the basics of how branch-weight theories are confirmed.

We close this section with two examples, intended to clarify and fix ideas by showing how this account works in two particular cases.

4.1 Example: The Unbiased Die

Recall the question with which we started: whether, on 24 tosses of a pair of fair dice, it is better to receive $1000 if a pair of sixes comes up at least once, or to receive $1000 if a pair of sixes never comes up. We now consider this question in a branching context.

There is a continuum of ways in which the dice can land, but we are interested only in which faces are up when the dice come to rest. We therefore partition the outcome space into the $36^{24} \approx 2.24 \times 10^{37}$ classes corresponding to distinct sequences of results. Suppose that these classes of outcomes are all regarded by our agent as equivalent, with respect to wagers—she will not change her estimation of the value of a wager on this experiment upon permutation of the payoffs associated with elements of the partition. Then she ought to prefer a wager **g** that pays $1000 on the branches on which a pair of sixes does not occur, and nothing on all branches on which this does occur, to a wager **f** with the payoffs reversed. Why? Because there are $35^{24} \approx 1.14 \times 10^{37}$ sequences of results on which a pair of sixes never occurs, and only $36^{24} - 35^{24} \approx 1.10 \times 10^{37}$ on which at least one pair of sixes does occur. The wager **f** can be converted via a permutation of payoffs into a wager **f'** in which the $1000 is received on $36^{24} - 35^{24}$ elements of our partition on which a pair of sixes does not occur, and nothing is paid on the remaining branches. By assumption, this does not change the value of the wager, and so **f** $\approx$ **f'**. We can obtain **g** from **f'** by giving a $1000 reward on each of the remaining branches—corresponding to approximately 4×10^{35} elements of our partition—on which a pair of sixes does not occur. If it is better for the agents on those branches to receive $1000 than to receive nothing, then, by P2, we should regard **g** as preferable to **f'**, and hence, to **f**.

4.2 Example: The Biased Die

Suppose that our agent has available to her the records of outcomes of a great many previous rolls of the die, and examination shows that, though one of them displays the behaviour expected of a fair die, the other has shown a 6 in a fraction of outcomes significantly higher than 1/6. On the ordinary view of a die toss, we would say that it should be possible for sufficient data of this sort to reverse her estimates of the values of the wagers **f** and **g**, and come to prefer **f** to **g**. It does not take a huge bias to reverse this preference. If, for example, one die is unbiased, and the other has a chance of 6/35 of showing a six on any given toss, then the chance that a pair of sixes shows up at least once in 24 tosses is approximately 0.501, and **f** is the marginally better wager.

On a branching view also, if our agent has exchangeable, non-dogmatic prior preferences about wagers on the dice tosses, then a sufficiently large number of tosses showing bias will lead her to prefer **f** to **g**. She regards the statistical evidence as informative about branch weights and concludes that, though there is a greater number of sequences of possible outcomes of 24 dice tosses in which a pair of sixes never occurs, the set of branches on which a pair of sixes comes up at least once has a higher total weight. That is, it is better to reward her successors on that set of branches, than on its complement.

A simpler example will give the flavour of this reasoning. Suppose that our agent is initially sure that a coin is either biased two-to-one in favour of heads, or two-to-one in favour of tails, with her degrees of belief evenly divided between these two alternatives. That is, she believes that the coin either produces, on each toss, branches with total weight 2/3 on which it lands heads and branches with total weight 1/3 on which it lands tails, or branches with total weight 1/3 on which it lands heads and total weight 2/3 on which it lands tails. Suppose that the coin is to be tossed twice, and that, after learning the result of the first toss, she will be given the choice between receiving $1000 if the second toss lands heads, and $1000 if the second toss lands tails. She resolves to bet on heads on the second toss if the first toss is heads, and on tails if the first toss is tails. She reasons as follows. If the coin is biased towards heads, then it is better to make the second bet on heads. On her strategy, this will happen on weight 2/3 of branches, with the wrong bet being made on weight 1/3. Similarly, if the coin is biased towards tails. Her estimation of a strategy is an epistemically weighted mean of its value if the coin is biased towards heads, and its value if the coin is biased towards tails. The strategy she has resolved to follow is the one with the highest expected value.

If she is to be coherent, and if she is to follow this strategy upon learning the outcome of the first toss, then she must revise her degree of belief about the branch weights via conditionalization. That is, an agent who sees heads on the first toss will have degree of belief 2/3 that the coin is biased towards heads. Of course, there will be branches on which our agent's successors are misled, and *decrease* their degrees of belief in the true hypothesis about branch weights. But on a higher weight of branches the agent's successors will have their belief-states improved.[6]

It is a consequence of updating beliefs about branch weights by conditionalization that agents on all branches take the results of experiments to favour

---

[6] This can be made precise: it can be shown, via the argument of Greaves and Wallace [2006], that, on any reasonable way of measuring the epistemic value of a belief-state, updating by conditionalization maximizes expected epistemic value. This epistemic-utility argument is complementary to the intertemporal-consistency defence of conditionalization given in the Appendix.

hypotheses that afford their own branches high weight, and so boost their degrees of belief in such hypotheses and lower their degrees of belief in hypotheses that afford their own branches low weight. The copy of our agent on each branch ends up believing that the set of branches that share the outcome that she has seen has high weight. Some of them will be mistaken, of course. But there will be a higher total weight of agents who have had their beliefs about branch weights altered in the direction of the truth, than of those who have been misled.

5 OBJECTIONS AND REPLIES

As our presentation above has tried to emphasize, there is a pervasive structural analogy between chance theories and branching-universe theories (and between chances and branch weights, and between possible worlds and branches). Correspondingly, many of the objections that might be raised against the proposed account of decision-making and/or belief-updating in the face of branching have equally compelling (or uncompelling) analogues in the chance case. This is important: we claim only that the Everett interpretation is *no worse off* than any other theory vis-à-vis the philosophy of probability, so any objection that applies equally to both cases will be irrelevant to the present project.

Before considering particular objections in any depth, we therefore summarize the analogy that we see between the two cases. It will be helpful to keep this analogy in mind in the discussion that follows because, if there is to be a branching-specific objection, it must take the form of a claim that the analogy presented here is incomplete in some *relevant* respect; in every case, our replies will claim that it is not.

Chance set-up (gamble)	Branch set-up (bramble)
Preferences between wagers go as maximizing expected utility, which is an average of utilities across . . .	
alternative possible outcomes	all branches
weighted by an α-function, and we call this α-function . . .	
a credence function.	a quasicredence function.[7]

. . . Continued

---

[7] In Greaves [2004] and Greaves [2007a], the term 'caring measure' was also used. It was applied to the measures over branches that lie to the future of a given branch in a given multiverse that one obtains by conditionalizing the agent's quasicredence function (that is, her α-function) on the self-locating proposition that she is currently on the branch in question in the multiverse in question. There is thus a 'caring measure' that coincides with the quasicredence function, and gives the agent's betting quotients, in the special case (and only in that case) in which the agent is sure which multiverse is actual and which branch in that multiverse she is on. (This comment is included only to clarify the relationship between the three papers in question; the concept of 'caring measure' plays no special role in the present paper.)

Chance set-up (gamble)	Branch set-up (bramble)
For an exchangeable sequence of experiments, the agent's α-function can be represented as a mixture of extremal exchangeable functions α_λ. The agent acts as if she believes that one of these extremal functions is objectively the best one to base decisions on (although, in general, she is not sure which), and her α-function is an epistemically weighted average of them. That is, she acts as if the α_λ's are candidates for being . . .	
objective chance distributions,	objective branch weights,
and the weighting (μ) of these that yields her α-function reflects her credences about which vector λ gives the right one. Updating by conditionalizing on results of experiments in the exchangeable sequence permits her to refine her credences about which α_λ is objectively best. Part of the content of quantum mechanics is the claim (which either is derivable from the nonprobabilistic part of the theory, or is an independent postulate of the theory), that . . .	
chance = |amplitude|^2.	branch weight = |amplitude|^2.
Call this the Born rule. Note that the Born rule is a substantive claim: left and right side of this equation have independent meanings (the left implicitly defined by decision theory, the right by quantum mechanics). Moreover, it is an empirically testable claim. Conditionalizing on results of experiments in the exchangeable sequence will cause the agent's credences about the values of . . .	
chances	branch weights
to become peaked about the observed relative frequency. If the observed relative frequency is close to the value calculated from the Born rule, it will raise credence that quantum mechanics is correct; if it is far from the Born rule value, it will lower credence that quantum mechanics is correct.	
There are possible worlds in which . . .	There will certainly be branches on which . . .
anomalous statistics occur. A frequentist *analysis* of . . .	
chance	branch weight
is untenable: one cannot hold that . . .	
'the chance of *E* is *x*' . . .	'the branch weight of *E* is *x*' . . .
just means 'the long-run frequency of *E* will be *x*' because, for any *E* and *x*,	
it is possible that . . .	there will be some branches on which . . .
the long-run frequency of *E* is *not x*. Relative frequencies are connected only evidentially with . . .	
chances. In anomalous-statistics possible worlds,	branch weights. On anomalous-statistics branches,

agents are misled: they rationally lower their credence in the theory that is in fact true. Still,

the possible worlds in which this occurs have a low total chance.	the branches on which this occurs have a low total weight.
There is no available updating policy that guarantees that agents raise their credence in the true theory . . .	
in *every* possible world.	on *every* branch.
It therefore makes sense that the conditionalization strategy recommended is the optimal one:	
low-chance . . .	low-weight . . .
events don't count for much in the evaluation of wagers.	

We now consider and reply to eight foreseen objections to the account proposed in the present paper.

Objection 1: Branch weights are not probabilities. Reply: we do not claim that they are. The claim is, rather, that, given reasonable constraints on an epistemic agent's preferences between wagers, she will act as if she believes that there are physical branch weights, analogous to physical chances, that can be estimated empirically in the same way that chances are, and that observation of events to which a theory assigns high branch weight boosts rational credence in a branching theory in the same way that observation of events to which a theory assigns high chance boosts rational credence in a chance theory.

Objection 2: The decision-theoretic account is all about the behaviour of rational agents; this is (surely) irrelevant to matters of physics, and so cannot supply the Everett interpretation with an acceptable account of *physical* probability.

We do not accept that the behavior of a rational decision-maker should play a role in modeling physical systems.   (Gill [2005])

The reply to this has two parts. The first is that there is a clear sense in which, in order to model a physical system, one does not need to invoke considerations of rationality, and that this remains true in Everettian quantum mechanics. The second (and deeper) point is that—the first point notwithstanding—considerations of rationality have always played a role, and indeed must play a role, in the *confirmation* of physical theories, so it is no objection to the approach outlined above that it brings rationality considerations into the discussion of the confirmation of Everettian quantum mechanics. Let us explicate each of these two points in turn.

First, the sense in which the modelling of physical systems is silent on issues of rationality. The point here is perfectly straightforward. According to quantum theory (Everettian or otherwise), one models a physical system by ascribing to it a quantum state—a vector in, or density operator on, some Hilbert space.

On the Everettian account of measurement, after a measurement there will exist a multiplicity of branches; the quantum state of the universe will be a superposition of the states of these branches, with some particular set of complex coefficients (amplitudes). Here we have, in outline, a physically complete account of the situation before and after measurement, and nothing has been said about rationality.

Now let us move on to the second point: that considerations of rationality *must* be relevant to theory confirmation. The point can be seen abstractly as follows. The question under consideration—when we are talking about theory confirmation—is that of *which physical theories it is rational to believe* (or have significant degree of belief in, or have significant degree of belief in the approximate truth of, etc.), given the evidence we in fact have. This is a question about a *relation* between physical theories on the one hand, and rationality on the other. It should then be of no surprise that, in answering the question, we need to consider the theory of rationality, as well as our various candidate theories of physics.

The point can be made more vivid by considering a more concrete case. Let us put the issue of branching-universe theories aside for the moment. Suppose that we have a physical theory, call it T, that is irreducibly stochastic. (T can be thought of as, for example, a dynamical collapse theory along the lines proposed by GRW et al.) Consider some fixed experimental set-up, A. Suppose the way that A is to be modelled in terms of T (including initial conditions) is always the same. Then, according to T, there are a number of possible outcomes for the experiment A: $s_1, \ldots, s_n$. T also assigns chances to the various possible outcomes: $p_1, \ldots, p_n$ for $s_1, \ldots, s_n$ respectively. But now suppose that these so-called 'chances' are unrelated to considerations of rationality. In particular, suppose that there is no rationality constraint to the effect that the experimenter, insofar as she believes T, should bet at odds given by $p_1, \ldots, p_n$ on the outcome of the experiment; and that there is no rationality constraint with the consequence that, if in a long run of repetitions of the experiment she observes relative frequencies that approximately match the single-case chances predicted by T, and that no other available theory has this so-called 'virtue', then she should increase her degree of confidence in the theory over its rivals. Under these suppositions, *the 'chances' ascribed by the theory would have become altogether idle*: for all practical and theoretical purposes, we would be no better off than if our theory merely said that such-and-such a range of outcomes was *possible*, and ascribed the various possibilities no chances at all. In particular, the evidential connection between theoretical single-case chances (on the one hand) and observed relative frequencies (on the other) can be made to reappear only by admitting the connection between chances and rational belief revision.

Why is this point often missed? In our view, the explanation is the prevalence of (a) a frequentist analysis of chance and (b) a falsificationist account of confirmation—both of which accounts are importantly defective. On the combination of these two (defective) accounts, one reasons as follows. First, one takes it that one knows perfectly well what to do with predictions of the form 'the chance of E is x', without touching on issues of rationality: such predictions *just mean* (according to frequentism) that in a long run of repetitions of the experiment, the relative frequency of E will be approximately x. Second, one notes (as a consequence) that if the observed relative frequency deviates significantly from the theory's single-case chance, then something has happened that the theory predicted would not happen, and hence has been falsified; if, on the other hand, there is approximate agreement (and, perhaps, the prediction was a 'risky' one), then the theory has been confirmed (or perhaps 'corroborated').

The deficiencies of frequentism and falsificationism are well known. To repeat: a probabilistic theory does *not* predict, *categorically*, that the observed relative-frequency *will* approximately match the theoretical single-case chance; what it predicts is that (in a sufficiently long run of experiments) this matching will be observed *with probability close to one*. So the first assertion in the above frequentist-deductivist account is false. And when one replaces 'the theory predicts that the observed relative frequency will approximately match the theoretical single-case chance' with 'the theory ascribes probability close to one to the proposition that the observed relative frequency will approximately match the theoretical single-case chance', the second step in the above account develops a glaring hole: if the observed relative frequency deviates significantly from the theory's single-case chance, then something has happened that the theory *ascribed low probability to*, but this is perfectly consistent with the theory's being true, so the theory has not been falsified.

The would-be deductivist is then tempted to patch up the account with a principle to the effect that, if something happens that the theory deemed *sufficiently improbable*—say, to which the theory ascribed probability less than some threshold p_{thresh}—then the theory is to be regarded as effectively falsified. But this patching-up will not work either: for every way the observations could turn out (including relative frequencies that approximately match the theoretical single-case chances), there will be some description under which 'those observations' were astronomically improbable (such as the particular ordered sequence of outcomes observed).

To escape from this quagmire, one must move to something more closely resembling a Bayesian account of theory confirmation. But then, if one really wants to be precise about the details, one is up to one's elbows in rationality constraints—on belief-updating, and on the connection between conditional credences and chances (the principal principle). The account we have given is just the extension to the branching case of this standard Bayesian account. To be sure, one can, for the purposes of most discussions of physical theory, avoid *explicit* discussion of rationality. One can simply help oneself to a particular consequence of the Bayesian theory: the principle (CC) stated in Section 2. The same thing can be done in the Everettian case: one can simply help oneself to the principle (CW) stated above. This, too, obviates the need to write several paragraphs on decision theory before drawing evidential conclusions from a laboratory experiment. But it is a myth that the foundations of these confirmation-theoretic principles are independent of the theory of rational belief and decision. Our task in this paper has been to provide the foundation for the principle (CW): it is only for this reason that our discussion has been more explicitly rationality-theoretic than that found in the average physics text.

Objection 3: The decision-theoretic account presented here shows that agents must attach *some* decision-theoretic weights to branches, but it does not show that these weights must equal those given by the Born rule. There is a sense in which this is correct, and a sense in which it is not.

The observation made by the 'objection' is correct in the sense that we have not supplied a 'derivation of the Born rule' from the pre-existing part of the theory. That is, we have not supplied an a priori proof that betting quotients for outcomes, conditional on the truth of Everettian quantum mechanics *stripped* of any explicit postulate about the relationship of *branch weights* to (say) the amplitude-mod-squared measure over branches, must, on pain of irrationality, be those given by the Born rule. That is, we have not made the claim that is made by Deutsch [1999] and Wallace [2003, 2007]. However, as we will now explain, we do not take this to be ground for any objection to our account.

The status of derivations of Born-rule weights within Everettian quantum mechanics is (at least prima facie) similar to the status of Gleason's theorem and related results[8] concerning probability in quantum mechanics. They show that certain assumptions lead to Born-rule probabilities (or weights). The assumptions used as premises in such proofs are not beyond question. At most, such proofs show that Born-rule chances/ignorance-probabilities/branch weights are the only ones that fit naturally with, or, perhaps, are definable in terms of, the *existing* structure of quantum mechanics. It is an open question what the significance of this is: whether, chance/ignorance-probability/weight predictions *should* be thus definable in terms of the structure that is already present in the theory prior to the introduction of chances/ignorance-probabilities/weights. Further, it is conceivable that the answer to this question could turn out to be different in a chance or an ignorance-probability theory than in a branching-universe theory. For example, Wallace [2007, section 6] can be understood as arguing that in the branching case decision-theoretic branch weights must be definable in terms of the structure of the theory, but that the analogous claim for chances or ignorance probabilities is not true; meanwhile, the existence of the subject of non-equilibrium statistical mechanics, of the work of Valentini et al. on Bohmian mechanics 'out of quantum equilibrium', (see e.g., Pearle and Valentini [2006])

---

[8] A nice recent example is the Zurek derivation of the Born rule from envariance, Zurek [2005]. Barnum [2003] has shown how to turn this proof into one that takes no-signalling as its main premise.

show (for whatever this is worth) that the principle that probabilities be definable in terms of pre-existing structure is in fact flouted by some (non-branching) theories taken seriously by working scientists.

Perhaps probabilities and/or weights must be derivable from pre-existing structure; perhaps this issue plays out differently in branching and non-branching theories; perhaps not. We have no further comment to offer on these issues. Fortunately, such issues are irrelevant to our central claim. If chances/ignorance-probabilities/branch weights must be definable in terms of the existing structure of the theory, then the Born rule *seems* to be the only option, in either a branching or a non-branching version of quantum mechanics. In that case, had we consistently observed non-Born frequencies, we would have been compelled to abandon quantum theory altogether. If (on the other hand) there is no requirement that weights be definable in terms of existing structure, then two versions of Everettian quantum mechanics that agree on everything but the branch weights but ascribe non-Born weights to branches are, for the purposes of theory confirmation, distinct theories, just as Bohmian mechanics in and out of quantum equilibrium are distinct theories. In this case, had we observed non-Born frequencies, we would have had more latitude; it is the package as a whole, branches (or possible histories) plus branch weights (or chances, or ignorance-probabilities) that is confirmed or disconfirmed; it might have been open to us to retain a core of quantum theory, but to adopt a different rule for the chances/ignorance-probabilities/branch weights.

As things have turned out empirically, however, this is all largely irrelevant: we have observed Born frequencies and so, whether or not there are other coherent theories out there that otherwise agree with Everettian quantum mechanics but postulate non-Born branch weights, the theory with Born-rule branch weights has been empirically confirmed (and any candidate theories with non-Born branch weights have been disconfirmed). This is the sense in which the objection is *in*correct: our account *does* have the consequence that—whether or not there exists a satisfactory 'derivation of the Born rule' from the pre-existing part of the theory—rational agents who observe long runs of Born-rule frequencies will increase their degree of belief that the weights of future branches are those given by the Born rule.

Objection 4: There are branches on which non-quantum statistics are observed.
Hemmo and Pitowsky write:

Even for agents like us, who observed up to now finite sequences which a posteriori seem to conform to the quantum probability [i.e. the Born rule], adopting the quantum probability as our subjective probability for future action is completely arbitrary, since there are future copies of us who are bound to observe frequencies that do not match the quantum probabilities. (Hemmo and Pitowsky [2007 p.348])

The inference from the existence of branches to the arbitrariness of adopting Born-rule probabilities as guides to future choice requires some explanation. The argument seems to be something like this. On the ordinary account, evidence from past relative frequencies provides grounds for believing, if not with certainty then at least with high degree of belief, that future relative frequencies will be similar. But, on an Everettian account, there are no grounds for such belief, and we are in fact *certain* that relative frequencies will deviate arbitrarily far from Born-rule weights on *some* future branches. In the absence of an account on which observation of past frequencies is evidence that it is better, in some sense, to adopt Born-rule probabilities as guides to future actions, these past observations are irrelevant to future action.

We claim to have supplied such an account. Theorem 2 shows that updating beliefs about branch weights by conditionalizing on observed data is preferred to any other strategy. In worlds like ours, provided only that the agent regards the sequences of experiments in question as repeatable, this leads to beliefs that the Born-rule branch weights are at least approximately correct. It is, therefore, not arbitrary.

If our agent has priors on which a sequence of experiments is exchangeable, and if these are non-dogmatic, then she will treat past experience as relevant to future action. Of course, an agent might have priors that are such that the result of one experiment is never relevant to that of another, and so be unable to learn from experience which betting strategies are better than which other. But she would be an agent who could not do science. We are not aware of any reason for thinking that the sort of assumption that entails learning from experience is any less reasonable in the branching than in the non-branching case.

Objection 5: According to the Everett interpretation, what the observer learns when she observes a measurement outcome is only self-locating information. This cannot possibly be relevant to theory confirmation. The idea here is as follows. Consider an agent who is about to perform some quantum measurement with n possible outcomes $O_1, \ldots, O_n$. Conditional on the proposition that the Everett interpretation is true, this agent is *certain*, for each value of i from 1 to n, that there will be some future branch on which O_i occurs, and some future copy of herself on that branch. The measurement is then performed. A later copy of our agent looks at the apparatus in her lab, and observes that, on her branch, some particular outcome O_j occurred. Then (the thought runs) *conditional on the truth of Everettian quantum mechanics, the information she has acquired is purely self-locating* — she knew all along (conditional on the truth of Everettian quantum mechanics) that there would be such a copy of herself, and now she has merely observed that indeed there is. Therefore (the objection continues), she cannot possibly have learned anything that is evidentially relevant to the truth of Everettian quantum mechanics. (The thought is related to that raised in objection 4.)

Let us put aside the awkwardness ('knew conditional on the truth of Everettian quantum mechanics', etc.) required to state the sense in which the information is 'purely self-locating'. The key mistake on which the above objection rests is the idea that information that is 'purely self-locating', in the sense that it does not rule out any possible worlds, is necessarily also evidentially irrelevant to *de dicto* propositions (i.e., that it cannot, under rational belief-updating, result in the redistribution of credences between possible worlds).

Such a principle cannot be sacrosanct; there are in any case many known counterexamples, independent of the Everett interpretation. Consider, for example, the prisoner in a lighted cell, who knows that it is six o'clock in the evening and that the light in her cell will be switched off at midnight if she is to be hanged at dawn. Some significant amount of time passes, and the light stays on; the prisoner rationally becomes more confident that she will live another 24 hours. But nothing that she has learnt *rules out* the possibility that she will be hanged at dawn: she remains uncertain as to whether or not midnight has really passed. For a second (more familiar, but also more controversial) example, consider Sleeping Beauty: the two most common analyses of Beauty's case, the 'thirder' (Elga [2000]) and 'halfer' (Lewis [2001]) analyses, agree that on learning that it's Monday, Beauty acquires evidence that the coin landed heads, despite the fact that her being awake on Monday is *consistent* both with heads and with tails.

Our account is one according to which this (anyway non-sacrosanct) principle is routinely violated: information about the outcomes of experiments (in the possible world in which, and/or on the branch on which, the agent is now located) is a type of information that, even in the highly idealized cases in which it becomes purely self-locating (i.e. cases in which the agent is certain that some branching-universe theory is correct), *is* evidentially relevant to *de dicto* propositions. Furthermore, we are aware of no well-motivated alternative account of belief-updating that renders it evidentially *ir*relevant. (The methodological point implicit in this reply is that it is often more reliable first to work out which global belief-updating strategies are candidates for rational status, and afterwards to draw conclusions about which sorts of information can be evidentially relevant to which sorts of propositions, than vice versa.)

Objection 6: Decision-making is incompatible with deterministic physics. This issue is an old one. If the underlying physical dynamics is deterministic, then the decision that our agent is going to make is already determined by the present state of the universe, together with the dynamical laws. It is an illusion, according to this objection, that she has any decision at all to make.

In reply, it should first be mentioned that this does not differentially affect our account, but applies equally well to any deterministic physical theory. Nor does a move to an indeterministic physics help; making my actions partly a matter of chance does not address the concerns behind this objection.

Fortunately, we do not have to consider the age-old problem of freedom of the will here. Our axioms concern *rational* preferences between wagers. It makes sense to have such preferences, and to evaluate them as rational or irrational, independently of questions of our ability to act on the basis of such evaluations.

We frequently evaluate our own actions (often, negatively) in cases in which, due to weakness of the will, we are unable to act in the way that we judge to be best.

Objection 7: Preferences between wagers is nonsensical in a branching universe. On the branching-universe view, *all* payoffs corresponding to non-null outcomes are actually paid to agents on the corresponding branches. This is certainly a departure from the usual way of thinking about wagers. Some readers may find themselves at sea when contemplating such a scenario, and it may seem that we have no clear ideas about what preferences between branching wagers might be reasonable.

We claim that the situation is not so grim. For one thing, there do seem to be some clearly defensible principles regarding rational preferences between branching wagers. A wager that pays a desirable payoff on all branches is surely preferred to a wager that pays nothing on all branches. If, on every outcome, the payoff paid by f is at least as desirable as that paid by g, then f is at least as desirable as g.

If we accept that preferences between wagers makes sense in the branching case, and accept also that there are principles that reasonable preferences between branching wagers ought to satisfy, then the question still arises whether the axioms we have laid down are reasonable constraints on rational preference in the branching case.

Objection 8: The decision-theoretic axioms are *not* as defensible in the branching as in the non-branching case. This is the most serious objection. If the axioms are accepted for preferences between wagers on a branching scenario, then confirmation of theories that posit branch weights proceeds in a manner entirely parallel to confirmation of theories that posit chances.

One occasionally comes across the following idea: since decision-making conditional on the assumption that the Everett interpretation is true is decision-making under conditions of certainty, 'the' decision theory for such decision scenarios is trivial (meaning: it consists merely of the requirement that preferences be total and transitive, that is, our axiom P1).[9]

If intended as an objection to the account defended in this paper, this point would beg the question entirely. One can, of course, write down both trivial and non-trivial decision theories, both for decision-making in the face of indeterminism and for decision-making in the face of branching. The fact that decision-making in the face of branching had not been seriously considered (and hence no non-trivial decision theories for that case advocated) prior to 1999 is irrelevant; the question is which decision theories are *reasonable*. Our claim is that the non-trivial decision theory we have outlined is no less reasonable in application to the branching case than in its long-accepted application to the

---

[9] Mostly this suggestion has been made in conversation; however, see also Wallace [2002, section 3.2].

structurally identical indeterministic case. A non-question-begging objection in this area must give a *reason* for thinking that, structural identity notwithstanding, the decision-theoretic axioms that we have discussed, while reasonable in the indeterministic case, ought not to be applied to branch set-ups in the way that we have advocated.

Such a reason must involve a difference between chance set-ups and branch set-ups. The most obvious difference is that, on a branching scenario, all outcomes actually occur, whereas, in the non-branching scenario, only one outcome is actual. Moreover, if a given payoff is paid only on a class of outcomes with low chance, our agent can be reasonably certain that that payoff will not be the one that is paid. In the branching scenario, the corresponding payoff is sure to be paid, albeit on a class of outcomes with low weight.

Why might this difference be relevant? We will not explore the full range of possible reasons here; we discuss only the one that, in our own opinion, poses the most serious prima facie problem for our account. It is this. The fact that all outcomes actually occur supplies a sense, *possibly* relevant, in which preferences between wagers on branch set-ups ('brambles', as we called them, following Barry Loewer, in the above table) are analogous to questions of distributive justice.[10] Now, the representation we obtain from our axioms P1–P6 is one in which alternative distributions of payoffs are judged according to a weighted mean of utilities on all branches. Though some, namely utilitarians, accept that judgements regarding distributive justice, no less than questions of rational decision under uncertainty, are to be addressed in this way, there are of course dissenters from such a view. Rawls, for example, argues for a maximin rule, which seeks to maximize the well-being of those that are worst off (Rawls [1999]). Someone who accepts such a view for questions of distributive justice could still think that rational strategies for prudent decision-making under uncertainty conform to the axioms. If she regards preferences among brambles as relevantly similar to questions of distributive justice rather than to preferences among gambles, she will then accept our axioms for chance set-ups, but be wary of a rule for *branch* set-ups that ranks wagers according to a weighted mean of payoffs on branches.

Two replies can be made to this objection. The first is that even if preferences among brambles is relevantly similar to questions of distributive justice, still the representation theorem discussed in Section 3 will do the epistemic work we have claimed it can do in the branching context. The second is that it is at least far from obvious that the similarity in question *is* relevant. We will set out these two replies in turn.

The first reply runs as follows. We concede (for the sake of argument) that preferences among brambles are relevantly similar to questions of distributive justice, but we claim that, *for a suitably restricted class of decisions*, even questions of distributive justice are suitably treated using the weighted-average formula.

---

[10] See Huw Price's contribution in Chapter 12.

The point is the following: all that is required, for the representation theorem to go through, is that there be at least two payoffs, one of which is strictly preferred to the other. This means that we do not need to make the strong claim that our axioms apply even when some of the brambles among which the agent is choosing assign a *terrible* outcome to some branches, or in general when the utility differences between branches are large. (This is relevant because part of the intuition underlying the objection is that there are some things that we ought not do to *anyone*, no matter how great the benefit to others might be.) Suppose, then, that we restrict our attention to preferences between wagers involving only the payoffs, a: receive one chocolate doughnut, and b: receive two chocolate doughnuts, with b strictly preferred to a. The trivial wager $I(b)$ is strictly preferred to the trivial wager $I(a)$. All other wagers in this restricted class are ties on the maximin rule, since they share the same worst outcome. It does not seem reasonable to simply be indifferent between all such wagers; if f and g coincide except on a non-null class F of outcomes, on which f awards a and g awards b, then, surely, g is to be strictly preferred to f. We claim that axioms P1–P6 are reasonable ones for breaking maximin ties within this restricted class of wagers. And, if this is accepted, then we still get a representation on which values of wagers are represented as a weighted mean of utilities on branches. For the confirmation-theoretic purposes of this paper, we do need to claim that a rational agent should always have the same preferences when faced with a bramble or with a corresponding gamble. Perhaps a case can be made for this strong claim; but we need not take the analogy so far.

The second reply, rather than conceding the objector's point and arguing that it is not damaging, challenges the point itself, as follows. It is far from obvious that Everettian decision-making is *relevantly* analogous to distributive justice, rather than to decision-making under classical uncertainty, in cases (if any such there be) in which the correct decision procedures for the latter two situations diverge. To be sure, as we noted above (and as our objector emphasizes), brambles and distributive justice problems share the attribute that all candidate reward-recipients are actual. But, since any two scenarios are similar in some respects and dissimilar in others, the existence of *some* criterion effecting this grouping is trivial. There are, of course, many other (more or less natural) criteria that would group brambles and gambles together while excluding distributive justice problems, and still others that would group gambles and distributive justice problems together while excluding brambles. (An example of the former type of criterion is: are the candidate reward recipients future copies of the decision-making agent?[11] An example of the latter is: was the scenario in question discussed prior to 1950?)

---

[11] Following a panel discussion at the Perimeter Institute conference during which one of us raised this point, Simon Saunders suggested an alternative that is probably more to the point: Is there any interaction between the recipients?

What is required, in order to assess the relevance or otherwise of appeals to distributive justice, is a careful exploration of precisely which differences between scenarios of classical uncertainty and those of distributive justice are responsible for the divergence in recommended policies; only once we have carried out such an explanation can we know to which the Everettian case it is relevantly analogous. This project has not (to our knowledge) been carried out, and lies beyond the scope of the present paper. But in the meantime, it is at least plausible that the fairness-based intuitions that motivate deviations from maximization of expected utility in the case of distributive justice are grounded in issues of trust and power dynamics, present in a complex community of distinct and interacting agents but absent in the case of brambles, and have nothing at all to do with the mere fact that all candidate recipients are actual. (When deciding whether to increase contributions to one's pension fund or blow the extra money on an expensive holiday next year, one doesn't worry about whether or not one's allocation of resources between one's next-year self and one's older self is *fair*, despite the fact that both are [timelessly] actual. It is very interesting to ask why not, and we do not know the answer; but the datum is clear.) It is thus at least plausible that the analogy to cases of distributive justice is irrelevant.

Let us turn to P7. This axiom *seems* non-controversial: unless our agent has cause to re-evaluate her earlier judgements about preferences between wagers, she should continue to employ the updating strategy that, on her initial preferences, she deemed the best. This, as we have seen, is equivalent to updating by conditionalization. The following objection, however, can be raised to P7 in the branching case.[12]

The account defended in this paper has the post-branching agent adopting the updating strategy ranked highest by the pre-branching agent. But (the objection runs) our post-branching agent *knows* that, if the Everett interpretation is true, then her interests now are not the same as the interests of her pre-branching self—the latter's interest was to maximize average utility across branches according to the measure of importance of those branches, whereas the former's interest is to maximize utility on whichever branch she is in fact now on. And (the objection continues) intertemporal consistency criteria—such as P7—can have the status of rationality constraints only if the agent-stages concerned believe that they have the same interests as their temporal counterparts. Therefore (the objection concludes), P7 is a rationality constraint for chance set-ups but not for branch set-ups.

Before replying, let us illustrate the objection by elaborating on the sort of example that it suggests. We imagine a situation in which, prior to a sequence of two coin flips, our agent (call her Alice$_0$) weighs options and decides whether,

[12] We are grateful to Tim Maudlin for raising a similar objection to a predecessor of the position defended in this paper, and for extensive discussion.

in her estimation, her successors on each of the post-flip branches should prefer wager **f** or wager **g** on the outcome of the second flip. Now let Alice$_1$ be a successor on one of the branches after the first coin flip, but suppose that Alice$_1$ has not yet learned the outcome of this first coin toss. We imagine that Alice$_1$ opts not to take the advice of Alice$_0$, on the grounds that her interests are different. However, since the situation is meant to be one of the sort for which updating by conditionalization would be required in the non-branching case, we must stipulate that Alice$_1$ still endorses Alice$_0$'s judgements as appropriate for Alice$_0$'s situation. (Recall that P7 restricts the intertemporal consistency requirement to 'pure learning experiences'.) Alice$_1$'s reasoning must then be: 'Alice$_0$, in formulating her advice, was concerned with maximizing mean benefit across all branches to *her* future. But I'm only concerned with myself and with my future branches. (And, by the way, I see no reason for my self-locating credences concerning which branch I am now on to bear any particular relationship to Alice$_0$'s estimates of the relative importance of branches.) Though Alice$_0$ was right, given her concerns, to recommend that I choose **f**, it would be better for me, with my concerns, to choose **g**.'

We claim that Alice$_1$ should, rather, accept Alice$_0$'s advice on whether to take **f** or **g**. The reason is that there is a relevant sense in which the interests of Alice$_0$ and Alice$_1$ are the same: they both ultimately aim to maximize actual payoff averaged with respect to the actual branch weights. But given that they do not know the actual branch weights, the preferences of each of them over wagers are given by maximizing average payoff with respect to their respective credences about the branch weights. Now, the actual branch weight of a given payoff on a wager on the second coin flip is the same downstream of Alice$_1$ as it was downstream of Alice$_0$, and Alice$_1$ knows this. And Alice$_1$ has gained no new information about the branch weights; she is, with respect to branch weights, in a 'pure learning situation' in which nothing has been learned. If she endorses Alice$_0$'s credences, she should therefore retain the same credences about branch weights, and hence the same preferences among wagers. Hence Alice$_1$ will endorse Alice$_0$'s recommendations about what she should do upon learning the result of the experiment. The subsequent learning of the outcome of the experiment involves no branching, merely a gain in knowledge, so there is no room for a supposed change of interest to alter her judgements about what she should do, upon learning which sort of branch she is on.

Someone might accept P1–P7 and nevertheless insist on a more egalitarian treatment of measurement outcomes, continuing to bet at even odds on the outcome of a coin toss, even in the face of a string of tosses in which heads predominate. After all, the argument goes, there will be a copy of me on the *H* branch, and a copy of me on the *T* branch; ought not I be fair, and treat both of these copies equally? This amounts to rejecting P8, which is meant to exclude dogmatism of this sort.

This is reminiscent of Laplaceanism, and the reply is similar. Counting branches is not as simple as that. After all, there are many ways in which the coin can land heads, and many ways in which it can land tails. A wager that makes the payoff depend only on heads or tails imposes a partition on the outcome space. There is no necessity, and no compelling reason, why this partition, rather than some other, must be treated so that copies of me in one class hold the same weight in my deliberations as copies of me in the other class. Just as, in applications of probability conceived in the ordinary way, Laplace's definition of probability must be supplemented by judgements of *which* classes of events are to be judged equipossible, so too would an egalitarian approach to preferences between branching wagers require a judgement of which partitions of an experiment's outcome space are to be afforded equal weight. What we suggest is the same for the branching case as for the non-branching case: let experience be your guide.

6 CONCLUSIONS

Everettian quantum mechanics ascribes weights to branches. We have outlined an account according to which rational agents use these weights as if they were chances in evaluating bets that may give different payoffs on different branches, and the occurrence of events to which the theory ascribed a weight higher than the average chance-or-weight ascribed by rival theories increases rational degree of belief in the Everettian theory. That is, on this account, branch weights play both the decision-theoretic and the confirmation-theoretic role that chances play. We have argued that this account is no less defensible than the structurally identical account according to which chances, in an indeterministic theory, have similar decision-theoretic and confirmation-theoretic relevance. It follows from the same decision-theoretic axioms, via the same representation theorems; and, we claim, the axioms are no less plausible under our suggested interpretation in branching contexts than they are under the familiar interpretation in non-branching contexts. If correct, this solves the prima facie evidential problem that the Everett interpretation seemed to face.

Acknowledgements

Would like to thank the organizers of, and participants in, two conferences at which this paper was presented—the 'Everett at 50' conference held in Oxford in July 2007, and the 'Many Worlds at 50' conference held at Perimeter Institute in September 2007—and the participants in a seminar at the University of Western Ontario in June 2007. Special thanks are due to Sona Ghosh, Bill Harper, Barry Loewer, Tim Maudlin, Simon Saunders, and David Wallace.

Appendix:

Experiments A, B are to be performed in succession. Our agent is to be informed which element of a partition $\{D_i^A \mid i = 1, \ldots, n\}$ of S^A that the outcome of experiment A falls into, and then offered a choice between wagers $\mathbf{f} = [E_j^B \to f_j]$, $\mathbf{g} = [E_j^B \to g_j]$ on the outcome of B, where $\{E_j^B \mid j = 1, \ldots, m\}$ is a partition of S^B. A *strategy* consists of a choice, for each i, of $\mathbf{f}$ or $\mathbf{g}$ as the preferred wager on B upon learning that the outcome of A is in D_i.

Define the strategy ϕ by

$$
\phi_i = \begin{cases}
\mathbf{f}, & \text{if } \sum_j a(D_i^A \cap E_j^B)\, u(f_j) \geq \sum_j a(D_i^A \cap E_j^B)\, u(g_j) \\[2ex]
\mathbf{g}, & \text{if } \sum_j a(D_i^A \cap E_j^B)\, u(f_j) < \sum_j a(D_i^A \cap E_j^B)\, u(g_j)
\end{cases}
$$

Let $\overline{\phi}_i$ be the opposite strategy: if ϕ_i is $\mathbf{f}$, $\overline{\phi}_s$ is $\mathbf{g}$, and vice versa. ϕ's choices are such that, for each i,

$$
\sum_j a(D_i^A \cap E_j^B)\left(u(\phi_{ij}) - u(\overline{\phi}_{ij})\right) \geq 0.
$$

We will say that ϕ strictly prefers ϕ_i to $\overline{\phi}_i$ iff

$$
\sum_j a(D_i^A \cap E_j^B)\left(u(\phi_{ij}) - u(\overline{\phi}_{ij})\right) > 0.
$$

We will show that:

i). For any strategy ψ, $\psi \preceq \phi$.
ii). If, for some i, ϕ_i is strictly preferred to $\overline{\phi}_i$, then, for any strategy ψ that disagrees with ϕ's choice on i, $\psi \prec \phi$.

Let ψ be any strategy. $\phi \prec \psi$ iff

$$
\sum_i \sum_j a(D_i^A \cap E_j^B)\, u(\phi_{ij}) < \sum_i \sum_j a(D_i^A \cap E_j^B)\, u(\psi_{ij}),
$$

or,

$$
\sum_i \sum_j a(D_i^A \cap E_j^B)\left(u(\phi_{ij}) - u(\psi_{ij})\right) < 0.
$$

There is no contribution to this sum from those i, if any, on which ϕ and ψ agree. When ϕ and ψ disagree, $\psi_i = \overline{\phi}_i$. For such i,

$$
\sum_j a(D_i^A \cap E_j^B)\left(u(\phi_{ij}) - u(\psi_{ij})\right) = \sum_j a(D_i^A \cap E_j^B)\left(u(\phi_{ij}) - u(\overline{\phi}_{ij})\right) \geq 0,
$$

and so,

$$\sum_i \sum_j \alpha(D_i^A \cap E_j^B)\left(u(\phi_{ij}) - u(\psi_{ij})\right) \geq 0,$$

or, $\psi \preceq \phi$. If, for any i,

$$\sum_j \alpha(D_i^A \cap E_j^B)\left(u(\phi_{ij}) - u(\overline{\phi}_{ij})\right) > 0$$

we have $\phi \prec \psi$ for any ψ with $\psi_i = \overline{\phi}_i$.

This gives us α_i^A up to an arbitrary scale factor. If we wish to normalize the updated α-function, so that $\alpha_i^A(S^B) = 1$, we have

$$\alpha_i^A(E_j^B) = \frac{\alpha(D_i^A \cap E_j^B)}{\alpha(D_i^A)}.$$

References

Albert, D. and B. Loewer [1988], 'Interpreting the many worlds interpretation', *Synthese* 77, 195–213.

Barnum, H. [2003], 'No-signalling-based version of Zurek's derivation of quantum probabilities: A note on "environment-assisted invariance, entanglement, and probabilities in quantum physics" '. Avaliable online at quant-ph/0312150.

Bernoulli, J. [2006], *The Art of Conjecturing*, Johns Hopkins University Press, Baltimore.

Borges, J.L. [1941], *El Jardín de Senderos que se Bifurcan*, Sur, Buenos Aires.

—— [1962], 'The garden of forking paths', in D.E. Yates and J.E. Irby (eds), *Labyrinths: Selected Stories & Other Writings*, New Directions, New York.

de Finetti, B. [1937], 'La prevision: ses lois logiques, ses sources subjectives', *Annales de l'Institute Henri Poincaré* 7, 1–68. English translation in de Finetti [1980].

—— [1980], 'Foresight: its logical laws, its subjective sources', in H.E. Kyburg and H.E. Smokler (eds), *Studies in Subjective Probability*, Robert E. Krieger Publishing Company, Huntington, New York. Translation of de Finetti [1937].

Deutsch, D. [1999], 'Quantum theory of probability and decisions', *Proceedings of the Royal Society of London* A455, 3129–37. Preprint available online at arXiv.org.

Diaconis, P. and D. Freedman [1980], 'De Finetti's generalizations of exchangeability', in R.C. Jeffrey (ed.), *Studies in Inductive Logic and Probability, Volume II*, University of Caifornia Press, Berkeley and Los Angeles.

Diaconis, P., S. Holmes, and R. Montgomery [2007], 'Dynamical bias in the coin toss', *SIAM Review* 49, 211–35.

Elga, A. [2000], 'Self-locating belief and the sleeping beauty problem', *Analysis* 60, 143–7.

Gill, R. [2005], 'On an argument of David Deutsch', in M. Schurmann and U. Franz (eds), *Quantum Probability and Infinite Dimensional Analysis: from Foundations to Applications. QP–PQ: Quantum Probability and White Noise Analysis, Volume 18*. World Scientific.

Greaves, H. [2004], 'Understanding Deutsch's probability in a deterministic multiverse', *Studies in History and Philosophy of Modern Physics* **35**, 423–56.

—— [2007a], 'On the Everettian epistemic problem', *Studies in History and Philosophy of Modern Physics* **38**, 120–52.

—— [2007b], 'Probability in the Everett interpretation', *Philosophy Compass* **2**, 109–28.

Greaves, H. and D. Wallace [2006], 'Justifying conditionalization: Conditionalization maximizes expected epistemic utility', *Mind* **115**, 607–32.

Hemmo, M. and I. Pitowsky [2007], 'Quantum probability and many worlds', *Studies in History and Philosophy of Modern Physics* **38**, 333–50.

Huygens, C. [1660], *Van Rekeningh in Spelen van Gluck*, Amsterdam.

Lewis, D. [1980], 'A subjectivist's guide to objective chance', in R.C. Jeffrey (ed.), *Studies in Inductive Logic and Probability*, Volume 2, pp.263–93, University of Caifornia Press.

—— [2001], 'Sleeping beauty: reply to Elga', *Analysis* **61**, 171–6.

Ore, O. [1960], 'Pascal and the invention of probability theory', *The American Mathematical Monthly* **67**, 409–19.

Pearle, P. and A. Valentini [2006], 'Generalizations of quantum mechanics', in G.N.J.-P. Francoise and T.S. Tsun (eds), *Encyclopaedia of Mathematical Physics*, Elsevier.

Rawls, J. [1999], *A Theory of Justice*, Harvard University Press, Cambridge, MA. Revised edition.

Saunders, S. [2005], 'What is probability?' in S.D.A. Elitzur and N. Kolenda (eds), *Quo Vadis Quantum Mechanics?* Springer, New York.

Savage, L.J. [1972], *The Foundations of Statistics*, Dover Publications, Inc, New York.

Skyrms, B. [1984], *Pragmatics and Empiricism*, Yale University Press, New Haven and London.

—— [1994], 'Bayesian projectibility', in D. Stalker (ed.), *Grue! : The New Riddle of Induction*, pp.241–62. Open Court, Chicago.

Smith, D. [1959], *A Source Book in Mathematics*, Dover Publications, New York.

Wallace, D. [2002], 'Quantum probability and decision theory, revisited'. Available online at philsciarchive.pitt.edu/archive/00000885/02/decarx.pdf.

—— [2003], 'Everettian rationality: defending Deutsch's approach to probability in the Everett interpretation', *Studies in History and Philosophy of Modern Physics* **34**, 415–39.

—— [2006], 'Epistemology quantized: Circumstances in which we should come to believe the Everett interpretation', *British Journal for the Philosophy of Science* **57**, 655–89.

—— [2007], 'Quantum probability from subjective likelihood: improving on Deutsch's proof of the probability rule', *Studies in History and Philosophy of Modern Physics* **38**, 311–32.

Zurek, W.H. [2005], 'Probabilities from entanglement, Born's rule $p_k = |\psi_k|^2$ from envariance', *Physical Review* **A 71**, 052105.

PART IV
CRITICAL REPLIES

10

One World Versus Many: The Inadequacy of Everettian Accounts of Evolution, Probability, and Scientific Confirmation

Adrian Kent

ABSTRACT

There is a compelling intellectual case for exploring whether purely unitary quantum theory defines a sensible and scientifically adequate theory, as Everett originally proposed. Many different and incompatible attempts to define a coherent Everettian quantum theory have been made over the past 50 years. However, no known version of the theory (unadorned by extra ad hoc postulates) can account for the appearance of probabilities and explain why the theory it was meant to replace, Copenhagen quantum theory, appears to be confirmed, or more generally why our evolutionary history appears to be Born-rule typical.

This article reviews some ingenious and interesting recent attempts in this direction by Wallace, Greaves–Myrvold and others, and explains why they don't work. An account of one-world randomness, which appears scientifically satisfactory, and has no many-worlds analogue, is proposed. A fundamental obstacle to confirming many-worlds theories is illustrated by considering some toy many-worlds models. These models show that branch weights can exist without having any role in either rational decision-making or theory confirmation, and also that the latter two roles are logically separate.

Wallace's proposed decision theoretic axioms for rational agents in a multiverse and claimed derivation of the Born rule are examined. It is argued that Wallace's strategy of axiomatizing a mathematically precise decision theory within a fuzzy Everettian quasiclassical ontology is incoherent. Moreover, Wallace's axioms are not constitutive of rationality either in Everettian quantum theory or in theories in which branchings and branch weights are precisely defined. In both cases, there exist coherent rational strategies that violate some of the axioms.

1 INTRODUCTION

1.1 Some Common Ground

Although I disagree with the Everettian contributors to this volume on some fundamental questions, I think they deserve much credit for developing some creative and interesting ideas and arguments, which have certainly helped advance our understanding of fundamental science. To elaborate on this, let me note some points on which I agree with many Everettians.

First, the Everettian programme had a sensible motivation. Everett [1957] asked, in effect, whether quantum theory really needs to be framed in such a way that the evolution of the wavefunction is governed by two different laws: generic unitary evolution together with the projection postulate when measurement takes place. It's a good question. Even if, rather than the projection postulate, quantum theory came equipped with a precise extra dynamical law implying the postulate as an approximation, it would be natural to ask if we really needed it. As it is, there is quite a compelling case for exploring whether we can make sense of purely unitary quantum theory.

Second, it *is* a sensible project to try to extract a physical ontology from a unitarily evolving quantum state vector, given a theory of the initial state or initial conditions, a Hilbert space defining a representation of position, momentum and other canonical operators, and a dynamical theory that expresses the Hamiltonian in terms of these operators. Whether the project succeeds in producing an ontology with the properties that Everettians fondly imagine is another question—but certainly one worth discussing. One still, strangely, sometimes hears the argument that it is illegitimate—a basic misunderstanding of quantum theory—even to examine the possibility of giving the state vector a direct physical interpretation. This seems to me simply unimaginative dogma. Everettians are right to insist that their programme should be judged on whether or not it works, not on whether it respects pre-Everettian quantum orthodoxies.

Third, neither the apparently fantastic nature of the Everettian worldview, nor the superficial conflict between postulating multiple independent mutually inaccessible worlds and Occam's razor, are entirely compelling arguments against the Everett programme. One needs to consider Everettian ideas in the context of other attempts to make sense of quantum theory, and in detail. One of the great intellectual challenges of theoretical physics is to find a mathematically elegant, universally applicable, Lorentz covariant, scientifically adequate version of quantum theory that supplies a well-defined realist ontology. If the Everett programme really could produce a well-defined Lorentz covariant physical ontology that adds little or no arbitrary structure to the mathematics of quantum theory, and that reproduces all the scientific successes of Copenhagen quantum theory within its domain of validity, it would have solved this fundamental problem. Given the present alternatives, we would, I think, at that point, have to consider it seriously as a possible account of reality.

Now, in fact, I think that the Everett programme fails in these ambitions, for reasons explained below. I am also optimistic (Kent [2010]) that we can find simpler one-world versions of quantum theory that have all the aforementioned virtues and none of the problems that afflict, and I think ultimately doom, the Everett programme. But I see no way to make either conclusion so transparently true as to eliminate the need for argument.

Fourth, it matters—it is scientifically important to understand—whether the Everettian programme can possibly succeed. If Everettians really could produce a theory of reality with all the proclaimed virtues, it would clearly weaken (though not eliminate) the motivation for other attempts at solving the quantum reality problem—just as finding a consistent quantum theory of gravity would weaken (though not eliminate) the motivation for looking for others. Conversely, if, as I argue, the Everettian programme has fairly definitely failed, then the problem of finding a viable formulation of quantum theory applicable to closed quantum systems looms rather large among the concerns of theoretical physics. The failure of the Everett programme adds to the likelihood that the fundamental problem is not our inability to interpret quantum theory correctly but rather a limitation of quantum theory itself. If so, my guess is that we most likely won't find an adequate cosmological theory so long as we assume that quantum theory is universally valid—so we should be looking for possible signals of the failure of quantum theory applied to the universe. Likewise, if so, quantum interference quite likely breaks down somewhere between the microscopic and the macroscopic—so we should be working harder to characterize the most promising types of experiment to test this.

Everettian ideas have been around for 50 years, and influential for at least the past 30. Yet there has never been a consensus among theoretical physicists either that an Everettian account of quantum theory can be made precise and made to work, or that the Everettian programme has been comprehensively refuted. These questions are quite central to the future of theoretical, experimental, and observational physics. We need to resolve them and move forward.

1.2 Everett's Elusive Essence

'When he died, his heirs found nothing save chaotic manuscripts. His family, as you may be aware, wished to condemn them to the fire; but his executor—a Taoist or Buddhist monk—insisted on their publication.'

'We descendants of Ts'ui Pên,' I replied, 'continue to curse that monk.
Their publication was senseless. The book is an indeterminate heap of
contradictory drafts.'

(Jorge Luis Borges, 'The Garden of Forking Paths' [1948])

. . . so crowded with . . . empty sophistication that it is extremely difficult
to perceive the simple errors at the basis. It is like fighting the hydra—cut
off one ugly head, and eight formalizations take its place.

(P.K. Feyerabend, 'How to Defend Society Against Science' [1975])

After 50 years, there is no well-defined, generally agreed set of assumptions and
postulates that together constitute 'the Everett interpretation of quantum theory'. Far from it: Everett [1957, 1973], DeWitt [1973], Graham [1973], Hartle
[1968], Geroch [1984], Deutsch [1985], Deutsch [1999], Saunders [Chapter 6
in this volume], Barbour [2001] (partly inspired by Bell [1987], though Bell's aim
was not to inspire), Albert-Loewer [1988], Coleman [1994], Lockwood [1996],
Wallace [Chapters 1 and 8], Vaidman [2002, Chapter 20], Papineau [Chapter
7], Greaves [2004], Greaves and Myrvold [Chapter 9], Gell-Mann and Hartle
[1993 and Chapter 2], Zurek [Chapter 13] and Tegmark [Chapter 19], among
many others, have offered distinctive and often fundamentally conflicting views
on what precisely one needs to assume in order to get the Everett programme
off the ground, and what precisely an Everettian (or, some say, post-Everettian)
version of quantum theory entails.

I am primarily interested here in contrasting realist 'one-world' and 'many-
worlds' accounts of quantum theory. By *one-worlders*, I mean those who aim to
find a version of quantum theory in which quantum experiments really have only
one outcome, we really have only one version of our future selves at any future
time, and some intrinsic randomness in nature determines which outcome occurs
and which future self is realized from among the range of possibilities defined
by the theory. For example, within its domain of validity, the Copenhagen
interpretation of quantum theory is a one-world theory. By *many-worlders*, I
mean those who share Everett's view that a unitarily evolving quantum state
vector should be interpreted as directly representing reality, and the future
versions of ourselves that observe different outcomes of quantum experiments
should be interpreted as equally real.

So, I will not discuss here attempts at 'post-Everettian' interpretations like
those of Gell-Mann and Hartle [1993] and Zurek [Chapter 13], which fall
into neither camp, and seem—despite much critical probing—unclear on, or
uncommitted to taking a stance on, precisely what, if anything, in the theory
corresponds to objective external reality. Extensive critiques of Gell-Mann and
Hartle's approach can, however, be found elsewhere (Dowker and Kent [1995,
1996], Kent [1996, 1997, 1998a,b, 2000]).

My main focus is on the recent attempts by Wallace [Chapters 1 and 8], Greaves and Myrvold [Chapter 9], and, to a lesser extent, Papineau [Chapter 7] and Saunders [Chapter 6], to define, analyse, and test realist many-worlds interpretations. These authors offer different, and on some points mutually inconsistent, approaches, but nonetheless share enough common perspectives to be considered together. Their papers include some very interesting and creative arguments, which raise important scientific questions. However, I will argue below that none of their approaches produces a scientifically adequate version of quantum theory.

Shadowing these discussions is the spectre of the 'many-minds interpretation' set out some time ago by Albert and Loewer [1988]. Essentially everyone, including Albert and Loewer, agrees that the many-minds interpretation, while logically consistent and in accord with the data, is utterly unsatisfactory, since it adds to the Everettian formalism a collection of ad hoc postulates which not only are (even by Everettian standards) fantastic, but also undercut the motivation for taking Everett seriously, namely that it purports to explain how to make sense of quantum theory without adding extra equations or interpretational postulates. So, no one—certainly no one represented in this volume—wants to be a many-minder. And here lies the problem: it seems to me (and to others—see in particular Albert's contribution in Chapter 11) that, at various points in their arguments, Saunders, Wallace, Greaves–Myrvold and Papineau tacitly—and, since they reject the many-minds interpretation, illegitimately—appeal to many-minds intuitions. Indeed, at least in the first three cases, it seems to me that if one fleshed their ideas out into a fully coherent and complete interpretation, one would necessarily arrive either at the many-minds interpretation or something even worse. I will elaborate on this below.

Of course, these discussions crucially turn on our understanding of what counts as scientifically adequate. The idea that reality contains many essentially independent quasiclassical worlds corresponding to different possible cosmological and experimental outcomes clearly isn't, per se, susceptible to logical refutation. That isn't at issue. The key question, to my mind—and I think modern Everettians, including the authors considered here, generally agree—is whether we can find an appealingly simple version of quantum theory in which a realist many-worlds ontology is essential (i.e. there is no equally simple one-world variant) and which (at minimum) replicates all the scientific successes of one-world quantum theory (i.e. quantum theory including some form of the projection postulate, or some principle from which it can, approximately and within a suitable domain of validity, be derived). I believe that we can't. In particular, it seems to me that the Everettian programme has not produced and cannot produce a scientifically adequate alternative account that reproduces the standard one-world account of probabilistic inferences derived from quantum theory—despite the ingenious recent attempts of contributors to this volume.

Some commentators sympathetic to the Everettian programme (for example Papineau [Chapter 7] and Greaves–Myrvold [Chapter 9]) argue that a double standard is at work here: that criticisms of the Everettian programme's attempt to account for the appearance of probability can and should equally well be applied to the standard understanding of the role of probability in one-world versions of quantum theory, and indeed of probabilistic scientific statements in general. To respond to this point, I consider below some fundamental differences between randomness (or apparent randomness) in one-world quantum theory and its purported Everettian analogue, and point out what seem to me irresolvable problems with the latter.

2 ONE-WORLD THEORIES AND PROBABILITY

Copenhagen quantum theory is a one-world version of quantum theory: any given experiment or quantum event has a number of possible outcomes, but only one actual outcome. Some other non-Everettian variants and modifications of quantum theory, such as de Broglie–Bohm theory and dynamical collapse models, similarly randomly select from many possible physical evolutions, and can be (and usually are) interpreted as defining a unique quasiclassical world. The consistent histories approach (Gell-Mann and Hartle [1993]), if combined with an (alas unknown) suitable set selection rule, would also lead naturally to a one-world interpretation, in which reality is described by one randomly chosen history from the selected set. And these by now venerable contenders certainly don't exhaust the possible options (Kent [2010]). My aim here is not to advocate a specific one-world version or variant of quantum theory, or to assess the current state of the art, but rather to compare and contrast one-world and many-world accounts of probability. For that purpose, let us suppose, for the sake of argument, that we have to hand a particular one-world theory that implies that, while the universe could have evolved in a (presumably very large) number of different ways, one quasiclassically evolving world—the one we observe—was randomly selected.

One-world versions of quantum theory, together with hypotheses about the initial conditions and unitary evolution, predict the probabilities of our experimental results and observations. We test the theory and these hypotheses by checking whether the results are of a form we would typically expect given the predicted probabilities. In practice, pretty much everyone agrees on the methodology of theory confirmation, at least sufficiently so that, for example, everyone agrees that, within the domain of validity of Copenhagen quantum theory, the Born rule is very well confirmed statistically. However, there is much less agreement on how, or even whether, we can make sense of fundamentally probabilistic physical theories. What exactly, if anything, does it mean to say that the probability of the universe turning out the way it did was 0.00038?

Everettian authors have stressed this last point lately. We should not, they argue, apply different standards to one-world and many-worlds quantum theory.

If our account of standard probability applied to one-world quantum theory is suspect, or incomplete, or involves ad hoc postulates, we cannot reasonably reject an alternative many-worlds account on the grounds that it runs into difficulties that might, on close analysis, turn out to be precisely analogous.

There are several possible responses for one-worlders here. One response is to try to defend or buttress or further develop frequentism, or another standard account of standard probability. A second is to try to point out some insuperable problems with many-worlds accounts of probability, and thus make the case that, whatever difficulties one-world quantum theory might run into, many-worlds quantum theory cannot possibly be satisfactory. A third is to argue that the difficulties that many-worlders face in dealing with probability are worse than—not, as claimed, precisely analogous to—those faced by one-worlders.

I think the first of these options is worth pursuing. I think too that the second and third lines of argument are valid, and I will develop them later. But, in this section, I want to make a separate point. I want to suggest a non-standard account in which the scientific space usually occupied by one-world probabilistic theories is filled instead by deterministic theories with a large amount of theoretically unspecified data. This allows us to compare, verify, and falsify theories, and to recover essentially all of current science, without assigning a fundamental role to probability per se. Convinced believers in a chancy world might regard this as a useful fallback position pending a fully satisfactory explanation of standard probability. It might, alternatively, be seen as an account with enough attractions of its own that it could be preferable to any standard account involving probability. Either way, it offers a way of making scientific sense of one-world quantum theory that has no many-worlds analogue.

Consider a probabilistic theory T, and suppose for simplicity that it predicts a finite set of probabilistic events, labelled by the index i, each with finitely many possible outcomes x_i^j, labelled by the index $j \in J_i$, for which it predicts non-zero probabilities p_i^j. For simplicity, we also suppose for the moment that the possible outcomes for any given event, and their probabilities, are independent of the outcome of any other event. We say two events i and i' are of the same type, according to T, if the sets $\{p_i^j\}$ and $\{p_{i'}^j\}$ are identical. Let $B = \{0, 1\}$, $B^* = \{\emptyset, 0, 1, 00, 01, \ldots\}$ be the set of finite binary strings, and B^r the set of length r binary strings. Let $n = \Pi_i |J_i|$ be the size of the list of possible sets of outcomes, which we write as $N = \{1, 2, \ldots, n\}$.

A length r code for the outcomes is any surjective map $C : B^r \to N$. Given such a code, we can define an alternative probabilistic theory T^C by stipulating that a binary string b in B^r is randomly chosen from the uniform distribution, and that the outcomes are given by $C(b)$. By taking r sufficiently large, and choosing C so that $|\{b : C(b) = i\}| \approx 2^r p(i)$ for each $i \in N$, we can find theories T^C whose probability assignments are arbitrarily close to those of T.

A length r subcode for the outcomes is any map (not necessarily surjective) $C : B^r \to N$. Again, given a subcode, we can define a probabilistic theory T^C

as above: here T^C may assign zero probability to some outcomes for which T assigns non-zero probability.

We can define another type of theory from the triple (T^C, C, r): a theory that simply states that the data will be those predicted by T^C and C given some length r binary string as input, and makes no prediction about the binary string. We call this theory $D(T^C, C, r)$, using D to emphasize that we now regard the theory as deterministic. One might view the binary string in $D(T^C, C, r)$ as playing a role analogous to a constant of nature in a deterministic theory: its value is not fixed by the theory, and can only be determined empirically. In this case, even if the map C is injective, determining the entire string would require observing every random event in the universe.

Now, on the view that there is a unique 'correct' fundamentally probabilistic theory of nature T, each probabilistic theory of the form T^C must be either equivalent to T (which is possible only if the probabilities p_i^j are all dyadic), or else incorrect (though possibly a good approximation to T). Note though that, given a finite set of data, many other probabilistic theories besides T, including some of the form T^C, will be consistent with the data. Indeed, we would generally expect some theories T' to fit the data better than T, in the sense that the same sets of events are of the same type according to T and T', and the probabilities $p_i^{\prime j}$ are closer than p_i^j to the observed relative frequencies for events of the same type. If we nonetheless regard T as likelier to be correct than T', it must be for reasons other than purely empirical—presumably on grounds of elegance or simplicity. And if we maintain that there is a unique correct fundamental theory, it seems to follow that the correct theory is determined by a set of probabilities $\{p_i^j\}$ not determined by the physical universe (although perhaps very well approximated by relative frequencies of physical events).

Here's an alternative view. It may be, if not meaningless, then at least unnecessary, to appeal to the idea of a unique correct fundamentally probabilistic theory of nature, or even to define probability as a fundamental physical concept. Instead of considering probabilistic theories T^C, we can compare deterministic theories $D(T^C, C, r)$ against one another and against the data. In evaluating these theories, we use the criteria of simplicity and elegance. These criteria have no precise mathematical definition. They include judgements about the form of T^C and C, as well as the parameter r (which is a precise measure of complexity for the part of the theory defined by the unknown binary string). In saying that one theory $D(T^C, C, r)$ is our best current theory—or perhaps that our best descriptions of nature are given by a class of similar such theories—we mean that we cannot presently find a substantially simpler and more elegant theory that fits the data. The stronger metatheoretic hypothesis that a theory $D(T^C, C, r)$ is, up to approximate equivalence, *the* best theory of nature implies that, given all the physical data in the universe, one would not be able to find a simpler, more elegant, compelling theory.

This could be made more quantitative by formalizing the discussion within the context of a fixed model of computation, for instance a (classical) Turing machine. (This is why we have chosen to consider theories with unknown binary strings, although of course bases other than binary could also be used.) Here, a theory is a program for generating a mathematical representation of the complete set of physical data. A theory with unknown data is a program that requires an unknown input string of stipulated length. The theory's simplicity depends, *inter alia*, on both the length of the program and the length of the required input string. Each of these is a natural simplicity parameter. The halting time of the program is another significant parameter, which gives one way of quantifying the elegance of a theory.

In principle, within a fixed computation model, it's possible to carry out an exhaustive search of all theories with total length $\leq L$ that halt after $\leq N$ steps. Thus, in principle, given all the physical data, one can test the hypothesis that $D(T^C, C, r)$ is the best theory among all those whose program and input strings satisfy given length bounds, and which satisfy other stipulated simplicity and elegance constraints, that halt after any given finite time, relative to a fixed computation model.

Thus, instead of talking about a probabilistic physical theory that produces a random set of physical data, we can consider a deterministic physical theory whose definition includes a set of predetermined but a priori unknown physical data, together with the metatheoretic hypothesis that this description is essentially algorithmically incompressible. If we learn empirically that the data are in fact significantly compressible, then this hypothesis is refuted, and we may replace the theory by a more economical one.

It should be stressed that these measures of simplicity and elegance are by no means intended to be an exhaustive list. For example, another elegance criterion is given by the principle of scientific induction, which suggests that we should prefer a theory that suggests that a hitherto apparently fair coin will continue to be apparently fair over one that suggests that it will henceforth always come up heads, even though the latter theory requires a shorter input string (and so is simpler by one of the above measures).[1] Comparing scientific theories generally involves a wide and arguable variety of quantitative and qualitative simplicity and elegance criteria, and nothing in this account alters that: the aim here is only to propose a different treatment of apparent randomness when comparing theories.

2.1 Example: Reinterpreting a Fair Coin

For example, in a universe with an apparently random process that apparently mimics a fair coin and produces a large number N of apparently independent

[1] I thank Jonathan Barrett for this point and this example.

outcomes, our metatheoretic hypothesis might suggest that we cannot find a simpler correct theory than one that states that the length N binary string is essentially algorithmically incompressible. If, in fact, the string S turns out to consist of $0.01N$ zeros and $0.99N$ ones, we can certainly generate a more economical theory, and this hypothesis is refuted.

According to the standard account of probabilistic theories, if a probabilistic theory PT says that zeros and ones are equiprobable and independently generated, the outcome S is extremely improbable, but not logically impossible. The theory PT is thus not logically refuted by the outcome S. In practice we would reject it—but, without a fundamentally satisfactory account of probability, it is hard to give a completely satisfactory justification for doing so.

In our alternative account, however, no such problem arises. Our hypothesis predicts that a given physical data set is essentially incompressible—where 'essentially' incorporates some judgements about the trade-offs between small gains in compression of the data set and simplicity and elegance in other aspects of the theory. If the data set turns out to be a string such as S that is significantly compressible, so that we can fit the data by a simpler theory, the hypothesis is falsified and the original theory replaced.

2.2 Example: Reinterpreting a Biased Coin

Now consider a universe with an apparently random process that apparently mimics a coin with bias $p > \frac{1}{2}$ towards zero and produces N apparently independent outcomes. We can then produce theories that state that the length N binary string is compressible to $H(p)N + o(N)$ bits. For example, a theory which says that the length N string will contain between $pN - 10\sqrt{N}$ and $pN + 10\sqrt{N}$ zeros has the required compression, since we can binary code all such strings in a code of length $H(p)N + o(N)$. Clearly there are many somewhat similar such theories—the string contains between $pN - 9\sqrt{N}$ and $pN + 9\sqrt{N}$ zeros, between $pN - 11\sqrt{N}$ and $pN + 11\sqrt{N}$ zeros, and so on. On this view of scientific accounts of apparently random data, that's the best one can hope for: generically, no single clearly optimal theory will emerge. However, we can hypothesize that theories of roughly this length are essentially the best possible—i.e. that the string cannot be compressed to significantly shorter than $H(p)N$ bits—and *this* hypothesis is testable and falsifiable.

Again, these theories deterministically reproduce predictions that the standard probabilistic theory says hold with probability very close (but not equal) to one. They exclude some very low probability events which would, if realized, in practice persuade almost everyone that the probabilistic theory was wrong, even though their occurrence is logically consistent with the theory.

2.3 Conclusion

According to this account, we should consider one-world quantum theory as
a theory which requires a binary string as input, and consider it alongside the
metatheoretic hypotheses that (a) there is no significantly more compressed
description of the data obtained from quantum experiments than that given
by encoding them in binary, using a coding that would produce an approximately uniform distribution over binary strings if the data were probabilistically
generated via the Born rule, (b) the data can indeed be thus described. If one
of these hypotheses turns out to be incorrect—if, for example, the data in all
Bell experiments consistently show significantly greater violations of the CHSH
inequality than quantum theory predicts—then we must find a better theory.
Conversely, the theory logically (not merely with high probability) implies that
we will see no consistent regularities in our experimental data that would, on the
usual account, be highly improbable.

Among the scientific virtues of this account, as I see it, are its explicitness
about the provisional nature of our theories, and its undogmatic sidestepping of
the problem of giving a fundamental meaning to probability. It recognizes the
possibility that random-seeming data may turn out to have a simpler description.
It recognizes too that, if we find consistent regularities that a probabilistic theory
says are highly improbable, then we should and will feel impelled to produce a
better theory. At the same time, it stays silent on the question of whether random-
seeming physical data are genuinely randomly generated in some fundamental
sense, and hence avoids the need to explain what such an assertion could really
mean and how we could be persuaded of its truth.

One-world quantum theory, read in this way, allows us to draw logical
inferences about the physical world. It predicts—it is not merely consistent with
the fact—that there will be no regularities in the data of a type that would allow
for a significantly simpler theoretical description. If that prediction turns out to be
wrong, the theory is refuted. Interpreted thus, one-world quantum theory can be
read as a well-formulated scientific theory, in a way that allows a straightforward
account of scientific confirmation and refutation. If we assume that it is correct, we
have an explanation for the apparent fact that our evolutionary and experimental
histories contain no regularities that would be inexplicably improbable according
to the Born rule. To the extent that the project outlined above can be fleshed
out and succeeds—and I am optimistic that it can and does—proponents of
one-world quantum theory can rest relatively easy on the question of randomness.

3 TOY MANY-WORLDS THEORIES AND THEIR USES

If we knew of probability theory *only* through its use in Copenhagen quantum
theory—if we had no familiarity with coin tosses, dice rolls, noise, or any other

effectively unpredictable classical systems—we would probably be (even more) deeply confused about the nature of both quantum theory and probability. I suspect this is the cause of much of the continuing confusion over many-worlds quantum theory: discussions need simultaneously to grapple with the quite unfamiliar concept of many branching worlds and the specific peculiarities of Everettian quantum theory.

This motivates defining some simpler many-worlds theories. Another reason for doing so is that some key Everettian ideas—for example, Greaves and Myrvold's attempt [Chapter 9] at an account of many-worlds theory confirmation—can really only sensibly be discussed if we can consider a class of many-worlds theories, not just the single example of Everettian quantum theory. Readers may initially find the form of the following theories a little intellectually unsettling, but I recommend persevering: they shed a great deal of light on Everettian arguments.

Let me stress right away that these are not perfect models for Everettian quantum theory. That is, in fact, part of the point: they allow us to separate out general claims about rationality and theory confirmation in multiverse theories from claims that rely on specific features of quantum theory. In particular, they allow us to see why Greaves–Myrvold's account of many-world theory confirmation doesn't work.

3.1 Some Toy Multiverses

The following toy multiverses are all classical, in the sense that the state of any branch at any time is defined by a classical physical theory, and they all have a definite branching structure.

Consider, first, the branching multiverse CBU_1, which includes conscious inhabitants, and also includes a machine with a red button on it and a tape emerging from it, with a sequence of numbers on it, all in the range 0 to $(N - 1)$. Whenever the red button is pressed in some universe within the multiverse, that universe is deleted, and N successor universes are then created. All the successors are in the same classical state as the original (and so, by hypothesis, all include conscious inhabitants with the same memories as those who have just been deleted), except that a new number has been written onto the end of the tape, with the number i being written in the i-th successor universe.

Suppose, too, that the multiverse's inhabitants believe that something like this is indeed happening. The numbers on the tape play a significant role in their society. In particular, it is quite common to place bets on future numbers, and social mores ensure that such bets are always honoured. Of course, since one's own universe will be destroyed before the next number is written, placing such a bet means—they correctly believe—redistributing resources amongst one's successors. Some inhabitants may find reasons for preferring some redistributions over others. We need not discuss yet precisely what these reasons and preferences (both of which may be different for different inhabitants) may be.

It might be helpful to imagine that the universes are being run on a simulator by technologically advanced beings, who simply end one simulation whenever the red button is pressed, and then start simulating the successor universes from the appropriate initial states. We will sometimes assume that the inhabitants, indeed, believe this to be the case.

Suppose, further, that some of the inhabitants of CBU_1 have acquired the theoretical idea that the laws of their multiverse might attach *weights* to branches, i.e. a number p_i is attached to branch i, where $p_i \geq 0$ and $\sum_i p_i = 1$. They may have different theories about how these weights are defined: for instance, that the weights are always $\{p_i\}$, that they are always $\{q_i\}$, that they vary over time according to some rule, and so on. As it happens, though, these theories are all incorrect: there are no weights attached to the branches. To be clear: this is not to say that the branches have equal weight. Nor are they necessarily physically identical aside from the tape numbers. They may perhaps be distinguished by other features: for example, if they are simulations, they may be simulated by different hardware or software. However, any such differences do not yield any natural quantitative definition of branch weights. There is just no fact of the matter about branch weights in this multiverse.

The multiverse CBU_2 is similar to CBU_1. In this universe, there *are* indeed numbers attached to the branches, but the way they are attached means that they should (by our lights, and also by the inhabitants', if only they understood the full picture) have no significance to any decisions that the inhabitants make about bets/redistributions. For instance, we could extend the simulation idea, and imagine that the technologically advanced beings simply choose, on whim, to write the number p_i somewhere inconspicuous in the simulation of successor universe i, in such a way that it has no effect on the inhabitants, and that it has no other significance.

The multiverse CBU_3 is similar to CBU_2. However, this time the numbers attached to the branches by the physical theory are attached in such a way that it can be plausibly argued that they *could* reasonably play a significant role in the decisions that the inhabitants make about bets/redistributions. For instance, we could imagine that when the technologically advanced beings create successor universes, they create not just one successor corresponding to each outcome i, but a number of distinct successor universes, all identical apart from their outcome values, and the number containing outcome i is proportional to the weight p_i. (We assume here the p_i are rational numbers.)

3.2 Some Possible Strategies

Consider an inhabitant of any of the above multiverses, who believes that the weight p_i is attached to the outcome i. Suppose they are offered a variety of bets that give their successor a good G_i in a universe in which outcome i obtains, and they (the original inhabitant) attach utility U_i to this good. We suppose the U_i are finite real numbers, not necessarily positive (the goods may be bads); and, of course, both G_i and U_i depend on the bet.

How might they proceed to evaluate and rank such bets? *Weight-sensitive* inhabitants believe that branch weights exist and should play a role in their betting preferences. *Weight-indifferent* inhabitants may also believe that the physical theory attaches weights, but if so, do not believe that they are of any relevance to a rational betting strategy. (Such an inhabitant might, for example, believe that they live in a multiverse like CBU_2.) Among their options is to mimic the strategy of a weight-sensitive inhabitant, except that they treat all branch weights as equal. By this means, given any weight-sensitive strategy, we can define a corresponding weight-indifferent strategy. Here are some examples of weight-sensitive strategies:

- The *mean utilitarian* ranks bets according to the value of $\sum_i p_i U_i$.
- The *Price–Rawlsian*'s dominant concern (Price [Chapter 12], Rawls [1971]) is with the welfare of their least satisfied future self. They rank bets first according to $\min(U_i)$, and then some list of tie-breaking criteria. To be definite, let's say their next criterion is the value of $\sum p_j$, summed over all j such that $U_j = \min(U_i)$, followed by $\min(U_j: U_j \neq \min(U_i))$, and so on.
- The *future self elitist*'s dominant concern is that the best possible version of their future self should be realized somewhere; they have little interest in mediocre future selves, whom they regard as losers. Their bet rankings are thus dominated by $\max(U_i)$, and they break ties using the mirror image of the Price–Rawlsian's criteria.
- The *rivalrous future self elitist* takes things one stage further. Not only do they identify their interests exclusively with those of their best possible future self, but they regard that self as in competition with the others, and feel happiest—all else being equal—if that competition is won by as large a margin as possible. They rank bets first by $\max(U_i)$, then by $\max(U_i) - \max(U_j: U_j \neq \max(U_i))$, and so on.
- The *median utilitarian*'s dominant concern is for median utility. Reordering the index labels so that $U_1 \leq U_2 \leq \ldots \leq U_n$, let j be such that $\sum_{i=1}^{j-1} p_i < \frac{1}{2}$ and $\sum_{i=1}^{j} p_i \geq \frac{1}{2}$: they rank bets first according to the value of U_j. (They also have some tie-breaking criteria: one option is to break ties by considering the mean utility.)
- The *x-percentile utilitarian*'s dominant concern is for the utility of the future self ranked at $x\%$ in the distribution. They proceed like the median utilitarian, with $\frac{1}{2}$ replaced by $\frac{x}{100}$. The Price–Rawlsian, median utilitarian, and future self elitist are all special cases.
- The *future self democrat* believes her preference should be that which would result from a democratic vote among her future selves. Given a finite list of

possible bets, for each value of x, she asks herself how she would order her preferences among the bets, if she knew that she would become the future self ranked at the x-percentile of the elected bet. (The answer might be that her future self's voting preference would always be dominated by its own welfare under this hypothesis, but it need not: it depends whether she cares about the welfare of contemporaneous versions of herself in other branches.) She then tallies the votes, integrating over x using branch weight measure, and using, say, a single transferable vote system. The winner of the vote is her preferred bet. If the election is tied, she has more than one equally preferred bet.

• An example of a *future self distribution engineer* is someone who seeks to maximize an expression of the form

$$\sum_i f_1(U_i)p_i + \sum_{ij} f_2(U_i, U_j)p_ip_j + \ldots, \tag{1}$$

where the U_i are the utilities of future branches with weights p_i, and f_n is some given joint function of n variables.[2]

3.3 Many-Worlds Rationality Reconsidered in Toy Models

If you do what you've always done, you'll get what you always got.

(variously attributed)

According to Wallace [Chapter 8] and Greaves–Myrvold [Chapter 9], we should *define* rational behaviour in a multiverse via axioms generalizing those proposed by Savage [1972] in order to justify using the standard calculus of probabilities and utility functions for rational decisions in a single world in which future events are uncertain.

3.3.1 Savagean Rationality in One World

Savage, engagingly and rather admirably, presented his approach to rationality in the presence of (one-world) uncertainty

. . . in a tentative spirit, for I realize that the serious blemishes in it apparent to me are not the only ones that will be discovered by critical readers (Savage [1972], p.5).

Everettian neo-Savageans [Chapters 8, 9], as I read them, seem rather less self-critical—puzzlingly so, since applying Savagean decision theory to Everettian quantum theory raises many new questions without solving any of the old ones. This raises some general worries, which are developed to some extent elsewhere

---

[2] There are more general possibilities.

in this paper, but might also be taken in other directions. First, if Savage's axioms are, in fact, unable to give a completely satisfactory account of ideal rational behaviour in the presence of one-world uncertainty, it seems very unlikely that a completely satisfactory axiomatic treatment of many-worlds rationality can be produced by generalizing them. Second, giving a satisfactory account of ideal rational behaviour in the presence of one-world uncertainty (or some many-worlds generalization) may in any case not be enough. (For one thing, we are not ideal rational agents. For another, as Albert [Chapter 11] has eloquently stressed, there is a crucial difference between showing that one can find a rational justification for behaving as though the world were a certain way and showing that the world actually is that way.) Third, however far Savage can or cannot guide rational agents in one uncertain world, it isn't obvious that his programme generalizes *at all* to many-worlds theories in general or to Everettian quantum theory in particular.

3.3.2 Many-Worlds Rationality According to Greaves–Myrvold

Let me now focus on Greaves–Myrvold's axioms, which are intended to apply to general many-worlds theories, and so can straightforwardly be considered within the toy models described above. I will consider later Wallace's arguments, which are framed for the special case of Everettian quantum theory, and for the moment simply note that their logic suggests the same conclusions here as Greaves–Myrvold's.

In Greaves–Myrvold's view, the mean utilitarian's strategy is rationally justifiable, and the others are branded irrational, since they violate one or more of the axioms. For example, the future self elitist and the Price–Rawlsian violate their continuity postulate, P6, the median utilitarian violates P2, the future self democrat violates transitivity, P1a, and the rivalrous future self elitist the dominance postulate P3.

However, the fact is that each of these strategies is well defined and has a coherent motivation (and many other such examples could also be constructed). To brand them irrational seems to me itself irrational dogma. Even the most contentious case, the rivalrous future self elitist, has a coherent, if ungenerous, philosophy of life in the multiverse and a rational strategy for implementing it. Note too that some of these strategies have arguable theoretical advantages over the mean utilitarian strategy. For instance, one can be an x-percentile utilitarian, or a future self democrat (if they are *purely self-concerned*, in the sense that each future self's preferences among options are completely determined by the implications for its own welfare), without having to quantify the utility of the possible outcomes: one needs only a preference ordering. This is arguably advantageous, since even if one accepts Savage's postulates (Savage [1972]) and, hence, the conclusion that one's preferences must be defined by some utility function, it may be difficult or even impractical to compute the relevant function for general outcomes, and yet relatively easy to identify preferences among any finite list of outcomes.

In short, Greaves–Myrvold's postulates only express in more abstract form a preference for being a mean utilitarian—i.e., for one possible choice among many. Their postulates are plausible possible prescriptions for rational behaviour when considering the welfare of a population of future selves, but also logically inconsistent with other plausible prescriptions. This shouldn't come as a complete surprise: after all, Arrow's celebrated impossibility theorem (Arrow [1950]) taught us that plausible decision theoretic principles for populations may be inconsistent.

Granted, the one-world counterparts of some of these strategies may look peculiar, but one can consistently accept the many-worlds strategies as rational and reject their one-world counterparts. As Price [Chapter 12] has persuasively argued, many-worlds agents can offer reasoned justifications for their strategies that aren't available to their one-world counterparts. The many-worlds future self elitist knows that his best possible future self will be an *actual* future self, while his one-world counterpart doesn't. The many-worlds future self democrat knows that there really will be a population of future selves who have preferences among the betting choices, while her one world counterpart knows there won't be; and so on.

One could, of course, adopt a weaker position. One could take Greaves–Myrvold's and Wallace's accounts of rationality as simply suggesting a possible attitude one *might* adopt to life in an Everettian multiverse, an attitude defined by a set of rules which are consistent and have some pleasant mathematical features but which are not meant to constitute a dogma. On this liberal reading, Greaves–Myrvold's preferred strategy could be termed 'rational', in the sense of being well defined and internally consistent, without denying the existence of other equally rational strategies. The problem is that abandoning any claim of uniqueness also removes the purported connection between theoretical reasoning and empirical data, and this is disastrous for the programme of attempting to interpret Everettian quantum theory via decision theory. If Wallace's arguments are read as suggesting no more than that one can consistently adopt the Born rule if one pleases, it remains a mystery as to how and why we arrived at the Born rule empirically. If Greaves–Myrvold's arguments are read as merely suggesting a possible attitude one might choose to take about testing and confirming many-worlds theories, one's left to investigate how many other equally valid attitudes there might be, and whether they mightn't—disastrously—imply the confirmation of inconsistent theories from the same data.

3.4 Rationality and Feasibility

Consider now a rather more complicated multiverse, CBU_4. Here, the universes are definitely being simulated by technologically advanced beings, and the inhabitants know it. They also know that, after the red button is pressed, there is a list of outcomes i, and that the list is indeterminately long (and possibly infinite). They know too that, for each i, some number of successor universes containing outcome i will be created. They do not know the number of successors there will be of each type: these vary for each i, and vary each time the button is pressed, at the whim of the simulators. What they do know—because, let's say, the simulators have credibly promised them—is that numbers playing the role of additive weights, following certain rules, will be written inconspicuously into each simulated universe. Thus, if their universe has the number x written in it, and the red button is pressed, and there are n_i successor universes with outcome i, these successors will have numbers of the form xq_i^j written into them, where the label j runs from 1 to n_i, $q_i^j \geq 0$, and

$$\sum_j q_i^j = p_i. \tag{2}$$

Here the p_i are known to be constants (i.e. they take the same value each time the button is pressed), with $p_i \geq 0$, and $\sum_i p_i = 1$. The inhabitants know the values of a finite set of the p_i, those with index $i \in I$, where the sum $\sum_{i \in I} p_i < 1$.

What the inhabitants would *like* to do—what they feel rationality would mandate they do if they could—is express betting/distribution preferences that value each successor universe equally. But they can't—they don't know how many successors will be created for any given i, nor do they know how long the list of possible outcomes i is. Nor can they express betting/distribution preferences that value each outcome equally, regardless of the number of successor universes containing it—again, they don't know how long the list of possible outcomes is.

What they *can* do is express betting/distribution preferences, for bets on the known possible outcomes, treating the known values of p_i as probability weights. Doing so is equivalent to treating a successor universe with the number y written into it as having an importance proportional to y—a rule which can be consistently applied, despite their ignorance about the number of successors of each type, because of equation (2). So, they have a consistent, feasible strategy available to them. Moreover, if they want to assign a measure of importance to each individual universe, and they want the importance they assign to the set of universes containing outcome i to be independent of the number of such universes, this is the *only available* rule. Nonetheless, it doesn't seem to have a fundamental rational justification. The numbers written into the universes happen to follow convenient bookkeeping rules, but they have no significance: there is no fundamental *reason* to treat the numbers as a measure of importance of their universes.

From this, I think we should conclude two things, to be borne in mind when we come to consider Wallace's arguments. It can make perfect sense, in a multiverse theory, to say that there exists a rational optimal strategy that is inaccessible to the agents in that multiverse. Conversely, the fact that a strategy is available does not per se make it rationally compelling, even if it is the unique available strategy satisfying some pleasant consistency properties: rational compulsion also needs rational justification, which may or may not exist.

4 WHY MANY-WORLDS THEORY CONFIRMATION DOESN'T WORK

Everettian quantum theory is essentially useless, as a scientific theory, unless it can explain the data that confirm the validity of Copenhagen quantum theory within its domain—unless, for example, it can explain why we should expect to observe the Born rule to have been very well confirmed statistically. Evidently, Everettians cannot give an explanation that says that all observers in the multiverse will observe confirmation of the Born rule, or that very probably all observers will observe confirmation of the Born rule. On the contrary, many observers in an Everettian multiverse will definitely observe convincing *disconfirmation* of the Born rule. Nor can one look at Everettian quantum theory and conclude that any given observer in the multiverse will probably observe confirmation: the theory has no notion of standard probability available to even make sense of any such claim. And if the theory doesn't explain the data, the data don't support the theory.

There seems to be no good way around this, and if so, then that's the end of Everettian quantum theory as a serious contender: a theory with no predictive power *should* lose the scientific competition against theories that predict what we actually see. However, Greaves and Myrvold [Chapter 9] have offered an attempt at a solution, by giving a general account purporting to explain why agents who take seriously the possibility of many-worlds theories can use observational data to confirm particular theories and refute others. Their account is illuminating, and raises some very interesting questions about many-worlds theories. Ultimately, though, it seems to me that it does not show, as claimed, the possibility of explaining our observations from a many-worlds theory and thus confirming one many-worlds theory against another. Rather, it highlights some apparently insuperable problems that prevent us from doing so. As Greaves and Myrvold's arguments are set out in Chapter 9, in this discussion I will simply summarize the implications of their confirmation algorithm in toy models, and point out the problems that arise.

4.1 The Problem of Inappropriate Self-Importance

It suffices to consider very simple many-worlds theories, containing classical branching worlds in which the branches correspond to binary outcomes of definite experiments. Consider thus the *weightless multiverse*, a many-worlds theory of type CBU_1, in which the machine produces only two possible outcomes, writing 0 or 1 onto the tape. Recall that in CBU_1 there is no fact of the matter about weights attached to the branches containing 0 outcomes and 1 outcomes, although the inhabitants think there may be. This is the many-worlds analogue of an indeterministic one-world theory containing a sequence of binary experimental outcomes which are not only not determined but also not governed by any probabilistic law. Suppose now that the inhabitants begin a series of experiments in which they push the red button on the machine a large number, N, times, at regular intervals. Suppose too that the inhabitants believe (correctly) that this is a series of independent identical experiments, and moreover—this is not essential, but simplifies the discussion—believe this *dogmatically*: no pattern in the data will shake their faith. Suppose also that they believe (incorrectly) that their multiverse is governed by a many-worlds theory with unknown weights attached to 0 and 1 outcomes, identical in each trial, and seek to discover the (actually non-existent) values of these weights by following Greaves–Myrvold's learning algorithm.

After N trials, the multiverse contains 2^N branches, corresponding to all N possible binary string outcomes. The inhabitants on a string with pN zero and $(1-p)N$ one outcomes will, with a degree of confidence that tends towards one as N gets large, tend to conclude that the weight p is attached to zero outcome branches and weight $(1-p)$ is attached to one outcome branches. In other words, everyone, no matter what outcome string they see, tends towards complete confidence in the belief that the relative frequencies they observe represent the weights.

Let's consider further the perspective of inhabitants on a branch with pN zero outcomes and $(1-p)N$ one outcomes. They do not have the delusion that all observed strings have the same relative frequency as theirs: they understand that, given the hypothesis that they live in a multiverse, *every* binary string, and hence every relative frequency, will have been observed by someone. So how do they conclude that the theory that the weights are $(p, 1-p)$ has nonetheless been confirmed? Because, following Greaves–Myrvold's reasoning, they have concluded that the weights measure the *importance* of the branches for theory confirmation. Since they believe they have learned that the weights are $(p, 1-p)$, they conclude that a branch with r zeros and $(N-r)$ ones has importance $p^r(1-p)^{N-r}$. Summing over all the branches with pN zeros and $(1-p)N$ ones, or very close to those frequencies, thus gives a set of total importance very close to 1; the remaining branches have total importance very close to 0. So, on a set of branches that dominates the importance measure, the theory that the weights are (very close to) $(p, 1-p)$ is indeed correct. All is well! By definition, the important branches are the ones that matter for theory confirmation. The theory is indeed confirmed!

The problem, of course, is that this reasoning applies equally well for all the inhabitants, whatever relative frequency p they see on their branch. All of them conclude that their relative frequencies represent (to very good approximation) the branching weights. All of them conclude that their own branches, together with those with identical or similar relative frequencies, are the important ones for theory confirmation. All of them thus happily conclude that their theories have been confirmed. And, recall, all of them are wrong: there are actually no branching weights.

4.1.1 Comparison With the One-World Case

It's illuminating to compare the case of an inhabitant of the analogous one-world universe, in which pressing the red button produces either a 0 or a 1 on the tape but there is no law, either deterministic or probabilistic, governing these outcomes. After N experiments in which he sees pN zeros and $(1 - p)N$ ones, he tends towards confidence in the theory that zeros have probability p and ones have probability $(1 - p)$.

Let us again restrict attention to theories—in this case probabilistic one-world theories—that dogmatically assume the experiments are identical and independent. Among such theories, the selected theory does indeed characterize, better than all its competitors, all the relevant data in the universe—i.e., all the outcomes of the N experiments. Of course, further data could change that conclusion. But, so long as we consider only the relevant data, it's something of a puzzle to pin down whether it's wrong to adopt the theory *pro tem*, and if so precisely why. Is there a physically meaningful sense in which a universe that *looks* as though it contains data resulting from a sequence of independent identical coin tosses with a probability p of outcome zero is distinct from one that *does* contain such data? And if so, how precisely should we characterize the distinction?

On the view of physical randomness discussed in Section 2, the answer to the first question is no. In any case, however one answers the questions, it seems that any possible error here must be subtler than and distinct from the error highlighted above in the many-worlds case. In the many-worlds case, recall, all observers are aware that other observers in worlds with other data must exist, but each is led to construct a spurious measure of importance that favours their own observations against the others', and this leads to an obvious absurdity. In the one-world case, observers treat what actually happened as important, and ignore what didn't happen: this doesn't lead to the same difficulty.

4.1.2 Numbers in the Sky

Consider next the *decorative weight multiverse*, a type CBU_2 variant of the weightless universe. This universe has a constant of nature fixed by the technologically advanced beings, a real number p, with $0 < p < 1$. As before, whenever the red button is pressed in a simulated universe, that universe is deleted, and successor universes with outcomes 0 and 1 written on the tape are initiated. This time, the technologically advanced beings also write the numbers p and $(1 - p)$ in an inaccessible part of the skies of the 0 and 1 successor universes, respectively. These numbers are visible to the inhabitants, but have no other physical significance.

There is thus a formal sense in which distinct weights are attached to the 0 and 1 branches. However, by hypothesis, these weights are decorative: there are no rational grounds for assigning them any fundamental physical meaning or any role in constraining rational actions. We can thus run through a discussion of theory confirmation precisely parallel to that for the weightless multiverse.

This illustrates again that the mere fact that Born weights are mathematically defined in Everettian quantum theory does not per se justify assigning them any role in theory confirmation. They could be merely decorative.

4.2 Separating Caring Weights from Theory Confirmation

To investigate further, it's helpful to consider branching world models in which there *are* weights attached to the branches, in such a way that the weights could plausibly be regarded as important for making rational decisions. I want here to tell specific stories about the weights, in order to illustrate a crucial distinction between two possible definitions of importance.

4.2.1 The Replicating Multiverse

Consider first the *replicating multiverse*, a multiverse of type CBU_3 with a machine like the one above, in which the branches arise as the result of technologically advanced beings running simulations. Whenever the red button is pressed in a simulated universe, that universe is deleted, and successor universes with outcomes 0 and 1 written on the tape are initiated. Suppose, in this case, that each time, the beings create *three* identical simulations with outcome 0, and just one with outcome 1. From the perspective of the inhabitants, there is no way to detect that outcomes 0 and 1 are being treated differently, and so they represent them in their theories with one branch each. In fact, though, given this representation, there is an at least arguably natural sense in which they ought to assign to the outcome 0 branch three times the importance of the outcome 1 branch: in other words, they ought to assign branch weights $(\frac{3}{4}, \frac{1}{4})$.

They don't know this. But suppose, as before, that they believe that there are unknown weights attached to the branches, and follow the Greaves–Myrvold procedure for identifying those weights. What happens now? After N runs of the experiment, there will actually be 4^N simulations—although in the inhabitants' theoretical representation, these are represented by 2^N branches. Of the 4^N

simulations, almost all (for large N) will contain close to $\frac{3N}{4}$ zeros and $\frac{N}{4}$ ones. These simulations will contain inhabitants who, following Greaves–Myrvold, believe that they have confirmed that the branch weights (in their own theoretical representation, which remember contains only 2^N branches) are very close to $(\frac{3}{4}, \frac{1}{4})$. They believe too that the weights define an importance measure on the branches: a branch with r zeros and $(N - r)$ ones has importance (very close to) $(\frac{3}{4})^r (\frac{1}{4})^{N-r}$. They thus conclude that their weight assignment will be confirmed on a set of branches whose total importance is close to 1.

Now, I think I can see how to run some, though not all, of an argument that supports this conclusion. The branch importance measure defined by inhabitants who find relative frequency $\frac{3}{4}$ of zeros corresponds to the counting measure on simulations. If we could argue, for instance by appealing to symmetry, that each of the 4^N simulations is equally important, then this branch importance measure would indeed be justified. If we could also argue, perhaps using some form of anthropic reasoning, that there is an equal chance of finding oneself in any of the 4^N simulations, then the chance of finding oneself in a simulation in which one concludes that the branch weights are (very close to) $(\frac{3}{4}, \frac{1}{4})$ would be very close to one. Turning that around, the theory that the branch weights are $(\frac{3}{4}, \frac{1}{4})$ would then imply that, with high probability, one should expect to see relative frequency of zeros close to $\frac{3}{4}$. There would indeed then seem to be a sense in which the branch weights define which subsets of the branches are important for theory confirmation.

It seems hard to make this argument rigorous. In particular, the notion of 'chance of finding oneself' in a particular simulation doesn't seem easy to define properly. Still, we have an arguably natural measure on simulations, the counting measure, according to which most of the inhabitants will arrive at (close to) the right theory of branch weights. That might perhaps be progress.

4.2.2 *The Qualia Enhancing Multiverse*

But consider now the *qualia enhancing multiverse*, again a multiverse with the same type of machine, in which the branches arise in the way we've previously considered, as the result of technologically advanced beings running simulations. Whenever the red button is pressed in a simulated universe, that universe is deleted, and successor universes with outcomes 0 and 1 written on the tape are initiated. This time, though, the beings create just one simulation with outcome 0, and one with outcome 1, but devise their simulations so that the qualia—the mental sensations—of the inhabitants in the outcome 0 simulation are three times as intense. As before, from the perspective of the inhabitants, there is no way to detect that outcomes 0 and 1 are being treated differently, and so they represent them in their theories with one branch each.

There is, again, an arguably natural sense in which they ought—if they were aware of the rules of their multiverse—to assign to the outcome 0 branch three times the importance of the outcome 1 branch: in other words, they ought to assign branch weights $(\frac{3}{4}, \frac{1}{4})$. Recall, pleasure and pain in outcome 0 branches have tripled in intensity. The welfare of successors on outcome 0 branches is felt more intensely, and in that sense it matters more.

Let me deal here with three possible objections:

(a) It might be argued that qualia enhancement should be analysed differently, as an example of an unannounced alteration in utility functions: the actual payoff of winning a bet with outcome 0 is three times the expected payoff, since the inhabitants don't expect any qualia enhancement. Certainly it *could* be analysed in this way. But this reflects an arbitrary choice that always needs to be made in many-worlds theories. (Precisely the same argument could be made in the case of the replicating multiverse, for example.) The statement that one branch is N times as important as another can always be recast as a statement that utilities on the first branch are rescaled by N relative to those on the second. So, we can legitimately analyse qualia enhancement as an effect altering the relative importance of branches, and it's interesting to do so, as this lets us test general propositions about the confirmation of theories attaching importance to branches.

(b) The reader may not believe that there is a sensible account of experience involving qualia, or that intensifying qualia makes any sense. Never mind. It's just a useful device to make a point about branch measures. It could be formulated in another way: we could suppose that the simulators arrange that all bets have payoffs with three times the expected utility on outcome 0, while erasing the relevant bits of the inhabitants' memories so that they're not aware that the payoff tripled.

(c) One might also worry that inhabitants in an outcome 0 branch would notice that the intensity of their qualia has just tripled. For the sake of the argument, we must assume not. Insofar as the notion of qualia enhancement makes sense, this seems reasonable: their memories will triple in intensity along with everything else.

Suppose, once again, that the inhabitants believe that there are unknown weights attached to the branches, and follow the Greaves–Myrvold procedure for identifying those weights. What happens now? After N runs of the experiment, there will be 2^N simulations—now correctly represented by 2^N branches in the inhabitants' many-worlds theory. The simulations will contain inhabitants who, following Greaves–Myrvold, believe that they have confirmed that the branch weights are very close to $(p, 1 - p)$, because their observed relative frequency is $p = r/N$, for each r in the range $0 \leq r \leq N$. They believe that the weights define an importance measure on the branches: a branch with r zeros and $(N - r)$ ones has importance (very close to) $(p)^r (1 - p)^{N-r}$. They thus conclude that their weight assignment will be confirmed on a set of branches whose total importance is close to 1. Now, in one sense, the inhabitants whose observed relative frequency

$p = 3/4$ are a special case. Their inferred importance measure equals the natural importance measure defined by qualia intensity. And if we weight the branches by this importance measure, it is the case, by the same calculation as before, that, on a set of branches with total measure close to one, the inhabitants end up with (very close to) the 'right' branch weights, $(\frac{3}{4}, \frac{1}{4})$.

But wait! If we count the simulations, the inhabitants who arrive at weights $(\frac{3}{4}, \frac{1}{4})$ are a tiny minority. Almost everyone arrives at the wrong branch weights—and, as in our earlier example, almost everyone arrives at a measure of importance according to which branches with (very close to) their observed relative frequency are the important ones. By the natural simulation counting measure, theory confirmation has spectacularly failed.

What these last two examples show is that there are two distinct senses, which Greaves–Myrvold and Wallace fail to separate, in which a branch weight might possibly be said to be a measure of importance. It could be said to be a 'caring measure', if there is some reason to care differently about the welfare of successors on different branches. And it could be said to be, for want of a better term, an 'explanatory counting measure', if there is some reason to think that we are likelier to *find ourselves* on some branches rather than others—or some other argument to show that a branching theory which predicts the observed relative frequencies (or other data) on a set of branches of high explanatory counting measure thereby explains them. What we've seen is that the first property doesn't necessarily imply the second, and it's the second that is needed for an adequate account of branching theory confirmation.

Couldn't a many-worlds theorist then simply *postulate* the existence of an explanatory counting measure? (And perhaps also postulate that a caring measure exists and equals the explanatory counting measure?)

A preliminary remark: even postulating a caring measure—which *has* been proposed (Papineau [Chapter 7]) in the Everettian literature—already seems a very strange manoeuvre. Physical theories can certainly give reasons for rational agents to perform certain actions if they have certain goals. But what's envisaged here is a theory that *by fiat* imposes a constraint on rational behaviour. I'm not clear—and at least some Everettians (e.g. Saunders [Chapter 6]) seem to share this worry—that this makes any sense, either as an idea about physics, or about rationality.[3]

In any case, when it comes to postulating an explanatory counting measure, one should be clear: the proposal is that a many-worlds theory *defines*, by fiat, without any attempt at further justification, whose observations matter and whose may be neglected, when it comes to testing and confirming the theory. The theory defines its own—highly non-standard—criteria for deciding whether or not it is a scientific success.

---

[3] See Appendix B for further discussion.

One could play this sort of game, of course, even in one world. For example, Alice could define a theory that includes—as a postulate, with no further explanation—the principle that everyone who agrees with her observations and her theoretical interpretation is important for theory confirmation, and everyone else is negligible. She could then announce, after checking with the important people, that her theory is confirmed. This would be self-consistent, and maybe politically adept, but it wouldn't be science.

It's no more scientifically respectable to declare that we can, without further justification, confirm Everettian quantum theory by neglecting the observations made on selected low Born weight branches. A Pavlovian association of low Born weight with small probability—illegitimately carried over from one world quantum theory—may perhaps lend an aura of greater respectability, but in Everettian quantum theory the Born weight is simply a number attached to branches. It has no intrinsic relevance to theory confirmation, and unless we add further structure to the theory, we cannot justify assigning it any such role.

Note again the contrast here with the one-world case: one-world probabilists do not pick and choose which observations are to be used for theory generation or confirmation.

4.3 Many-Worlds Confirmation: Conclusion

To explain how we could come to confirm Everettian many-worlds quantum theory it is not enough to note that we have Born weights to hand and so can automatically give them a confirmation-theoretic role. As the decorative weight multiverse illustrates, branch weights can be simply irrelevant to theory formation and confirmation.

Nor can Wallace's arguments for treating the Born weights as a caring measure suffice, even if we take Wallace's result at face value. As the qualia enhancing multiverse illustrates, a caring measure is not necessarily an explanatory counting measure.

Thus, the most sympathetic (though unauthorized) translation of Greaves–Myrvold's account of many worlds and confirmation that I can find requires us to add structure that *justifies* the existence of an explanatory counting measure. This requires interpreting Everettian quantum theory, along with competing many-worlds theories, as modelled by versions of the replicating multiverse, with branches constantly being deleted, and successor branches created. We need to postulate that the number of simulations or realizations of a given branch at a given time is proportional to the branch weight, and to assume that it is rational to treat all realizations as equally valuable. We need also to postulate something like an anthropic principle that tells us that, in some sense that needs to be properly defined, the chance of finding ourselves in one of a given class of realizations at a given time is proportional to the number of realizations in the class.

This, if it could be made rigorous, would suggest something resembling the objectively determinist 'momentary minds' version of Albert–Loewer's many-minds interpretation (Barbour [2001], Bell [1987], Albert and Loewer [1987]), in which the minds exist only instantaneously, with no continuous identity extending over time. This isn't a picture I find easy to take seriously. As I read them, none of the Everettian contributors to the present volume would wish to defend this account—and yet it seems very closely aligned with some of their intuitions. Let me close here by inviting readers to see if they can find a better way of rigorously justifying Greaves' gloss (Greaves [2004]):

But since we have a measure over our successors, we can, if we find it intuitive, talk of 'how much successor' sees spin-up. I have a preference for my spin-down successor to receive chocolate, rather than my spin-up successor, because there is more of the former; more of my future lies that way. Thus, I think, Lockwood's (1996) talk of a 'superpositional dimension', and/or Vaidman's (1998, 2001) suggestion that we speak of the amplitude-squared measure as a 'measure of existence', are somewhat appropriate (although we are not to regard lower-weight successors as less real, for being real is an all-or-nothing affair—we should say instead that there is less of them).

5 FUZZINESS, RATIONALITY, AND DECISION THEORY IN MANY WORLDS

Two of the most interesting recent developments in the Everettian literature, in my view, have been the attempt to argue for an intrinsically fuzzy emergent quasiclassical ontology (Wallace [Chapter 1]) and (as already discussed) the attempt to reinterpret Born weights via a many-worlds version of decision theory (Wallace [Chapter 8]). Interesting, but flawed—each project has deep problems, and they appear to be based on inconsistent premises.

5.1 Fuzziness and Its Limitations

Granted, as Wallace [Chapter 1] notes, viable higher-level scientific theories can and do, indeed, supervene on more fundamental theories. Objects in those theories need not have any unique and precise definition in terms of fundamental concepts: there is, indeed, no unique, natural, precise, chemical characterization of a tiger.

Nonetheless, there is a very strong reason for seeking (Kent [1990]) a precise mathematical formulation of the intuition that many branching worlds emerge from unitary quantum theory—or else a precise mathematical formulation of some other structure consistent with Everettian ideas—namely, that it is not at all clear that, without such a formulation, we have a well-defined scientific theory to discuss. (This, it seems to me, is why both Everettians (DeWitt and

Graham [1973], Deutsch [1985]) and critics (Bell [1987], Albert and Loewer [1988]) have often attempted to find mathematical structures that might explain the notion of branching.) The alternative strategy, proposed by Wallace [Chapter 1], of trying to interpret the implications of a fundamentally mathematical theory in terms of higher level fuzzily defined constructs carries a very obvious danger—namely, a retreat into vagueness and hand-waving on points where precision really is required. It's hard to run a serious argument (pro or con), let alone prove a rigorous theorem, if one doesn't, in the end, know quite what one's talking about.

5.2 Fuzzy Minds

A case in point is Wallace's appeal to functionalist intuitions in trying to give an account of the mind states of agents in Everettian quantum theory. Readers are, I think, owed a much more precise explanation of what, actually, is supposed to follow from this, since some rather crucial points appear to turn on unspecified details.

For instance, on this account, do distinct mind states necessarily correspond to orthogonal quantum states? If so, wouldn't this account necessarily supply us with a preferred orthogonal decomposition of the unitarily evolving quantum state? And wouldn't this, *pace* Wallace [Chapter 1], allow a precise definition of a relevant branching structure after all?

Wallace places great emphasis on the lack of a unique natural definition of a quasiclassical branch, and hence the impossibility of agents formulating a rational strategy based on counting distinct future branches. But it's at least as relevant to examine whether our account of mind states supplies a natural definition of a future self, and whether it might be possible for agents to formulate a rational strategy based on counting distinct future selves? Can't an agent identify successor selves as distinct if and only if they have distinct mind states, ascribe to distinct successors a branching history corresponding to that recorded in their memories, and use *those* data to define a rational strategy for taking account of their welfare? (These points are pursued further in Appendix A.)

On the other hand, if non-orthogonal quantum states *could* correspond to distinct mind states, how would we even begin to connect quantum theory with even the appearance of probabilities? Quantum theory gives no general rule to calculate a probability of a transition from an unknown state belonging to one fuzzily defined set of states (corresponding to mind state A) to an unknown state belonging to another (corresponding to mind state B). But that's what we'd need to calculate, in principle, in order to obtain a number corresponding to the apparent probability of arriving at state B when starting in state A. Maybe one could cook up such a rule, and then explain how the Born rule emerges as an approximation under suitable circumstances—but it's not obvious how, and this would certainly be going beyond quantum theory as presently understood.

Both options thus lead to serious, perhaps insuperable, difficulties.

5.3 Can Precise Preferences Arise in a Fuzzy Ontology?

Another very basic worry about Wallace's programme is its equivocation over mathematical rigour. Everything in Wallace's ontology that's relevant to rational decisions—including agents, the quasiclassical branches they inhabit, the branch states, and the branch Born weights, and the distinction between microstates and macrostates—is intrinsically fuzzily defined [Chapter 1]. There is, on Wallace's account, no precise fact of the matter about the different quasiclassical states that would result after a bet on a quantum experiment, nor about the Born weights of the branches corresponding to those quasiclassical states. And this isn't merely because quantum theory doesn't supply a unique natural definition of elementary branches and branching events: the *total* Born weight of all the quasiclassical branches describing a spin-up outcome of a Stern–Gerlach experiment isn't precisely defined either.

Now, to be sure, the total weight *is* supposed to be approximately defined. We are supposed, on Wallace's account, to be able to say that it's in a range of the form $R = (p - \epsilon, p + \epsilon)$, where ϵ is very small, and p thus represents an approximate total Born weight.[4] But we're not supposed to be able, on this account, to reduce ϵ to zero: below some level of precision, it becomes unavoidably arbitrary, just a matter of taste in your choice of branch definition, whether you take the total weight as $p_1 \in R$ or $p_2 \in R$.

And yet Wallace's decision-theoretic programme postulates that each rational agent should have a *precisely specified* and *complete* preference ordering among a very large class of possible unitary maps that produce different possible future global states. Where could such a preference ordering possibly come from? The ordering is supposed to be agent-dependent. Physics doesn't equip rational agents with some personal preference ordering on global states: they have to arrive at their preferences by introspection and reasoning. If one accepts Wallace's conclusions, the only ultimately relevant quantities are branch weights and the agent's personal utilities for macrostates (whose existence is supposed to follow given the preference ordering axioms). But even a super-agent who finds that they can calculate the former and can identify the latter by pure introspection would find these quantities only fuzzily defined—so that, in comparing some pairs (U_1, U_2) of actions on a given state $|\psi\rangle$, however hard

---

[4] That all ambiguities in total weights of quasiclassical outcomes are necessarily very small seems plausible and is what Wallace expects. Given the level of conceptual imprecision in discussing the emergence of quasiclassical structures from unitary quantum theory, though, it is hard to be certain even of this.

they try and however carefully they analyse the alternatives, they wouldn't be able to identify a reliable preference, *not* because the resulting global states are precisely equivalent, but because their difference is fuzzily ambiguous. On some views, $U_1|\psi\rangle$ would seem very slightly preferable; on others, $U_2|\psi\rangle$ would. In Wallace's notation (Wallace [Chapter 8]) for preference orderings, neither $U_1 \succeq_\psi U_2$ nor $U_2 \succeq_\psi U_1$ would hold in all ways of looking at the situation. Nor does it seem legitimate to postulate that $U_1 \sim_\psi U_2$ must hold in such cases. One can imagine the possibility of a sequence $(U_1, \ldots, U_n)$ such that no preference can reliably be identified between $U_i|\psi\rangle$ and $U_{i+1}|\psi\rangle$, for $i = 1, \ldots, (n-1)$, but nonetheless setting $U_i \sim_\psi U_{i+1}$ violates transitivity, since $U_1 \succ_\psi U_n$ *does* hold no matter what view the agent adopts of the fuzzy facts.

We are not, in any case, super-agents, and can only read Wallace's arguments as prescriptions for ideal rationality rather than descriptions of our real-world behaviour. None of us in fact has a complete and precise preference ordering among the relevant unitaries. Wallace, in effect, is telling us that we should ideally adjust our reasoning and our behaviour so as to be consistent with some complete preference ordering. But how? There is no natural algorithm available: any choice will involve uncountably many arbitrary decisions on pairs of preferences.[5] And why? Given that no choice of ordering will have any intelligible justification, even after the entire analysis is complete, how can there be a rational compulsion to make some choice (even if, counterfactually, it were practical)?

Here, it seems to me, Wallace's prescription runs into essentially the same difficulties that he identifies in other ways of thinking about Everettian branching. One *could*, in principle, find some (perhaps ad hoc) prescription defining a branching structure for the unitarily evolving state vector, and one could then use this structure to define a rational Born-rule-independent strategy based on branch counting. Wallace accepts that such a strategy is not logically inconsistent, but argues that it is likely to be difficult to implement in practice (because defining a precise branching structure is difficult) and hard to justify in principle (because the definition seems to require ad hoc choices). Both objections apply—arguably with at least equal force—to the Wallace programme.

This also reinforces the point that the case for Wallacean rationality cannot possibly rely on the lack of any practical alternative strategy. A very practical alternative is to follow whatever combination of instinct and reasoning evolution provided us before we became aware of Everettian quantum theory. Altering that strategy so as to comply rigorously with Wallace's axioms isn't practical; even coming close to doing so may not be. To be persuaded that we ought to try, we would need to be rationally persuaded not only that we should ideally be Wallaceans, but also that there is a practical method which allows us to become

---

[5] See chapter 4 of Savage [1971], where Savage makes a related point in criticizing his own approach.

closer to being Wallaceans, and that we will be better off if we employ this method.[6]

In short, given Wallace's account of a fuzzy ontology, there seems a strong reason to doubt Wallace's most basic postulate of rationality, R1, which states that rational agents have a *complete* (or *connected*) preference ordering on the unitary operations available to them at any state $|\psi\rangle$. No actual agent in a fuzzy Everettian ontology will ever be able to arrive at such an ordering in practice. Moreover, even if they had infinite computational power, fixing an ordering would require making a very complicated ad hoc choice which can have no complete rational justification. Yet, without R1, the purported derivation of the Born rule (Wallace [Chapter 8]) fails at the first step.

It's not clear to me that there is any fix for this, but let me comment briefly on two possible responses.

First, one might perhaps try weakening the postulate R1 to suggest that agents have, or should aspire to have, a preference ordering that *approximates* a complete ordering, in the hope of then proving that their policy should approximate Born-weighted mean utilitarianism. One problem with this is that one would need first to find and justify a suitable definition of approximation applied to preferences between pairs of unitary operations. As these are unquantified binary relations, it doesn't seem obvious that any suitable definition exists.

Second, one might consider the desperate resort of *postulating* a total ordering as part of the physical theory. But even that surely isn't available here. The orderings, recall, are agent-dependent, and even the most postulate-happy Everettian would surely recoil from requiring that *fundamental physical laws* specify independently, agent by agent, the preferences of every agent instantiated in nature.

Trying to formulate a rigorous decision theory for preferences in a fuzzy ontology may thus be rather like trying to build a skyscraper on mud.

5.4 Circularity of the Wallace Programme?

Zurek [Chapter 13] flags another worry about the logical relation between the two parts [Chapters 1, 8] of Wallace's programme, namely an apparent circularity. Wallace envisages a fuzzy quasiclassical ontology arising as the result of mathematical regularities observable within components of the unitarily evolving universal wavefunction. These regularities are supposed, in a realistic cosmological model, to arise through the decoherence of classical variables and to be defined by what Gell-Mann and Hartle [1993] term a quasiclassical domain, in which, for example, operators approximately quantifying local mass

---

[6] And to run such an argument, Wallaceans would, *inter alia*, need to find suitable precise definitions of 'closer' and 'better off'.

densities approximately follow classical equations of motion with probability close to one. Here the probability for a history defined by a sequence of operators $P_1(t_1), \ldots, P_n(t_n)$ is given by the decoherence functional

$$\mathrm{Tr}(P_n(t_n) \ldots P_1(t_1)\rho_{\mathrm{initial}}P_1(t_1) \ldots P_n(t_n)). \qquad (3)$$

In other words, the ontology is *defined* by applying the Born rule. Even if one could show, as Wallace claims, that agents defined within that ontology are rationally justified in using the Born rule as a calculus for decisions, it would seem incorrect to portray this argument as a *derivation* of the Born rule within Everettian quantum theory. Wallace's argument should rather be understood as attempting to show something weaker: that the Born rule re-emerges as output (albeit, to be fair, in an interesting and non-obvious way) if assumed as input. Even if correct, this would leave open the possibility that there are many different consistent and essentially inequivalent ways of defining ontologies that include distinct types of agents for whom different rational decision calculi can be established. It would thus fail to explain whether and (if so) why our own decision calculus should be based on the Born rule. It would also leave open the questions as to whether and (if so) how agents in some consistently defined Everettian ontology can arrive at the rational decision calculus appropriate to their ontology.

5.5 Problems with Born-Weighted Mean Utilitarianism

Wallace [Chapter 8], developing earlier ideas of Deutsch [1999], partly in response to criticisms (e.g. Barnum et al. [2000]) of the latter, then goes on to argue that from a few simple and purportedly natural axioms we can prove that rational agents who believe themselves to be in a universe described by many-worlds quantum theory are rationally required to (a) have a utility function that quantifies the value they assign to possible future quasiclassical events, (b) act so as to maximize their Born-weighted mean utility.

As we just saw, Wallace's first postulate, R1, seems to run into a fundamental obstacle, since neither Born weights nor quasiclassical histories (and thus their utility) are precisely defined in his ontology, and without R1 the decision-theoretic argument, which, *inter alia*, implies the existence of a utility function, fails. Moreover, even for an agent who *has* a utility function applicable to all relevant quasiclassical histories, the strategy of maximizing Born-weighted mean utility is not well defined. For a real world agent in state ψ there will generally be available unitaries U_1 and U_2 for which it's a matter of arbitrary definitional choice whether U_1 or U_2 produces higher Born-weighted mean utility.

There's a further practical problem, which isn't apparent in simple models of many-worlds experiments but is a serious worry in realistic applications. To be a rigorous Born-weight mean utilitarian in the real world, one must allow for the possibility of small Born weight branches with extreme negative or positive utility. The mean Born-weighted utility of a bet that, with Born weight close to 1, involves small utility gains or losses, is radically altered if it also creates, say, 10^{-25} Born weight branch of utility -10^{30}. Now, the Deutsch–Wallace–Savage arguments imply no bounds on agents' utility functions. It seems unlikely that any a priori argument can supply one, since pure rationality imposes no bound on utility functions—and in practice, for example, there seems to be no generally agreed lower bound on the utility cost assigned to the destruction of the Earth or similar catastrophes (Kent [2004]). A rigorous real world Born-weight mean utility calculation thus typically requires very careful analysis of small weight branches. In fact, even ensuring that the sum defining the mean utility *converges* requires careful analysis of small weight branches: consider, for example, the possibility of a set of branches of weight 2^{-n} and utility -3^n for all integers $n \geq N$.

Practically speaking, the best that real world agents are likely to be able to do is first simplify their model, by excluding events below some weight threshold, and then estimate a Born-weighted mean utility within that model—with no assurance that the estimated mean utility is close to the true mean utility (if indeed the latter exists).[7] This needs emphasizing, since much of Wallace's case against alternative rational strategies is based on the claim that they are ill-defined or impractical or both. Actually, as we will see, alternative strategies can sometimes be rather *better* defined and *more* practical than Born-weight mean utilitarianism.

5.6 Everettian Many-Worlds Rationality Reconsidered

5.6.1 General Remarks on Life in a Multiverse

It seems prima facie surprising to claim that mathematical analysis could show that Born-weight mean utilitarianism, or any other strategy, is the unique rational way of optimizing the welfare of one's own, and other people's, many future selves in a multiverse. After all, human parents are faced with the not entirely disanalogous question of how to take into account the welfare of their genetic descendants in (most of us assume) a single world, and it's a notoriously complex problem. People generally care not only about their descendants' present welfare, but also about their expected future welfare after our death. They can, and sometimes do, frame guiding rules of thumb to arbitrate between competing claims on their resources—for instance, to divide their estate equally among their children,

---

[7] Given the fuzzy ontology, we should more accurately say 'to any possible assignment of the imprecisely defined value of the true mean utility'. For the sake of readability, we take this qualification as read in what follows.

or to divide it according to their need. They take into account their children's relationships with one another, with others, and with society. They tend to care about immediate descendants more than distant ones, in a way that generally follows no well-defined formula. Evolutionarily developed instincts also impel a more general concern for our genes and those of the species. This concern probably cannot be precisely codified, but we can often find principles with which they are roughly aligned and which roughly characterize the behaviour they motivate. For instance, some species' instinctive behaviour might be roughly modelled as aiming to maximize an individual's expected number of descendants after 10^2 years. Some humanists' aims might be modelled as aiming to maximize the survival probability of the human race (and its genetic successors) over the next 10^9 years.

Some of these principles require impossible calculations to implement precisely, but can nonetheless legitimately be regarded as rational aims. If we adopt them, we commit ourselves to trying to satisfy them as best we can. In general, they imply conflicting courses of action. No one, I think, would seriously claim that any one of them is uniquely rationally preferable to all the others. We just make decisions as best we can, imperfectly guided by logic, sometimes perhaps trying our best to optimize quantities that we know we cannot properly calculate. And we always did: before we were capable of rational reflection, evolution equipped us to muddle through, sometimes following one rule of thumb, sometimes another. That's life. Why should we expect evolution or rationality to have equipped us any better when faced with the bewilderingly underdetermined imperative to care about our and everyone else's quantum descendants in a hypothetical multiverse?

5.6.2 Alternative Born-weight-Sensitive Strategies

Suppose, for the sake of the discussion, that we can somehow ignore the fuzziness of the ontology. Suppose that we have an agent faced with a finite number of choices j, each of which will create quasiclassical branches (although not a unique quasiclassical branching structure) with well-defined utilities U_i^j, in such a way that the set S_i^j of branches with the same utility U_i^j has a well-defined total Born weight p_i^j, and that the sums $\mu^j = \sum_i p_i^j U_i^j$ are finite.[8]

Consider again some of the strategies listed in Section 3.2. The x-percentile utilitarian, for $0 < x < 100$, always has a well-defined strategy, as does the future self democrat. The future self elitist and Price–Rawlsian's strategies are defined provided that $\max_j \sup_i (U_i^j)$ and $\max_j \inf_i (U_i^j)$, respectively, are defined. These will always hold true if the indexing set $I \ni i$ is finite. They need not hold true

---

[8] Without these assumptions, the mean utilitarian's strategy isn't defined, in which case Wallace's argument has failed.

if the branch utilities are unbounded above or below (possibilities which are not usually considered by Everettians, and which perhaps might be excluded by assumption, but are possibilities nonetheless).

As a practical matter, unless low-Born-weight extreme-utility branches can be excluded, the future self elitist and Price–Rawlsian may have difficulty optimizing their strategies, even if an optimal strategy exists, since calculating $\sup_i(U_i^j)$ or $\inf_i(U_i^j)$ requires analysing low-Born-weight branches that realize, or converge towards, the extreme utility values. This is also a problem—which may be easier or harder, depending on the details—for the mean utilitarian. Generically, it should not be a significant problem for the x-percentile utilitarian (for most x, say $1 < x < 99$), assuming the utility function is generically well behaved over the range, since the utility at the x-th percentile is then relatively insensitive to small perturbations of x, and so the calculation is relatively insensitive to the details of low-Born-weight extreme-utility branches. It should also generally not be a problem for a purely self-concerned future self democrat, who would generally hope to be able to attain a majority decision without counting the votes from low-Born-weight extreme-utility branches.[9]

5.6.3 Some Other Strategies

The *Gell-Mann–Hartle aesthete* fixes a particularly pretty quasiclassical consistent set S, which she uses to define a way of counting branches containing her future selves.[10] Her quantum ontology is Everettian: she agrees that her selected set has no fundamental physical significance. However, she thinks that one needs *some* way of weighting future selves and that this one is as rationally defensible as Born-rule-weighting or any other, and more aesthetically pleasing.

The *value teleologist* fixes a particular cosmological final density matrix ρ_f, whose spectrum does not include zero. In considering whether or not to accept a generalized bet on a quantum experiment, or indeed making any decision dependent on a quantum event, he uses pre- and post-selection, with some standard theory of the initial cosmological conditions defining the initial state ρ_i, and with ρ_f defining the final state, in order to calculate the probabilities of the future worlds corresponding to the possible outcomes (Aharonov et al. [1964], Gell-Mann and Hartle [1994]). He bets as if these were the actual probabilities. This is not because he believes they are—he believes in deterministic unitary quantum mechanics and so doesn't think probabilities are fundamental, and in any case his physical theory is a standard cosmological theory with initial state ρ_i and no post-selection on ρ_f. However, for aesthetic or existential reasons, his interest in future events is conditional on the chosen final state post-selection.

[9] Of course, in both these last two cases, it *could* still be a problem if the numbers so conspire.
[10] I thank Hans Westman and Ward Struyve for suggesting this example.

5.6.4 *Wallace's Rationality Postulates*

We have noted already that branch weight, branch macrostate, branch microstate, and reward are all only fuzzily defined in Wallace's ontology [Chapter 1], and that this gives strong reason to doubt Wallace's ordering axiom $R1$. It casts doubt too on whether the availability axioms $A3-A5$ and the rationality axioms $R3$, $R5$ even have a precise definition.

Wallace's diachronic consistency axiom, R2, is violated by the x-percentile utilitarian strategy, among others. Now, to be fair, one can find examples where the two conflict which illustrate some motivation for diachronic consistency. Consider the possibility of being offered N dollars per unit time to stand in a radiation field, with a risk p of lethality per unit time. An x-percentile utilitarian who considers this offer will generally find their response depends on the timescale over which they regard their decisions as binding: it could seem a good offer considered as valid for the next second, and then good again for each successive second, but a bad offer if they have to make a single decision about whether to accept for the next hour.

Yet, even in this rather unusual example, the motivation for x-percentile utilitarians is still clear when x is close to 0 or 100, and their actual strategy is intended as a practical approximation to their ideal strategy of Price-Rawlsianism or future self elitism. The Price–Rawlsian will decline unless the total risk is zero; the future self elitist will accept unless the survival probability is zero. Note too that even here x-percentile utilitarianism *is* a well-defined strategy once a timescale for decisions is fixed. In the more normal circumstance of separated discrete decisions, x-percentile utilitarianism seems both rationally defensible and practical, which suggests that the diachronic consistency axiom is less rationally compelling than Wallace argues.

Another reason to doubt R2, it seems to me, is that, *pace* Wallace's comment [Chapter 8]—

In the presence of widespread, generic violation of diachronic consistency, agency in the Everett universe is not possible at all.

—diachronic consistency actually *is*, strictly speaking, generically violated in real-world decisions. A Savagean or Wallacean rational agent, recall, has to be equipped with a utility function as well as a probability measure for outcomes. Rationality is silent on the precise form of the utility function. If we have one, it reflects our current values. These generally change over time, as we do, partly as a result of decisions we have previously taken, whose outcomes affect us in ways that we cannot reliably predict beforehand: our own natures are too complex and too opaque to us, and we also change in response to our environment, which is also complex and unpredictable. The best it seems to me that one might hope to say of diachronic consistency in real-world decisions is that pretty often, in the short term, it approximately holds—which clearly isn't a strong enough assumption to prove an interesting decision-theoretic representation theorem.

Elga's proposal that Everettians might have a rational preference for future self diversity[11] also seems pertinent here, as does the case for rationally preferring future society diversity. (Why *not* exploit the scope for political compromise by causing society to evolve in different ways along different branches?) In both cases, it seems to me, contra Wallace [Chapter 8], diachronic consistency can be rationally violated. I can consistently believe now that it's a good thing that the global state should include future copies of me as a king and a beggar, while knowing that, if I ever find myself a beggar, I would strive to become a king if I could.[12] From the perspective of my future beggar self, the unpleasantness of finding that *he* is the beggar outweighs the satisfaction of knowing that diversity was achieved. From my present perspective, the prospect of diversity nonetheless remains appealing.

As Wallace himself notes earlier in his discussion:[13]

... to make a copy of myself and send him off to do a dangerous or disagreeable task—and ... to take actions designed to prevent him shirking that task ... is not *irrational*.

Indeed—and this remains true if the task, for which I have a strong present desire, is to ensure future self diversity. The fact that my future selves will never interact makes no difference to the rational justification.

Turning briefly to other postulates:

(R3) Microstate indifference, R3, can be violated by value teleologist strategies, among others.

(R4) Continuity, R4, is violated by x-percentile utilitarian strategies, among others.[14]

(R5) Branching indifference, R5, is violated by Gell-Mann–Hartle aesthete strategies, among others.

5.6.5 Summary

Wallace argues that strategies other than mean utilitarianism turn out, on closer inspection, either to be not rigorously defined, completely impractical, or to violate criteria such as diachronic consistency that allegedly define the very essence of rationality. The last two claims—impracticality and violation of rational essentials—surely require mathematical underpinning and justification, if they are to have any possible relevance to what is presented as a rigorous

---

11 See the discussion of 'The variety rule' in Wallace [Chapter 8].
12 Or perhaps, considering the uneasiness of crown-wearing heads, vice versa.
13 In Section 5 of Chapter 8.
14 Wallace [Chapter 8] notes and discusses the special case of the Price–Rawlsian.

mathematical argument. For example, an account of practicality needs some complexity criteria for rational agent computations; one could then at least discuss the empirical justification for the proposed criteria and whether and when they actually distinguish mean utilitarianism from other strategies. Similarly, if one accepts that diachronic consistency is generically violated and sometimes grossly violated in the real world, an account of its role in decisions needs to quantify and compare the degree of violation implied by different strategies in different circumstances. At present, though, the arguments for diachronic consistency and those concerning practicality rest only on very debatable verbal intuitions.

As for lack of rigorous definition, there seems to be a danger of a double standard, whereby the fuzziness of the ontology is used to point out difficulties for alternative strategies (though in fact it also causes difficulties for mean utilitarianism), while the arguments for mean utilitarianism are justified in the context of toy models in which a precise definition of a branching structure can be found (in which case many strategies other than mean utilitarianism can be precisely defined). The case has not been made that mean utilitarianism is well defined or practical in Wallace's fuzzy Everettian ontology, in which the mean utility of a strategy can at best only be fuzzily defined. One can imagine examples in which it either fails to be finite or is impractical to estimate—and it seems hard to exclude the possibility that these features often apply in the real world. One can also easily construct examples in which other strategies are easier either to approximate or to implement precisely.

Wallace's rationality postulates, likewise, are hard to motivate in Wallace's fuzzy Everettian ontology, where they are ultimately intended to apply, but where they are difficult, perhaps impossible, to define precisely. They generally appear, in any case, possible but uncompelling guides for rational agents. Where defined and practical, mean utilitarianism is certainly a rationally defensible strategy, with some mathematically convenient properties. But, like other critics (Albert [Chapter 11], Price [Chapter 12]) I am far from persuaded that, if I were an Everettian, I should or would be a Born-weighted mean utilitarian.

6 AGAINST SUBJECTIVE UNCERTAINTY

One of the stranger claims in the recent Everettian literature is the suggestion, first made by Saunders ([Chapter 6], [1998], Saunders and Wallace [2008], Wallace [2006]), that Everettian quantum theory, although deterministic, nonetheless has a natural probabilistic interpretation that can be found not by amending the theory or by adding further postulates, but simply by—somehow—analysing the experience and linguistic usages of agents, that is, creatures like ourselves, in an Everettian universe. In support of this claim are offered highly technical and controversial arguments concerning the philosophy of language. It seems to me simply a mistake, an exercise in wish fulfilment, to think that anything of significance to fundamental physics could turn on such questions, as though waving the magic wand of linguistic philosophy over a unitarily evolving state vector could somehow conjure up a probability measure and a sample space.[15]

Consider Wallace's succinct summary (Wallace [2006]) of the argument:

[The argument for subjective uncertainty] may be summarised as follows: in ordinary, non-branching situations, the fact that I expect to become my future self supervenes on the fact that my future self has the right causal and structural relations to my current self so as to count as my future self. What, then, should I expect when I have two or more such future selves? There are only three possibilities:

1. I should expect abnormality: some experience which is unlike normal human experience (for instance, I might expect somehow to become both future selves).

2. I should expect to become one or the other future self.

3. I should expect nothing: that is, oblivion.

Of these, (3) seems absurd: the existence of either future self would guarantee my future existence, so how can the existence of more such selves be treated as death? (1) is at least coherent—we could imagine some telepathic link between the two selves. However, on any remotely materialist account of the mind this link will have to supervene on some physical interaction between the two copies—an interaction which is not in fact present. This leaves (2) as the only option, and in the absence of some strong criterion as to which copy to regard as 'really' me, I will have to treat the question of which future self I become as (subjectively) indeterministic.

This is a false trichotomy. Consider an (obviously simplified) Everettian description of an experiment in which an agent Alice, initially in brain state $|0\rangle_A$, observes a system in a quantum superposition $\sum_{i=1}^{2} c_i|i\rangle_S$, where $|1\rangle_S$ and $|2\rangle_S$ correspond, say, to the up and down states of a spin 1/2 particle, and becomes entangled:

$$|0\rangle_A \sum_{i=1}^{2} c_i|i\rangle_S \rightarrow \sum_{i=1}^{2} c_i|i\rangle_A|i\rangle_S. \qquad (4)$$

Here $|i\rangle_A$ is Alice's brain state after observing the system in state i, for $i = 1, 2$.

Now, as an aside, we actually *should* take possibility (3) seriously, for two reasons. First, our conclusions ought to be based on empirical evidence rather than prejudice. We do not know that Everettian quantum theory is actually correct; we do not have a good theory of how consciousness is attached to quantum states; we do not know that we or any other agents have ever been

---

[15] Incidentally, this issue has also divided Everettians: Papineau [1997, 2003] and, at one point, Greaves [2004], have also argued that subjective uncertainty is not to be had in Everettian quantum theory.

in a superposition of macroscopically distinct brain states. We thus do not know whether, if we were able to place an agent in such a superposition, they would experience anything—nor, if so, what. Second, there's a coherent view of Everettian quantum theory in which we are continually being replaced by multiple copies of future selves. On this view, even if we assume that superposed selves have individual experiences, *we* will experience nothing in future (though our various future selves will).

The more immediately pertinent point, though, is that if we do take Everettian quantum theory seriously, it says, indeed, that Alice becomes entangled in a macroscopic superposition. A coherent way of describing this, which respects the link between brain states and mind states, is that just as materially she becomes several future selves, her mind becomes several *disjoint, non-interacting* future minds, with no telepathic link: i.e. option (1) without Wallace's misleading gloss.

The one description that seems obviously wrong, given the rules of the game Wallace sets out, is option (2): this really *is* an account of mind that supervenes on something not present in the physics, namely a probabilistic evolution law taking brain state $|0\rangle_A$ to one of the states $|i\rangle_A$.

The dangers of attaching some fuzzy theory of experience to Everettian quantum theory provoke two comments:

First, the fact that we don't have a good theory of mind, even in classical physics, doesn't give us a free pass to conclude anything we please. That way lies scientific ruin: *any* physical theory is consistent with *any* observations if we can bridge any discrepancy by taking on arbitrary assumptions about the link between mind states and physics. We should, rather, be all the more cautious and tentative in offering any conclusion.

Second, the fact that at present no theory of mind can be expressed *purely* mathematically doesn't remove the obligation to strive to express one's ideas in mathematics *as far as possible*. Adorning Everettian quantum theory with extra assumptions expressed in words—for instance, as arguments in linguistic philosophy—without equations doesn't alter the fact that one's making extra assumptions: it merely makes them more vaguely expressed.

Consider[16] Saunders' exposition [Chapter 6]:

Consider a concrete example. Alice, we suppose, is about to perform a Stern–Gerlach experiment; she understands the structure of the apparatus and the state preparation device, and she is convinced EQM is true. In what sense does she learn, post-branching, something new? The answer is that *each* Alice, post-branching, learns something new (or is in a position to learn something new)—each will say something (namely, 'I see the outcome is spin-up (respectively, spin-down), and not spin-down (respectively, spin-up)') that Alice prior to branching cannot say. It is true that Alice, prior to branching, knows that this is what each successor will say—but still she herself cannot speak in this way. . . . The implication of this line of thought is that, appearances notwithstanding,

---

[16] My remarks here follow Albert's lucid discussion [Chapter 11].

prior to branching Alice does *not* know everything there is to know. What is it she does not know? I say 'appearances notwithstanding' for of course in one sense (we may suppose) Alice does know everything there is to know: she knows (we might as well assume) the entire corpus of impersonal, scientific knowledge. But what that does not tell her is *just which person she is—or where she is located*—in the wave-function of the universe.

But equation (4) suggests that there is no meaning to this question before the experiment.[17] Nothing in the mathematics corresponds to 'Alice, who will see spin up' or 'Alice, who will see spin down'. On the left we have 'Alice, before the experiment'; on the right we have 'Alice, who has seen spin up' and 'Alice, who has seen spin down'. If one wants to postulate an 'Alice, who will see spin up', well, one can—but one should then include her in the mathematics. One could, for instance, start with a postulate of the form:

(P) the probability that A's mind ends up believing that spin is up is $|c_0|^2$ and the probability that A's mind ends up believing that spin is down is $|c_1|^2$.

This—Albert and Loewer's 'single mind view' (Albert and Loewer [1988])— gives only one sentient future Alice. To introduce a collection of present Alices who in future will experience each of the different possible experimental outcomes, one could, instead, follow Albert and Loewer in postulating a continuum of Alice minds of which a proportion $|c_0|^2$ will see spin-up. One could, in short, adopt the many-minds interpretation. I am not persuaded that there is a legitimate alternative formulation of Saunders' account.

Appendix A: Further Comments on Counting Descendants

Wallace places great stress on the fact that there is no unique natural definition of a quasiclassical branch in an Everett universe, and so no way of counting the number of branches with any given feature. For example, a naive analysis of a quantum experiment with three possible outcomes might suggest that a single branch, pre-experiment, divides into three, post-experiment. But this, Wallace stresses, neglects the fact that quantum interactions take place very frequently in time and densely in space, outside our control. A careful attempt to quantify quasiclassical branches would show many branches splitting into many more during the lifetime of the experiment; however, there is no unique natural definition that would allow us to pin these numbers down. Hence, it is argued, there is no way of implementing the naive idea of using branch counts to define a rational strategy—an approach which would, if it worked, be a coherent alternative to the Born-rule-dependent strategy, and so refute the claim that the latter is the unique rational strategy.

There is indeed no known natural way of characterizing and counting branches. It is worth reconsidering, however, whether there may nonetheless be a natural way for an

---

[17] The same is true in more realistic models.

agent about to observe an experiment to characterize and hence count his descendants, by considering their memory states after the experiment.

Consider first a simple model of an observer, apparatus, and quantum state which evolve unitarily during an experiment so that

$$|O_0\rangle\,|A_0\rangle\,|Q_0\rangle \;\rightarrow\; \sum_{j=1}^{3} a_j |O_j\rangle\,|A_j\rangle\,|Q_j\rangle, \tag{A1}$$

where the first state is at time 0 and the second at time t, the state $|A_j\rangle$ is the apparatus state registering outcome j and $|O_j\rangle$ is the observer state having observed the apparatus registering outcome j. The agent in state $|O_0\rangle$ can reasonably say that he will have three successors $|O_j\rangle$ at time t, corresponding to his three future distinct brain states.

Now consider a more detailed model in which we include an environment, and suppose that

$$|O_0\rangle\,|A_0\rangle\,|Q_0\rangle\,|E_0\rangle \;\rightarrow\; \sum_{j=1}^{3}\sum_{i=1}^{n_j} a_{ij}|O_j\rangle\,|A_{ij}\rangle\,|Q_j\rangle\,|E_{ij}\rangle. \tag{A2}$$

Again, the sum over j represents the three possible observer states. The sums over i represent decompositions into orthogonal quasiclassical branches, and the number of terms n_j in each sum depends on an arbitrary choice of definition of quasiclassical branch from among many possible definitions. Since this is supposed to be a model of the same experiment, we have that $\sum_i |a_{ij}|^2 = |a_j|^2$ for $j = 1,2,3$. The agent in state $|O_0\rangle$ has no unique natural way of characterizing the branches. However, he could consistently view each of the three components, containing the state $|O_j\rangle$, for $j = 1,2,3$, as representing precisely one successor.[18] After all, he is interested in his successors' welfare, and this is determined by their mind states; at this point in the analysis, at least, it is not affected by the state of the rest of the universe.

It might be objected that we still have not taken sufficiently into account the pervasiveness of environment-induced quantum interactions, which will presumably also be taking place within the agent's brain during the experiment. A more detailed model still would replace the sums on the right-hand side of (A2) by sums including at least small components of a variety of different agent mind states, corresponding to different brain states that arise through zapping by stray cosmic rays and other quantum effects.

A related objection is that this way of counting successors leads to ambiguities when sequences of experiments are carried out. Suppose that the agent will carry out a second experiment, with two possible outcomes, if he observes outcome 1 in the first experiment, but not otherwise. After the first experiment but before the second, it appears that he has three successors, whose welfare he should value equally. After the second experiment, it appears that he has four successors, whose welfare he should again value equally. But

---

[18] He is not rationally compelled to accept this view. The point is that, in this model, successor counting is mathematically well defined and hence it's *possible* to use it to define a rational strategy.

since two of these descend from one of the original successors, and each 'inherit' any resources he 'bequeaths' to that successor, the two ways of counting lead to different allocation strategies—i.e. they disagree on which 'bets' he should be willing to accept on the experiment.

To these objections, however, the agent could make several responses.

First, that his policy is not incoherent, but merely so far incompletely specified. In the case of the two experiments, he does care equally about the welfare of his three successors between the experiments, and he also cares equally about the welfare of his four successors after the experiments. To formulate a more precise policy, he would need to set out some way of trading off the welfares of different successors at different times: for example, by summing the time-integrated welfares over each distinct lifetime segment. That this may become complicated doesn't imply that the aim is not rational. Indeed, we face very similar problems in worrying about the well-being of our genetic descendants. It is perfectly rational to value the welfare of all of your children equally, and also perfectly rational to value the welfare of all of your as yet unborn grandchildren equally, in both cases *ceteris paribus*. However, finding a rational asset allocation policy that respects both these preferences may require some further policy decisions, complicated calculations, and predictions of uncertain future events.

Second, that he would indeed take into account the welfare of all his successors, including those whose mind states differ because of environmentally induced interactions, if he could. Here again, he can maintain that he has a rational policy in principle, albeit one that he cannot fully implement in practice because of the impracticality of carrying out the relevant calculations. And again, he can say that the latter caveat does not detract from the rationality of his goal.

Third, that a consistent strategy for weighting his concern for the welfare of successors could be defined by considering the branching structure recorded in their own memory states. In the example above, successors who experience two experiments in succession remember that fact, and this distinguishes them from successors who experience only one experiment. It's logically consistent—and not obviously any more absurd than any Born-weight-dependent many-worlds strategy—to assign the former caring weight 1/6 and the latter caring weight 1/3.

To these responses, Everettians might in turn object that there is no natural way, even in principle, of characterizing and counting all the possible mind states of successors of an agent exposed to real-world environmental interactions. But if the Everettian case eventually turns on *this* point, then the objection to rational strategies based on counting successors ultimately arises from an intrinsic vagueness in the quasifunctionalist theory of mind attached to the quantum formalism by Wallace et al., not, as claimed, from the vagueness in the notion of a quasiclassical branch. It seems a most uncomfortable defence of a purportedly fundamental theory to say that it is not well enough developed for us to be able to assess whether or not one of the key arguments advanced in its favour is valid.

Appendix B: Further Comment on Physical Laws of Rational Compulsion

What could it possibly mean to believe that the *laws of physics per se rationally compel* a particular behaviour for rational agents in a branching multiverse? The idea here, to be clear, is not merely the truism that the laws of physics imply significant facts about the world which rational agents might, or even must, sensibly take into account. It is that there are basic postulates, on an equal footing with other physical laws, that state by fiat that a particular type of behaviour is rationally compulsory for rational agents. I don't think I know what this can mean—the idea of such a law isn't consistent with my understanding of either physics or rationality—but the idea is definitely in play in some discussions of many-worlds theories in this volume. Papineau [Chapter 7] proposes an axiom of this type, and Greaves–Myrvold [Chapter 9] consider how to (purportedly) confirm theories including such axioms. My impression is that many Everettians share Greaves' view (Greaves [2004]) that resorting to such a postulate would be, at least, an adequate fall-back should Wallace's [Chapter 8] and other arguments not hold up.

Here's a point that seems not to have been considered in the Everettian literature. If we *were* to take seriously the idea that physical axioms can rationally compel rational beings to act in a certain way—by fiat, without further justification—then we must also take seriously the possibility that these rationally compelling axioms can take unfamiliar forms. For instance, in our branching universe, there's no reason to restrict to axioms that require rational preferences to be given by the ordering of values of expressions of the form $\sum_i p_i U_i$, where U_i is the agent's utility for the outcome on branch i and the p_i are positive branch weights satisfying the Kolmogorov axioms. There's nothing *logically* inconsistent about postulating laws with negative or complex p_i, or preference orderings given by general joint functions $f(p_i, U_i)$, or indeed any other mathematical structure one cares to dream up.[19]

Such laws would generally violate Savage's axioms and perhaps other cherished intuitions about rational behaviour. But once one enters the strange game of postulating *physical* laws *defining* rational behaviour in multiverses, one needn't restrict one's postulates to intuitions developed in an attempt to provide a foundation for decision theory in a single chancy universe. Everettians who miss this point seem to me like hypothetical 17th-century theorists who learn Hooke's law, come up with the idea that one can postulate abstract physical force laws that define forces between objects unmediated by springs, but then still maintain that these laws necessarily have to set force proportional to separation. Their boldness is inconsistently selective: an abstract law need not be constrained by the details of the concrete model that inspired it.

Of course, Everettians who think it makes sense to postulate laws of rational compulsion still have the option of basing their postulates on one-world probability theory, and specifically on optimizing Born-rule-weighted average utility. But one needs to be clear that, prima facie, this is an arbitrary choice from a very large range of possibilities. Along with everything else in this peculiar game, that choice seems to lack justification. Moreover, as the above analysis of Greaves–Myrvold's account of confirmation applied to the weightless universe shows, allowing arbitrary choices of laws of rational

---

[19] Indeed, one could imagine an exotic story about inverted qualia and hence reversing of utilities on some branches, which justifies giving them negative weights. And perhaps, some might argue, the complex quantum amplitudes defining the path integral should be interpreted as directly defining rational constraints on an agent's preferences for the entire set of paths defining a hypothetical future unitary evolution.

compulsion means not only that many mutually inconsistent choices can be postulated in the same multiverse but also that each of them can be, by their own lights, confirmed.

Appendix C: A Possible Empirical Distinction between Many-Worlds and One-World Quantum Theory

Finally, suppose, notwithstanding all the arguments above, that we arrive at an Everettian theory that, while perhaps ad hoc and unattractive, is coherent—for example, some version of the many-minds interpretation (Albert and Loewer [1988]). It is generally believed that, without very advanced technology which allows the re-interference of macroscopically distinct branches, such a theory will necessarily be empirically indistinguishable from Copenhagen quantum theory.

The following argument against this conclusion relies on anthropic reasoning and also on the hypothesis that species may evolve a consistent preference for or against higher population expectation over higher survival probability. Anthropic reasoning is notoriously tricky to justify, and we may anyway not necessarily have evolved demonstrable consistent preferences one way or the other, so the argument may not necessarily have practical application. Nonetheless, it does show in principle that evolutionary evidence could make many-worlds theories more or less plausible.

Consider a simple model of two species A and B, both of which begin with population P and are offered, each year, the option of doing something that depends on a quantum event and carries a 0.5 probability of extinction and a 0.5 probability of trebling the species population. Suppose that, if they reject the option, their population remains constant, as it does in between these decisions. Species A is risk-averse, and so always declines the option. Species B is risk-tolerant, and instinctively driven to maximize expected population, and so always accepts.

Now let N be a large integer. After N years, if one-world quantum theory is correct, species A will have population P, and species B will have either population 0 (with probability $(1 - (\frac{1}{2})^N)$) or population 3^N (with probability $(\frac{1}{2})^N$). In other words, species B will almost surely be extinct. If these are the only two species, and you are alive in the N-th year, almost certainly you belong to species A.

If many-worlds quantum theory is correct, species A still has population P in all branches. Species B has population 0 in branches of total Born weight $(1 - (\frac{1}{2})^N)$, and population 3^N in branches of total Born weight $(\frac{1}{2})^N$. Now, if anthropic reasoning is justifiable here, and you are alive in the N-th year, almost certainly you belong to species B. (There are $(\frac{3}{2})^N$ times as many minds belonging to species B as to A after N years.)

In other words, there is a sense in which long-run evolutionary success is defined by different measures in one-world and many-worlds quantum theory. If anthropic reasoning were justifiable, then one could in principle infer whether one-world or many-worlds quantum theory is likelier correct by seeing whether one belongs to a Born-weighted expected population maximizing species or to a risk-averse species that seeks to maximize its Born-weighted survival probability. Readers may thus wish to consider whether their species has evolved a coherent strategy of either type.[20]

Acknowledgements

I am very grateful to Jonathan Barrett for many thoughtful and constructive comments on, and criticisms of, a preliminary version of the manuscript, as well as other very valuable conversations, and also to Hilary Greaves and David Wallace for patiently tolerating and helpfully engaging with my critical probing. Thanks too to David Albert, Harvey Brown, Jeremy Butterfield, Chris Fuchs, Lucien Hardy, Graeme Mitchison, Wayne Myrvold, David Papineau, Huw Price, Simon Saunders, Tony Short, John Sipe, Rob Spekkens, and Tony Sudbery for some valuable conversations. This research was partially supported by a grant from The Foundational Questions Institute (fqxi.org) and by Perimeter Institute for Theoretical Physics. Research at Perimeter Institute is supported by the Government of Canada through Industry Canada and by the Province of Ontario through the Ministry of Research and Innovation.

References

Aharonov, Y., B. Bergman, and J. Leibowitz, [1964], 'Time symmetry in the quantum process of measurement', *Physical Review* B **134**, 1410–16.

Albert, D. [2010], 'Probability in the Everett picture', this volume.

Albert, D. and B. Loewer [1988], 'Interpreting the many-worlds interpretation', *Synthese* 77, 195–213.

Arrow, K.J. [1950], 'A difficulty in the concept of social welfare', *Journal of Political Economy* **58**, 328–46.

Barbour, J. [2001], *The End of Time*, Oxford University Press, Oxford.

Barnum et al., [2000], 'Quantum probability from decision theory?', *Proceedings of the Royal Society of London* A456, 1175–82.

Bell, J.S. [1987a], 'Quantum mechanics for cosmologists', in *Speakable and Unspeakable in Quantum Mechanics*, Cambridge University Press, Cambridge.

——— [1987b], 'The measurement theory of Everett and De Broglie's pilot wave', in *Speakable and Unspeakable in Quantum Mechanics*, Cambridge University Press, Cambridge.

Borges, J.L. [1948], 'The garden of forking paths', *Ellery Queen's Mystery Magazine* (August 1948); reprinted in *Collected Fictions: Ficciones*, by J.L. Borges, Viking, New York (1999).

Coleman, S. [1994], 'Quantum mechanics in your face', lecture video archived at http://media.physics.harvard.edu/video/index.php?id=SidneyColeman_QMIYF.flv.

Deutsch, D. [1985], 'Quantum theory as a universal physical theory', *International Journal of Theoretical Physics* **24**, 1–41.

——— [1999], 'Quantum theory of probability and decisions', *Proceedings of the Royal Society of London* A455, 3129–37.

---

[20] Unfortunately, I suspect mine has not.

DeWitt, B. [1973], 'Quantum mechanics and reality', in *The Many-Universes Interpretation of Quantum Mechanics*, B. DeWitt and N. Graham (eds), pp.155–66 and 167–218, Princeton University Press, Princeton.

Dowker, F. and A. Kent [1995], 'Properties of consistent histories', *Physical Review Letters* 75, 3038–41.

—— [1996], 'On the consistent histories approach to quantum mechanics', *Journal of Statistical Physics* 82, 1575–646.

Everett, H. [1957], ' "Relative state" formulation of quantum mechanics', *Reviews of Modern Physics* 29, 454–62. Reprinted in *The Many-Worlds Interpretation of Quantum Mechanics*, B. DeWitt and N. Graham (eds), pp.141–9, Princeton University Press, Princeton (1973).

—— [1973], 'The theory of the universal wavefunction', in *The Many-Worlds Interpretation of Quantum Mechanics*, B. DeWitt and N. Graham (eds), pp.3–140, Princeton University Press, Princeton.

Feyerabend, P.K. [1975], 'How to defend society against science', *Radical Philosophy* 11, 277–83.

Gell-Mann, M. and J.B. Hartle [1993], 'Classical equations for quantum systems', *Physical Review* D 47, 3345–82. Available online at arXiv: gr-qc/9210010.

—— [1994], 'Time symmetry and asymmetry in quantum mechanics and quantum cosmology, in *Physical Origins of Time Asymmetry*, J.J. Halliwell, J. Perez-Mercader, and W. Zurek (eds), Cambridge University Press, Cambridge.

Geroch, R. [1984], 'The Everett interpretation', *Noûs* 18, 617–33.

Graham, N. [1973], 'The measurement of relative frequency', in *The Many-Universes Interpretation of Quantum Mechanics*, B. DeWitt and N. Graham (eds), pp.229–53. Princeton University Press, Princeton.

Greaves H. and W. Myrvold [2010], 'Everett and evidence', this volume.

Greaves, H. [2004], 'Understanding Deutsch's probability in a deterministic multiverse', *Studies in History and Philosophy of Modern Physics* 35, 423–56. Available online at philsci-archive.pitt.edu/archive/00001742/.

Hartle, J.B. [1968], 'Quantum mechanics of individual systems', *American Journal of Physics* 36, 704–12.

—— [2010], 'Quasiclassical realms', this volume.

Kent, A. [1990], 'Against many-worlds interpretations', *International Journal of Modern Physics* A5, 1745–62.

—— [1996], 'Quasiclassical dynamics in a closed quantum system', *Physical Review* A54, 4670–75.

—— [1997], 'Consistent sets yield contrary inferences in quantum theory', *Physical Review Letters* 78, 2874–77.

—— [1998a], 'Quantum histories', *Physics Scripta* T76, 78–84.

—— [1998b], 'Reply to Griffiths and Hartle', *Physical Review Letters* 81, 1982.

—— [2000], 'Quantum histories and their implications', in *Relativistic Quantum Measurement and Decoherence, Lecture Notes in Physics* 559, 93–115, H.-P. Breuer and F. Petruccione (eds), Springer-Verlag, Berlin.

—— [2004], 'A critical look at risk assessments for global catastrophes', *Risk Analysis* 24, 157–68.

—— [2010], 'A solution to the quantum reality problem', in preparation.

Lockwood, M. [1996], ' "Many minds" interpretations of quantum mechanics', *British Journal for the Philosophy of Science* 47, 159–88.

Papineau, D. [1997], 'Rational decisions and the many minds interpretation of quantum mechanics', *The Monist* 80, 97–117; reprinted in *The Roots of Reason: Philosophical Essays on Rationality, Evolution and Probability*, pp. 212–39, Oxford University Press, Oxford (2003).

—— [2010], 'A fair deal for Everettians', this volume.

Price, H. [2010], 'Decisions, decisions, decisions: can Savage salvage Everettian probability?', this volume.

Rawls, J.S. [1971], *A Theory of Justice*, Harvard University Press, Cambridge, Mass.

Saunders, S. [1998], 'Time, quantum mechanics, and probability', *Synthese* 114, 405–44. Available online at arXiv.org/abs/quant-ph/0112081.

—— [2010], 'Chance in the Everett interpretation', this volume.

Saunders, S. and D. Wallace [2008], 'Branching and uncertainty', *British Journal for the Philosophy of Science* 59, 293–305. Available online at philsci-archive.pitt.edu/archive/00003811/.

Savage, L. [1972], *The Foundations of Statistics*, Dover, New York.

Tegmark, M. [2010], 'Many worlds in context', this volume.

Vaidman, L. [2002], 'Many worlds interpretations of quantum mechanics', *Stanford Encyclopedia of Philosophy*, available online at http://plato.stanford.edu/entries/qm-manyworlds/.

—— [2010], 'Time symmetry and the many-worlds interpretation', this volume.

Wallace, D. [2006], 'Epistemology quantized: circumstances in which we should come to believe in the Everett interpretation', *British Journal for the Philosophy of Science* 57, 655–89. Available online at philsci-archive.pitt.edu/archive/00002839.

—— [2010], 'Decoherence and ontology', this volume.

—— [2010], 'How to prove the Born rule', this volume.

Zurek, W.H. [2010], 'From relative states to quantum jumps, Born's rule, and objective reality', this volume.

11

Probability in the Everett Picture

David Albert

1 THE PROBLEM

Let me start off by reminding you of what I take to be the simplest and most beautiful and most seductive way of understanding what it was that Everett first decisively put his finger on 50 years ago.

There is supposed to be a problem with the linear, deterministic, unitary, quantum-mechanical equations of motion. And that problem—in its clearest and most vivid and most radical form—runs as follows: the equations of motion (if they apply to everything) entail that in the event that somebody measures (say) the x-spin of an electron whose y-spin is initially up, then the state of the world, when the experiment is over, is with certainty going to be a superposition, with equal coefficients, of one state in which the x-spin of the electron is up and the measuring-device indicates that that spin is up and the human experimenter *believes* that that spin is up, and another state in which the x-spin of the electron is down and the measuring-device indicates that that spin is down and the human experimenter *believes* that that spin is down. And superpositions like that—on the standard way of thinking about what it is to be in a superposition—are situations in which there is no matter of fact about what the value of the x-spin of the electron is, or about what the measuring-device indicates about that value, or about what the human experimenter *believes* about that value. And the problem with that is that we know—with certainty—by means of direct introspection, that there is a matter of fact about what we believe about the value of the x-spin of an electron like that, once we're all done measuring it. And so the superposition just described can't possibly be the way experiments like that end up. And so the quantum-mechanical equations of motion must be false, or incomplete. Or that (at any rate) is the conventional wisdom.

And it was Everett who first pointed us in the direction of a scientifically realist strategy for resisting that conventional wisdom. It was Everett who first remarked that the problem rehearsed above is in fact *ill-posed*. It was Everett who first argued (more particularly) that precisely the same linearity of the quantum-mechanical equations of motion which gives rise to the troubling superpositions of brain-states described above also radically undermines the *reliability* of the sorts of introspective reports with which those superpositions are supposed to be incompatible! And the suggestion here—the intriguing possibility here—was that perhaps there is nothing wrong with the equations of motion, and nothing incomplete about them, after all.

And there have been a host of questions, ever since, about how to go on from there.

And I want to focus on one particularly difficult such question in my remarks here: the question of how to make sense—in the context of the sort of picture of the world that Everett seems to be suggesting—of all of the apparently indispensable quantum-mechanical talk of *probabilities* (and the talk I have in mind here includes our use of probabilities as a guide to life, and as a component of explanation, and as a tool of confirmation, and so on).

There is a simple and straightforward and perfectly obvious worry here—a worry that will be worth putting on the table, at the outset, in two slightly different forms:

1) Everettian pictures of the world are apparently going to have no room in them for *ignorance about the future*. The worry is that the Everettian picture of the world—whatever, precisely, that picture is going to turn out to be—is going to be completely deterministic, and (moreover) it is going to impose none of the sorts of ignorance of the initial conditions that allow us to make sense of probabilistic talk in deterministic theories like classical statistical mechanics and Bohmian mechanics.

2) Everettian pictures of the world are apparently not going to be susceptible to confirmation or disconfirmation by means of experiment—or not (at any rate) by means of anything even remotely like the sorts of experiments that we normally take to be confirmatory of quantum mechanics. Why (for example) should it come as a surprise, on a picture like this, to see what we would ordinarily consider a low-probability string of experimental results? Why should such a result cast any doubt on the truth of this theory (as it does, in fact, cast doubt on quantum mechanics)?

2 FIRST PASSES

Everybody's first unreflective reaction to these worries is to think of the probability in question here as the probability that the real me or the original me or the sentient me ends up, at the conclusion of a measurement, on this or that particular branch of the wavefunction. And the serious discussion of these questions gets underway *precisely* with the realization that—in so far as the Everett picture is committed to the proposition that quantum-mechanical wavefunctions amount to metaphysically complete descriptions of the world—such thoughts make no sense. (Many-minds? Dualism? Incompleteness of the wavefunction? Preposterous!)

Here's an idea: suppose that we measure the x-spin of each of an infinite ensemble of electrons, where each of the electrons in the ensemble is initially prepared in the state $a|x\text{-up}\rangle + \beta|x\text{-down}\rangle$. Then it can easily be shown that in the limit as the number of measurements already performed goes to infinity, the state of the world approaches an eigenstate of the frequency of (say) up-results, with eigenvalue $|a|^2$. And note that the limit we are dealing with here is a perfectly concrete and flat-footed limit of a sequence of vectors in Hilbert space, not a limit of probabilities of the sort that we are used to dealing with in applications of the probabilistic law of large numbers. And the thought has occurred to a number of investigators over the years (Sidney Coleman, and myself, and others too) that perhaps all it *means* to say that the probability that the outcome a measurement of the x-spin of an electron in the state $a|x\text{-up}\rangle + \beta|x\text{-down}\rangle$ up is $|a|^2$ is that if an infinite ensemble of such experiments were to be performed, the state of the world would with certainty approach an eigenstate of the frequency of (say) up-results, with eigenvalue $|a|^2$. And what is particularly beautiful and seductive about that thought is the intimation that perhaps the Everett picture will turn out, at the end of the day, to be the only picture of the world on which probabilities fully and flat-footedly and not-circularly *make sense*. But the business of parlaying this thought into a fully worked-out account of probability in the Everett picture quickly runs into very familiar and very discouraging sorts of trouble. One doesn't know what to say (for example) about finite runs of experiments, and one doesn't know what to say about the fact that the world is after all very unlikely ever to be in an eigenstate of my undertaking to carry out any particular measurement of anything.

3 DECISION THEORY

There has lately been a very imaginative and very intriguing and very intensely discussed third pass at all this—due (among others) to David Deutsch and David Wallace and Hilary Greaves and Simon Saunders—which exploits the formal apparatus of decision theory. And that third pass is going to be my main topic here.

3.1 The Fission Picture

Let me start off with what seems to me to be the clearest and most straightforward and most radical version of this idea, the version of Greaves and (I think) of Deutsch as well, the version that David Wallace refers to as the "fission program".

The point of departure here is to eschew any talk of probabilities whatsoever—to acknowledge frankly that in a world like this (this being a deterministic world, with none of the relevant ignorance of initial conditions) there are none. The situation is (rather) this: the Schrödinger-evolution is the complete story of the evolution of the world. Every branch (in the appropriate basis) supports an actual experience of the observer. Every quantum-mechanically possible outcome of a measurement occurs with certainty. What I should rationally expect, on undertaking a measurement, is to see *all* (sometimes people prefer to say 'each') of its possible outcomes, with certainty. In every sequence of similar measurements, all of the possible *frequencies* occur, and are experienced by the observer, with certainty. Period.

And now the following question is raised: imagine that this is actually the way the world is, and suppose that our preferences are given (we want, say, to maximize our financial holdings, and we don't care about anything else)—how is it rational for us to *act*; what is it rational for us to *decide*?

And there is a whole collection of arguments to the effect that the square of the absolute value of the coefficient of this or that particular branch is going to play precisely the same formal role in rational deliberations about how to act in a world like that as the probability of this or that particular state of affairs plays in such deliberations in the analogous genuinely chancy world. There is a whole collection of arguments (that is) to the effect that square-amplitudes in the many-worlds interpretation are going to play precisely the same role in decision-theory as probabilities do in chancy theories.

And the thought is that that's all we need—the thought is that that exhausts the role the probabilities play in our lives.

How, precisely, should we think of these square-amplitudes, on the fission picture, if the sort of argument described above succeeds? What they certainly are *not* (remember) are *probabilities*. Greaves thinks that what these sorts of arguments establish, if they succeed, is that rational agents must treat the square-amplitude as what she calls a 'caring measure'—a measure of the degree to which we *care* about the situation on this or that particular branch. Our goal in making decisions, then, is to maximize the average over all the branches of the product of how well we do on a branch and the degree to which we *care* about that branch.

And I have two distinct sorts of worries about this strategy. One worry has to do with whether or not these arguments actually succeed in establishing the particular point that they are *advertised* as establishing—whether or not, that is, these arguments actually succeed in establishing that the square of the absolute value of the coefficient of this or that particular branch is going to play precisely the same formal role in rational deliberations about how to act in a world like that as the probability of this or that particular state of affairs plays in such deliberations in the analogous genuinely chancy world. And the other has to do with whether or not establishing that would amount to anything along the lines of a solution to the puzzle about probabilities in the many-worlds interpretation.

Let me talk about the second of these worries, which is the more abstract and more general of the two, first.

The worry here is that the question at which this entire program is aimed, the question out of which this entire program arises, seems like the wrong question. The questions to which this program is addressed are questions of what we would do if we believed that the fission hypothesis were correct. But the question at issue here is precisely whether to believe that the fission hypothesis is correct! And what needs to be looked into, in order to answer that question, has nothing whatever to do with how we would act if we believed that the answer to that question were 'yes'. What needs to be looked into, in order to answer the question of whether to believe that the fission hypothesis is correct, is the *empirical adequacy* of that hypothesis. What needs to be looked into, in order to answer the question of whether to believe the fission hypothesis is correct, is whether or not the truth of that hypothesis is explanatory of our empirical experience. And that experience is of certain particular sorts of experiments having certain particular sorts of outcomes with certain particular sorts of frequencies—and not with others. And the fission hypothesis (since it is committed to the claim that all such experiments have all possible outcomes with all possible frequencies) is *structurally incapable* of explaining anything like that.

The decision-theoretic program seems to act as if what primarily and in the first instance stands in need of being explained about the world is why we *bet* the way we do. But this is crazy! Even if the arguments in question here were to succeed, even, that is, if it could be demonstrated that any rational agent who believed the fission hypothesis would bet just as we do, that would merely show that circumstances can be imagined, circumstances which—mind you—are altogether different from those of our actual empirical experience, circumstances in which the business of betting on X has nothing whatsoever to do with the business of guessing at whether or not X is going to occur, in which, as it happens, we would bet just as we do now. For us, on the other hand, the business of betting on X has everything in the world to do with the business of guessing at whether or not X is going to occur. And the guesses we make are, in the best cases, a rational reaction to what we see—the guesses we make are (in the best cases) informed by the frequencies of relevantly similar occurrences in the past—and it is those frequencies, and not the betting behaviors to which they ultimately give rise, which make up the raw data of our experience. It is those frequencies, or at any rate the *appearance* of those frequencies, and not the betting behaviors to which they ultimately give rise, which primarily and in the first instance stand in need of a scientific explanation. And the thought that one might be able to get away *without* explaining those frequencies or their appearances, the thought that one might be able to make some sort of an end run around explaining those frequencies or their appearances—which is the central thought of the decision-theoretic strategy—is, when you think about it, mad.

There's a sleight of hand here—a bait and switch. What we need is an account of our actual empirical experience of frequencies. And what we are promised (which falls entirely short of what we need) is an account of why it is that we bet as we do. And what we are given (which falls entirely short of what we were promised) is an argument to the effect that if we held an altogether different set of convictions about the world from the ones we actually hold, we would bet the same way as we actually do. And, to top it all off—and this brings me to the first and more concrete and more technical of the two worries I mentioned above—the argument itself seems wrong.

Let's look (then) at the actual details of these arguments.

They all start out by asking us to consider the following set of circumstances:

Let $|a\rangle|payoff\rangle$ represent a state of the world in which I am given a certain sum of money and the state of the rest of the world is $|a\rangle$, and let $|\beta\rangle$ $|no\ payoff\rangle$ represent a state of the world in which I am given no money and the state of the rest of the world is $|\beta\rangle$. And suppose that my only interest is in maximizing my wealth—suppose (in particular) that I am altogether indifferent as to whether $|a\rangle$ or $|\beta\rangle$ obtains.

Then—on pain of irrationality—the utility I associate with the state

$$\frac{1}{\sqrt{2}}\ |a\rangle|payoff\rangle\ +\ \frac{1}{\sqrt{2}}|\beta\rangle\ |no\ payoff\rangle$$

must be equal to the utility I associate with the state

$$\frac{1}{\sqrt{2}}\ |a\rangle|no\ payoff\rangle\ +\ \frac{1}{2}|\beta\rangle\ |payoff\rangle$$

because those two states differ *only* in terms of the roles of $|a\rangle$ and $|\beta\rangle$, and I am, by stipulation, indifferent as to whether $|a\rangle$ or $|\beta\rangle$ obtains.

Or that, at any rate, is what these arguments claim. And that claim goes under the name of 'equivalence' in the literature. And it turns out that once the hypothesis equivalence is granted, the game is over. It turns out that supposing equivalence amounts to supposing that whatever it is that plays the functional role of probabilities can depend on nothing other than the quantum-mechanical amplitudes. And it turns out to be relatively easy to argue from there to the conclusion that these 'functional probabilities' can be nothing other than the absolute squares of the amplitudes.

And the worry is that all of the initial plausibility of this hypothesis seems, on reflection, to melt away. For suppose that I adopt a caring measure which (contra the equivalence hypothesis) depends on the difference between $|a\rangle$ and $|\beta\rangle$. Suppose (for example) that I decide that the degree to which it is reasonable for me to care about what transpires on some particular one of my future branches ought to be proportional to how fat I am on that branch—the thought being that since there is more of me on the branches where I am fatter, those branches deserve to attract more of my concern for the future. Would there be something incoherent or irrational or unreasonable in that? Would it somehow make less sense for me to adopt my fatness as a caring measure than it would for me to adopt the absolute square of the quantum-mechanical amplitude as a caring measure?

Let's think it through: Hilary Greaves has pointed out that a caring measure which depends *exclusively* on how fat I am is probably not going to work, since the coherence of a measure like that is going to depend on there being some perfectly definite matter of fact about exactly how many branches there *are*—and there are very unlikely to turn out to be any facts like that. But this is easily remedied by replacing the naive fatness measure, the one that depends *exclusively* on how fat I am, with a slightly more sophisticated one: let the degree to which I care about what transpires on a certain branch, then, be proportional to how fat I am on that branch multiplied by the absolute square of the amplitude associated with that branch.

It has sometimes been suggested that moving from the standard quantum-mechanical square-amplitude caring measure to the sophisticated form of the fatness caring measure is (as a matter of fact, when you get right down to it, notwithstanding superficial appearances to the contrary) not really a case of changing my caring measure *at all*, but, rather, a case of changing my *preferences*, a case of deciding that I want not only to be rich, but to be fat as well. It has sometimes been suggested (to put it slightly differently) that my adopting such a measure would somehow be inconsistent with the claim—or somehow irrational in light of the claim—that I am as a matter of fact entirely indifferent as to whether I am fat or thin. But this is a mistake. I can perfectly well have no preference at all when faced with a choice between two different non-branching deterministic future evolutions, in one of which I get fat and in the other of which I get thin, and at the same time be very eager to arrange things—when I am faced with an upcoming branching-event—so as to ensure that things are to my liking on the branch where I am fatter. What would explain such behavior? What would make sense of such behavior? Precisely the conviction I mentioned above, that where branching-events are concerned—but *only* where branching-events are concerned—there is more of me to be concerned about on those branches where I am fatter. In the non-branching cases, no such considerations can come into play, since in those cases the entirety of me, fat or thin, is on the single branch to come.

So, although my present concern about the overall well-being of my descendants is in general going to involve my caring a great deal about their relative fatnesses, those same fatnesses are going to be no concern at all of those descendants themselves. And, once again, there is nothing worrisome or paradoxical or mysterious in that. In cases of branching, after all, there is no reason at all why my interests at t_1 vis-à-vis the circumstances of my descendants at t_2 should coincide with the interests of any particular one of my descendants at

t_2 vis-à-vis his circumstances at t_2. In cases of branching, my concerns now are going to embrace *the entire weighted collection* of my descendants, whereas *their* concerns are going to embrace only their individual *selves*—and *their* descendants.

David Wallace has pointed out that acting in accord with a fatness caring measure might sometimes prove difficult in practice—it might, for example, involve my trying to anticipate, even taking measures to try to control, how much I am going to eat on this or that future branch. Now, difficulties like that could presumably be reduced, or perhaps even eliminated altogether, by other choices of a caring measure—but the more important point is that the existence or non-existence of such difficulties seems altogether irrelevant to the question of what it is reasonable for me to care about. It hardly counts as news, after all, that it can sometimes be difficult to bring about the sorts of situations that we judge the most desirable—but it would be absurd to pretend that those situations are any less desirable for that. Forget for the moment about quantum mechanics, and about branching, and about chances, and consider the business of making a decision. Consider the business of tracing the consequences of my acting in such-and-such a way, at such-and-such a moment, all the way out to the end of time, in the face of a classical-mechanical sort of determinism. Consider how easy it might be to imagine that my going to movie A rather than movie B tonight might result in the deaths of millions of innocent people over the next several hundreds of thousands of years. We just don't know. We can't know. The calculations are utterly and permanently and inescapably beyond us. Ought we to pretend (then) that we don't care how many millions live or die as a result of what we do? Ought we to come to understand that as a matter of fact it doesn't *matter* to us how many millions live or die as a result of what we do? Of course not! We do the best we can, with what we have, to bring about what we want. And what we find we cannot do is occasion for sadness, and for resignation, and not at all for concluding that it was somehow irrational to want that in the first place!

The worry, then, is that equivalence turns out to be false. And more than that: the worry is that it is in fact not one whit less rational—in the face of the conviction that something like the fission picture is true—for me to operate in accord with the sophisticated fatness measure than it is for me to operate in accord with the Born measure. And if *that* is true, then there can be no *uniquely* rational way of operating under such convictions at all, and the whole argument falls apart.

[Let me pause for just a minute here to mention a very *different* argument—an argument due to W. Zurek, an argument that makes crucial and imaginative use of the *locality* of the fission picture—to the effect that any rational agent who believes in the fission picture has got to bet as if the probability of the outcome of a measurement of, say, the z-spin of an electron which is initially a member of any maximally entangled *pair* of electrons is equal to one half.

The argument goes like this: suppose that, when the singlet state obtains, we apply an external field to the electron in question, the electron (that is) whose z-spin is to be measured, which has the effect of flipping the electron's z-spin. The application of that field (Zurek argues) must necessarily exchange whatever credences the agent in question assigns to z-spin-up and z-spin-down. But note that, at this point, the original maximally entangled state can be recovered by means of a second application of precisely the same external field to the other electron. And note that it follows from the locality of the fission picture that this application of an external field to this second electron can have no effect at all the agent's credences concerning the outcomes of upcoming measurements on the first electron.

So the situation is as follows: the first application of the external field must exchange the agent's credences in z-spin-up and z-spin-down for electron number 1. But once that first application is done, there is something we can do to electron number 2, something which the locality of the fission picture guarantees can *have no effect whatever* on the agent's credences in z-spin-up and z-spin-down for electron number 1, which must nonetheless (because it fully restores the original quantum state of the two electrons) fully restore the agent's original credences in z-spin-up and z-spin-down for electron number 1. So the agent's original credences in z-spin-up and z-spin-down for electron number 1 must have been such as to be unaffected by exchanging them, which is to say that they must, all along, have been $\frac{1}{\sqrt{2}}$.

The trouble with this argument—and this was precisely the trouble with the hypothesis of equivalence—is that it takes for granted that an agent's credences in the outcomes of upcoming measurements on this or that physical system can depend on nothing other than the *quantum state* of that system just prior to the measurements taking place. And that's what we have just now learned isn't right. The reader should at this point have no trouble at all in confirming for herself (for example) that the fatness measure provides just as straightforward a counterexample to Zurek's argument as it does to the hypothesis of equivalence.]

Let's back up and come at all this again, from a slightly different angle, through the question of confirmation.[1]

Advocates of the fission hypothesis have argued that there is a very straightforward generalization of the standard Bayesian technique of confirmation which can be applied to branching and to stochastic theories alike, and (moreover) that such an application shows how the fission hypothesis is confirmed by our empirical experience of the world to *precisely* the same degree, and by *precisely* the same evidence, as a stochastic theory governed by the Born rule is.

---

[1] It seems to me a distinct weakness, a distinct artificiality, of the decision-theoretic discussions of probabilities in the Everett interpretation, that those discussions are always at such pains to separate the consideration of probabilities as a guide to action from the consideration of probabilities as a tool of confirmation. The goal should surely be a simple, unified account of probability-talk which makes it transparent how it is the Everettian probabilities play *both* those roles!

And if all this is right, then much of what I have been saying here must somehow be wrong. Hillary Greaves has pointed out, for example, that if all this is right, then there just cannot be an objection to the fission hypothesis to the effect that that hypothesis cannot 'explain' the frequencies of the outcomes of the standard quantum-mechanical experiments, since, after all, it is *precisely* those frequencies that are confirmatory of the fission hypothesis!

But it turns out that the application of Bayesianism to the fission hypothesis that investigators like Greaves and Myrvold have in mind is much less standard, and much less straightforward, and much less innocent, than advertised. And this will be worth explaining carefully. And it will be best to start at the beginning.

Any sensible strategy for updating one's credence in this or that scientific theory in light of one's empirical experience of the world has presumably got to be a matter of considering how well or how poorly that experience bears out the proposition that the theory in question is true. Any sensible strategy for updating one's credence in this or that scientific theory in light of one's empirical experience of the world has presumably got to be a matter of evaluating how well or how poorly the theory does in predicting what is going to happen.

In the standard Bayesian account of confirmation, for example, we decide how to update our credence in hypothesis H in light of some new item of evidence E, by evaluating what the probability of E would be if H were true—and our credence in H goes *up* (other things being equal) if E is the sort of occurrence that H counts as likely.

Put that beside the discussions I mentioned above of Bayesian confirmation (or maybe neo-Bayesian confirmation, or maybe faux-Bayesian confirmation) in the fission picture. The way Greaves and Myrvold set things up, what needs to be evaluated there is not what the probability of E would be if H were true, but (rather) how we would bet on E if we believed that H were true—and what they recommend is that those of my descendants who witness E ought to raise their credence in H (other things being equal) if E is the sort of occurrence that I would have bet on if I believed that H were true. But remember—and this is the absolutely crucial point—that deciding whether or not to bet on E, in the fission picture, has *nothing whatsoever* to do with guessing at whether or not E is going to occur. It *is*, for sure. And so is $-E$. And the business of deciding how to bet is just a matter of maximizing the payoffs on those particular branches that—for whatever reason—I happen to care most about. And if one is careful to keep all that at the center of one's attention, and if one is careful not to be misled by the usual rhetoric of 'making a bet', then the epistemic strategy that Greaves and Myrvold recommend suddenly looks silly and sneaky and unmotivated and wrong.

There surely is, on the other hand, a perfectly legitimate question of how one ought rightly to proceed; there is a perfectly legitimate question of what epistemic strategy one ought rightly to adopt, once the possibility of branching is taken into account. And dispensing with the suggestion of Greaves and Myrvold leaves that question altogether unanswered.

Suppose, then, that we update our credences in accord with the standard Bayesian prescription, where the probability of E on H is stipulated to be the probability that E would occur if H were true. That won't work either. Greaves has very rightly pointed out that a policy like that would unreasonably *favor* a fission picture, that if we adopt a policy like that, our credence in the fission picture must rise and rise, *no matter what we see*, compared to any other theory, since every possible outcome of every possible measurement has probability 1 on that picture.

Let's try another tack. Suppose I have a device D which prints out numerals on a tape, and I am considering a number of theories about how this device works. Consider, to begin with, the following two theories: (1) The device, once each second, prints out one of the numerals $1-10$ with probability $1/10$ for each particular numeral; (2) The device, once each second, prints out all of the numerals $1-10$. And suppose that the only empirical access I have to what is printed on the tape is by means of a measuring device M, which operates as follows: when I press the button on M, M prints out one of the numerals that are on the D-tape. If there is more than one numeral on the D-tape, M it prints out one of those, and there are no rules whatsoever, neither strict nor probabilistic that bear on the question of which one (the largest, say, or the one on the left, or the one in the middle, or the third one, or whatever) it prints. So the information I get by pressing this button on M is that the output numeral appears on the D-tape, and nothing more. And theory (2) is going to associate no probability whatever with any particular output of the M-device. And so there is going to be nothing at all to feed into the Bayesian updating formula. And so measurements with M can have no effect whatever on our relative credences in theories (1) and (2).[2]

And all of this strikes me as exactly analogous to the epistemic situation of an observer who remembers a certain string of experiments coming out a certain particular way, and is wondering how those memories ought rightly to affect his comparative credences in chancy versus fission understandings of quantum mechanics.

All of the above complaining takes it for granted, of course, that chances and frequencies and rational degrees of belief are all very intimately tied up with one another. And advocates of the fission picture are constantly reminding us that we have no clear and perspicuous and uncontroversial *analysis*, as yet, of the links between chances and frequencies and rational degrees of belief. And they

---

[2] Note that if we add another theory to the mix, a theory in which D prints out one of the numbers from 1 to 10 but with some other, non-uniform, probability-distribution over them, then our credences in the two chancy theories can perfectly well evolve as a result of measurements with M, but *not* our credence in the all-the-numerals theory.

are fond of insisting that in the absence of such an analysis, all of the above complaints against the fission picture are guilty of an implicit and unjustified double standard. The thought, I take it, is that the absence of such an analysis somehow makes it clear that the chance of E can have no more to do with questions of whether or not E is going to occur than the caring measure of E does, that the absence of such an analysis somehow makes it clear that caring measures can be no less fit to the tasks of explanation and confirmation than chances are.

But all this strikes me as wildly and almost wilfully wrong.

The point of a philosophical analysis of chance is not to establish *that* chances are related to frequencies, but, among other things, to show precisely *how* chances are related to frequencies. And if it should somehow become clear that such an analysis is impossible, if it should somehow become clear that no such relationship exists, then the very idea of chance will have been exposed as nonsense, and the project of statistical explanation will need to be abandoned. Period.

But things are unlikely ever to get that bad. It's true, of course, that there is as yet no *uncontroversial* analysis of the links between chances and frequencies and rational degrees of belief. Philosophy, after all, is hard. But there are plenty of smart people at work on the problem, and there are already proposals on the table which are very promising indeed, and it seems profoundly misleading to act as if our best guess at present is that there are no such links at all.

And it goes without saying that nothing like that is true, that nothing like that could *ever* be true, of the fission picture. The fission picture starts out—after all—precisely by denying that there is any determinate fact about the frequency with which such-and-such a quantum-mechanical experiment has such-and-such an outcome—and so the possibility of explaining frequencies like that is out of the question from the word 'go'.

3.2 The Uncertainty Picture

Let's back up to the beginning, one last time.

There is a very basic and obvious and straightforward worry about whether or not *anything* along the lines of an Everettian picture of quantum mechanics can possibly make sense of the standard quantum-mechanical talk about probabilities—a worry I mentioned at the outset of my remarks here—which runs as follows: talk about the probability of this or that future event would seem to make no sense unless there is something about the future of which we are uncertain, and there seems to be no room for any such uncertainty in the context of anything along the lines of an Everettian picture.

The strategy of the fission picture, of course, is to bite down hard on just that particular bullet, and to propose a way of trying to get along *without* probabilities. But there are more polite approaches on offer as well. There are a number of attempts in the literature at dressing things up in such a way as to take away the ground of the sort of worry I just described. There are (more particularly) a number of attempts in the literature at analyzing the semantics of locutions of the form, 'I am uncertain about the outcome of this upcoming experiment' in such a way as to make it plausible—notwithstanding initial appearances to the contrary—that such locutions can amount to sensible descriptions of the epistemic situations of observers in a branching universe. And the idea, I take it, is that such an analysis, if it succeeds, will make it possible to think of the argument from equivalence as fixing not merely a *caring* measure, but a measure of genuine probability.

Now, there are subtle and interesting questions about whether or not any of these semantic analyses actually succeed—and there are subtle and interesting questions about whether or not any merely semantic analysis can *possibly* succeed—in making it plausible that locutions like 'I am uncertain about the outcome of this upcoming experiment' can amount to sensible descriptions of the epistemic situations of observers in a branching universe. But in so far as our purposes here are concerned, all such questions can safely be put to one side—because even if one or another of the above-mentioned semantic analyses should succeed, it is apparently going to be fatal for the uncertainty picture, just as it was fatal for the fission picture, that the argument from equivalence fails.

There is an idea of Lev Vaidman's for introducing an altogether familiar and pedestrian sort of uncertainty into the Everett picture—an uncertainty that involves none of the fancy semantical footwork we have just been discussing. And it has been suggested here and there in the decision-theoretic literature that Lev's idea might afford yet *another* way of understanding the argument from equivalence as fixing a measure of probability—but suggestions like that are manifestly going to suffer precisely the same fate, for precisely the same reasons, as the fission picture and the uncertainty picture, and Lev himself seems to have something altogether different in mind. Lev seems to think of himself as having discovered the sort of uncertainty that can make room in an Everettian universe for the introduction of a new, free-standing, fundamental quantum-mechanical law of chances.

Suppose (says Lev) that I make the following arrangements: I arrange to be put to sleep. And once I am asleep a measurement of the x-spin of an initially y-spin up electron is carried out. And if the outcome of that measurement is 'up' (or rather: in the *branch* where the outcome of that measurement is 'up') my sleeping body is conveyed to a room in Cleveland. And if the outcome of that measurement is 'down' (or rather: in the *branch* where the outcome of that measurement is 'down') my sleeping body is conveyed to an identical room in Los Angeles. And then I am awakened. It seems right to say (then) that on being awakened I am simply, genuinely, flat-footedly, familiarly, uncertain of what city I am in.

The trouble with Lev's uncertainty is that it seems altogether avoidable, and that it comes too late in the game. The uncertainty we *need*—the uncertainty that quantum mechanics imposes on us—is something not to be bypassed, something that comes up whether or not we go out of our way to keep ourselves in the dark about anything, something that comes up no matter what pains we may take to know all we can; something that comes up not after experiments are over but before they get started.

Now, I think Lev is likely to respond to this last complaint—the complaint that his uncertainty comes up too late in the game—like this: suppose that the observer in the scenario described above knows all there is to know about the state of the world, and about the equations of motion, before the experiment gets underway. Among the things she knows, then, is that *every single one* of her descendants is with certainty going to be uncertain, on being awakened, about what city they are in. Moreover, that uncertainty does not arise in virtue of those descendants having *forgotten* anything—they know everything that the pre-measurement observer knows! And so if there is something about the world of which these descendants are uncertain, the pre-measurement observer must have been uncertain about that thing too!

And the trouble here—it seems to me—has to do with the locution 'about the world'. The fact is that there is nothing whatsoever about the objective metaphysical future of the world of which the pre-measurement observer is uncertain. Period, end of story. The questions to which that observer's descendants do not have answers are questions which can only be raised in indexical language, and only from perspectives which are not yet in the world at all before the measurement has been carried out. Completely new uncertainties do indeed come into being, on this scenario, once the measurement is done. But those uncertainties have nothing whatever to do with objective metaphysical features of the world, and they are not the sorts of uncertainties that can only arise by means of *forgetting*. Those new uncertainties have to do, rather, with the coming into being of completely new *centers of subjectivity*, from the standpoints of which completely new and previously unformulable indexical questions can come up.

12

Decisions, Decisions, Decisions: Can Savage Salvage Everettian Probability?

Huw Price

1 FABULOUS AT 50?

Our 40s are often a fortunate decade. Understanding ourselves better, resolving old problems, we celebrate our 50th year with new confidence, a new sense of purpose. So, too, for the Everett interpretation, at least according to one reading of its recent history. Its fifth decade has resolved a difficulty that has plagued it since youth, indeed since infancy, the so-called problem of probability. Critics have long maintained that the Everett view cannot make sense of quantum probabilities, in one or both of two senses: either it cannot make sense of probability at all, in a world in which 'all possibilities are actualized'; or, at best, it cannot explain why probability should be governed by the Born rule.

According to the optimistic view just mentioned, these problems have been sorted out, in the past decade, by an approach due to David Deutsch [1999]. Deutsch argues not only that an analogue of decision under uncertainty (of the kind traditionally associated with probability) makes sense in an Everett world; but also that under reasonable assumptions, the betting odds of a rational Everettian agent should be constrained by the Born rule. In one important respect, Deutsch argues, probability is actually in *better* shape in the Everettian context than in the classical 'one-branch' case. If so, the Everett view turns 50 in fine form indeed.

Deutsch's argument has been reformulated and clarified by David Wallace [2002, 2003, 2006, 2007]. Wallace stresses the argument's reliance on the distinguishing symmetry of the Everett view, viz. that all possible outcomes of a quantum measurement are treated as equally real. The argument thus makes a virtue of what is usually seen as the main obstacle to making sense of probability in this context. Further important contributions have been made by several other writers, such as Simon Saunders [1998, 2005] and Hilary Greaves [2004,

2007a, 2007b]. The result is a fascinating collection of papers, of great interest to researchers in philosophy of probability, as well as the foundations of quantum mechanics.

But does it work? In this paper, presenting the pessimistic face of turning 50, I want to argue that it does not. I'll argue in particular that the distinguishing symmetry of the Everett view (the fact that it treats all outcomes of a quantum measurement as equally real) is the downfall rather than the saviour of the Deutsch–Wallace (DW) argument (or the *Oxford* approach, as I'll sometimes call it). For Everett's new ontology provides something new for agents to have preferences *about;* and this, as we'll see, makes it inevitable that rational Everettian agents need not be Deutschian agents. (In other words, rationality does not require that their decision behaviour be modelled by an analogue of classical decision under uncertainty, with the Born weights playing the role of subjective probability.) And I'll argue that this is no mere abstract possibility. Once the loophole is in view, it is easy to see why we might prefer not to be Deutschian agents, at least in some respects, if we believed we lived in an Everett world.

Little in this paper is new. As I'll note, the main point is made by Greaves herself (though I think she underestimates its significance). But the package has not previously been assembled in this form, so far as I know.

2 A MISSING LINK?

It is often suggested that the problem of probability in the Everett interpretation is analogous to, and perhaps no worse than, a problem long recognized in the classical case. An early and forceful version of this claim is that of David Papineau [1996]. Papineau considers what he terms the decision-theoretical link:

The Decision-Theoretical Link. We base rational choices on our knowledge of objective probabilities. In any chancy situation, a rational agent will consider the difference that alternative actions would make to the objective probabilities of desired results, and then opt for that action which maximizes objective expected utility.   [1996 p.238])

'Perhaps surprisingly,' Papineau continues, 'conventional thought provides no agreed further justification [for this principle]':

Note in this connection that what agents want from their choices are desired *results,* rather than results which are objectively *probable* (a choice that makes the results objectively probable, but unluckily doesn't produce them, doesn't give you what you *want*). This means that there is room to ask: *why* are rational agents well advised to choose actions that make their desired results objectively probable? However, there is no good answer to this question . . . Indeed many philosophers in this area now simply take it to be a primitive fact that you ought to weight future possibilities according to known objective probabilities in making rational decisions. . . . It is not just that philosophers can't agree on the right justification; many have concluded that there simply isn't one.   [1996 p.238]

Applying this to the many-worlds case, Papineau suggests that the Everettian is therefore entitled simply to *assume* that rational decision in a branching universe is constrained by certain physical magnitudes postulated by quantum mechanics:

[T]he many minds view should simply stipulate that the quantum mechanical coefficients . . . provide a decision-theoretic basis for rational decisions. As to a justification for these stipulations, the many minds theory can simply retort that it provides as good a justification as conventional thought does for treating its probabilities similarly—namely, no good justification at all.   [1996 pp.238–9]

Supporters of the DW approach claim that if their argument succeeds, we actually get more than this: we get a *proof* that rational decision in an Everett world is properly constrained by the Born rule. But if the proof doesn't quite go through, then no matter—we are still no worse off than in the classical case.

In my view, this comparison both exaggerates the difficulty in the classical case, and misrepresents and hence underestimates the difficulty in the Everettian case. I want to argue that so long as we keep in mind—as emphasized by the *pragmatist* tradition in philosophy of probability[1]—that probability properly *begins* with decision under uncertainty, there isn't any pressing mystery about Papineau's decision-theoretic link. (There may be a problem for some views of probability, but those views are far from compulsory—and the problem itself counts against them, in so far as it is a problem.) On the other side, the crucial issue for the Everettians is in a sense *prior* to probability: it is the problem whether decision under uncertainty—or some suitable analogue—makes any sense in the many-worlds case. This problem is prior, in particular, to the question as to whether we might take it to be a brute fact that rational Everettian decision is constrained by the Born rule; for until it is clear that there is something coherent to constrain in this way, such an assumption is premature, if not question-begging. The true problem in this vicinity for the Everett view—somewhat misleadingly labelled as the problem of probability, and with no analogue in the classical case—is the problem about uncertainty.

To give a sense of the pragmatist viewpoint, imagine creatures who make maps of their surroundings, marking the positions of various features of practical significance: blue dots and lines where they find the liquids they drink and wash in, for example, and green shopping baskets where they find things to eat.

Is there any mystery, or primitive assumption at work, in the fact that the mapmakers use these maps as a guide to where they can drink, wash and eat?

---

[1] Aka the *subjectivist* tradition, though I think there are good reasons to prefer the former term.

No, for that's precisely what the maps are *for*. (To understand the map is to understand that this is how it is *used*, as a Wittgensteinian might say.) The mapmakers might find other interesting questions in the vicinity, of course. Why are blue lines correlated with contour lines in such a distinctive way? Is there is any unified *physical* account, either of the stuff they drink and wash in, or of the stuff they eat? And perhaps most interestingly, what is it about the mapmakers and their environment that explains why this mapmaking practice is both *possible* and *useful*, to the extent that it is? But these are not the practical puzzle about why someone who adopts the map should use it to guide her drink-seeking and food-seeking behaviour: to adopt the map *is* to take it as a practical guide.

A pragmatist regards probabilistic models in this same practical spirit. They are maps to guide us in making decisions under uncertainty in particular domains. The previous map said, 'Go *here* if you need a wash or a drink, *there* for something to eat.' This map says, 'Use *this number* if you need a credence—i.e. if you have to make a decision with imperfect knowledge about something that matters.' If a probability map just *is* a practical guide to decision under uncertainty, it isn't a mystery why it should be used for exactly that purpose. There's no primitive assumption, and no decision-theoretic 'missing link'. Again, there might be other interesting questions in the vicinity—e.g., about whether, or why, decision-theoretic 'blue lines' are correlated with something else on our maps—but these are not the practical puzzle about why probability properly guides action.[2]

The pragmatist thus applauds the Oxford approach for beginning with the question of rational decision, but asks it to allow that classical probability does the same. With that concession in place, the issues line up as they should. There's no essential primitive assumption in the classical case, to which Everettians can appeal in the spirit of 'we do no worse than that'. And the crucial question, the crux of the so-called problem of probability, is whether the Everett picture allows any analogue of decision-theoretic uncertainty—any analogue of *credence*, in effect.[3]

---

[2] Would it matter is there wasn't any unified story about 'something else' that correlated with decision theoretic probability—if 'probability' turned out to be more like 'food' (where there isn't any interesting natural kind, apparently), than like 'liquid for washing and drinking' (where there is a natural kind, at least to a first approximation)? If so, why? Obviously there is (much) more to be said on this matter, but I hope the food example goes some way to undermining the rather simplistic conception of the options that tends to characterize objections to the pragmatist viewpoint. Wallace, for example, says, 'Whilst it is coherent to advocate the abandonment of objective probabilities, it seems implausible: it commits one to believing, for instance, that the predicted decay rate of radioisotopes is *purely* a matter of belief' [2002 p.15, my emphasis]. Would abandonment of the view that 'food' picks out a natural kind commit us to believing that it is *purely* a matter of taste? Surely there's room for some subtlety here—room for a bit of natural science, in effect, that begins by asking why the concepts concerned ('food' or 'probability') should be useful for creatures in our circumstances, to the extent that they are useful (and anticipates an answer that refers both to aspects of the environment and to aspects of ourselves)?

[3] To the satisfaction of what *need* could Everettian probability be our guide, as it were?

3 RATIONAL CHOICE FROM SAVAGE TO DEUTSCH

In a recent survey article, Hilary Greaves provides a concise summary of the classical model of rational decision theory due to Savage, and of its modification by Deutsch in the Everettian case. Here I follow and appropriate Greaves' exposition, compressing further where possible:

> Decision theory is a theory designed for the analysis of rational decision-making under conditions of uncertainty. One considers an agent who is uncertain what the *state* of the world is: for example, she is uncertain whether or not it will rain later today. The agent faces a choice of *acts:* for example, she is going for a walk, and she has to decide whether or not to take an umbrella. She knows, for each possible state of the world and each possible act, what will be the *consequence* if the state in question obtains and the act in question is performed: for example, she knows that, if she elects not to take the umbrella and it rains, she will get wet. The agent is therefore able to describe each of her candidate acts as a function from the set of possible States to the set of Consequences. (Greaves [2007b p.113])

As Greaves explains, Savage then shows that given certain plausible rationality constraints, one may prove a 'representation theorem':

> [F]or any agent whose preferences over acts satisfy the given rationality constraints, there exists a unique probability measure p on the set of States, and a utility function U on the set of Consequences (unique up to positive linear transformation), such that, for any two acts A,B, the agent prefers A to B iff the expected utility of A is greater than that of B. Here, the expected utility of an act A is defined by:

$$EU(A) := \sum_{s \in S} p(s) \cdot U(A(s)).$$

Deutsch modifies this classical apparatus by replacing the set of States with the set of branches that will result from a specified quantum measurement, and the set of Consequences with, as Greaves puts it 'things that happen to individual copies of the agent, *on particular branches*'. An Act is still a function from States to Consequences, but in this new sense of Consequence: in other words, it is an assignment of Consequences or rewards to branches defined by measurement outcomes. Then, by 'imposing a set of rationality constraints on agents' preferences among such quantum games',

> Deutsch is able to prove a representation theorem that is analogous in many respects to Savage's: the preferences of a rational agent are representable by a probability measure over the set of States (branches) for every possible chance setup, and a utility function on the set of Consequences (rewards-on-branches), such that, for any two Everettian acts A,B, the agent prefers A to B iff $EU(A) > EU(B)$. (Expected utility is defined via the same formal expression as above.) [2007b p.115]

3.1 Wallace's Version of Deutsch's Approach

How does an analogue of uncertainty enter the picture? Since much hangs on this issue, I'll reproduce a clear account of the relevant move from Wallace [2007 pp.316–17], in a formulation of the axioms in which uncertainty is 'on the surface'.

[D]efine a *likelihood ordering* as some two-place relation holding between ordered pairs $\langle E, M \rangle$, where M is a quantum measurement and E is an event in $\mathcal{E}_M$ (that is, E is a subset of the possible outcomes of the measurement). We write the relation as $\succeq$:

$$E|M \succeq F|N$$

is then to be read as 'It's at least as likely that some outcome in E will obtain (given that measurement M is carried out) as it is that some outcome in F will obtain (given that measurement N is carried out)'. We define $\simeq$ . . . as follows: $E|M \simeq F|N$ if $E \succeq M \succeq F|N$ and $F|N \succeq E|M$ . . .

We will say that such an ordering is *represented* by a function Pr from pairs $\langle E, M \rangle$ to the reals if

1. $\Pr(\emptyset|M) = 0$, and $\Pr(\mathcal{S}_M|M) = 1$, for each M.
2. If E and F are disjoint then $\Pr(E \cup F|M) = \Pr(E|M) + \Pr(F|M)$.
3. $\Pr(E|M) \geq \Pr(F|N)$ iff $E|M \succeq F|N$.

The ordering is *uniquely represented* iff there is only one such Pr.

The subjectivist programme then seeks to find axioms for $\succeq$ so that any agent's preferences are uniquely represented. . . . [I]n a *quantum-mechanical* context we can manage with a set of axioms which is both extremely weak . . . and fairly simple. To state them, it will be convenient to define a *null event*: an event E is *null with respect to* M (or, equivalently, $E|M$ is null) iff $E|M \simeq \emptyset|M$. (That is: E is certain not to happen, given M). If it is clear which M we're referring to, we will sometimes drop the M and refer to E as null *simpliciter*.

We can then say that a likelihood ordering is *minimally rational* if it satisfies the following axioms:

Transitivity $\succeq$ is transitive: if $E|M \succeq F|N$ and $F|N \succeq G|O$, then $E|M \succeq G|O$.
Separation There exists some E and M such that $E|M$ is not null.
Dominance If $E \subseteq F$, then $F|M \succeq E|M$ for any M, with $F|M \simeq E|M$ iff $E - F$ is null.

This is an extremely weak set of axioms for qualitative likelihood . . . Each, translated into words, should be immediately intuitive:

1. Transitivity: 'If A is at least as likely than B and B is at least as likely than C, then A is at least as likely than C.'
2. Separation: 'There is some outcome that is not impossible.'

3. Dominance: 'An event doesn't get less likely just because more outcomes are added to it; it gets more likely iff the outcomes which are added are not themselves certain not to happen.'

3.2 Comments

Wallace is right, of course, that any reasonable notion of qualitative likelihood should be expected to satisfy these axioms. But to characterize the target is not yet to hit the bullseye. The axiom system provides no answer to the puzzle as to what such a notion of subjective likelihood could amount to, in the Everettian context, where there is no classical uncertainty. So if we assume these axioms as the basis for a quantum version of the representation theorem, we are *assuming*, rather than *demonstrating*, that there is some non-trivial Everettian analogue of decision-theoretic uncertainty. (We know that there is a *trivial* measure of likelihood available in the Everett world: it assigns equal and maximal likelihood to all results not certain not to happen. But this measure doesn't satisfy Dominance, for it fails the requirement that Wallace glosses as: 'An event . . . gets more likely iff the outcomes which are added are not themselves certain not to happen.')

I'm not suggesting that Wallace is confused about these points, of course. On the contrary, he and others have discussed at length what the required analogue of uncertainty might be. (More on this below.) Nevertheless, I think it is worth stressing that the availability of an appropriate notion of uncertainty doesn't *emerge from* the DW argument, but is *presupposed by it*. So if we sceptics are challenged to say which of these axioms we disagree with, we have at least the following answer: we're sceptical about any axiom—e.g., Wallace's Dominance, for one—that presupposes an analogue of uncertainty. And we'll remain sceptical, until our opponents convince us that they have a notion that will do the job.

4 THE QUEST FOR UNCERTAINTY

There are two main proposals as to how we might make sense of the required notion of decision-theoretic uncertainty, or likelihood. The first turns on the claim that an Everettian agent properly feels genuine though *subjective* uncertainty, as to which branch she will find herself in, after a quantum measurement. The second argues that the required analogue of uncertainty isn't really any sort of uncertainty at all, but rather what Greaves calls a 'caring measure'—a measure of how much 'weight' a rational agent should give to a particular branch.

Greaves [2004] herself has offered some robust and (to me) fairly persuasive criticisms of the subjective uncertainty approach. I want to raise an additional

difficulty for this view, not mentioned (so far as I know) in the existing literature.
It, too, seems to suggest that Greaves' approach is the right approach, if anything
is—though it also contains the seeds of an objection to her view, I think. It turns
on what I hope are some uncontroversial observations about the role of subjects
and survival in rational decision theory.

4.1 Decision Theory Doesn't Care About Personal Identity

First, rational decision theory depends on the fact that agents care about
outcomes, but not on whether these outcomes are things that the agent herself
experiences. They are things the agent has preferences *about*—her *loci of concern,*
as I'll say—but her own future experiences are just one possible class of such
things.

Thus consider a case in which the relevant outcome is that a child is happy
at a specified time in the future. What is revealed by my betting preferences
is the utility—for *me, now*—of the child's being happy, not the utility for
the child (now or in the future). Similarly if we replace the child by some
future person-stage of me, what's revealed by my betting preferences *now* is the
utility *for the present me* of the future me being happy, not the utility *for* the
future me.

In general, then, it is entirely inessential to classical subjective decision theory
that its subjects *have* future selves, at the time of the relevant Consequences.
We can imagine a race of short-lived creatures for whom all decisions would be
like deathbed decisions are for us—we prefer that there should be some futures
rather than others, and act so as to maximize our expected utility, in the light
of uncertainty, of a future that we ourselves shall not experience. Indeed, we
can imagine a view of the metaphysics of personal identity according to which
this is inevitably our own situation: there is no genuine trans-temporal personal
identity, according to this view, and our decisions cannot but concern a future
that we ourselves will never see. In such a case, of course, there is simply no
place for subjective uncertainty about which future we will experience. Properly
informed by our metaphysical friends, we *know* that we will experience none
of them. No matter—our classical decision theory takes all this in its stride.
(The relevant uncertainty is just the agent's uncertainty as to which future is
actual.)

I've emphasized these points to highlight a respect in which the subjective uncertainty version of Deutschian decision theory seems inevitably less
general than its classical ancestor. It needs to take sides on some heavy-
duty metaphysical issues about personal identity, so as to rule out the view
according to which there is no such thing as genuine survival. And even
with the help of such metaphysics, it only applies, at least directly, to cases
in which the agent does survive—for it is only in these cases that the relevant notion of subjective expectation makes sense. On my deathbed, I expect

nothing for tomorrow—Everett-embarrassing Lewisian loopholes to one side, at least![4]

4.2 What Does Survival Mean, and Why Does it Matter?

There are other difficulties lurking nearby. For what does *survival* mean? Consider a classical analogy. Suppose, modifying the above example, that I am making a decision which will affect the happiness of a group of children. There is no uncertainty involved. I have several options, each with different consequences for the happiness of each of the children. I act so as to maximize *my* utility, which (by assumption) depends on nothing other than (my beliefs about) the resulting happiness of the children. There are many ways my utility function might vary over the N-dimensional space representing the degree of happiness of each of the N children, but suppose I am a simple-minded consequentialist—then my utility just goes by the total happiness, summed over the group of children.

Now suppose that initially I plan to make my choice in the belief that I will not live to see the outcome, but then learn that I am going to be reincarnated as one of the children. Does survival make any difference to my calculation? In particular, does it provide a role for an ascription of a *probability* to the matter of which child I shall be?

There are two ways it can fail to do so. One possibility is that I believe that I will find myself in a child's body with the same 'global' perspective as before—the same concern for the welfare of the group of children as a whole. If so, then although survival means that I will live to enjoy the utility of my choice, it makes no difference to the calculation. My reincarnated self will have the same utility function as I do now, whichever child he turns out to be. Another possibility is that 'I' will find 'myself' with the preferences of the child in question, in which case 'I' will be happy or sad according as he is happy or sad—the fate of the other children won't come into it. But unless I take this *now* as a reason to abandon my even-handed approach to the future welfare of the children—more on this possibility in a moment—then again, survival makes no difference. (After all, I knew already that there would be a child with that perspective.) In either case,

---

[4] See Lewis 2004. The claim I make here about the generality of the subjective uncertainty view might be challenged. Wallace himself says:

'Subjective' should not be taken too literally here. The subjectivity lies in the essential role of a particular location in the quantum universe (uncertainty isn't visible from a God's-eye view). But it need not be linked to first-person expectations: 'there will be a sea battle tomorrow' might be as uncertain as 'I will see spin up'. [2007 p.314]

I don't have space to pursue this point here, but it seems that for a view not linked to first-person expectations, the subjective uncertainty approach spends an inordinate amount of time discussing issues of personal identity over time. My point is that these issues are essentially irrelevant to classical subjective decision theory, for which the only 'I' who matters is the 'I' at the time of decision.

then, there's no role for a probability in my calculation, despite the survival. In other words, if reincarnation doesn't break the symmetry of my original attitude to the children then it is irrelevant; if it does break the symmetry, but only from the moment of reincarnation, then it is not survival, in any sense that matters here.

In the classical case there's a third possibility: the news about reincarnation might break the symmetry at the beginning, in the sense that I will make my initial utility calculation in a different way altogether: I'll base it on the future happiness of one child, weighted by my credence that that child is the future me.

Notice that although this last option makes sense in the classical case—we can intelligibly suppose that the symmetry is broken in this way—it would be absurd to suggest that rationality *requires* me to do it this way. Rationality requires that I act now on the basis of what matters to me now. This might depend on what happens to me in the future, in so far as we can make sense of that notion, but it need not do so: rationality doesn't impose that outlook. And it doesn't take *my* survival to break the symmetry, either. We can get the same result by adding the supposition that only one of the group of children—I don't know which one—will survive to the relevant time in the future. In this case, too, I may well need a decision theory that can cope with uncertainty. But that need stems from the fact that the world breaks a symmetry among a class of things that I care about, not from the fact (if it is a fact) that one of those things is a future version of 'me'.

Now transpose this to an Everett case, in which I am facing a choice between gambles on a quantum measurement. I believe that there are multiple real future branches, one (or at least one) for each possible measurement outcome. The different gambles produce different outcomes for individuals whose welfare I care about, in each of the branches. Does it matter whether any of those individuals are 'me'? As in the previous case, there are two senses in which it seems not to matter: the first in which the future individual concerned has the same 'global' concerns as I have, and simply takes himself to be enjoying the net global utility which I could already foresee with certainty; and the second in which the future individual is no longer the me that matters, for present purposes, in that his concerns have become more selfish than mine: he only cares about what happens in his branch.

Is there a third sense, a symmetry-breaking sense, such that my initial decision should now become a weighted sum over a range of possibilities? The subjective uncertainty view claims that there is—that we can make sense of the symmetry being broken 'subjectively', though not objectively. (The whole point of the Everett view is that it is not broken objectively.) My point is that even if this were so—and even if it could conceivably be relevant to the decision-maker, who clearly doesn't occupy any of these subjective stances, to the exclusion of any of the others—it couldn't be a *rationality* requirement that decisions be made on that basis. If an Everettian agent's present utilities depend on his view of the welfare of occupants of multiple future branches, then—just as in the case of the children—it is irrelevant which of those branches he takes himself (subjectively!) to occupy.

One lesson of the move to the case in which only one child would survive was that in that case, uncertainty was forced on us by the ontology—uncertainty about which of our various loci of concern (the children) would be 'actual' (alive), at the relevant time in the future.[5] The shift to the Everett world takes us in the other direction. It removes the uncertainty, by rendering equally real all of an agent's previous ('classical') loci of concern. As in the case of the children, it couldn't be a requirement of rationality that we put the uncertainty back in, even if we did have a notion of subjective uncertainty that could do the job.

5 TWO CHEERS FOR THE CARING MEASURE

These considerations were intended to supplement the arguments that Greaves, in particular, has offered against the subjective uncertainty approach. It is a great advantage of Greaves' caring measure approach that it does take seriously the new 'global' perspective of an Everettian agent. Moreover, although Greaves herself formulates the view in terms of an agent's care *for her own future descendants*, this seems inessential. If the approach works, then caring measures are a rational discounting factor for any source of utility from future branches, whether it concerns the welfare of the agent's own descendants or not. (This inoculates the approach from issues about personal identity, survival, and the like—those matters are simply irrelevant, in my view, for the reasons above.)

So why only two cheers for the caring measure? Because (I claim) its victory over the subjective uncertainty view turns on a point that proves its own Achilles' heel. Once the new *global* viewpoint of the Everettian agent is on the table, it turns out to count against Greaves' proposal, too, in two senses. First, and more theoretically, it shows us how to make sense of an Everettian agent who, although entirely rational in terms of his own utilities, fails to conform to the Deutschian model.[6] Second, and more practically, it leads us in the direction of reasons why we ourselves might reasonably choose to be such agents, if we became convinced that we lived in an Everett world. Thus I want to make two kinds of points:

1. The new ontology of the Everett view introduces a new locus of possible concern. The utilities of a rational Everettian agent might relate directly to the global ontology of the Everett view, and only indirectly, if at all, to 'in-branch'

---

[5] The other lesson was that it isn't relevant whether the child in question is 'me', unless my present preferences make it so.

[6] As I noted, Greaves [2004 pp.451–2] is well aware of this possibility.

circumstances. (For such an agent uncertainty enters the picture, if at all, only in a classical manner.) The DW argument rules out such agents by fiat, in effect, by assuming that utilities are an 'in-branch' matter.

2. Even if we restrict ourselves to agents whose global preferences do show some 'reasonable' regard for the welfare of their in-branch descendants, it doesn't follow that the global preference should be to maximize a Born-weighted sum of in-branch utilities. This isn't the only option, and there are at least two kinds of reasons for thinking is should not be the preferred option: one objects to *summing,* and the other to *weighting.*

5.1 The Threat of Globalization

One of the lessons of the previous section was that 'Where goes ontology, there goes possible preference'. Decision theory places no constraints on what agents care about, other than that it be *real.*[7] The new ontology of the Everett view—the global wavefunction itself—thus brings in its wake the possibility of an agent who cares about *that.* Hence the challenge, in its most general form: by what right do we assume that the preferences of Everettian agents are driven by 'in branch' preferences *at all?*

This may seem a trivial point. After all, even in the classical case we can imagine agents whose preferences are so non-specific that uncertainty disappears, from their point of view. (All I care about is that something happens tomorrow.) This doesn't seem to lessen the importance of the theory for less uninteresting agents. Isn't a similar move possible in the Everettian case? Granted, the caring measure approach doesn't apply non-trivially to any possible agent, but if it applies to a large and interesting class of agents, isn't that good enough? Can't we just restrict ourselves by fiat to the case of agents whose preferences do depend on what happens in future branches?

Unfortunately for the DW argument, no such restriction seems likely to get it off the hook at this point, unless it is so strong as to be question-begging. For the argument's own prescription for rational choice between quantum games—'Choose the option that maximizes expected utility, calculated using the Born weights'—is *itself* a rule for choosing between future global wavefunctions. To adopt such a rule, then, is to accept a principle for choice at this global level—at which, obviously, there is neither uncertainty nor any analogue of uncertainty. (The caring measure is already rolled up inside this global rule.)

But this means that if someone already has a *different* preference at the global level—in particular, a *different* way of ranking wavefunctions according to what goes on in the branches they entail (we'll meet some examples in a

---

[7] It would be interesting to explore the subtleties of this restriction in the case of modal realities, variously construed, but that would take us a long way astray.

moment)—then the DW argument has nothing to say to them. Rationality may dictate choice in the light of preference, but it doesn't dictate preference itself. (Recall our principle: 'Where goes ontology, there goes possible preference.') *Pace* Greaves [2004 p.452], this argument has no analogue in the classical case, because in that case there's no ontology at the 'global' level—only the 'local' ontology, and then uncertainty about that.[8] By actualizing the epistemic *possibilia*, the Everett view introduces a new locus of possible preference, and hence leaves itself vulnerable to this challenge.

It seems to me that at a minimum, this argument establishes that the DW argument cannot rest on a principle of *rationality*. It shows us how to imagine agents whose preferences are such that it would clearly be *irrational* for them to follow the DW prescription, in an Everett world—given their preferences, rationality requires them to make different choices about the global wavefunction.[9]

It seems to me that the best response to this objection would be to fight back at the level of global preferences—to try to show that a preference for anything other than maximizing a Born-weighted sum of in-branch payoffs would be (in some sense) *unreasonable*. My next goal is to show that this fight will be an uphill battle: in some respects, such anti-Deutschian global preferences look very reasonable indeed. Reasonable folk like us might well conclude that there are better ways to care for the welfare of future branches.

5.2 Choices About Group Welfare

The considerations I have in mind of are two kinds:

1. An MEU model seems in some respect just the *wrong kind of model* for this kind of decision problem, which concerns the welfare of a group of individuals (i.e. the inhabitants of multiple future branches). In the classical case, there

---

[8] Could someone sufficiently realist about probability find a classical analogue in the possibility of an agent who cared *directly about the probabilities,* so that probability itself provided their (classical) analogue of additional ontology provided by the Everett view? As a pragmatist, I'm inclined to say that there is a potential analogy here, and so much the worse for some kinds of realism about probability. For present purposes, however, let me just emphasize two points of disanalogy. First, the extra ontology of the Everett view is in no sense optional—on the contrary, it is the heart of the physical theory itself. Second, what this extra ontology consists in, *inter alia,* is a whole lot more of the kind of things that ordinary people care about anyway (as opposed, so to speak, to something that only a metaphysician could love). The suggested analogy, by contrast, would fly in the face of the fact noted by Papineau, in the passage we quoted above: '[W]hat agents want from their choices are desired *results,* rather than results which are objectively *probable* (a choice that makes the results objectively probable, but unluckily doesn't produce them, doesn't give you what you *want*).'

[9] I note in passing another reason for thinking that the DW argument might be vulnerable in this way, viz., that the argument itself appeals to attitudes to global states at a crucial point. The principle Wallace calls Equivalence requires *as a matter of rationality* that an agent should rank two Acts equally, if they give rise to the same global state. In relying on this principle, the argument can hardly afford to be dismissive about the idea of preferences for global states.

are well-known difficulties for weighted sum approaches to the welfare of groups of individuals, and I want to argue that similar considerations seem to apply in the Everett case.

2. In so far as an MEU model is appropriate, there is no adequate justification for discounting the interests of low-weight individuals. I'll make this point by asking why we feel entitled to discount low-weight alternatives in the classical case, and arguing that this justification has no analogue in an Everettian framework (whether via a caring measure or subjective uncertainty).

Concerning both points, it will be helpful to have a vivid example in mind, to guide (or pump) our intuitions.

6 LEGLESS AT BONDI

Suppose I'm swimming at Bondi Beach, and a shark bites off my right leg. Saved from the immediate threat of exsanguination, I'm offered a wonderful new treatment. In the hospital's new cloning clinic (CC), surgeons will make a (reverse) copy of my left leg, and attach it in place of the missing limb. As with any operation there are risks: I might lose my left leg, or die under anaesthetic. But if these risks are small, it seems rational to consent to the procedure.

On my way to the CC, however, I learn a disturbing further fact: it is actually a *body* cloning clinic (BCC). The surgeons are going to reverse-copy all of me, remove the good right leg from the clone, and attach it to me. So I get two good legs at the expense of a legless twin. (This poor chap wakes up, complaining, in another ward—and is handed the consent form, on which 'he' accepted the risk that he would find himself in this position.)

This new information seems to make a difference. I'm a lot less happy at the thought of gaining a new leg at this cost to someone else—especially someone so dear to me!—than I was at gaining it merely at the cost of risk to 'myself'. What's more, my discomfort seems completely insensitive to considerations of subjective uncertainty or branch weight. It isn't lessened by the certainty—as I told the story—that it would be 'me' who gets the legs; or, apparently, by the information that the other ward is in another branch of the wavefunction, with very low weight.

Perhaps these intuitions simply need to be stared down, or massaged away. Perhaps I would lose them, once fully acquainted with the meaning of probability and uncertainty, or their analogues, in an Everettian context. Perhaps. But if so, a case needs to be made out. The significance of the example is that it suggests at least two ways in which such a case might be weak—two ways in which, at least arguably, a reasonable decision rule in an Everett world would not follow a Born-weighted MEU model.

7 BRANCHING AND DISTRIBUTIVE JUSTICE

Decision in the Everett world concerns the welfare of a *group* of future individuals—one's future descendants, or more generally one's 'objects of concern', in all future branches. This fact is highlighted by Greaves' approach, but clearly true for the subjective uncertainty approach as well. (As we saw, subjective uncertainty might make it *possible* to be selfish, but can't make it rationally *obligatory*. And my disquiet in the Bondi example isn't offset by certainty that it's me who gets the legs.) The challenge is that rational decision in such a context—a context involving group welfare—seems in some respects fundamentally different from any weighted sum model.

As I noted, this is a familiar point in discussions of distributive justice. The problem with a weighted-sum allocation of goods to a group is that it always permits a large cost to one individual to be offset by small gains to others. A principle of maximizing such a sum thus conflicts with plausible principles of justice: 'Pleasure for some should not knowingly be gained at the cost of pain for another.'

The plausibility of some such principle is highlighted by the famous Trolley Problem, in which we are asked whether it would be acceptable to kill one person—e.g. in one well-known version, a fat man sitting on a bridge, who might be pushed into the path of a runaway tram, thus killing him but saving the lives of several passengers—in order to benefit others. To a remarkable degree, ordinary humans seem to agree that this would not be morally acceptable.[10]

If we approach the choices facing Everettian agents in this frame of mind, viewing them as problems of group welfare, then some attractive decision principles might include the following:

1. First, do no harm: we should try to establish a baseline, below which we don't knowingly allow our descendants to fall (at least not simply for the sake of modest advantage to others).

---

[10] Another version, even closer to Legless at Bondi, is Transplant:

Imagine that each of five patients in a hospital will die without an organ transplant. The patient in Room 1 needs a heart, the patient in Room 2 needs a liver, the patient in Room 3 needs a kidney, and so on. The person in Room 6 is in the hospital for routine tests. Luckily (for them, not for him!), his tissue is compatible with the other five patients, and a specialist is available to transplant his organs into the other five. This operation would save their lives, while killing the 'donor'. There is no other way to save any of the other five patients. . . . [W]ith the right details filled in, it looks as if cutting up the 'donor' will maximize utility, since five lives have more utility than one life. If so, then classical utilitarianism implies that it would not be morally wrong for the doctor to perform the transplant and even that it would be morally wrong for the doctor not to perform the transplant. Most people find this result abominable. They take this example to show how bad it can be when utilitarians overlook individual rights, such as the unwilling donor's right to life. (Sinnott-Armstrong [2006]).

2. Second, unequal 'good luck' seems a lot less objectionable than unequal 'bad luck'. An Everettian might be inclined to play lotteries, for example, accepting that most of his descendants will make rather trivial sacrifices, in order to make it certain that one will hit the jackpot.
3. Third, we might trade off 'quantity' for 'quality', as in quantum Russian roulette (Squires [1986 p.72]).

This isn't intended as a definitive list—it is far from clear that there could be such a thing. It is simply intended to illustrate that agents who approach Everettian decision problems in the spirit of Greaves' caring measure, with the welfare of descendants uppermost in their minds, need not see matters entirely, or primarily, in terms of an MEU model. There are other intuitive appealing decision principles available in such cases, of a much more qualitative nature. An argument for the unique reasonableness of the Deutschian global preference would need to convince us that these contrary intuitions are guilty of some deep error.

7.1 Objections to the Distributive Justice Analogy

7.1.1 What About the Axioms of Rationality?

One objection appeals to the plausibility of the decision-theoretic axioms. Granted, there are other possible preferences for the global state, other than a preference for maximizing the Born-weighted sum of in-branch utilities. (Granted, too, perhaps, that this is an important difference, in principle, from the classical case.) But doesn't the intuitive appeal of the axioms still provide a sense in which such a preference would be *unreasonable?*

There are two ways to counter this objection, a direct way and an indirect way. The direct way is to give examples of ways in which intuitively reasonable global preferences can conflict directly with some of the axioms in question. Again, the analogy with classical problems of distributive justice is likely to prove helpful. Suppose I have a choice of leaving $1000 to each of my seven children, or $1100 to six of them and $1000 to the last. The latter choice *dominates* the former, in that no child is worse off and some are better off. But is it a better choice? Many of us would say that it is worse, because it is unjust.[11]

The indirect way to counter the objection is to point out that the appeal to the axioms makes assumptions about an agent's utility function that simply beg the question against the possibility of agents of the kind relevant here: agents who have preferences (of a non-MEU form) about the global state. As I've explained, the Oxford approach amounts to trying to mandate a particular global preference as a principle of rationality. What's wrong with an attempt to appeal to the

---

[11] For some different axiom-busting considerations, see Lewis [2005 section 4].

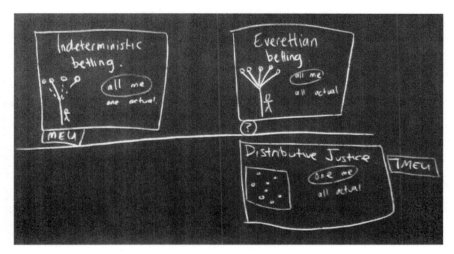

Figure 1. Greaves on Everett and Distributive Justice

axioms to rule irrational an alternative global preference is that the axioms make assumptions about utilities, about what an agent cares about, that are simply inapplicable to the kind of agent my objection has in mind—an agent for whom the payoff lies in an Everettian version of distributive justice, for example.

7.1.2 A Difference Between Everett and Distributive Justice?

In a panel discussion at the Many Worlds@50 meeting, Greaves [2007c] suggested a way of drawing a distinction between Everettian decisions and those faced in classical problems of distributive justice. Her idea (see Fig. 1) is that there are two dimensions on which we may compare and contrast the Everettian situation to one-branch decision under uncertainty, on one side, and cases of distributive justice on the other. One dimension—the one emphasized in my objections—tracks the number of individuals involved. In this dimension, we have one recipient in cases of classical uncertainty, and many recipients both in Everett and in cases of distributive justice. The other dimension—the one that Greaves recommends that we should emphasize instead—tracks the issue as to whether all the individuals involved are 'me' (i.e. the agent). Greaves suggests that it is the fact that all the beneficiaries are 'me' that unifies classical uncertainty and Everett; and the fact that not all are 'me' that makes distributive justice different (so that MEU doesn't apply, in those cases).

As I've set things up above, however, we've already taken the 'me' out of the picture, for other reasons. (The basic reason was that only 'me' relevant to classical subjective decision theory is the 'me at the time of choice'.) So it simply isn't available to do the work that Greaves wants it to do, in the present context.[12]

7.1.3 *What About Weight?*

Finally, it might be objected that the distributive justice challenge simply ignores the branch weights—the crucial disanalogy between the Everett case and classical problems of distributive justice is precisely that in the former case, it is not true that all 'concernees' are created equal, at least from the decision-maker's perspective. After all, that's what weight *is*, on Greaves' view: a measure of rational degree of *care*. Don't we simply beg the question against the view, if we insist on the analogy with classical distributive justice?

This brings us to a crucial issue. What *kind* of consideration could it be, in the Everett picture, that would make it rational to give more consideration to some of our loci of concern than others, in the way proposed. I'll approach this issue by considering the analogous question in the classical context.

8 WHY SHOULD WEIGHT MATTER?

8.1 The Credence-Existence Link

Consider the initial version of Legless at Bondi (only my leg is cloned). Suppose I survive the operation, as planned, with two good legs. Why don't I care about the unfortunate 'possible me', whom the operation left with no legs at all? Because, happily, he doesn't exist. (I have no duty of care to a Man Who Never Was.)

Why didn't I care about him (much) before the operation, when I wasn't certain that he didn't exist? Because I was *very confident,* even then, that he wouldn't exist. And in the limit, being very confident equates simply to believing. At least to a first approximation, then, we can say that I ignored him because I believed—rightly, as it turned out—that he didn't exist.[13] In the classical case, then, the justification for giving preference to higher weighted alternatives goes something like this. We give (absolute) preference to actual things over 'merely possible' versions of the same things, and the weights simply reflect our degrees of confidence about which things are actual. (Call this the *credence-existence link*.)[14]

---

[12] Could we replace it, for Greaves' purposes, with my notion of 'locus of concern'? No, for it simply isn't true, of classical distributive justice problems, that there is only one person in the group whose welfare is 'really' my concern. (Recall the case of arranging one's affairs to benefit a group of children.)

[13] And note that if my belief had proved false, I might be culpable for that, but not for what I did in the light of it—given what I believed, I was right to ignore him.

[14] True, there are some subtleties about the notion of belief as a limit of credence. One way to unpack the idea a little further, sticking closely to the pragmatic understanding of credence, goes

This doesn't work in the Everett case: if well-weighted pleasure comes at the expense of low-weighted pain, then it is *real* pain, despite its weight. What Greaves needs is a measure which has the same *normative* implications as 'confidence in non-existence', while being entirely compatible with 'certainty of existence'.[15] That's what an analogue of classical uncertainty would look like.

8.2 What Alternative is There?

Greaves suggests that there is no realistic alternative to a caring measure based on the Born weights. But we need to distinguish two questions. First, is there any rival MEU approach—i.e., any rival to the Born measure, if Everettian decision theory is agreed to an analogue of classical decision under uncertainty? And second, is there any rival to the MEU approach itself? We need to answer the second question, before we can take refuge in a negative answer to the first.

I've suggested that there do seem to be some non-MEU alternatives, modelled in part on problems of distributive justice. It is no answer to this challenge, obviously, to claim that there are no alternatives *within* the MEU framework. Moreover, Greaves herself notes another possible positive answer to the second question, viz., the nihilist option, that rational action is simply incoherent in the Everett world:

[I]t could turn out that there were no coherent strategy for rational action in an Everettian context. . . . As Deutsch notes, this possibility cannot be ruled out of court:

It is not self-evident . . . that rationality is possible at all, in the presence of quantum-mechanical processes—or, for that matter, in the presence of electromagnetic or any other processes.   (Deutsch [1999 p.3130]) (Greaves [2004 p.432])

The 'what alternative is there?' plea can hardly have much force, presumably, while nihilism waits in the wings (drawing strength from the difficulties

---

something like this. In practice, any assignment of degrees of belief has a resolution limit—an 'epistemic grain', beneath which differences are of no practical significance. At this limit, low probability equates to zero, for the purposes at hand. I act *just as if* I believe that the low-weight alternatives don't exist.

[15] Greaves proposes that existence might come by degrees, though reality doesn't:

Lockwood's . . . talk of a 'superpositional dimension', and/or Vaidman's . . . suggestion that we speak of the amplitude-squared measure as a 'measure of existence', are somewhat appropriate (although we are not to regard lower-weight successors as *less real*, for *being real* is an all-or-nothing affair—we should say instead that there is *less* of them).   (Greaves [2004 p.30])

For my part, I can't see how such a distinction could make a difference here. Why should my concern for another's welfare depend on whether he *exists* (and hence on the degree to which he exists, on this view), but not simply on whether he is *real*? And in any case, the kind of 'Do no deliberate harm' principles which weight needs rationally to trump, to meet the objections above, are insensitive to quantity in the sense of frequency. What could make them sensitive to quantity in the sense of degree of existence?

that the DW approach has in explaining why weight should have normative significance).[16]

Greaves notes that the most commonly suggested rival to the Born measure is what she calls *egalitarianism*, or *naive counting*: a view that tries to treat all branches equally. Following Wallace, however, Greaves claims that this view turns out to be incoherent, in a decoherence-based version of the Everett picture, because the number of branches is not well defined:

> Why does naive counting break down in the decoherence approach? The core of the problem is that naive counting . . . presupposes the existence of a piece of structure that is not in fact present in the theory.   [2007b p.120]

However, I think that Greaves and others have failed to distinguish two versions of egalitarianism, only one of which is aptly called naive counting (and subject to this objection).[17] The version Greaves has in mind is a view that accepts that rational decision in the Everett world should properly be an analogue of classical decision under uncertainty, but simply proposes a rival measure—a measure that in the finite case in which there are N branches, gives each branch a weight of $1/N$. Grant that this turns out to be incoherent, for the reason Greaves describes.

The rival form of egalitarianism ('outcome egalitarianism') simply rejects the attempt to make numerical comparisons, treating all non-null *outcomes* as having the same kind of claim to be taken into account. Thus it rejects the idea that there is any sort of comparative weight to be associated with outcomes, with normative significance, whether counting-based or not. Is this view incoherent? If so, it doesn't seem to be because the number of *branches* is ill-defined in a decoherence-based version of the Everett interpretation—this view embraces that lesson.

9 CONCLUSION

Some brief conclusions. The 'Why does weight matter?' challenge has not been met. It is far from clear that there aren't compelling alternatives, or at least supplements, to a Deutschian decision policy, for an agent who believes that she lives in an Everett world. (Some of these alternatives are motivated by considerations analogous to those of distributive justice, but perhaps not all—if quantum Russian roulette appeals, it is for other reasons.) It is true that these considerations fall a long way short of a unified Everettian decision policy, and

---

[16] In the light of Section 5.1, nihilism can only be the view that there is no *preferred* rational strategy. As we saw, suitable preferences about the global state can certainly determine a rational strategy (without uncertainty). So nihilism needs to be understood as the view that there is no rational constraint on global preferences themselves.

[17] Cf. Lewis's [2005 pp.14–15] distinction between the Average Rule and the Sum Rule.

in particular, do little to stave off the larger threat of nihilism. But without an explanation as to why weight should have normative significance, Deutschian decision theory does no better in this respect. Most importantly, there seems little prospect that a Deutschian decision rule can be a constraint of rationality, in a manner analogous to the classical case. The fundamental problem rests squarely on the distinctive ontology of the Everett view: on the fact that it reifies what the one-branch model treats as mere *possibilia*, and hence moves its own decision rule into the realm of 'mere preference'. I conclude that the problem of uncertainty has not been solved.

Acknowledgements

My interest in these issues dates from a characteristically lucid talk by David Wallace in Konstanz in 2005, which introduced me to the basics of the DW approach. It was further encouraged by the opportunity to discuss them with Hilary Greaves in Sydney in 2006. I'm grateful to the organizers of the Many Worlds@50 Conference at the Perimeter Institute for inviting me to take part in such a fascinating meeting, and to Wayne Myrvold, David Papineau and other participants there, for further discussions. I'm also very much indebted to Guido Bacciagaluppi, Jenann Ismael, and especially Peter Lewis, for helpful conversations in Sydney about previous versions of this material. And I'm grateful to the Australian Research Council and the University of Sydney, for research support.

Bibliography

Deutsch, D. [1999], 'Quantum theory of probability and decisions', *Proceedings of the Royal Society of London* **A455**, 3129–37.

Greaves, H. [2004], 'Understanding Deutsch's probability in a deterministic multiverse', *Studies in the History and Philosophy of Modern Physics* 35, 423–56.

—— [2007a], 'On the Everettian epistemic problem', *Studies in History and Philosophy of Modern Physics* 38, 120–52.

—— [2007b], 'Probability in the Everett Interpretation', *Philosophy Compass* 2(1), 109–28.

—— [2007c], 'Comments in panel discussion', *Many Worlds@50 Conference*, Perimeter Institute, 24 September 2007. [Video and audio available at http://pirsa.org/07090076/].

Lewis, D. [2004], 'How many lives has Schrödinger's cat?' *Australasian Journal of Philosophy* **82**, 3–22.

Lewis, P. [2005], 'Probability in Everettian quantum mechanics'. Available online at philsci-archive.pitt.edu/archive/00002716/.

Papineau, D. [1996], 'Many minds are no worse than one', *British Journal for the Philosophy of Science* 47, 233–41.

Saunders, S. [1998], 'Time, quantum mechanics, and probability', *Synthese* **114**, 373–404.

Saunders, S. [2005], 'What is probability?' in *Quo Vadis Quantum Mechanics?*, A. Elitzur, S. Dolev, and N. Kolenda (eds), Springer-Verlag.

Sinnott-Armstrong, W. [2006], 'Consequentialism', in *The Stanford Encyclopedia of Philosophy* (Winter 2006 Edition), Edward N. Zalta (ed.). Accessible online at http://plato.stanford.edu/archives/win2006/entries/consequentialism/.

Squires, E. [1986], *The Mystery of the Quantum World*, Bristol, Adam Hilger.

Wallace, D. [2002], 'Quantum probability and decision theory, revisited'. Available online at quant-ph/0211104.

—— [2003], 'Everettian rationality: defending Deutsch's approach to probability in the Everett interpretation', *Studies in the History and Philosophy of Modern Physics* **34**, 415–39.

—— [2006], 'Epistemology quantized: circumstances in which we should come to believe in the Everett interpretation', *British Journal for the Philosophy of Science* **57**, 655–89.

—— [2007], 'Quantum probability from subjective likelihood: improving on Deutsch's proof of the probability rule', *Studies in History and Philosophy of Modern Physics* **38**, 311–32.

Transcript: Probability

Wallace

It's a theory of Dennett's that when you get an intuition pump, what you should do is play with it, twiddle the dials, vary the details, and then see what difference that makes, and I think that's quite instructive when we think about David [Albert]'s example [of the fatness rule]. Let's suppose I've got a situation where I'm considering buying a quantum lottery ticket for a large amount of money. And if I win the ticket—and let's say the weight for me winning is 0.5—then I go on holiday in Mauritius, and if I don't win the ticket I don't go on holiday. And I might do the calculation and think, well, it looks like, on the expected utilities, it's not worth me buying the ticket. But, I know what I'll do, when I'm in Mauritius I'll eat lots. I'll eat like a pig in Mauritius; my fatness on the Mauritius branch is going to shoot up and now it's worth me going to Mauritius, and now it's worth me buying a ticket. But hold on, I like to be fit and bronzed on holiday and I know that once I get to Mauritius I'm not going to want to put on weight at all. So what I'll do is, I'll pay some money to hire a minder and the minder will come to Mauritius with me and the minder will enforce the fact that I put on weight, put on fatness that is.

And of course when I get there I'll be annoyed about the minder, so I'd better brief the minder very carefully on my foibles to make damned sure I'm not able to escape from him.

OK, so I get that far, then I think some more, and I think, oh dear, if I end up not getting the ticket I might get depressed, I might go bankrupt, other bad financial things might happen and I know that my successor who doesn't win the ticket is going to be very worried about that. And I know that when I get miserable, I comfort eat and so I put on weight, and that's going to lower the utility of buying the ticket too, so I'm going to hire another minder to make sure that actually I do not comfort eat, and of course I have to brief this minder. And I know that my future self is going to hire a minder to make the first minder go away, and I've got to stop that guy.

. . . So I'm committed to this massive game of conspiratorial action against my own future selves and I'm committed to second-guessing all the things that I'm going to do in the future. And this is after two instances of branching—but branching is happening quintillions of times a second.

So that's the kind of way in which I think those sorts of objections, while they're well worth having on the table to see what happens, ultimately don't work. And it makes vivid, if you like, what was in the arguments with which I—and earlier in different ways David [Deutsch]—argued for equivalence or an equivalent principle. It kind of shows why, even if equivalence is not a priori obvious as a rationality principle, the other principles we derive it from—like the requirement of temporal consistency—are kind of obvious.

Albert

Of course, the choice of fatness in this example was partly to be funny. So, one can make up stories in which my life might become complicated if I were to adopt that caring measure. I have two things to say about that. First, I can make it depend on features of the world, physical features of the world not involving the amplitudes, that are less malleable than that so as to make these complications a little less threatening. But second, the bigger point seems to be: I'm not sure how the observation that it would be complicated to follow a certain decision procedure impacts on the worry that such a procedure poses for a claim or an argument or a theorem to the effect that there's a correct procedure—that some other procedure is the one a rational agent would have to adopt under those circumstances. That's all I want to say about the fatness argument for now.

Albert (to Saunders)

This is not a critique, this is a confession that I don't really understand what's at stake in these semantical discussions. If the discussions really are supposed to have semantical but not fundamental metaphysical import, if they're just supposed to show how speaking in a certain way about a fundamental metaphysical structure on which we're all agreed could make sense, OK, then I don't know why I'm not justified in thinking about these problems about probability and so on and so forth in just *ignoring* them, in just saying, 'Look, I understand the other structure, I understand the fundamental metaphysical structure, I think I see how things go there.' It just seems like it could be held to be completely irrelevant to that discussion that there is another coherent way of mapping this onto our ordinary language. So, that's one level of confusion I may have.

The second level of confusion I may have is: even if the semantical discussion turns out to have more teeth than I'm understanding, and even if the semantics somehow establishes not merely that there's a coherent way of uttering 'I am ignorant' or not an insane way of uttering 'I am ignorant', but that I really am ignorant in some sense, it's still hard to see, given the fixed metaphysical structure underneath it, how learning what I learn after the branching is going to be a way of learning about the fundamental metaphysical structure of the world, in a way that's going to allow me to distinguish between two dynamics that, say, assign different amplitudes to the same branches. OK, that's an unfairly long and multi-part question.

Saunders

To your first point, it seems that when you say 'metaphysics' I have to translate that as 'physics'.

Albert

Yeah, fine.

Saunders

And what I think you come back to as a way of rejecting this semantics and so forth is still a metaphysical view by my lights, and it's a metaphysical view, I think, which is essentially a stage-theorist view. I think you're a Siderian stage theorist in the Everettian case.

Albert

Simon, look, I'm wondering about things like how this semantics system is suddenly making it clear that the amplitudes explain the frequencies.

Saunders

Oh sure, but that's coming to the second question. On the second point, what I'm discovering by doing experiments and so forth is what happens in my world. That's what I'm discovering. I hope my world is typical, it may not be typical, I may get deviant outcomes; but that's the language in which the amplitudes are explanatory—they quantify 'typicality'.

Wallace (to Albert)

I think your second point is your first point again. If the semantic argument goes through then I think what we've established is that the ordinary-language-stated set of rules we have in our theory of confirmation map onto the theory in a different way than the way you might have expected. If they map onto that theory in this way then our existing confirmation theory is appropriate to confirm the Everett interpretation.

Loewer (comment)

I want to look at the way the principal principle works, and I think actually the best way for me to do it is to look at the list given by Hilary and Wayne (pp. 287–9). Here's what they've done on these pages. They've taken the branch weight case and the chance case, and they've played out a complete analogy between the two. As far as confirmation is concerned, chances and objective branch weights and credences and subjective weights are treated as completely on a par: that's the idea of the list. What I would like to do is to put another column there, which represents another, rather particular view about chance, namely David Lewis's view. Probably most people here are familiar with it. I'm going to say very quickly what it is. Lewis makes a connection between what actually happens in the world and what the chances are, let me call them the L-chances. That's tighter than the kind of chances that Wayne was talking about. It's tighter in this way: what the chances are is built into the laws, and the laws are given by the theory which best summarizes all the facts about the world, all the fundamental facts about the world. Here Lewis has an idea of there being a theory of the world which best combines simplicity and informativeness. That's what's meant by 'best summary'. And one way to be very informative while staying simple is to introduce a notion of probability, a probabilistic notion. Then you have to say something about how that notion, of how those probability claims, inform us about the world, and the idea is that a principle connecting them to the world will tell you how it's informing about the world. The principal principle is a way of making that connection. So the Lewis account has the principal principle, understood this way (this isn't quite the way Lewis understood it), built right into it. OK, so imagine that it's in the list there.

If you put it in the list there too, I want to point out some differences (they also have similarities) between what's said under the Lewis account, and in particular one difference under the chance account and the branch weight account. As far as the Lewis account is concerned, it agrees with everything that is said under chances. So Lewis chances satisfy the first thing, the theory assigns chances to possible worlds; the second, about updating; the third, there are possible worlds in which anomalous statistics occur. The next one is that a frequentist analysis of chance is untenable. That's right for Lewis, the simple frequentist analysis is not Lewis's, but Lewis's is closer to an actual frequentist account because what the Lewis chances are can supervene on the whole history, the actual facts of the world. So whereas the sort of primitivist propensities account of chance makes chances something over and above the ordinary facts of the world, Lewis's account makes them supervenient on the categorical facts in the world. Similarly, on the branch weight case the amplitudes, which are going to be identified with the objective weights, those over and above the branching structure—the categorical structure. (You might think of the branch weights as themselves additional categorical structure: I'll come to that in just a second.)

In the case of propensity chances and chances in Lewis's sense, but not in the case of the Greaves–Myrvold branch weights, the chances are built into the laws. The laws are the explanatory structure in the world. Some people don't find the Lewis laws the appropriate thing to do explaining—that's a different argument—but the Lewis account does build the chances and the laws in together. In the branch case that's not so, and I'm thinking that this gets at David Albert's worry that we're not really getting explanations from the branch weights of why it is that we obtain various frequencies on the various branches; why we don't get explanations for the weights either.

Next two lesser points. One is that the Lewis account as I explained it provides an understanding, a kind of rationale, for this principle connecting these Lewis chances, facts of the actual world, to the degrees of belief about the facts of the actual world. As has been emphasized by Wayne and Hilary, and by David Papineau earlier, this is just accepted as a primitive principle within the branching account just as within the chancy account. I think Lewis's account is better off there. For the second point, I was struck by the fact that Lewis's account simply doesn't apply to the many-worlds account, to the many-worlds ontology, because you have all of these frequencies—all of the branches which have all of the frequencies. If you're going to give a simple informative account of all that it would be the Schrödinger equation. Nothing that corresponds to probabilities would even show up there.

Maudlin (to Greaves and Myrvold)

There's a very fundamental puzzle on the Everett picture which is easy to state which just gets lost, which goes: as far as decision theory goes, this is decision theory under certainty, and that's easy. It's not a problematic thing. Under certainty you have various options, each option has an outcome, you have a preference order, that's fine, you take the one you like. You go back to David Papineau's question, if you want money, why do you bet to maximize expected utility? Answer: because I know what's going to give me money. There's no puzzle about why I'm doing something other than going for what I want.

Now, here's a situation, let me just spell it out. There's a kind of Schrödinger device; I'm going to prepare an x-spin-up electron, I'm going to measure the spin near x-spin but not quite that. I've got two boxes, if it comes out one way one box will be filled with deadly gas, if it comes out the other way the other box will be filled with deadly gas. I have Kitty, I love Kitty, and I have two choices in front of me: I can put Kitty in Box One or I can put Kitty in Box Two. From an Everett point of view it looks like I know what's going to happen in each case; there's no uncertainty. I put Kitty in Box One, I end up with two equal-caring branches, one with dead Kitty, the other with Kitty surviving. If I put Kitty in the other box, I'll also have two branches with equal-caring measures. Now, here's my fundamental puzzle. The branches decohere, what we're told is all that matters is the structure, and the structure doesn't change. If I pump up and down the relative amplitudes of the waves, the structure doesn't change, and you've told me that as an Everettian all I care about is structure. So the difference between putting Kitty in Box One or putting Kitty in Box Two, the only difference is the relative amplitudes of the branches, and that doesn't make any difference to the relevant *structure* of the branches, so as far as I'm an Everettian, I don't care whether dead Kitty ends up on a high amplitude branch or on a relatively low amplitude branch. And I don't understand: Hilary can come along and say maybe I have a further, unexplained, primitive desire that I care about high-amplitude Kitty more than low-amplitude Kitty; but that seems to be in direct conflict with David Wallace's talk, which said what I care about is structure, when here the structure's the same.

Greaves

Let me first make a brief comment about the last thing you said: that changing relative amplitudes of branches doesn't change 'the structure'. It's true that an adjustment to relative amplitudes doesn't affect the in-branch structure of either of the branches, but such an adjustment does affect the structure of the overall state. So I don't think there's any conflict between saying on the one hand (and as David does) that what the Everettian cares about is structure, and on the hand (as Wayne and I do) that agents should care about amplitudes.

Now let me go back to the part of your question that I think is more important: the objection that we are considering decision problems under conditions of certainty, and that 'the' decision theory for such problems is trivial.

The short reply to this objection is that talking of 'the' decision theory applicable to a given case begs the question. There are multiple theories available; our argument has urged a particular choice of decision theory, distinct from the one you have in mind.

To elaborate: Even for cases of decision-making under uncertainty, we could, if we wanted, write down a trivial decision theory. Such a trivial decision theory would say that the rational agent's preferences over gambles have to be transitive, and would impose no further constraints. We can also write down a non-trivial decision theory for such cases; that's what Savage did. Similarly, in the Everettian case, we could, if we wanted, write down a trivial decision theory; such a decision theory would merely say that a rational agent's preferences over Everettian 'brambles' (as Barry calls them) have to be transitive. I take it this is the decision theory that you have in mind when you say that decision theory for cases like this is 'easy', and that you just 'take the preference order you like'. But we could also do what Wayne and I have been proposing, which is to write down a non-trivial decision theory for preferences over brambles, structurally isomorphic to the non-trivial decision theory that Savage writes down for preferences over gambles. With all these decision theories available, the objection can't be that 'the' decision theory for branching cases is trivial. We have both a trivial and a non-trivial theory on the table, and the question is whether or not the stronger (non-trivial) theory is defensible. Wayne's and my claim is that the non-trivial theory is just as defensible in the branching case as Savage's non-trivial theory is in the non-branching case.

Barbour (to Greaves and Myrvold)

In Everett's wonderful paper he makes two fantastic suggestions. One, he picks up Einstein's argument that every theory should carry its complete explanation, its interpretation, in its bones. And it seems to me that that has failed in what you're presenting because you've added the branch weights as a new concept. I'm looking for a theory which will really do what Everett said. And the other thing Everett did in proposing his theory was to help people try to create quantum cosmology. Now, since Everett did that quite a lot of work has been done on quantum cosmology, and one of the strongest hints that comes out of that is that it is not really appropriate to think about time at all when you think about quantum cosmology. If we talk about a universal wavefunction, if we're talking about the wavefunction of the universe, there are lots of indications that that will be absolutely static. It will be a time-independent solution. And I believe that within this framework one can still get an explanation of the arrow of time, and of Born statistical weights, rather in the manner that John Bell explains them in his paper. I think Bell gave the perfect explanation of (Everettian) quantum mechanics more or less in its entirety, in this timeless sense, but then he finished up by saying that you find that solipsists after all also have life insurance. And with that remark he seemed to completely abandon it, and then died before he could try and justify what he said.

   A very important part of this story that has not been explained—not in any of the discussions I've heard—is the arrow of time and the low entropy state. This I think is an essential part of the story. And Jim is quite right I think—he said it years ago and we are very much of a like mind—we should be thinking of interpreting quantum mechanics much more like geology, where we try to find an explanation of existing records, including multiple experiments which have shown all these outcomes. We should be explaining the records as something static, trying to find the theory to explain that.

Greaves

I only have a comment on the first bit: about the idea that we want a theory to 'carry its interpretation in its bones', that is, roughly, to entail its own interpretation. Your thought is that Everett was supposed to fulfil this desire, but that this seems to be spoiled when we add the theoretical term 'branch weight' that is not definitionally equivalent to 'amplitude-mod-squared'.

   I think that to get a theory that entails its own interpretation, we would need the arguments that David Deutsch and David Wallace have been giving. If their arguments are sound, then the quantum state entails particular branch weights. Wayne and I were trying to remain neutral in this paper on the question of whether or not those arguments work and to say, look: if you think that the Deutsch–Wallace arguments are successful then you're going to think that, within the class of theories that agree on the quantum state, there's only going to be one rule for the branch weights, namely the rule according to which branch weights are the amplitudes squared. If you think those arguments don't work, you're going to think there's a larger space of possibilities, that the branch weights could be the amplitudes squared or they could also be one or other of these other functions. Our thought was that we can remain neutral on the issue of whether or not those arguments work because, at the end of the day, it's not going to make a difference given that in fact we've observed Born frequencies: it's going to be the 'branch weight equals amplitude squared' version of the theory that gets confirmed. But I think you're right, if you really want an 'in the bones' interpretation you have to commit to the Deutsch–Wallace claim. It's consistent with our programme but we didn't commit to it in this paper.

Albert (to Greaves and Myrvold)

I have two brief questions. The first is, I think, along the lines of what Tim was saying, but put a little differently, and put specifically with regard to the question of confirmation, of updating. I'm completely puzzled along the following lines. Suppose somebody believes in Everett, they believe in fission. They know that their job is to choose which branching they'd like. If they do this they get this branching, if they do that they get that branching. Forget about the general worries Barry raised, which I agree with, along the lines: what could you learn from where you end up by branching, since you knew in advance everything was going to be there?—Forget about that general worry. Here's a much more specific worry: whether you learn anything from that or not, what you for sure don't learn anything about is what branchings are going to result from what choices you make in the future. So why would you be tempted to update or to change which branchings you prefer in any way, based on what's happened before? You know exactly, completely independent of what's happened before, what branching is going to occur if you do this, and what branching is going to occur if you do that. Your job is to choose which branching you prefer, but you're clearly not going to learn anything about that, because it's just given to you deterministically by the theory; what could this updating ceremony possibly do for you?

Myrvold

OK, suppose you're flipping a coin and you know for certain that you're going to get that kind of branching structure, and you're offered a choice of a dollar on heads on the second flip or a dollar on tails on the second flip, so there are two wagers. Now suppose you're uncertain about what physical theory's right. You have two theories about branch weights, one has two-thirds for heads and one-third there, and theory two has the branch weights the other way round. And you're going to be offered the choice of the wagers after the first flip. So, T1 says each branching gives two-thirds heads and one-third tails and T2 says it is one-third heads and two-thirds tails. If I had to make a choice here I'd probably be indifferent between the two. But if I wait until I see the first one and then make a choice, I'd rather have the people who see heads go for heads subsequently, and the people who see tails go for tails, because that maximizes the weight of the payoff. And so if I say that beforehand, and then I actually have the agent do that after the first flip the agent is going to be acting as if she has . . .

Albert

I still don't understand the rationale for why the agent would do that.

Myrvold

Suppose you know for certain that it's two-thirds weight heads and one-third weight tails; you'd rather get a dollar on the heads branch than a dollar on the tails branch. Here's what I think: given our representation theorem, what it means for you to have those beliefs about branch weights *is* for you to prefer a dollar on the heads branch to a dollar on the tails branch. In this I do not identify the weights with degrees of belief. Here's the theorem, if your preferences between wagers satisfy our axioms . . .

Albert

Yes, but this is a question about why they plausibly would—OK, we'll continue this—I have one other quick question. This isn't really so much a question as to highlight something you guys have already said, but that I think may come as a surprise to people and seems to me a little damaging to your case. So, you want to be free of the earlier Deutsch–Wallace arguments which I was criticizing in my talk, which select a unique weight, in the quantum-mechanical case. And you want to say, look, we're going to learn what the weights are, we're going to learn what the relationship between the weights and the amplitudes is by seeing how our experiments come out. I think it's worth emphasizing that to the extent that you take that line you're distinguishing between two theories both of which have the exact same quantum state evolution. And the distinction between the two theories—and this is moreover a distinction that you think you can empirically distinguish between by looking at the frequencies—you're distinguishing between two theories which make completely identical claims about the quantum state evolution but differ in the claims they make about the relationship between the amplitudes and the weights. You have a line in your paper where you say: at least for the purposes of confirmation, these are going to count as distinct theories. That seems to me a terribly heavy burden for your view to carry, that is: either these weights are additional physical facts about the world, in which case you've already given up the main goal of Everett which is to see the wavefunction as the whole story, or, if they're not *physical* facts about the world, they're some other kind of fact which you think you're confirming or disconfirming by observing these frequencies. I think this is a bizarre situation to be in.

So the dialectical situation is as follows. If the Deutsch–Wallace proofs succeed, and there is only one coherent way to associate weights with the branches then you're back in their programme and a lot of this gets a lot easier but I've already tried to raise considerations against those. To the extent that you want to hang independently of those, to the extent to which you *don't* want your view to depend on those proofs, then you're positing something extra about the world, either physical or non-physical, in the relation between the amplitudes and the branch weights, that seems very strange.

Greaves

I agree this is a bit puzzling, but notice it's also a puzzle that the non-Everettian has. We're not claiming to resolve all philosophical worries about probability here.

Brown to Myrvold

As I understand the logic of this material, you're learning from experience in the way that we normally do, but making clear its rational basis. But you seem to suggest—correct me if I'm wrong—that if you take this line, then in particular the equivalence condition, or assumption, looks very natural. But if it looks very natural only because you have learned from experience and know the Born rule is empirically correct, doesn't it make the Born rule theorem redundant?

Myrvold

Here's my attitude towards the Deutsch–Wallace theorem: It has a status similar to Gleason's theorem in that they show that the only branch weights that fit nicely with quantum mechanics without adding extra structure are the Born-rule weights. OK, so we've got this theory, quantum mechanics, and the only natural branch weights we can get out of quantum mechanics are the Born-rule weights and you want to know whether that theory is right or some other theory that might posit different branch weights as right, and then you have to compare the observed relative frequencies with the calculated branch weights in the two theories. So a theory with different branch weights would be a different theory to quantum mechanics.

Wallace (comment)

David Albert brought up the evidence problem, I think, very clearly, but I think perhaps the shape of Everettian responses to this haven't been made completely clear. I think there have been two families of responses available, one being driven by Simon and myself and one being driven by Hilary and more lately by Wayne. And complicating matters is the fact that Simon and I broadly buy Hilary and Wayne's response, and Hilary broadly is at least sympathetic to Simon's and mine, or at least some of the premises of it, so I don't think we should regard these responses as in conflict. And this is great because it ought to be the case that when things are right one *ought* to be able to work out that they're right by more than one means. It's when you've got positions that can only be argued for by a single tortuous route, that you want to get worried about whether they're really getting at the truth.

But just to say, here's how Simon and my response is supposed to go: The reason we're worried—and Tim put this very clearly—is that it looks as if the framework for decision-making, for confirmation, and for evaluation of theories just goes across hopelessly badly into the Everett interpretation. In particular, it is deterministic, so shouldn't we use deterministic decision theory, and what is there in deterministic decision theory except transitivity? Also, how do we judge the validity of theories, and collect data, when all the data is there in the branching structure?

Essentially the point of what Simon and I are saying, and the response that he'd and I'd make to David [Albert]'s worries, is essentially this: By saying, 'look, our entire framework fits very badly into the Everett interpretation' we're conflating two things. We're conflating our ordinary pre-theoretic grip of how to think about theory and confirmation, which scientists have been doing fine for ages and which people doing informal science have been doing for ages, and which we can express in informal, intuitive statements of natural language, with the way those ideas translate and map and fit into the fundamental metaphysical picture, 'the book of the world' as John called it. And the claim is that when we take the things we say about the way we all use probability ordinarily, and try to map them onto some set of statements about third-person accessible God's-eye view facts, about propositions which take truth values on possible maximal states of the universe, all sorts of frameworks like that, then because we might have the wrong semantics and metaphysics, the way to do that mapping is more up for grabs than we realized.

So when we say things like, for instance, that there's no point carrying out an experiment when we know the answer, that's just fine inside our pre-theoretic framework: we shouldn't carry out experiments when we know the answer. But it's a mistake to move from that to the claim that we shouldn't carry out experiments when we're in possession of a theory according to which the results of the experiment are predictable from a God's-eye view with certainty. That move incorporates a lot of metaphysical assumptions about the single universe. If the single universe picture is right then those assumptions are just great, but if it's not they're not great at all; it would beg the question to assume they're right and then apply them in the wrong way to the many-universe picture.

So that's the response to the David criticism: it is true that this doesn't affect our metaphysics, it only affects our semantics and non-metaphysical aspects of our worldview. But some of the aspects it affects are the rules of the game in our own theory of confirmation and decision. Anyway, that's how our programme is supposed to go.

How does Hilary's programme go? I think the way it goes is perhaps only partially visible in the presentation that Wayne gave on that project so I'd like to say a little bit more about it, and Hilary can correct me if I get it wrong. Hilary's framework, as I see it, is saying 'never mind these semantic moves, we'll accept that agents need to accept the possibility that they should have non-zero credence in the Everett interpretation and non-zero credence in things other than the Everett interpretation, and we'll work out how they should update those credences when they receive data.' So she's assuming the Bayesian epistemological framework, rightly or wrongly but very respectably and very mainstreamly, and she's saying there has to be a version of that framework that lets us work this way because if not, it's not a scandal for Everettians, it's a scandal for everyone. Because unless we're prepared to say the Everett interpretation is logically impossible, that we must somehow have to give no credence to it at all, then we have to have a manageable epistemic framework which applies to agents with some credence in the interpretation, even if it's ten to the power minus 35. And if our first stab at that framework says that all data we get massively confirms the Everett interpretation, that's not a problem for the Everett interpretation, that's an urgent problem for our theory of confirmation that needs to be dealt with, Everett or not-Everett. And I take it that what she does is construct a framework of that form which genuinely is unified and which includes, as a sub-component in it, how we handle these various branching situations, but it doesn't presuppose those branching situations. And within that unified framework, she establishes on what seem to be very reasonable assumptions, that we would indeed rationally increase our credence in the Everett interpretation given the sort of data that we do in fact receive from quantum mechanics. That's how I see that programme. It's been said at various points that it's not at all obvious how it is, if the world is Everettian, that these statistics confirm it. Absolutely, on Hilary's premises, it's not at all obvious. It's hard work, and here's the work, here's how it goes.

Pitowsky (to Wallace)

I'm speaking now about the Deutsch–Wallace and similar but classical derivations, de Finetti and Savage and the like. In the latter case there's a very clear distinction between what they call rules of rationality which are justified by things like the Dutch-book argument and so on, and what they consider in say de Finetti's case as prior probability, or in the Savage case a 'small world' model, in which you just set certain possibilities to have probability zero, in your prior. These possibilities are not going to get higher probability later on. I think some of the assumptions that are coming into your derivations are actually part of what traditionally, within such derivations, are assumptions about the prior probability. Just to give one example, suppose that there's a person who is considering a particular experiment at MIT, say a measurement of spin, where the Born rule applied to the wavefunction just gives half and half. But he thinks, for some reason, that the outcome spin-up is somehow higher because of the level of the stock exchange in New York and that's his prior; so he violates Born's rule, and he's violating non-contextuality of probability, and he's certainly not being rational in the general sense of what we mean by rationality, but he's not violating any rule of what is called rationality in probability theory. In your derivations you should at least be clear what you take as part of the rationality principles and what you take as some sort of empirical judgement of the kind that is usually made. I'm not saying that these empirical judgements are not justified; I think they are justified but I think that their justification requires a prior notion of probability—because after all we're talking about priors—and this may be circular. This is my challenge here.

Wallace

I think you're being unduly kind by saying it's unclear in my paper [about which assumptions are rationality principles]. I think it's crystal clear, but it might be completely wrong. They're all supposed to be rationality principles, they're not supposed to be empirical at all. And I think you'll just say that's wrong, they should not be counted as rationality principles. But David came up with these ideas and I'd like to hear what he thinks.

Deutsch (to Pitowsky)

First of all there's decision-theoretic rationality which doesn't extend as far as ruling out that the stock exchange could affect . . .

Pitowsky

Right, exactly, that's my point—

Deutsch

But then there's a wider conception that one might call scientific rationality—physicalism, let's say—where you want to say that your view of what is actually happening in a particular place only depends on the physical variables at that place, in the Heisenberg picture, and not on the variables in other places. Now that—I think we probably do need to make several assumptions like that. For me, I don't care how many assumptions we make, so long as they're not *probabilistic* assumptions. So long as we start off with something that isn't talking about probability and end up saying that the rational agent behaves as if there were probabilities. So long as that happens, I don't mind those additional assumptions, and in the case that you've mentioned they are, I would call them, scientific rationality not . . .

Albert

But David, if you mean that literally, why not just assume that the rational agent will behave as if the square amplitudes are probabilities and get it over with?

Deutsch

Well that *is* a probabilistic assumption.

Albert

No, no. *As if*. The rational agent will maximize utility by the following formula including the squared amplitudes, period.

Deutsch

The game is, you start off with assumptions, as many assumptions as you like, so long as you can make them if you didn't know about observers bifurcating—by the way, it doesn't cover in my opinion other types of probability which don't involve the observer bifurcating—so then the question is, do those principles, and you can have as many of them as you like so long as they apply, so long as they can be rationally justified, in the case where observers *don't* bifurcate; and then you add the additional fact that observers do bifurcate according to these laws, and then use the same principles and see what they then imply. That was the game.

Pitowsky

I think—that was my point, that these are probabilistic assumptions, because these are assumptions about the prior; what you are giving initial probability zero. Initially the probability is zero, so they are assumptions about probability. Within this framework of rational decision theory, these are exactly the assumptions about the prior probabilities that you set up before the game begins. You can stand aside and say, well, I mean, these are just general scientific assumptions; I wouldn't disagree with that, I'm just saying that, within this framework, they have probabilistic content. And actually because of Gleason's theorem—technical details apart—because of Gleason's theorem, you are actually justified in assuming that the square of the amplitude is the probability.

Hemmo

I want to continue what Itamar just said, and that will relate to Hilary and Wayne's paper. Given what we think, that the assumptions we are talking about are assumptions about probability, in the quantum picture, the question immediately arises: can you really argue—can you get to these assumptions—from empirical data? Just consider in the branching picture the situation in which you live on an anomalous branch; I mean there must be such a you, and you're going to come up with assumptions about priors which are not quantum mechanical—and then of course you're going to reject quantum mechanics. It seems to me there's no sense in which one could say that you are objectively wrong in such a picture, wrong in a sense which is related to the features of the world in which you live.

Pitowsky

There's no sense in which you're unlucky, that's the point.

Wallace

Thinking about this project in the light of these objections: it's nice to think: ah, we Everettians solved probability (and, ultimately I do buy this way of thinking). But if you'd like a more conciliatory starting-point route, ask: What do you need to put in to get probability out in Everett, compared to what you need to put in to get it out in a non-Everettian framework. In a non-Everettian framework it looks like you just have to put in probability itself. Maybe, maybe, maybe you can do better, and it would be great if that's the case, but I don't think so, it doesn't look that way at the moment. What do you need to put into Everett? Things that seem to be much more plausible as axioms of rationality than the probability axiom itself. Are they logically true? Probably not. Are they reasonable assumptions about rationality? I'd be inclined to say yes. Are they probabilistic? I'd say no. So I don't think it's entirely fair, certainly in the version of the argument I presented, for Itamar to feel that the assumptions going in are probabilistic. Maybe they're unreasonable, but I don't think they're probabilistic. In this particular case, for instance, the assumption going in is that the agent cares only about the state of the multiverse that is generated by whatever process he's considering. So, for instance, unless he actually has independent reasons for caring about the New York stock exchange, he should be indifferent about any transformation in the state of the stock exchange. And notice one of the assumptions going in there, and it is an assumption, is that the details of the short-period history that get us from before the bets are taken till after the result of the bet, including, for instance, in this fanciful example, the quite dramatic rearrangement of the entire New York stock exchange, are events the agent ought to be rationally indifferent to. Is that a logically true assumption? Of course not. Is it a reasonable assumption? I'd say yes. Is it weaker than what's going on in classical accounts? I'd say yes. Is it probabilistic? I'd say no.

Myrvold

A moment ago David Deutsch stated a principle of scientific rationality, something like: your probabilities only depend on whatever physical variables there are. OK, I take that as a principle of rationality. To actually get something like equivalence out we also need some kind of substantive physical claim about what kind of physical variables there are. And I don't think you have any problem with getting probability assignments out of symmetry considerations—I'm going to require my probability assignments to depend only on certain variables . . . if you call that an assumption about priors I would agree. But let me just make the point that it can't be the case that equivalence follows from principles of pure rationality alone; it follows from—and this sometimes gets obscured, but I don't think either David is claiming otherwise—it follows from traditions of rationality plus certain claims about what is and isn't in the ontology of the world.

Let me just give an example, Antony is going to go look for evidence of non-equilibrium matter and, if Everettian quantum mechanics is true, there will be branches on which Antony finds stuff that everyone is going to take as evidence about non-equilibrium matter, and we're going to award Antony the Nobel prize, we're going to all think that quantum mechanics isn't quite right, that some other theory is right, and we'd be rational to do so, and I think that—I *know* that—David Wallace would say we'd be rational to do so, even if in fact the case is that we're misled on that branch and Everettian quantum mechanics is right.

Wallace

Absolutely. Unlucky, wrong. But rational.

Hemmo

So even if you find evidence in the physical world which refutes Everettian quantum mechanics you would still say what you just said and that Everett's theory is still true and we are misled.

Wallace

That's not true, that's not the case. And that's not what Wayne's saying either.

PART V

ALTERNATIVES TO MANY WORLDS

13

Quantum Jumps, Born's Rule, and Objective Reality

Wojciech Hubert Zurek

ABSTRACT

This brief guide describes three insights into the transition from quantum to classical that are based on the recognition of the role of the environment. I start with a minimalist derivation of preferred sets of states. This breaking of the unitary symmetry of the Hilbert space yields—without the usual tools of decoherence—quantum jumps and *pointer states* consistent with those obtained via *einselection*. Pointer states obtained this way define events without appealing to Born's rule for probabilities, which can be now derived from *envariance*—a symmetry of entangled quantum states. With probabilities at hand one can analyze information flows in the course of decoherence. They explain how classical reality can arise from the quantum substrate by accounting for the objective existence of the einselected states of quantum systems through the redundancy of pointer state records in their environment—through *quantum Darwinism*. Taken together, and in the right order, these three advances (which fit well within Everett's relative states framework, but do not require 'many worlds' per se) extend the *existential interpretation* of quantum theory.

1 INTRODUCTION: THREE QUANTUM QUESTIONS

It is instructive to revisit the 'relative state interpretation' set out 50 years ago by Hugh Everett III [1957a,b] and re-evaluate it with the basic axioms of quantum theory (as abstracted, for example, from Dirac [1958]) at hand. This is one way to motivate exploring the effect of the environment on the state of the system. (Of course, a complementary motivation based on a non-dogmatic reading of Bohr [1928] is also possible.)

This note is not a review (such review is already available; see Zurek [2007a]). Rather, I aim at a brief 'annotated guide' to some of the recent results, primarily of the Los Alamos group. And I mean here 'a guide'. This is no substitute for the original papers.

The basic idea we shall pursue here is to accept a relative state explanation of the 'collapse of the wavepacket' by recognizing, with Everett, that observers perceive the state of the 'rest of the Universe' *relative* to their own state, or—to be more precise—relative to the state of their records. This allows quantum theory to be universally valid. (This does *not* mean that one has to accept a many-worlds ontology; see (Zurek [2007a]) for discussion.)

Everett explains the perception of collapse. However, the relative state approach raises three questions absent in Bohr's Copenhagen interpretation [1928] which relied on the independent existence of an *ab initio* classical domain. Thus, in a completely quantum universe one is forced to seek sets of preferred, effectively classical but ultimately quantum states that define branches of the universal state vector, and allow observers to keep reliable records. Without such a **preferred basis** relative states are 'too relative', and the relative state approach suffers from *basis ambiguity*.

In my view, this issue was—over the past quarter century or so—settled by decoherence (Zurek [1981, 1991], Paz and Zurek [2001], Joos et al. [2003], Zurek [2003a], Schlosshauer [2004, 2007]). Environment-induced decoherence shows that in open quantum systems—that is, in systems interacting with their environments—certain quantum states retain stability, while their superpositions quickly decay into mixtures of pointer states. This is known as einselection—a nickname for environment—*in*duced super*selection*.

Einselection can account for the preferred sets of states, and, hence, for Everettian 'branches'. But this is achieved at a price that would have been unacceptable to Everett: the usual practice of decoherence is based on averaging (as it involves reduced density matrices defined by a partial trace). This means that one is using Born's rule to relate amplitudes to probabilities. But, as emphasized by Everett, Born's rule should not be postulated in an approach that is based on purely unitary quantum dynamics. Thus the universal validity of quantum theory raises the issue of (b) **the origin of Born's rule** $p_k = |\psi_k|^2$ which—following the original conjecture [1926]—is simply postulated in textbook discussions.

Last but not least, even preferred quantum states defined by einselection are still quantum. Therefore, they cannot be found out by initially ignorant observers through direct measurement without getting disrupted. Yet, states of macroscopic systems in our everyday world seem to exist objectively—they can be found out by anyone without getting perturbed. (This ability to find out an unknown state is in fact an operational definition of 'objective existence'.) So, if we are to explain the emergence of everyday classical reality, we need to identify (c) **the quantum origin of objective existence**.

We shall do that by dramatically upgrading the role of the environment: in decoherence theory the environment is in effect the collection of degrees of freedom where quantum coherence (and, hence, phase information) is lost. However, it was recently proposed that, in 'real life', in our quantum universe, the role of the environment is far more significant (see e.g. Zurek [2000, 2003a, 2009]): it is in effect a witness to the state of the system of interest, and a communication channel through which a vast majority of the information reaches us, the observers. This new view of the environment is the subject of the theory of quantum Darwinism.

2 QUANTUM POSTULATES AND RELATIVE STATES

It is best to start our discussion from well-defined solid ground. To this end, we have extracted the list of quantum postulates that are explicit in Dirac [1958], and at least implicit in many quantum textbooks.

The first two postulates are 'purely quantum':

(i) *The state of a quantum system is represented by a vector in its Hilbert space* $\mathcal{H}_S$.
(ii) *Evolutions are unitary (e.g., generated by the Schrödinger equation).*

These two axioms provide an essentially complete summary of the mathematical structure of the theory. They are sometimes (DeWitt [1971], DeWitt and Graham [1973]) supplemented by a 'postulate (o)' that a state of a composite quantum system is represented by a vector in the tensor product of the Hilbert spaces of its components. We cite it here for completeness, and note that physicists differ in assessing how much of postulate (o) follows from (i). We shall not be distracted by this minor issue, and move on to where the real problems are. Readers can follow their personal taste in supplementing (i) and (ii) with whatever portion of (o) they deem necessary.

Using (i) and (ii), and suitable Hamiltonians, one can calculate everything that can be calculated in quantum theory. Yet such calculations would only be a mathematical exercise—one can predict nothing from their results. What is missing in (i) and (ii) is a connection with physics—a connection with measurements. A way to establish correspondence between abstract state vectors in $\mathcal{H}_S$ and laboratory experiments (and/or everyday experience) is needed to relate quantum mathematics with the real world. The task of establishing this correspondence starts with the next axiom:

(iii) *Immediate repetition of a measurement yields the same outcome.*

Axiom (iii) is idealized (it is hard to devise such non-demolition measurements, but in principle it can be done). Yet it is also uncontroversial. The very notion of a 'state' is based on predictability, i.e., something like axiom (iii): the most rudimentary prediction is a confirmation that the state is what it is known to be. Moreover, a classical equivalent of (iii) is taken for granted, so there is no clash with our classical intuition here.

Axiom (iii) ends the uncontroversial part of the list of postulates. In particular, in contrast to classical physics (where an unknown state can be found out by

an initially ignorant observer) the very next quantum axiom limits attributes of
the state:

(iv) *Measurement outcomes are limited to an orthonormal set of states (eigenstates of
     the measured observable). In any given run of a measurement an outcome is just
     one such state.*

This *collapse postulate* is certainly controversial. To begin with, in a completely
quantum universe it is inconsistent with the first two postulates: starting from
a general pure state $|\psi_S\rangle$ of the system (axiom (i)), and an initial state $|A_0\rangle$ of
the apparatus $\mathcal{A}$, and assuming unitary evolution (axiom (ii)) one is led to a
superposition of outcomes:

$$|\psi_S\rangle|A_0\rangle = \left(\sum_k a_k|s_k\rangle\right)|A_0\rangle \Rightarrow \sum_k a_k|s_k\rangle|A_k\rangle, \qquad (1)$$

which is in contradiction with, at least, a literal interpretation of the 'collapse'
anticipated by axiom (iv).

This part of the problem with postulate (iv) was settled by Everett. But perhaps
the most significant and disturbing implication of (iv) is that quantum states
do not exist—at least not in the objective sense to which we are used to in the
classical world. The outcome of the measurement is typically *not* the pre-existing
state of the system, but one of the eigenstates defined by the measured observable.
This clashes with the classical idea of what the state should be. Whatever the
quantum state is, 'objective existence' independent of what is known about it is
clearly not one of its attributes. Some even go as far as to claim that quantum
states are simply a description of the information that an observer has, and have
essentially nothing to do with 'existence'. I think this is going too far—after all,
there are situations when a state can be found out, and (iii) postulates that its
existence can be confirmed. But, clearly, (iv) limits the 'quantum existence' of
states to situations that are 'under the jurisdiction' of postulate (iii) (or slightly
more general situations where the pre-existing density matrix of the system
commutes with the measured observable).

This difficulty with postulate (iv) has been appreciated since Bohr [1928] and
von Neumann [1932]. Yet, (iv) captures what happens in the laboratory, and (at
least before Everett) it was often cited as an indication of the ultimate insolubility
of the 'measurement problem'.

To resolve the clash between the mathematical structure of the theory and
subjective impressions of what happens in real world measurements, one can
accept—with Bohr—the primacy of experience. The inconsistency of (iv) with
the core of the quantum formalism—(i) and (ii)—can then be blamed on
the nature of the apparatus. According to the Copenhagen interpretation the
apparatus is classical, and, therefore, not subject to the quantum principle of

superposition (which follows from (i)). Moreover, its evolution need not be unitary. So, collapse can happen on the 'lawless' quantum–classical border.

Taken literally, this quantum–classical duality is a challenge to the unification instinct of physicists. One way of viewing decoherence is to regard einselection as a mechanism that replaces literal quantum–classical duality with an effective classicality that suspends the validity of the quantum principle of superposition in a subsystem while upholding it for the composite system that includes the environment (Zurek [1991, 2003a]).

The alternative to Bohr's approach proposed by Everett was to abandon the literal view of collapse and recognize that a measurement (including appearance of the collapse) is already described by Eq. (1) providing quantum states are regarded as relative. Once the observer is included in the wavefunction, one can consistently interpret the consequences of such correlations. The right-hand side of Eq. (1) contains all the possible outcomes, so the observer who records outcome #17 perceives the branch of universe that is consistent with that event reflected in his records. This view of the collapse is also consistent with axiom (iii); re-measurement by the same (non-demolition) device yields the same state.

But this relative state view of the quantum universe suffers from a basic problem: the principle of superposition (the consequence of axiom (i)) implies that the state of the system or of the apparatus after the measurement can be written in infinitely many unitarily equivalent basis sets in the Hilbert space of the pointer of the apparatus (or of the observer's memory cell). So,

$$\sum_k a_k |s_k\rangle |A_k\rangle = \sum_k a'_k |s'_k\rangle |A'_k\rangle = \sum_k a''_k |s''_k\rangle |A''_k\rangle = \ldots \qquad (2)$$

This is *basis ambiguity*. It appears as soon as—with Everett—one eliminates axiom (iv). The bases employed above are typically non-orthogonal, but in the relative state setting there is nothing that would preclude them, or that would favor, e.g., the Schmidt basis of $\mathcal{S}$ and $\mathcal{A}$ (the orthonormal basis that is unique, provided that the absolute values of the *Schmidt coefficients* in such a *Schmidt decomposition* of an entangled bipartite state differ).

In our everyday reality we do not seem to be plagued by such basis ambiguity problems. So, in our universe there is something that (in spite of (i) and the egalitarian superposition principle it implies) picks out preferred states, and makes them effectively classical. Axiom (iv) anticipates this.

In other words, before there is an (apparent) collapse in the sense of Everett, a set of preferred states—one of which is selected by (or at the very least, consistent with) the observer's records—must be somehow chosen. There is nothing in the writings of Everett that would even hint that he was aware of this question.

The obvious other issue concerns probabilities: how likely is it that, after I measure, my state will be, say, $|\mathcal{I}_{17}\rangle$? Everett was keenly aware of this question, and even believed that he solved it by deriving Born's rule. In retrospect, it is clear that the solution he proposed (as well as the solutions proposed by his followers, including DeWitt [1970, 1971, 1973], Graham [1973], and Geroch [1984]) did not accomplish as much as was hoped for, and did not amount to a derivation of Born's rule (see Squires [1990] and Stein [1984], and especially Kent [1990] for influential critical assessments).

In the textbook version of the quantum postulates, probabilities are assigned by adding another axiom:

(v) *The probability p_k of an outcome $|s_k\rangle$ in a measurement of a quantum system that was previously prepared in the state $|\psi\rangle$ is given by $|\langle s_k|\psi\rangle|^2$.*

This last postulate in our list fits very well with Bohr's approach to quantum foundations (and, especially, with postulate (iv)). However—as was noted by Everett—it is at odds with the spirit of the relative state approach. This does not mean that there is a mathematical inconsistency here: one can certainly use Born's rule (as the formula $p_k = |\langle s_k|\psi\rangle|^2$ is known) along with the relative state approach in averaging to get expectation values and the reduced density matrix. Everett's point was, rather, that Born's rule should be *derivable* from the other axioms of quantum theory.

3 QUANTUM ORIGIN OF QUANTUM JUMPS

The above discussion laid out the problem, or, rather, a set of problems. To restate them briefly, we need to derive (iv) and (v) from (i)–(iii). In particular, even when we accept Everett's relative state resolution of 'single outcomes' and 'collapse', we still need to find the preferred basis that is a part—indeed, the essence—of (iv). Moreover, we need to do it without appealing to Born's rule—without decoherence, or at least without its usual tools. Once we have it, we will also have a set of candidate *events*. Once we have events, we shall be able to pursue the issue of probabilities.

In my view, the *preferred basis problem* was settled by the characterization of environment-induced superselection (*einselection*), usually discussed along with decoherence. This is discussed elsewhere (Zurek [1981, 1982]). However, pointer states and einselection are usually justified by appealing to decoherence. Therefore, they come at a price that would have been unacceptable to Everett: decoherence and einselection employ reduced density matrices and trace, and so their predictions are based on averaging, and thus, on probabilities—on Born's rule.

Here we briefly survey the strategy and direct the reader to references where different steps of that strategy are carried out. In short, we describe how one should go about doing the necessary physics, but we only sketch what needs to be done. The requisite steps were carried out in the references we provide. We emphasize again that our discussion is incomplete, meant as a guide to the literature, and not a substitute.

Accounting for the quantum origins of our classical everyday world requires the solution of *several* problems. This calls for several ideas. Moreover, in order to avoid circularities, they need to be introduced in the right order. Much of the heat in various debates on the foundations of quantum theory seems to be generated by the expectation that a *single* idea should provide a complete solution. When this does not happen—even when there is progress, but there are still unresolved issues—the possibility that an idea responsible for this progress may be a step in the right direction—but that more than one idea, one step, is needed—is often dismissed.

Decoherence done 'in the usual way' (which, by the way, is a step in the right direction, in the understanding of the practical and even many of the fundamental aspects of the quantum–classical transition!) is not a good starting point for addressing the more fundamental aspects of the origins of the classical. In particular, decoherence is not a good starting point for the derivation of Born's rule. As the saying goes, there is no preacher like a reformed sinner. I previously proposed a derivation of Born's rule based on the symmetries—invariance of a state of the system under permutations of pointer states, 'events' obtained in the usual way from decoherence (Zurek [1998]). We have already noted the problem with this strategy: it courts circularity. It employs Born's rule to arrive at the pointer states by using reduced density matrix which is obtained through trace—i.e., averaging, which is where Born's rule is implicitly invoked (see e.g. Nielsen and Chuang [2000]). Therefore, using decoherence to derive Born's rule is at best a consistency check. While the above is a *mea culpa*, this circularity would also afflict other approaches, including proposals based on decision theory (Deutsch [1999], Wallace [2003], Saunders [2004]), as noted also by Forrester [2007] among others.

So one has to start the task from a different end. But the focus on the preferred basis as a first step turns out to be fruitful. To get anywhere—e.g., to define 'events' essential in the introduction of probabilities—we need to show how the mathematical structure of quantum theory (postulates (i) and (ii)) supplemented by the only uncontroversial measurement axiom (iii) (which demands immediate repeatability—and, hence, predictability—of idealized measurements) leads to preferred sets of states.

Surprisingly enough, this turns out to be simple. A line of reasoning (Zurek [2007b]) reminiscent of the 'no cloning theorem' (Wootters and Zurek [1982], Dieks [1982], Yuen [1986]) yields **(a) pointer states**—potential events (outcomes). Their stability is needed to establish an effectively classical domain within the quantum universe, and to define events such as measurement outcomes.

How to do that with minimal assumptions (postulates (i)–(iii) on the above list) is described in Zurek [2007a,b]. Here we recapitulate the basic steps. We assume that $|v\rangle$ and $|w\rangle$ are among the possible outcome states of $\mathcal{S}$, i.e.,

$$|v\rangle|A_0\rangle \implies |v\rangle|A_v\rangle, \tag{3a}$$

$$|w\rangle|A_0\rangle \implies |w\rangle|A_w\rangle. \tag{3b}$$

So far, we have employed postulates (i) and (iii). We now assume that the process described by Eq. (3) is fully quantum, so postulate (ii)—unitarity of evolutions—must also apply. Unitarity implies that the overlap of the states before and after must be the same. Hence:

$$\langle v||w\rangle(1 - \langle A_v||A_w\rangle) = 0. \tag{4}$$

This simple equality is the basis of our conclusions. Depending on the overlap $\langle v||w\rangle$ there are two possibilities. Let us first suppose that $\langle v||w\rangle \neq 0$ (but otherwise arbitrary). In this case, one is forced to conclude that the state of A cannot be affected by the process above. That is, the transfer of information from S to A must have failed completely, since $\langle A_v||A_w\rangle = 1$ must hold. The apparatus can bear no imprint that distinguishes between components $|v\rangle$ and $|w\rangle$ of the superposition $|\psi_S\rangle$, the prospective outcome states of the system. The attempted measurement was completely unsuccessful!

The second possibility is that $\langle v||w\rangle = 0$. This allows for an arbitrary $\langle A_v||A_w\rangle$, including a perfect record, $\langle A_v||A_w\rangle = 0$. Thus, outcome states must be orthogonal if—in accord with postulate (iii)—they are to survive intact a successful information transfer in general or a quantum measurement in particular, so that immediate re-measurement can yield the same result. The same derivation can be carried out for S with a Hilbert space of dimension $\mathcal{N}$ starting with a system state vector $|\psi_S\rangle = \sum_{k=1}^{\mathcal{N}} a_k|s_k\rangle$, where (as before)—a priori $\{|s_k\rangle\}$ need to be only linearly independent.

The straightforward derivation above leads to a surprisingly decisive conclusion: orthogonality of outcome states of the system is absolutely essential for them to exert distinct influences—to imprint even a minute difference—on the state of any other system while retaining their identity. The overlap $\langle v||w\rangle$ must be 0 *exactly* for $\langle A_v||A_w\rangle$ to differ from unity. Also, sloppy and accidental information transfers (e.g., to the environment in the course of decoherence) can define preferred sets of states providing that the crucial non-demolition demand of postulate (iii) is imposed on the unitary evolution responsible for the information transfer. A straightforward extension of the above derivation—addition of the environment $\mathcal{E}$ that would interact with the apparatus—provides a new perspective on decoherence (Zurek [2007a,b]).

It is important to emphasize that we are not asking for clearly distinguishable records (i.e., we are not demanding orthogonality of the states of the apparatus, $\langle A_v||A_w\rangle = 0$). Even with these rather weak assumptions one is still forced to conclude that *quantum states can exert distinguishable influences and remain unperturbed only when they are orthogonal*. We only used postulate (i)—the fact that when two vectors in the Hilbert space are identical then physical states they correspond to must also be identical.

The existence of sets of orthogonal states established above on the foundation of very basic (and very quantum) assumptions leads one to postulate the existence of observables through the inverse of the spectral theorem. Observables are associated with Hermitian operators. Consequences of measurements of observables that have the same eigenstates on the measured system are essentially identical. They must be—they obviously commute.

We conclude that the restriction to an orthogonal set of outcomes yields **(a) preferred basis**. Thus the essence of the collapse axiom (iv) need not be postulated! We established it here from the uncontroversial postulates (i)–(iii). We note that the preferred basis arrived at in this manner coincides with the basis obtained a long time ago via einselection (Zurek [1981, 1982]). It is just that we have arrived at this familiar result without implicit appeal to Born's rule, which is essential if we want to take the next step, and derive postulate (v).

Of course, we did not derive actual *collapse* of the wavepacket. Selection of one specific outcome is non-unitary, so one cannot hope to deduce it starting from postulate (ii). But, in a certain sense, we have come close. We deduced the necessity of a symmetry-breaking choice of a single *orthonormal* set of states from amongst various possible basis sets each of which can equally well span the Hilbert space of the system. This is all that is needed to exorcize basis ambiguity in the relative state setting.

4 PROBABILITIES FROM ENTANGLEMENT

The derivation of events allows, and even forces one, to enquire about their probabilities or—more specifically—about the relation between probabilities of measurement outcomes and the initial pre-measurement state. As noted earlier, several past attempts at the derivation of Born's rule turned out to be circular. Here we present the key ideas behind a circularity-free approach. Thus, we briefly recount some salient points of a recent derivation of Born's rule based on a symmetry of entangled states—on **(b) entanglement—assisted invariance** or **envariance**. The study of envariance as a physical basis of Born's rule started with Zurek [2003a, 2005], and is now the focus of several other papers (see, for example, Schlosshauer and Fine [2005], Barnum [2003], Herbut [2007]). The key idea is illustrated in Fig. 1. Symmetry of entanglement allows one to prove that equal amplitudes imply equal probabilities.

Envariance also accounts for the loss of the physical significance of local phases between Schmidt states (in essence, for decoherence). Thus, the eventual loss of coherence between pointer states can also be regarded as a consequence of quantum symmetries of the states of systems entangled with their environment. So, the essence of decoherence arises from symmetries of entangled states, and certain aspects of einselection (as we have seen it in the previous section) can be studied without employing the usual tools of decoherence theory (reduced density matrices and trace) that rely on Born's rule.

Envariance also allows one to justify the additivity of probabilities (Zurek [2005]), while the only generally accepted previous derivation of Born's rule by Gleason [1957] assumed it. This is a significant advance. In quantum theory the overarching additivity principle is the quantum principle of superposition. Anyone familiar with the double-slit experiment knows that probabilities of quantum states (such as the states corresponding to passing through one of the two slits) do *not* add, which in turn leads to interference patterns. Moreover, Gleason's theorem provides absolutely no motivation why the measure that obtains should have any physical significance—i.e., why it should be regarded as a probability. As illustrated in Fig. 1, the envariance approach has a transparent physical motivation.

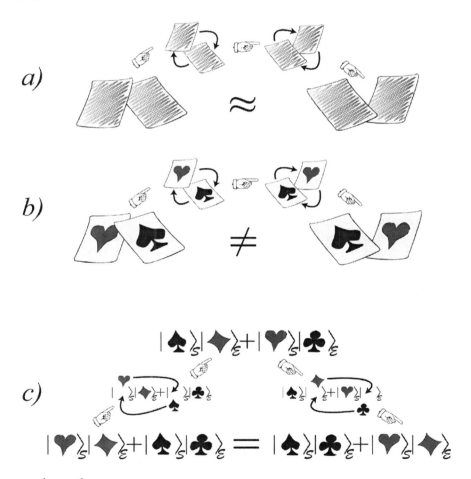

caption continues on next page

Fig. 1. Envariance is a symmetry of entangled quantum states. It allows one to demonstrate Born's rule using a derivation (Zurek [2003b, 2005]) that combines an old intuition of Laplace [1820] about symmetry and the origins of probability with quantum symmetries of entanglement. (a) Laplace's *principle of indifference* (illustrated with playing cards) aims to establish a symmetry from invariance of the state of the system under swaps. When a player does not know the face values of the cards, he will be indifferent—he will not care—whether the cards are swapped before he gets the one on the left. For Laplace, this indifference was evidence of invariance related to a (subjective) symmetry. It translated into a subjective assessment of the probability of a future outcome. Indifference implied *equal likelihood*, and led to the assignment of equal probabilities of the invariantly swappable alternatives. For the two cards above, subjective probability $p_\spadesuit = \frac{1}{2}$ would be inferred by an observer (player) who does not yet know their face value, but knows that one (and only one) of the two cards is a spade. This allows one to assign probabilities to outcomes—to future events—on the basis of symmetry established by invariance of a state under a transformation. When the probabilities of a set of elementary events are provably equal, this can be used to compute probabilities of composite events and thus to develop the theory of probability. Even the additivity of probabilities can be *established* (see, e.g., Gnedenko [1968]) given more elementary assumptions about the 'algebra of events'. This is in contrast to Kolmogorov's measure-theoretic axioms (which *include* additivity of probabilities). Thus, Laplace's symmetry-based approach yields probabilities where symmetries of elementary events (e.g., cards) under swaps are known. By contrast, Kolmogorov's theory is general, and does not deal with associating probabilities to elementary events in a specific system (physical or otherwise). It only deals with deriving probabilities of composite events from the arbitrarily assigned probabilities of elementary events, which only need to be positive and sum up to unity (Gnedenko [1968]). (b) From the point of view of a physicist, the problem with the 'principle of indifference' is its subjectivity. The actual physical state of the system (the two cards) will of course be altered by the swap. A related problem is that the assessment of indifference is based on ignorance. Therefore, as was argued, e.g., by supporters of the relative frequency approach (which is regarded by many as a more 'objective' foundation of the concept of probability) it is impossible to deduce anything useful (including probabilities) from ignorance. This is (in addition to subjectivity) a reason why equal likelihood (and, hence, Laplace's approach) is rightly regarded with suspicion as a basis of probability in physics. (c) Quantum entanglement allows one to use symmetries of entanglement to deduce objective probabilities starting with a perfectly known state. The relevant symmetry is the *entanglement*—assisted in*variance* or *envariance*. When a pure entangled state of a system $\mathcal{S}$ and of another system (which we call 'an environment $\mathcal{E}$', anticipating connections with decoherence) $|\psi_{\mathcal{SE}}\rangle = \sum_{k=1}^{N} a_k |s_k\rangle |\varepsilon_k\rangle$ can be transformed by $U_\mathcal{S} = u_\mathcal{S} \otimes 1_\mathcal{E}$ acting solely on $\mathcal{S}$, but the effect of $U_\mathcal{S}$ can be undone by acting solely on $\mathcal{E}$ with an appropriately chosen $U_\mathcal{E} = 1_\mathcal{S} \otimes u_\mathcal{E}$, $U_\mathcal{E}|\eta_{\mathcal{SE}}\rangle = (1_\mathcal{S} \otimes u_\mathcal{E})|\eta_{\mathcal{SE}}\rangle = |\psi_{\mathcal{SE}}\rangle$, it is envariant under $u_\mathcal{S}$. For such composite states one can rigorously establish that the local state of $\mathcal{S}$ remains unaffected by $u_\mathcal{S}$. Thus, for example, the phases of the coefficients in the Schmidt expansion $|\psi_{\mathcal{SE}}\rangle = \sum_{k=1}^{N} a_k |s_k\rangle |\varepsilon_k\rangle$ are envariant, as the effect of $u_\mathcal{S} = \sum_{k=1}^{N} \exp(i\phi_k)|s_k\rangle \langle s_k|$ can be undone by a *countertransformation* $u_\mathcal{E} = \sum_{k=1}^{N} \exp(-i\phi_k)|\varepsilon_k\rangle \langle \varepsilon_k|$ acting solely on the environment. This envariance of phases implies their irrelevance for the local states—in effect, it implies decoherence. Moreover, when the absolute values of the Schmidt coefficients are equal (as in (c) above), swapping states is also possible: a swap $|\spadesuit\rangle\langle\heartsuit| + |\heartsuit\rangle\langle\spadesuit|$ in $\mathcal{S}$ can be undone by a 'counterswap' $|\clubsuit\rangle\langle\diamondsuit| + |\diamondsuit\rangle\langle\clubsuit|$ in $\mathcal{E}$. So, it can be established rigorously that $p_\spadesuit = p_\heartsuit = \frac{1}{2}$ follows from the objective symmetry of such an entangled state. This proof of equal probabilities is based not on ignorance (as in Laplace's indifference) but on a perfect knowledge of the 'wrong thing'—of the global observable that rules out (via quantum indeterminacy) any information about complementary local observables. When supplemented by simple counting, this leads to Born's rule (Zurek [2003a,b, 2005]).

The presence of entanglement solves—with a single stroke—two different problems. It eliminates local phases (doing the job of decoherence). This leads to additivity. Moreover, when the Schmidt coefficients are equal, symmetries of entanglement force one to conclude that the probabilities must also be equal. The crux of the proof is that, after a swap on the system, the probabilities of the swapped states must be equal to the probabilities of their new partners in the Schmidt decomposition (which did not yet get swapped). But—when the coefficients are equal—a swap on the environment restores the original states. So the probabilities must be the same as if the swap never happened. These two requirements (that a swap exchanges probabilities, and that it does not exchange them) can be simultaneously satisfied only when probabilities are equal.

Getting rid of the phases is very important: swaps on isolated pure states will, in general, change them. For instance, $|\spadesuit\rangle + i|\heartsuit\rangle$, after a swap $|\spadesuit\rangle\langle\heartsuit| + |\heartsuit\rangle\langle\spadesuit|$, becomes $i|\spadesuit\rangle + |\heartsuit\rangle$, i.e., is orthogonal to the pre-swap state. The crux of the proof of equal probabilities was that the swap does not change anything *locally*. This can be established for entangled states with equal coefficients but—as we have just seen—is simply not true for a pure unentangled state of just one system.

It goes without saying that in the real world the environment will become entangled (in the course of decoherence) with the preferred states of the system of interest (or with the preferred states of the apparatus pointer). We have seen earlier how postulates (i)–(iii) lead to preferred sets of states. We have also pointed out that—at least in idealized situations—these states coincide with the familiar pointer states that remain stable in spite of decoherence. So, in effect, we are using the familiar framework of decoherence to derive Born's rule. Fortunately, as we have seen, it can be analyzed without employing the usual Born's rule-dependent tools of decoherence (reduced density matrix and trace). Thus, one can reach the goal of Everett without the danger of circularity.

In this brief summary we have only sketched how one can establish equality of probabilities for the outcomes that correspond to Schmidt states associated with coefficients that differ at most by a phase. This is not yet Born's rule. But it turns out that this is the hard part of the proof: once such equality is established, a simple counting argument (a version of that employed in Zurek [1998], Deutsch [1999], Wallace [2003], and Saunders [2004]) leads to the relation between probabilities and unequal coefficients (Zurek [2003b, 2005]).

5 QUANTUM DARWINISM, CLASSICAL REALITY, AND OBJECTIVE EXISTENCE

Monitoring of the system by the environment (the process responsible for decoherence) will typically leave behind multiple copies of its pointer states in $\mathcal{E}$. Pointer states are favored. Only states that can survive decoherence can produce information theoretic progeny in this manner (Ollivier et al. [2004, 2005]).

So only information about pointer states can be recorded redundantly. States that can survive decoherence can use the same interactions that are responsible for einselection to proliferate information about themselves throughout the environment.

This redundancy of selected information is the key to **(c) the quantum origin of objective existence**, which was listed in the introduction. Quantum Darwinism (Zurek [2003a, 2009]) allows observers to use the **environment as a witness**—to acquire information about pointer states indirectly, leaving the system of interest untouched and its state unperturbed. This is how objective existence—cornerstone of classical reality—arises in the quantum world. Observers can find out the state of the system without endangering its existence (which would be inevitable in direct measurements). Indeed, the reader of this text is—at this very moment—intercepting a tiny fraction of the photon environment by his eyes to gather all of the information he needs. This is how virtually all of our information is acquired. A direct measurement is not what we do. Rather, we count on redundancy, and settle for information that exists in many copies.

The existence of redundant copies of pointer states implies that observables which do not commute with the pointer observable are inaccessible. The simplest model of quantum Darwinism that illustrates this is a somewhat contrived arrangement of many (N) target qubits that constitute subsystems of the environment interacting via a *controlled not* (c-not) with a single control qubit S. As time goes on, consecutive target qubits become imprinted with the state of the control S:

$$(a|0\rangle + b|1\rangle) \otimes |0_{\varepsilon_1}\rangle \otimes |0_{\varepsilon_2}\rangle \cdots \otimes |0_{\varepsilon_N}\rangle \implies$$
$$(a|0\rangle \otimes |0_{\varepsilon_1}\rangle \otimes |0_{\varepsilon_2}\rangle + b|1\rangle \otimes |1_{\varepsilon_1}\rangle \otimes |1_{\varepsilon_2}\rangle) \cdots \otimes |0_{\varepsilon_N}\rangle \implies$$
$$a|0\rangle \otimes |0_{\varepsilon_1}\rangle \otimes \cdots \otimes |0_{\varepsilon_N}\rangle + b|1\rangle \otimes |1_{\varepsilon_1}\rangle \cdots \otimes |1_{\varepsilon_N}\rangle.$$

This simple dynamics creates multiple records of the logical basis of the 'pointer' states of the system in the environment. The existence of the preferred pointer basis that is untouched by the interaction is essential. As we have seen earlier, this is possible—such quantum jumps emerge from the purely quantum axioms (i)–(iii).

The mutual entropy between S and a subsystem $\mathcal{E}_k$ can be easily computed. As the k'th c-not is carried out, $I(S : \mathcal{E}_k)$ increases from 0 to:

$$I(S : \mathcal{E}_k) = H(S) + H(\mathcal{E}_k) - H(S, \mathcal{E}_k)$$
$$= -(|a|^2 \lg |a|^2 + |b|^2 \lg |b|^2).$$

Thus, each environment qubit $\mathcal{E}_k$ constitutes a sufficiently large fragment of $\mathcal{E}$ to supply complete information about the pointer observable of S. The interaction with the first c-not leads to complete decoherence of $\mathcal{S}$ in its pointer basis $\{|0\rangle, |1\rangle\}$. This illustrates the relation between decoherence and quantum Darwinism. It also shows that redundancy can continue to increase after coherence between pointer states is lost.

We have already noted the special role of the pointer observable. It is stable and, hence, it leaves behind information-theoretic progeny—multiple imprints, copies of the pointer states—in the environment. By contrast, complementary observables (e.g., the phase between the pointer states $\{|0\rangle, |1\rangle\}$) are destroyed by the interaction with a single subsystem of $\mathcal{E}$. They can in principle still be accessed, but only when *all* of the environment is measured. Indeed, because we are dealing with a quantum system, things are much worse than that: the environment must be measured in precisely the right (typically global) basis to allow for such reconstruction. Otherwise, the accumulation of errors over multiple measurements will lead to an incorrect conclusion and re-prepare the state and environment, so that it is no longer a record of the state of $\mathcal{S}$, and phase information is irretrievably lost.

As each environment qubit is a perfect copy of $\mathcal{S}$, redundancy in this simple example is eventually given by the number of fragments—that is, in this case, by the number of the environment qubits—that have (more or less) complete information about $\mathcal{S}$. In this simple case there is no reason to define redundancy in a more sophisticated manner. Such a need arises in more realistic cases when the analogues of c-not's are imperfect.

Quantum Darwinism was defined only recently. Previous studies of the records 'kept' by the environment were focused on its effect on the state of the system, and not on their utility. Decoherence is a case in point, as are some of the studies of the decoherent histories approach (Gell-Mann and Hartle [1998], Halliwell [1999]). The exploration of quantum Darwinism in specific models has only just started (Blume-Kohout and Zurek [2005, 2006, 2008]). We do not intend to review these results here in any detail. The basic conclusion of these studies is, however, that the dynamics responsible for decoherence is also capable of imprinting multiple copies of the pointer basis on the environment. Moreover, while decoherence is always implied by quantum Darwinism, the reverse need not be true. One can easily imagine situations where the environment is completely mixed, and, thus, cannot be used as a communication channel, but would still suppress quantum coherence in the system.

While much remains to be done, quantum Darwinism settles the issue of the origin of classical reality by accounting for all of the operational symptoms of objective existence in a quantum universe. While a single quantum state cannot be found out through a direct measurement, pointer states can leave multiple records in the environment. Observers can use these records to find out the state of the system of interest. They can afford to destroy photons while reading the evidence—the existence of multiple copies implies that other observers can access the information about the system indirectly and independently, and that they will all agree about the outcome. This is, I believe, how objective existence arises in our quantum world.

6 DISCUSSION: FREQUENTLY ASKED QUESTIONS

The subject of this paper has a long history (Wheeler and Zurek [1983]). As a result, there are different ways of talking, thinking, and writing about it. It is almost as if different points of view have developed different languages. As a result, one can find it difficult to understand the ideas, as one often has to learn 'the other language' used to discuss the same problem. This is further complicated by the fact that all of these languages use essentially the same words, but charged with a very different meanings. Concepts such as 'existence', 'reality', or 'state' are good examples.

The aim of this section is to acknowledge this problem and to deal with it to the extent possible within the framework of a brief guide. We shall do that in a way inspired by modern approach to languages (and to travel guides): rather than study vocabulary and grammar, we shall use 'conversations' based on a few 'frequently asked questions'. These FAQs are distilled from issues raised by the participants of the meeting or by the referees. The hope is that this exercise will provide the reader with some useful hints of what is meant by certain phrases. This is very much in the spirit of the 'travel guide', where a collection of frequently used expressions is often included.

FAQ #1: *What is the relation between 'decoherence' and 'einselection'?*

Decoherence is the process of the *loss of phase coherence* caused by the interaction between the system and the environment. Einselection is an abbreviation of 'environment-induced superselection', which designates *selection of a preferred set of pointer states* that are immune to decoherence. Decoherence will often (but not always) result in einselection. For instance, an interaction that commutes with a certain observable of a system will preserve eigenstates of that *pointer observable*, *pointer states* that are einselected, and do *not* decohere. By contrast, superpositions of such pointer states will decohere. This picture can be (and generally will be) complicated by the evolution induced by the Hamiltonian of the system, so that perfect pointer states will not exist, but approximate pointer states will be still favored—will be much more stable than their superpositions. There are also cases when there is decoherence, but it treats all the states equally badly, so that there is no einselection, and there are no pointer state. A perfect depolarizing channel (Nielsen and Chuang [2000]) is an example of such decoherence that does not lead to einselection. Section 3 of this paper emphasizes the connection between predictability and einselection, and leads to a derivation of preferred states that does not rely on Born's rule.

FAQ #2: *Why does the collapse axiom (iv) conflict with the 'objective existence' of quantum states?*

The criterion for objective existence used here is pragmatic and operational: finding out a state without prior knowledge is a necessary condition for a state to objectively exist (Zurek [2003a], Ollivier et al. [2004, 2005], Blume-Kohout and Zurek [2005, 2008]). Classical states are thought to exist in this sense. Quantum states do not: Quantum measurement yields an outcome—but, according to axiom (iv), this is one of the eigenstates of the measured observable, and not a pre-existing state of the system. Moreover, according to axiom (iii) (or collapse part of (iv)) measurement re-prepares the system in one of the eigenstates of the measured observable. A sufficient condition for objective existence is the ability of many observers to independently find out the state of the system without prior knowledge, and to agree about it. Quantum Darwinism makes this possible.

FAQ #3: *What is the relation between the preferred states derived using their predictability (axiom (iii)) in Section 3 and the familiar 'pointer states' that obtain from einselection?*

In the idealized case (e.g., when perfect pointer states exist) the two sets of states are necessarily the same. This is because the key requirement (stability in spite of the monitoring/copying by the environment or an apparatus) that was used in the original definition of pointer states in Zurek [1981] is essentially identical to 'repeatability'—the key ingredient of axiom (iii). It follows that when interactions commute with certain observables (e.g. because they depend on them), these observables are constants of motion under such an interaction Hamiltonian, and they will be left intact. For example, interactions that depend on position will favor (einselect) localized states, and destroy (decohere) non-local superpositions. Using a *predictability sieve* to implement einselection (Paz and Zurek [2001], Zurek [2003a], Schlosshauer [2007]) is a good way to appreciate this.

FAQ #4: *Repeatability of measurements, axiom (iii), seems to be a very strong assumption. Can it be relaxed (e.g. to include POVMs)?*

Non-demolition measurements are very idealized (and hard to implement). In the interest of brevity we have imposed a literal reading of axiom (iii). This is very much in the spirit of Dirac's textbook, but it is also more restrictive than necessary (Zurek [2007b]), and does not cover situations that arise most often in the context of laboratory measurements. All that is needed in practice is that the *record* made in the apparatus (e.g., the position of its pointer) must be 'repeatably accessible'. Frequently, one does not care about repeated measurements of the quantum system (which may be even destroyed in the measurement process). Axiom (iii) captures in fact the whole idea of a record—it has to persist in spite of being read, copied, etc. So one can impose the requirement of repeatability at the macroscopic level of an apparatus pointer with a much better physical justification than Dirac did for the microscopic measured system. The proof of Section 3 then goes through essentially as before, but details (and how far can one take the argument) depend on specific settings. This 'transfer of the responsibility for repeatability' from the quantum system to a (still quantum, but possibly macroscopic) apparatus allows one to incorporate non-orthogonal measurement outcomes (such as POVMs) very naturally: the apparatus entangles with the system, and then acts as an ancilla in the usual projective measurement implementation of POVMs (see e.g. Nielsen and Chuang [2000]).

FAQ #5: *Probabilities—why do they enter? One may even say that in an Everettian setting 'everything happens', so why are they needed, and what do they refer to?*

Axiom (iii) interpreted in relative state sense 'does the job' of the collapse part of axiom (iv). That is, when an observer makes a measurement of an observable he will record an outcome. Repetition of that measurement will confirm his previous record. That leads to the symmetry breaking derived in Section 3 and captures the essence of the 'collapse' in the relative state setting (Zurek [2007b]). So, when an observer is about to measure a state (e.g. prepared previously by another measurement) he knows that there are as many possible outcomes as there are eigenvalues of the measured observable, but that he will end up recording just one of them. Thus, even if 'everything happens', a specific observer would remember a specific sequence of past events that happened to him. The question about the probability of an outcome—a future event that is about to happen—is then natural, and it is most naturally posed in this 'just before the measurement' setting. The concept of probability does not (need not!) concern alternatives that already exist (as in classical discussions of probability, or some many-worlds discussions). Rather, it concerns future potential events one of which will become a reality upon a measurement.

FAQ #6: *The derivation of Born's rule given here and in* Zurek [2003b, 2005, 2007a,b, 2009], *and even the derivation of the orthogonality of outcome states uses scalar products. But the scalar product appears in Born's rule. Isn't that circular?*

The scalar product is an essential part of the *mathematics* of quantum theory. The derivation of Born's rule relates probabilities of various outcomes to amplitudes of the corresponding states using symmetries of entanglement. So it provides a connection between the mathematics of quantum theory and experiments—i.e. physics. Hilbert space (with the scalar product) is certainly an essential part of the input. And so are entangled states and entangling interactions. They appear whenever information is transferred between systems (e.g. in measurements, but also as a result of decoherence). All the derivations use only two values of the scalar product—0 and 1—as input. Both correspond to certainty.

FAQ #7 *How can one infer probability from certainty?*

Symmetry is the key idea. When there are several (say, n) mutually exclusive events that are a part of a state invariant under their swaps, their probabilities must be equal. When these events exhaust all the possibilities, the probability of any one of them must be $\frac{1}{n}$. In contrast to the classical case discussed by Laplace, the tensor nature of states of composite quantum systems allows one to exhibit *objective* symmetries (Zurek [2003b, 2005, 2007a]). Thus, one can dispense with Laplace's *subjective* ignorance (his 'principle of indifference'), and work with objective symmetries of entangled states. The keys to the derivation of probabilities are the proofs: (i) that phases of Schmidt coefficients do not matter (this amounts to decoherence, but is established without the reduced density matrix and partial trace, the usual Born-rule-dependent tools of decoherence theory) and (ii) that equal amplitudes imply equal probabilities. Both proofs (Zurek [2003a,b, 2005, 2007a]) are based on *en*tanglement-assisted in*variance* (or *envariance*). This symmetry allows one to show that certain (Bell-state-like) entangled states of the whole imply equal probabilities for local states. This is done using symmetry and certainty as basic ingredients. In particular, one relies on the ability to undo the effect of local transformations (such as a 'swap') by acting on another part of the composite system, so that the pre-existing state of the whole is recovered with certainty. Using envariance, one can even show that an amplitude of 0 necessarily implies probability 0 of the corresponding outcome.[1] One can also prove the additivity of probabilities (Zurek [2005]) using a modest assumption—the fact that probabilities of an event and its complement sum up to 1.

FAQ #8: *Why are the probabilities of two local states in a Bell-like entangled state equal? Is the invariance under relabeling of the states the key to the proof?*

Envariance is needed precisely because relabeling is not enough. For instance, states can have intrinsic properties that they 'carry' with them even when they get relabeled. Thus, a superposition of a ground and excited state, $|g\rangle + |e\rangle$, is invariant under relabeling, but this does not change the fact that the energy of the ground state $|g\rangle$ is less than the energy of the excited state $|e\rangle$. So there may be intrinsic properties of quantum states (such as energy) that 'trump' relabeling, and it is a priori possible that probability is like energy in this respect. This is where envariance saves the day. To see this, consider a Schmidt decomposition of an entangled state $|\heartsuit\rangle|\diamondsuit\rangle + |\spadesuit\rangle|\clubsuit\rangle$ where the first ket belongs to S and the second to $\mathcal{E}$. Probabilities of Schmidt partners must be equal, $p_\heartsuit = p_\diamondsuit$ and $p_\spadesuit = p_\clubsuit$. (This 'makes sense', but can be established rigorously, e.g. by showing that the amplitude of $|\clubsuit\rangle$ vanishes in the state left after a projective measurement that yields $\heartsuit$ on S.) Moreover, after a swap $|\spadesuit\rangle\langle\heartsuit| + |\heartsuit\rangle\langle\spadesuit|$, in the resulting state $|\spadesuit\rangle|\diamondsuit\rangle + |\heartsuit\rangle|\clubsuit\rangle$, one has $p_\spadesuit = p_\diamondsuit$ and $p_\heartsuit = p_\clubsuit$. But probabilities in the environment $\mathcal{E}$ (that was not acted upon by the swap) could not have changed. It therefore follows that $p_\heartsuit = p_\spadesuit = \frac{1}{2}$, where the last equality assumes (the usual) normalization of probabilities with $p(\texttt{certain event}) = 1$.

---

[1] This is because in a Schmidt decomposition that contains n such states with zero coefficients one can always combine two of them to form a new state, which then appears with the other $n-2$ states, still with the amplitude of 0. This purely mathematical step should have no implications for the probabilities of the $n-2$ states that were not involved. Yet, there are now only $n-1$ states with equal coefficients. So the probability w of any state with zero amplitude has to satisfy $nw = (n-1)w$, which holds only for $w = 0$ (Zurek [2005]).

FAQ #9: *Probabilities are often justified by counting, as in the relative frequency approach. Is counting involved in the enviariant approach?*

There is a sense in which the enviariant approach is based on counting, but one does not count the actual events (as is done in statistics) or members of an imaginary ensemble (as is done in the relative frequency approach) but, rather, the number of potential invariantly swappable (and, hence, equiprobable) mutually exclusive alternatives. Relative frequencies statistics can be recovered (very much in the spirit of Everett) by considering branches in which a certain number of events of interest (e.g. detections of $|\heartsuit\rangle$, $|1\rangle$, spin-up, etc.) has occurred. This allows one to quantify probabilities in the resulting fragment of the 'multiverse', with *all* of the branches, including the 'maverick' branches that proved so difficult to handle in the past (DeWitt [1971], DeWitt and Graham [1973], DeWitt [1970], Geroch [1984], Squires [1990], Stein [1984], Kent [1990]). They are still there (as they certainly have every right to be!) but appear with probabilities that are very small, as can be established using enviariance (Zurek [2003b, 2005]). These branches need not be 'real' to do the counting—as before, it is quite natural to ask about probabilities before finding out (measuring) what actually happened.

FAQ#10: *What is the 'existential interpretation'? How does it relate to the many-worlds interpretation?*

The existential interpretation is an attempt to let quantum theory tell us how to interpret it by focusing on how effectively classical states can emerge from within our universe that is 'quantum to the core'. Decoherence was a major step in solving this problem: it demonstrated that in open quantum systems only certain states (selected with the help of the environment that monitors such systems) are stable. They can persist, and, therefore—in that very operational and 'down to earth' sense—exist. The results of decoherence theory (such as einselection and pointer states) are interpretation independent. But decoherence was not fundamental enough—it rested on assumptions (e.g. Born's rule) that were unnatural for a theory that aims to provide a fundamental view of the origin of the classical realm starting with unitary quantum dynamics. Moreover, it did not go far enough: einselection focused on the stability of states in the presence of an environment, but it did not address the question of what states can survive measurement by the observer, and why. Developments described briefly in this 'guide' go in both directions. Axiom (iii) that is central in Section 3 focuses on repeatability (which is another symptom of persistence and, hence, existence). Events it defines provide a motivation (and a part of the input) for the derivation of Born's rule sketched in Section 4. These two sections shore up our 'foundations'. Quantum Darwinism explains why states einselected by decoherence are detected by the observers. Thus, it reaffirms the role of einselection by showing (so far, in idealized models, but see Zwolak, Quan, and Zurek [2009], Riedel and Zurek [2010]) that pointer states are usually reproduced in many copies in the environment, and that observers find out the state of the system indirectly, by intercepting fragments of the environment (which now plays a role of the communication channel). These advances rely on unitary evolutions and Everett's 'relative state' view of the collapse. However, none of these advances depends on adopting the orthodox many-worlds point of view, where each of the branches is 'equally real'.

7 CONCLUSIONS

In conjunction with Everett's relative state account of the wavepacket collapse, these three advances—a 'decoherence free' derivation of preferred pointer states (key to postulate (iv)), the envariant derivation of probabilities (postulate (v)), and quantum Darwinism—illuminate the relation of quantum theory with the classical domain of our experience. They complete the *existential interpretation* based on the operational definition of objective existence, and justify confidence in quantum mechanics as the ultimate theory that needs no modifications to account for the emergence of the classical.

Of the three advances mentioned above, we have summed up the main idea of the first (the quantum origin of quantum jumps), provided an illustration of the second (the envariant origin of Born's rule), and briefly explained quantum Darwinism. As noted earlier, this is not a review, but a guide to the literature. A more complete review of these advances, their interdependence, and their relation of decoherence is available (Zurek [2007a]). A concise summary of salient points is also at hand (Zurek [2009]).

Everett's insight—the realization that relative states settle the problem of collapse—was the key to these developments (and to progress in understanding fundamental aspects of decoherence). But it is important to be careful in specifying what exactly we need from Everett and his followers, and what can be left behind. There is no doubt that the concept of relative states is crucial. Perhaps even more important is the idea that one can apply quantum theory to anything—that there is nothing *ab initio* classical. But the combination of these two ideas does not yet force one to adopt a many-worlds interpretation in which all of the branches are equally real.

Quantum states combine ontic and epistemic attributes. They cannot be 'found out', so they do not exist as classical states did. But once they are known, their existence can be confirmed. This interdependence of existence and information brings to mind two contributions of John Wheeler: his early assessment of relative states interpretation (which he saw as an *extension* of Bohr's ideas) [1957], and also his 'It from Bit' program [1990] (where information was the source of existence).

This complementarity of existence and information was very much in evidence in this paper. Stability in spite of (deliberate or accidental) information transfer led to preferred pointer states, and is the essence of einselection. Entanglement deprives local states of information (which is transferred to correlations) and forces one to describe these local states in probabilistic terms, leading to Born's rule. Robust existence emerges 'It from Many Bits' through quantum Darwinism. The selective proliferation of information makes it immune to measurements, and allows einselected states to be found out indirectly—without endangering their existence.

Acknowledgements

I would like to thank Robin Blume-Kohout, Juan Pablo Paz, David Poulin, Hai-Tao Quan, and Michael Zwolak for stimulating discussions. This research was funded by DoE through an LDRD grant at Los Alamos.

References

Barnum, H. [2003], 'No-signalling-based version of Zurek's derivation of quantum probabilities: A note on "Environment-assisted invariance, entanglement, and probabilities in quantum physics" '. Available online at quant-ph/0312150.

Blume-Kohout, R. and W.H. Zurek [2005], 'A simple example of "quantum Darwinism": redundant information storage in many-spin environments', *Foundations of Physics* **35**, 1857–76.

—— [2006], 'Quantum Darwinism: entanglement, branches, and the emergent classicality of redundantly stored quantum information', *Physical Review* **A73**, 062310.

—— [2008], 'Quantum Darwinism in quantum Brownian motion: the vacuum as a witness', *Physical Review Letters* **101**, 2404C. Available online at arXiv:0704.3615.

Bohr, N. [1928], 'The quantum postulate and the recent development of atomic theory', *Nature* **121**, 580.

Born, M., [1926], 'The quantum mechanics of the impact process', *Zeitschrift der Physik* **37**, 863.

Deutsch, D. [1999], 'Quantum theory of probability and decisions', *Proceedings of the Royal Society of London* **A455**, 3129–37.

DeWitt, B. [1970], 'Quantum mechanics and reality', *Physics Today* **23**, No.9, 30–xx. Reprinted in *The Many-Worlds Interpretation of Quantum Mechanics*, B. DeWitt and N. Graham (eds), Princeton University Press (1973).

—— [1971], 'The many-universes interpretation of quantum mechanics', in *Foundations of Quantum Mechanics*, B. d'Espagnat (ed.), Academic Press, New York. Reprinted in *The Many-Worlds Interpretation of Quantum Mechanics*, B. DeWitt and N. Graham (eds), Princeton University Press (1973).

DeWitt, B. and N. Graham (eds) [1973], *The Many-Worlds Interpretation of Quantum Mechanics*, Princeton University Press, Princeton.

Dieks, D. [1982], 'Communication by EPR devices', *Physics Letters* **92A**, 271–2.

Dirac, P.A.M. [1958], *Quantum Mechanics*, Clarendon Press, Oxford.

Everett, H. [1957], ' "Relative state" formulation of quantum mechanics', *Reviews of Modern Physics* **29**, 454–62. Reprinted in *The Many-Worlds Interpretation of Quantum Mechanics*, B. DeWitt and N. Graham (eds), Princeton University Press (1973).

—— [1973], 'The theory of the universal wavefunction', in *The Many-Worlds Interpretation of Quantum Mechanics*, B. DeWitt and N. Graham (eds), Princeton University Press.

Forrester, A. [2007], 'Decision theory and information propagation in quantum physics', *Studies in the History and Philosophy of Modern Physics*, **38**, 815–31. Available online at arXiv:quant-ph/0604133.

Gell-Mann, M. and J.B. Hartle [1998], 'Strong decoherence', in *Proceedings of the 4th Drexel Conference on Quantum Non-Integrability: The Quantum-Classical Correspondence*, D.-H. Feng and B.-L. Hu, (eds), International Press of Boston, Hong Kong.

Geroch, R. [1984], 'The Everett Interpretation', *Noûs* **18**, 617–33.

Gleason, A.M. [1957], 'Measures on the closed subspaces of a Hilbert space', *Journal of Mathematics and Mechanics*, **6**, 885.

Gnedenko, B.V. [1968], *The Theory of Probability*, Chelsea, New York.

Halliwell, J.J. [1999], 'Somewhere in the universe: where is the information stored when histories decohere?', *Physical Review* **D60**, 105031.

Herbut, F. [2007], 'Quantum probability law from "environment-assisted invariance" in terms of pure-state twin unitaries', *Journal of Physics* **A40**, 5949–71.

Joos, E.H., D. Zeh, C. Kiefer, D. Giulini, J. Kupsch, and I.O. Stamatescu [2003], *Decoherence and the Appearance of a Classical World in Quantum Theory*, Springer, Berlin.

Kent, A. [1990],' Against many-worlds interpretations', *International Journal of Modern Physics* **A5**, 1745–62.

Laplace, P.S, [1820], *A Philosophical Essay on Probabilities*, trans. by F.W. Truscott and F.L. Emory, Dover, New York (1951).

Nielsen, M.A. and I.L. Chuang, [2000], *Quantum Computation and Quantum Information*, Cambridge University Press, Cambridge.

Ollivier, H. et al., [2004], 'Objective properties from subjective quantum states: environment as a witness', *Physical Review Letters* **93**, 220401.

—— [2005], 'Environment as a witness: selective proliferation of information and emergence of objectivity in a quantum universe', *Physical Review* **A72**, 42113.

Paz, J.-P. and A.J. Roncaglia [2009], 'Redundancy of classical and quantum correlations during decoherence', *Physical Review* **A80**, 42111.

Paz, J.-P. and W.H. Zurek [2001], 'Environment-induced decoherence and the transition from quantum to classical', in *Coherent Atomic Matter Waves, Les Houches Lectures*, R. Kaiser, C. Westbrook, and F. David, (eds), pp. 533–614, Springer, Berlin.

Riedel, C.J. and W.H. Zurek [2010], 'Quantum Darwinism of scattering radiation', available online at arXiv:1001.3419.

Saunders, S. [2004], 'Derivation of the Born rule from operational assumptions', *Proceedings of the Royal Society of London* **A460**, 1–18.

Schlosshauer, M. [2004], 'Decoherence, the measurement problem, and interpretations of quantum mechanics', *Review of Modern Physics* **76**, 1267.

—— [2007], *Decoherence and the Quantum to Classical Transition*, Springer, Berlin.

Schlosshauer, M. and A. Fine [2005], 'On Zurek's derivation of the Born rule', *Foundations of Physics* **35**, 197–213.

Squires, E.J. [1990], 'On an alleged "proof" of the quantum probability law', *Physics Letters* **A145**, 67–8.

Stein, H. [1984], 'The Everett interpretation: many worlds or none?', *Noûs* **18**, 635–52.

Von Neumann, J. [1932], *Mathematische Grundlagen Der Quantenmechanik*, translated by R.T. Beyer as *Mathematical Foundations of Quantum Mechanics*, Princeton University Press, 1955.

Wallace, D. [2003], 'Everettian rationality: defending Deutsch's approach to probability in the Everett interpretation', *Studies in the History and Philosophy of Modern Physics* **34**, 415–39. Available online at arXiv.org/abs/quant-ph/0303050.

Wheeler, J.A. [1957], 'Assessment of Everett's relative state formulation of quantum theory', *Reviews of Modern Physics* **29**, 463–65, reprinted in *The Many-Universes Interpretation of Quantum Mechanics*, B. DeWitt and N. Graham (eds), pp.155–66, Princeton University Press.

—— [1990], 'Information, physics, quantum: The search for links', p.3, in *Complexity, Entropy, and the Physics of Information*, W. H. Zurek (ed.), Addison Wesley, Redwood City.

Wootters, W.K. and W.H. Zurek [1982], 'A single quantum cannot be cloned', *Nature* **299**, 802–3.

Yuen, H.P. [1986], 'Amplification of quantum states and noiseless photon amplifiers', *Physics Letters* **113A**, 405.

Zurek W.H. [1981], 'Pointer basis of quantum apparatus: Into what mixture does the wavepacket collapse?', *Physical Review* **D24**, 1516–25.

—— [1982], 'Environment-induced superselection rules', *Physical Review* **D26**, 1862–80.

—— [1991], 'Decoherence and the transition from quantum to classical', *Physics Today* **44**, No. 10, 36–44. Available (in updated form) online at quant-ph/0306072.

—— [1998], 'Decoherence, einselection, and the existential interpretation (the Rough Guide)', *Philosophical Transactions of the Royal Society of London* **A356**, 1793–820. Available online as quant-ph/9805065.

—— [2000], 'Einselection and decoherence from an information theory perspective', *Annalen der Physik* (Leipzig) **9**, 855–64.

—— [2003a], 'Decoherence, einselection, and the quantum origins of the classical', *Review of Modern Physics* **75**, 715–75.

—— [2003b], 'Environment-assisted invariance, entanglement, and probabilities in quantum physics', *Physical Review Letters* **90**, 120404.

—— [2005], 'Probabilities from entanglement, Born's rule $p_k = |\psi_k|^2$ from envariance', *Physical Review* **A71**, 052105.

Zurek W.H. [2007a], 'Relative states and the environment: einselection, envariance, quantum Darwinism, and the existential interpretation'. Available online at
quant-ph/0707.2832.
—— [2007b], 'Quantum origin of quantum jumps: Breaking of unitary symmetry
induced by information transfer in the transition from quantum to classical', *Physical
Review* **A76**, 052110.
—— [2009], 'Quantun Darwinism', *Nature Physics*, vol. 5, pp. 181–8.
Zwolak, M., H.T. Quan, W.H. Zurek [2009], 'Quantum Darwinism in a hazy environment', *Physics Review Letters* **103**, 110402.

14

Two Dogmas About Quantum Mechanics

Jeffrey Bub and Itamar Pitowsky

ABSTRACT

We argue that the intractable part of the measurement problem—the 'big' measurement problem—is a pseudo-problem that depends for its legitimacy on the acceptance of two dogmas. The first dogma is John Bell's assertion that measurement should never be introduced as a primitive process in a fundamental mechanical theory like classical or quantum mechanics, but should always be open to a complete analysis, in principle, of how the individual outcomes come about dynamically. The second dogma is the view that the quantum state has an ontological significance analogous to the significance of the classical state as the 'truthmaker' for propositions about the occurrence and non-occurrence of events, i.e., that the quantum state is a representation of physical reality. We show how both dogmas can be rejected in a realist information-theoretic interpretation of quantum mechanics as an alternative to the Everett interpretation. The Everettian, too, regards the 'big' measurement problem as a pseudo-problem, because the Everettian rejects the assumption that measurements have definite outcomes, in the sense that one particular outcome, as opposed to other possible outcomes, actually occurs in a quantum measurement process. By contrast with the Everettians, we accept that measurements have definite outcomes. By contrast with the Bohmians and the GRW 'collapse' theorists who add structure to the theory and propose dynamical solutions to the 'big' measurement problem, we take the problem to arise from the failure to see the significance of Hilbert space as a new *kinematic* framework for the physics of an indeterministic universe, in the sense that Hilbert space imposes kinematic (i.e., pre-dynamic) objective probabilistic constraints on correlations between events.

1 OXFORD EVERETT

The salient difference between classical and quantum mechanics is the non-commutativity of the operators representing the physical magnitudes ('observables') of a quantum mechanical system—or, equivalently, the transition from a classical event space, represented by the Boolean algebra of (Borel) subsets of a phase space, to a non-Boolean quantum event space represented by the projective geometry of closed subspaces of a Hilbert space, which form an infinite collection of intertwined Boolean algebras, each Boolean algebra corresponding to a resolution of the identity: a partition of the Hilbert space representing a family of mutually exclusive and collectively exhaustive events.

Probabilities in quantum mechanics are, as von Neumann [2001 p.245] put it, 'uniquely given from the start' as a non-classical relation between events represented by the angles between the one-dimensional subspaces representing atomic (elementary) events in the projective geometry of subspaces of Hilbert space. If e and f are atomic events, the 'transition probability' (Born probability) between the events is:

$$\text{prob}(e, f) = |\langle e|f \rangle|^2 = |\langle f|e \rangle|^2 = \cos^2 \theta_{ef} \qquad (1)$$

The transition probability can be expressed as:

$$\text{prob}_e(f) = \text{Tr}(P_e P_f) \qquad (2)$$

where P_e and P_f are the projection operators onto the one-dimensional subspaces representing the events e and f, respectively. Uniqueness is shown by Gleason's theorem [1957]:[1] in a Hilbert space $\mathcal{H}$ of dimension greater than two, if $\sum_i \text{prob}(f_i) = 1$ for the atomic events f_i in each Boolean algebra generated by a partition of the Hilbert space into orthogonal one-dimensional subspaces, then the probabilities of events f represented by subspaces of $\mathcal{H}$ are uniquely represented as:

$$\text{prob}_\rho(f) = \text{TR}(\rho P_f) \qquad (3)$$

where P_f is the projection operator onto the subspace representing the event f and ρ is a density operator representing a pure state ($\rho = P_e$, for some atomic event e) or a mixed state ($\rho = \sum_i w_i P_{e_i}$).

It is assumed that the assignment of probabilities satisfies a condition that Barnum et al. [2000] call 'the noncontextuality of probability,' that the probability assigned to an event f depends only on f and is independent of the Boolean algebra to which the event belongs. Note that if 'f in context 1' and 'f in context 2' represented two distinct events, we could not represent the structure of quantum events as the projective geometry of subspaces of a Hilbert space: we would have to enlarge the structure.

---

[1] For von Neumann, uniqueness is a consequence of invariance under the unitary symmetries of the projective lattice representing events.

The question is: what do these 'transition probabilities' or 'transition weights' mean? The probabilities are probabilities of—*what*? Evidently, $|\langle e|f\rangle|^2$ does not represent the probability of a spontaneous transition from an event e to the event f. The textbook answer is that $|\langle e|f\rangle|^2$ represents the probability, for a system in the state $|e\rangle$ in which the event e has probability 1, of finding the event f in a measurement of an observable of the system, where the set of possible outcomes of the measurement generates a Boolean algebra, representing a partition of the Hilbert space containing the event f (but note, not the event e).

The textbook answer by itself, without adding anything more to the story of how these events are supposed to come about in a measurement process, is adequate only if we are content with an instrumentalist interpretation of the theory. Why? The structure of the quantum event space determines the kinematic part of quantum theory. This includes the association of Hermitian operators with observables, the Born probabilities, the von Neumann–Lüders conditionalization rule, and the unitarity constraint on the dynamics, which is related to the event structure via a theorem of Wigner (Uhlhorn [1963]). The transition from the state $|e\rangle$, in which the event e has probability 1, to the state $|f\rangle$, in which the event f has probability 1, with probability $|\langle e|f\rangle|^2$ in a measurement process is a non-unitary stochastic transition that is not described by the unitary dynamics. Since the probability of the event e was 1 before the measurement and is now, in the state $|f\rangle$ after the occurrence of the measurement outcome f, less than 1, there is a loss of information on measurement or—as Bohr put it—an 'irreducible and uncontrollable' measurement disturbance. Without a dynamical explanation of this measurement disturbance, or an analysis of what is involved in a quantum measurement process that addresses the issue (including, possibly, rejecting the 'eigenvalue-eigenstate rule'—the association of the outcome event f with the state $|f\rangle$—as in Bohm's theory or modal interpretations), the theory qualifies as an algorithm for predicting the probabilities of measurement outcomes, but cannot be regarded as providing a realist account, in principle, of how events come about in a measurement process.

This is the measurement problem. Proposed solutions to the problem, such as Bohm's 'hidden variable' theory or the GRW 'dynamical collapse' theory, add structure to the theory: particle trajectories in the case of Bohm's theory or a non-unitary stochastic dynamics for the 'primitive ontology' of the GRW theory: mass density in the GRWm version, or 'flashes' in the GRWf version (see Allori et al. [2007]). The Everett interpretation purports to solve the problem without adding any new structural elements to quantum mechanics.

The central claims of the Everett interpretation in the 'Oxford' version developed by Deutsch, Saunders, Wallace, Greaves, and others can be outlined as follows:

Ontology At the most fundamental level, what there *is* is described by the quantum state of the universe—so whatever is true or false is determined by the quantum state as the 'truthmaker' for propositions about the occurrence and non-occurrence of events.

Branching A family of effectively non-interfering or decoherent histories of coarse-grained events associated with relatively stable systems at the macrolevel emerges through the dynamical process of decoherence, as a consequence of the Hamiltonian that characterizes the dynamical evolution of the universal quantum state. With respect to the coarse-grained basis selected by decoherence, the quantum state decomposes into a linear superposition that can be interpreted as describing an emergent branching structure of non-interfering quasiclassical histories or 'worlds', identified with the familiar classical macroworlds of our experience, weighted by the Born probabilities. The alternative outcomes of a quantum measurement process are associated with different branches in the decomposition of the quantum state with respect to the decoherence basis. There is no fact of the matter as to the number of branches: the history space is a quasiclassical probability space that is inherently vaguely defined (appropriately so, given the vague specification of macroconfigurations). The coarse-graining of the event space can be refined or coarsened to a certain extent without compromising effective decoherence, and the decoherence basis can be unitarily transformed (e.g. rotated) over a certain range of transformations without compromising decoherence.

Uncertainty/caring There is a sense in which a rational agent on a branch, faced with subsequent branching, can be uncertain about the future (i.e., uncertain about 'which branch the agent will subsequently occupy'). Such an agent can have rational credences (degrees of belief that satisfy the axioms of probability theory) about the outcomes of quantum measurements, even though all outcomes occur on different branches. Alternatively, even without uncertainty, an agent faced with multiple futures will care about what happens on a branch, and so will have a 'caring measure' for decision-making that quantifies the extent of caring for different branches and satisfies the axioms of probability theory.

Probability To achieve a realist interpretation of quantum mechanics that solves the measurement problem, it suffices to postulate that an agent's credence function or caring measure conforms to the objective quantum mechanical weights of the different branches. In fact, it is possible to prove that this must be so, given standard rationality constraints on an agent's preferences, and a measurement neutrality assumption: that a rational agent is indifferent between two quantum wagers that agree on the quantum state, the observable measured, and the payoff function on the outcomes, i.e., the agent is indifferent between alternative measurement procedures; alternatively, the result follows from a related equivalence assumption: that a rational agent assigns equal credences to events that are assigned equal quantum weights. These additional assumptions can be justified as rationality constraints, but only on the Everett interpretation, in which all possible measurement outcomes occur, relative to different branches.

The Everettian aims to show that standard quantum mechanics can be understood as a complete theory in a realist sense—that the measurement problem does not reduce the theory to an instrument for the probabilistic prediction of measurement outcomes. The basic problem for the Everettian is to 'save the appearances', given the radical difference between our experience of a stable macroworld and the ontological assumption. The dynamics of decoherence yields an emergent weighted branching structure of quasiclassical histories at the macrolevel. So what has to be explained is how uncertainty or caring makes sense when all alternatives occur relative to different branches, and how the quantum weights—which are a feature of the quantum state, i.e., the ontology—are associated with the credence function or caring measure of rational agents. The measurement problem is the problem of explaining the apparently 'irreducible and uncontrollable disturbance' in a quantum measurement process, the 'collapse' of the wavefunction described by von Neumann's projection postulate. The Everettian's solution is to show how appearances can be saved by denying that there is any such disturbance, on the basis that no definite outcome is selected in a measurement—all outcomes are selected relative to different branches, according to the quantum theory. The appearance of disturbance on a single branch is a reflection of how the quantum weights are distributed in the emergent process of branching, and if we either assume or prove that our credence function or caring measure should conform to these weights, then we have an explanation for the appearance of disturbance in a realist interpretation of quantum mechanics as a complete dynamical theory.

Of course, everything hinges on whether the different components of the intepretation can be established satisfactorily, and there is now an extensive literature challenging and defending these claims, especially **uncertainty/caring** and **probability**. Here we simply list these components[2] and note that the claim is that the Everett interpretation solves the measurement problem on the basis of (i) the weighted branching structure of quasiclassical histories that emerges through the dynamical process of decoherence, (ii) an argument that rational agents can be uncertain or care differently about different futures in a branching universe, and (iii) the proposal that the credence function or caring measure of rational agents should conform to the weights of the branches. For the Everettian, the icing on the cake is that the interpretation yields a derivation of Lewis's principal principle: the identification of an objective feature of the world—the quantum weights—with the credence function or caring measure of rational agents, and hence the interpretation of the quantum weights as objective chances. But the cake itself, so to speak, is independent of this additional feature. (See Wallace [2006].)

In a previous publication (Pitowsky [2007]), one of us characterized debates about the foundations of quantum mechanics in terms of two assumptions or

---

[2] For a critique of **probability** by one of us, see Hemmo and Pitowsky [2007].

dogmas, and distinguished two measurement problems: a 'big' measurement problem and a 'small' measurement problem. The first dogma is Bell's assertion (defended in [1990]) that measurement should never be introduced as a primitive process in a fundamental mechanical theory like classical or quantum mechanics, but should always be open to a complete analysis, in principle, of how the individual outcomes come about dynamically. The second dogma is the view that the quantum state has an ontological significance analogous to the ontological significance of the classical state as the 'truthmaker' for propositions about the occurrence and non-occurrence of events, i.e., that the quantum state is a representation of physical reality. The 'big' measurement problem is the problem of explaining how measurements can have definite outcomes, given the unitary dynamics of the theory: it is the problem of explaining *how individual measurement outcomes come about dynamically.* The 'small' measurement problem is the problem of accounting for our familiar experience of a classical or Boolean macroworld, given the non-Boolean character of the underlying quantum event space: it is the problem of explaining the *dynamical emergence of an effectively classical probability space of macroscopic measurement outcomes* in a quantum measurement process.

The 'big' measurement problem depends for its legitimacy on the acceptance of the two dogmas. We argue below that both dogmas should be rejected, and that the 'big' measurement problem is a pseudo-problem. In a sense, the Everettian, too, regards the 'big' measurement problem as a pseudo-problem, because the Everettian rejects the assumption that measurements have definite outcomes, in the sense that one particular outcome, as opposed to other possible outcomes, actually occurs in a quantum measurement process. By contrast with the Everettians, we accept that measurements have definite outcomes. By contrast with the Bohmians and the GRW 'collapse' theorists who add structure to the theory and propose dynamical solutions to the "big" measurement problem, we take the problem to arise from the failure to see the significance of Hilbert space as a new *kinematic* framework for the physics of an indeterministic universe, in the sense that Hilbert space imposes kinematic (i.e. pre-dynamic) objective probabilistic constraints on correlations between events. By 'pre-dynamic' here, we refer to generic features of quantum systems, independent of the details of the dynamics (see Janssen [2007] for a similar kinematic-dynamic distinction in the context of special relativity). The 'small' measurement problem is resolved by considering the dynamics of the measurement process and the role of decoherence in the emergence of an effectively classical probability space of macroevents to which the Born probabilities refer (alternatively, by considering certain combinatorial features of the probabilistic structure: see Pitowsky [2007, section 4.3]).

In the following section, we list the essential features of the proposed information-theoretic interpretation, somewhat more extensively than our brief sketch of the Everett interpretation. Further discussion follows in a subsequent Commentary.

2 AN INFORMATION-THEORETIC INTERPRETATION OF QUANTUM MECHANICS

The elements of the information-theoretic interpretation we propose[3] can be set out as follows:

'No Cloning' The empirical discovery underlying the transition from classical to quantum mechanics is the discovery that chance set-ups behave differently than we thought they did. More precisely: there are information sources that cannot be broadcast—there is no universal cloning machine capable of copying the outputs of an arbitrary information source.

Kinematics Hilbert space as a projective geometry (i.e. the subspace structure of Hilbert space) represents a non-Boolean event space, in which there are built-in, structural probabilistic constraints on correlations between events (associated with the angles between events)—just as in special relativity the geometry of Minkowski spacetime represents spatio-temporal constraints on events. Certain principles characterizing physical processes motivate the choice of Hilbert space as the representation space for the correlational structure of events, just as Einstein's principle of special relativity and the light postulate motivate the choice of Minkowski spacetime as the representation space for the spatio-temporal structure of events. In the case of quantum mechanics, these principles are information-theoretic and include a 'no signaling' principle and a 'no cloning' principle. The structure of Hilbert space imposes kinematic (i.e. pre-dynamic) objective probabilistic constraints on events to which a quantum dynamics of matter and fields is required to conform, through its symmetries, just as the structure of Minkowski spacetime imposes kinematic constraints on events to which a relativistic dynamics is required to conform. In this sense *Hilbert space provides the kinematic framework for the physics of an indeterministic universe*, just as Minkowski spacetime provides the kinematic framework for the physics of a non-Newtonian, relativistic universe. There is no deeper explanation for the quantum phenomena of interference and entanglement than that provided by the structure of Hilbert space, just as there is no deeper explanation for the relativistic phenomena of Lorentz contraction and time dilation than that provided by the structure of Minkowski spacetime.

Dynamics The unitary quantum dynamics evolves the whole structure of events with probabilistic correlations in Hilbert space (in the Heisenberg picture), not the evolution from one configuration of the universe to another, i.e., not the evolution from one actual co-occurrence of events to a subsequent actual co-occurrence of events. This means that there can be a real change in the correlations between events at the microlevel without a change in the

---

[3] For related views, see Demopoulos [2009], Pitowsky [2003, 2007].

occurrence of events at the macrolevel (as in the evolution of a quantum system through the unitary gates of a quantum computer, prior to the final measurement).

Probability By Gleason's theorem, there is a unique assignment of credences conforming to the structural probabilistic constraints (the objective chances) of Hilbert space (see Pitowsky [2003]). These credences are encoded in the quantum state. So the quantum state is a credence function.

Information loss The salient principle marking the transition from classical to non-classical theories of information is the 'no cloning' principle: there is no universal cloning machine capable of copying the outputs of an arbitrary information source.[4] This principle entails a loss of information in a measurement process—an 'irreducible and uncontrollable disturbance'—*irrespective of how the measurement process is implemented dynamically*. The loss of information is to be understood, ultimately, as a kinematic effect of the non-classical quantum event space, just as Lorentz contraction is, ultimately, a kinematic effect in special relativity.

Completeness Conditionalizing on a measurement outcome leads to a non-classical updating of the credence function represented by the quantum state via the von Neumann–Lüders rule, which expresses the information loss on measurement. This updating is consistent with a dynamical account of the correlations between micro- and macroevents in a quantum measurement process. The Hamiltonians characterizing the interactions between microsystems and macrosystems, and the interactions between macrosystems and their environment, are such that certain relatively stable structures of events associated with the familiar macrosystems of our experience emerge at the macrolevel, forming an effectively classical probability space. This amounts to a consistency proof that, say, a Stern–Gerlach spin-measuring device or a bubble chamber behaves dynamically according to the kinematic constraints represented by the projective geometry of Hilbert space, as these constraints manifest themselves at the macrolevel. Such a consistency proof demonstrates the completeness of quantum mechanics. Given the 'no cloning' principle underlying the kinematics of Hilbert space, there is no further story to be told about how individual measurement outcomes come about dynamically (assuming we don't add structure to the theory, such as Bohmian trajectories or dynamical 'collapses'). Similarly, the dynamical explanation of relativistic phenomena like Lorentz contraction in terms of forces, insofar as the forces are required to be Lorentz covariant, amounts to a consistency proof. There is no further story to be told about Lorentz contraction, once it is shown how to provide a dynamical account consistent with the kinematic constraints of Minkowski geometry (assuming we don't add structure to the theory, such as the ether).

---

[4] More precisely, there is no universal broadcasting machine. See below.

Realism The possibility of a dynamical analysis of measurement processes consistent with the Hilbert space kinematic constraints justifies the information-theoretic interpretation of quantum mechanics as realist and not merely a predictive instrument for updating probabilities on measurement outcomes.

3   COMMENTARY

On the information-theoretic interpretation, the quantum state is a credence function, a bookkeeping device for keeping track of probabilities—the universe's objective chances—not the quantum analog of the dynamically evolving classical state understood as the 'truthmaker' for propositions about the occurrence and non-occurrence of events.

Conditionalization on the occurrence of an event *a*, in the sense of a minimal revision—consistent with the subspace structure of Hilbert space—of the probabilistic information encoded in a quantum state given by a density operator ρ, is given by the von Neumann–Lüders rule:[5]

$$\rho \to \rho_a \equiv \frac{P_a \rho P_a}{\mathrm{Tr}(P_a \rho P_a)} \tag{4}$$

where P_a is the projection operator onto the subspace representing the event *a*. That is, ρ_a is the conditionalized density operator, conditional on the event *a*, and the normalizing factor $\mathrm{Tr}(P_a \rho P_a) = \mathrm{Tr}(\rho P_a)$ is the probability assigned to the event *a* by the state ρ.

If we consider a pair of correlated systems, A and B, then conditionalization on an A-event, for the probabilistic information encoded in the density operator ρ_B representing the probabilities of events at the remote system B, will always be an updating, in the sense of a refinement.

For example, suppose the system A is associated with a three-dimensional Hilbert space $\mathcal{H}_A$ and the system B is associated with a two-dimensional Hilbert space $\mathcal{H}_B$. Suppose the composite system AB is in an entangled state:

$$\begin{aligned} |\psi^{AB}\rangle &= \frac{1}{\sqrt{3}}(|a_1\rangle|b_1\rangle + |a_2\rangle|c\rangle + |a_3\rangle|d\rangle) \\ &= \frac{1}{\sqrt{3}}(|a_1'\rangle|b_2\rangle + |a_2'\rangle|e\rangle + |a_3'\rangle|f\rangle) \end{aligned} \tag{5}$$

where $|a_1\rangle, |a_2\rangle, |a_3\rangle$ and $|a_1'\rangle, |a_2'\rangle, |a_3'\rangle$ are two orthonormal bases in $\mathcal{H}_A$, and $|b_1\rangle, |b_2\rangle$ is an orthonormal basis in $\mathcal{H}_B$. The triple $|b_1\rangle, |c\rangle, |d\rangle$ and the

---

[5] See Bub [1977] for a discussion.

triple $|b_2\rangle$, $|e\rangle$, $|f\rangle$ are non-orthogonal triples of vectors in $\mathcal{H}_B$.[6] The state of B (obtained by tracing over $\mathcal{H}_A$) is the completely mixed state $\rho_B = \frac{1}{2}I_B$:

$$\frac{1}{3}|b_1\rangle\langle b_1| + \frac{1}{3}|c\rangle\langle c| + \frac{1}{3}|d\rangle\langle d| = \frac{1}{3}|b_2\rangle\langle b_2| + \frac{1}{3}|e\rangle\langle e| + \frac{1}{3}|f\rangle\langle f| = \frac{I_B}{2} \quad (6)$$

Conditionalizing on one of the eigenvalues a_1, a_2, a_3 or a'_1, a'_2, a'_3 of an A-observable A or A' via (4), i.e., on the occurrence of an event corresponding to A taking the value a_i or A' taking the value a'_i for some i, changes the density operator ρ_B of the remote system B to one of the states $|b_1\rangle$, $|c\rangle$, $|d\rangle$ or to one of the states $|b_2\rangle$, $|e\rangle$, $|f\rangle$. Since the mixed state $\rho_B = \frac{1}{2}I_B$ can be decomposed as an equal weight mixture of $|b_1\rangle$, $|c\rangle$, $|d\rangle$ and as an equal weight mixture of $|b_2\rangle$, $|e\rangle$, $|f\rangle$, the change in the state of B is an updating, in the sense of a refinement of the information about B encoded in the state $|\psi^{AB}\rangle$, taking into account the new information a_i or a'_i. In fact, the mixed state $\rho_B = \frac{1}{2}I_B$ corresponds to an infinite variety of mixtures of pure states in $\mathcal{H}_B$ (not necessarily equal weight mixtures, of course). The effect at the remote system B of conditionalization on any event at A will always be an updating, in the sense of a refinement, with respect to one of these mixtures.[7] This is the content of the Hughston–Jozsa–Wootters theorem [1993]. It is what Schrödinger called 'remote steering' and is the basis of quantum teleportation, quantum dense coding, and other peculiarities of quantum information, including the impossibility of unconditionally secure bit commitment (see Bub [2006] for a discussion).

The effect of conditionalization at a remote system (the system that is not directly involved in the conditionalizing event) is then consistent with a 'no signaling' principle:

$$\sum_b p(ab|AB) \equiv p(a|AB) = p(a|A) \quad (7)$$

$$\sum_a p(ab|AB) \equiv p(b|AB) = p(b|B) \quad (8)$$

where a represents a value of A, and b represents a value of B. If conditionalization on the value of an A-observable changed the probabilities at a remote system B in a way that could *not* be represented as an updating in the sense of a refinement of the prior information about B expressed in terms of correlations between A-observables and B-observables (as encoded in the entangled state $|\psi^{AB}\rangle$), then conditionalization would allow instantaneous signaling between A and B. The

---

[6] The vectors in each triple are separated by an angle $2\pi/3$. For a precise specification of these vectors, see Bub [2007].

[7] Fuchs makes a similar point in Fuchs [2002].

occurrence of a particular sort of event at A—corresponding to a determinate value for the observable A as opposed to a determinate value for some other observable A'—would produce a detectable change in the B-probabilities, and so Alice at A could signal instantaneously to Bob at B merely by performing an A-measurement and gaining a specific sort of information about A (the value of A or the value of A').

The 'no signaling' principle is a special case of what Barnum et al. [2000] call 'the noncontextuality of probability', which can be expressed as a condition on the probabilities assigned to the eigenvalues of any two commuting observables $[X, Y] = 0$:

$$\sum_y p(xy|XY) \equiv p(x|XY) = p(x|X) \tag{9}$$

$$\sum_x p(xy|XY) \equiv p(y|XY) = p(y|Y) \tag{10}$$

This formulation of the non-contextuality of probability follows from the representation of an observable in terms of its spectral measure.[8] We obtain the 'no signaling' condition if we take $X = A \otimes I$ and $Y = I \otimes B$. Note that 'no signaling' is not specifically a relativistic constraint on superluminal signaling. It is simply a condition imposed on the marginal probabilities of events for separated systems, requiring that the marginal probability of a B-event is independent of the particular set of mutually exclusive and collectively exhaustive events selected at A, and conversely, and this might well be considered partly constitutive of what one means by separated systems.

To preserve the 'no signaling' principle, quantum probabilities must also satisfy a 'no cloning' principle: there can be no universal cloning machine, i.e., it is impossible to construct a cloning machine that will clone the output of an arbitrary information source. More precisely, there can be no universal broadcasting machine—no device that takes a probability distribution over an event space to a new probability distribution over a product space of events, where the marginal probability distributions over each factor space are the same as the original distribution. We will continue to use the term 'cloning' rather than 'broadcasting' because it is more intuitive and more familiar, but note that we have in mind copying the *outputs* of an information source, not the information source itself (defined by the probability distribution).

Suppose that a universal cloning machine were possible. Then such a device could copy any state in the orthogonal triple $|b_1\rangle$, $|c\rangle$, $|d\rangle$ as well as any state in

---

[8] Barnum et al. formulate non-contextuality as the requirement that the probability assigned to an event e depends only on e and is independent of the other events in each mutually exclusive and collectively exhaustive set of events $\{e_i\}$ containing e, i.e., that the probability of an event is independent of the Boolean subalgebra to which the event belongs.

the orthogonal triple $|b_2\rangle$, $|e\rangle$, $|f\rangle$. It would then be possible for Alice at A to signal to Bob at B. If Alice obtains the information given by an eigenvalue a_i of A or a'_i of A', and Bob inputs the system B into the cloning device n times, he will obtain one of the states $|b_1\rangle^{\otimes n}$, $|c\rangle^{\otimes n}$, $|d\rangle^{\otimes n}$ or one of the states $|b_2\rangle^{\otimes n}$, $|e\rangle^{\otimes n}$, $|f\rangle^{\otimes n}$, depending on the nature of Alice's information. Since these states tend to mutual orthogonality in $\otimes^n \mathcal{H}_B$ as $n \to \infty$, they are distinguishable in the limit. So, even for finite n, Bob would in principle be able to obtain some information instantaneously about a remote event.

More fundamentally, the existence of a universal cloning machine is inconsistent with the interpretation of Hilbert space as providing the kinematic framework for an indeterministic physics, in which probabilities (objective chances) are 'uniquely given from the start' by the geometry of Hilbert space. For such a device would be able to distinguish the equivalent mixtures of non-orthogonal states represented by the same density operator $\rho_B = \frac{1}{2} I_B$. If a quantum state prepared as an equal weight mixture of the states $|b_1\rangle$, $|c\rangle$, $|d\rangle$ could be distinguished from a state prepared as an equal weight mixture of the states $|b_2\rangle$, $|e\rangle$, $|f\rangle$, the representation of quantum states by density operators would be incomplete.

Now consider the effect of conditionalization on the state of A. The state of AB can be expressed as the bi-orthogonal (Schmidt) decomposition:

$$|\psi^{AB}\rangle = \frac{1}{\sqrt{2}}(|g\rangle|b_1\rangle + |h\rangle|b_2\rangle) \tag{11}$$

where

$$|g\rangle = \frac{2|a_1\rangle - |a_2\rangle - |a_3\rangle}{\sqrt{6}} \tag{12}$$

$$|h\rangle = \frac{|a_2\rangle - |a_3\rangle}{\sqrt{2}} \tag{13}$$

The density operator ρ_A, obtained by tracing $|\psi^{AB}\rangle$ over B, is:

$$\rho_A = \frac{1}{2}|g\rangle\langle g| + \frac{1}{2}|h\rangle\langle h| \tag{14}$$

which has support on a two-dimensional subspace in the three-dimensional Hilbert space $\mathcal{H}_A$: the plane spanned by $|g\rangle$ and $|h\rangle$ (in fact, $\rho_A = \frac{1}{2}P_A$, where P_A is the projection operator onto the plane). Conditionalizing on a value of A or A' yields a state that has a component outside this plane. So the state change on conditionalization cannot be interpreted as an updating of information in the sense of a refinement, i.e., as the selection of a particular alternative among a set of mutually exclusive and collectively exhaustive alternatives represented by the state ρ_A.

This is the notorious 'irreducible and uncontrollable disturbance' arising in the registration of new information about the occurrence of an event that underlies the measurement problem: the loss of some of the information encoded in the original state (in the above example, the probability of the A-event represented by the projection operator onto the two-dimensional subspace P_A is no longer 1, after the registration of the new information about the observable A or A'). If the registration of new information is the outcome of a measurement then, since the state change on measurement will have to be stochastic and non-unitary, it cannot be described by the deterministic dynamics of the theory, which must be unitary (for closed systems) for consistency with the Hilbert space representation of probabilities. A solution to the problem is generally understood to require amending the theory in such a way that the loss of information can be accounted for dynamically, and the quantum probabilities can be reconstructed dynamically as measurement probabilities. Then the quantum probabilities are not 'uniquely given from the start' as kinematic features of an appropriately represented event structure, i.e., they do not arise kinematically but are derived dynamically, as artifacts of the measurement process or of decoherence. Even on the Everett interpretation, where Hilbert space is interpreted as the representation space for a new sort of ontological entity, represented by the quantum state, and no definite outcome out of a range of alternative outcomes is selected in a quantum measurement process (so no explanation is required for such an event), probabilities arise as a feature of the branching structure that emerges in the dynamical process of decoherence.

From the perspective of the information-theoretic interpretation, the 'disturbance' involved in conditionalization is a kinematic phenomenon associated with the non-Boolean quantum event space. If there were no information loss in the conditionalization of quantum probabilities, then cloning would be possible, and equivalent mixtures associated with the same density operator would be distinguishable, in which case Hilbert space would not be an appropriate representation space for quantum events and their probabilistic correlations.[9] In the Appendix, we show that this follows directly from the 'no cloning' principle for a large class of theories. We prove that in this class of theories the 'no cloning' principle demarcates the boundary between classical theories and theories in which measurement involves an 'irreducible and uncontrollable disturbance'. It seems plausible, therefore, that this principle should play a central role in a derivation of the Hilbert space structure from information theory.

It is instructive here to recall Einstein's distinction between 'principle' theories, like the special theory of relativity, formulated in terms of the relativity principle and the light postulate (empirical regularities raised to the level of postulates),

---

[9] For the Everettian, there is the appearance of measurement disturbance on each branch, or rather, on 'most' branches, because there will always be some branches on which it appears that there is no measurement disturbance—and on these branches it will appear that cloning is possible.

and 'constructive' theories, like Lorentz's theory, formulated in terms of a rich ontology of objects like particles, fields, and the ether. Einstein compared thermodynamics as a principle theory ('no perpetual motion machines of the first and and second kind') to the kinetic theory of gases as a constructive theory (where the mechanical and thermal behavior of a gas is reduced to the motion of molecules, modeled as little billiard balls). He proposed special relativity as a kinematic replacement for Lorentz's dynamical interpretation of what we now refer to as Lorentz covariance, which he saw as unsatisfactory, not as a rival theory of matter and radiation. One might say that what eventually replaced Lorentz's theory was relativistic quantum theory. From this perspective, Minkowski spacetime is the constructive theory corresponding to Einstein's principle theory formulation of special relativity: it is a component of the kinematic part of the constructive theory of the constitution of matter provided by relativistic quantum theory. (See Balashov and Janssen [2003] pp. 331–2] for an account along these lines.)

In an article entitled 'How to Teach Special Relativity', John Bell [1987] considers the following puzzle: Three identical spaceships, A, B, and C, are at rest relative to one other, drifting freely far from other matter without rotation, with A equidistant from B and C. The spaceships B and C are connected by a fragile thread, which is just long enough to span the distance between them. On reception of a signal from A, the spaceships B and C start their engines and accelerate gently. Since B and C are assumed to be identical, with identical acceleration programs, they will have the same velocity and so remain separated by the same distance relative to A. When B and C reach a certain velocity, the thread breaks. The question is: why does the thread break? Note that the thread would not break under similar assumptions in a Newtonian universe.

The relativistic kinematic explanation goes along the following lines:

Let $F1$ be the inertial frame in which the spaceships A, B, C are *initially* at rest (and A remains at rest). In $F1$, the distance between B and C, as the spaceships begin to move and continue moving, remains the same as the initial resting distance. But the moving thread undergoes a Lorentz contraction in the direction of its motion in $F1$. The explanation, in $F1$, of why the thread breaks is just this: the thread breaks because it is contracting, and this contraction is resisted by the thread being tied to B and C, which maintain a distance apart greater than the contraction requires. The thread will break when B and C reach a sufficiently high velocity in $F1$ and the prevention of the Lorentz contraction produces sufficient stress to break the thread.

Let $F2$ be the inertial frame in which B and C are *finally* at rest again, after their engines have been shut off. From the perspective of $F2$, there is a different explanation for the thread breaking. In $F2$, the two spaceships B and C are decelerating, and eventually come to rest. However, they are not decelerating at the same rate (they would be if B and C were connected by a rigid rod). It is this difference in deceleration that is responsible for the stress in the thread, which eventually causes the thread to break.

To clarify further, one might consider two additional spaceships, E and F, identical to B and C, with identical acceleration programs, initially at rest in $F1$ (before B and C start their engines), with E adjacent to B, and F adjacent to C. Suppose E and F are connected rigidly, so that EF behaves like a rigid rod with the two spaceships as endpoints, initially at rest in $F1$. Suppose also that EF starts accelerating at the same time as B and C in $F1$, and that the rod connecting E and F is strong enough to remain rigid under the acceleration. Bell's characterization of the set-up requires that, in $F1$, the distance between B and C, as the spaceships begin to move and continue moving, remains the same as the initial resting distance. So, in $F1$, this distance will become greater than the distance between E and F, once the spaceships start moving, since EF will suffer a Lorentz contraction in the direction of its motion. In the explanation in frame $F1$, the thread breaks because it is contracting by as much as EF contracts. In the explanation in frame $F2$, B and C are not decelerating at the same rate—rather, the endpoints of EF are decelerating at the same rate—and this difference in deceleration, relative to the deceleration of EF, is responsible for the stress in the thread, which eventually causes it to break.

The explanations are frame-dependent, insofar as they involve elements that are frame-dependent notions in special relativity. However, the increasing stress in the thread that causes it to break, and the fact that the thread breaks when the stress exceeds the tensile strength of the thread, are frame-independent features common to all explanations. What Bell pointed out was that one ought to be able to provide an explanation for the thread breaking in terms of an explicit calculation of the forces involved, and the tensile strength of the thread. He suggests that such a dynamical explanation is a deeper or at least more informative explanation than the kinematic explanation. Harvey Brown's book *Physical Relativity* [2005] develops this theme.

In Bell's spaceship example, the dynamical explanation for the thread breaking in terms of forces, insofar as the forces are Lorentz covariant, shows the possibility of a dynamics consistent with the kinematics of special relativity. The only factor relevant to the thread breaking is the Lorentz contraction, a feature of the geometry of Minkowski spacetime which is quite independent of the material constitution of the thread and the nature of the specific interactions involved. Given Einstein's two principles, there is no deeper explanation for the thread breaking than the kinematic explanation provided by the structure of Minkowski spacetime.[10] The demonstration that a dynamical explanation yields the same result as the kinematic explanation sketched above amounts to a consistency proof that a relativistic dynamics—a dynamics that conforms to the structure of Minkowski spacetime—is possible.

---

[10] Harvey Brown's book [2005] presents an extended argument for the contrary view.

Jeffrey Bub and Itamar Pitowsky

If we take special relativity as a template for the analysis of quantum conditionalization and the associated measurement problem,[11] the information-theoretic view of quantum probabilities as 'uniquely given from the start' by the structure of Hilbert space as a kinematic framework for an indeterministic physics is the proposal to interpret Hilbert space as a constructive theory of information-theoretic structure or probabilistic structure, part of the kinematics of a full constructive theory of the constitution of matter, where the corresponding principle theory includes information-theoretic constraints such as 'no signaling' and 'no cloning.'[12] Lorentz contraction is a physically real phenomenon explained relativistically as a kinematic effect of motion in a non-Newtonian spacetime structure. Analogously, the change arising in quantum conditionalization that involves a real loss of information is explained quantum mechanically as a kinematic effect of *any* process of gaining information of the relevant sort in the non-Boolean probability structure of Hilbert space (irrespective of the dynamical processes involved in the measurement process). Given 'no cloning' as a fundamental principle, there can be no deeper explanation for the information loss on conditionalization than that provided by the structure of Hilbert space as a probability theory or information theory. The definite occurrence of a particular event is constrained by the kinematic probabilistic correlations encoded in the structure of Hilbert space, and only by these correlations—it is otherwise 'free'.

The Born weights are probabilities in a purely formal sense unless they are related to experience by some explicitly formulated principle. The cash value of the 'transition probability' $|\langle e|f\rangle|^2$ is that $|\langle e|f\rangle|^2$ represents the probability, in the state $|e\rangle$, of finding the outcome corresponding to the state $|f\rangle$ in a measurement of an observable of which $|f\rangle$ is an eigenstate. But if quantum mechanics is more than an instrument for predicting the probabilities of measurement outcomes, it must be possible, in principle, to locate structures that represent macroscopic measuring instruments and recording devices in Hilbert space, where the dynamical behavior of such structures is consistent with the kinematic information-theoretic (probabilistic) principles encoded in the structure of Hilbert space.

In special relativity one has a consistency proof that a dynamical account of relativistic phenomena in terms of forces, like the breaking of the thread in Bell's spaceship example, is consistent with the kinematic account in terms of the structure of Minkowski spacetime. An analogous consistency proof for quantum mechanics would be a dynamical explanation for the effective emergence of classicality, i.e., Booleanity, at the macrolevel, because it is with respect to the

---

[11] See Brown and Timpson [2007] for a contrary view.

[12] While the 'no cloning' principle demarcates classical from non-classical theories, we require some further principle or principles to recover Hilbert space and exclude 'superquantum' theories for which the correlation of entangled states violates the Tsirelson bound for quantum states, while conforming to the 'no signaling' constraint. See Barnum et al. [2006, 2007].

Boolean algebra of the macroworld that the Born weights of quantum mechanics have empirical cash value.

In classical mechanics, taking a Laplacian view, one can consider the phase space of the entire universe, in principle. The classical state, represented by a point in phase space that evolves dynamically, defines a two-valued homomorphism on the Boolean algebra of (Borel) subsets of phase space, distinguishing events that occur at a particular time from events that don't occur. In this sense, the classical state is the 'truthmaker' for propositions about the occurrence or non-occurrence of events, for all possible events.

Similarly, in quantum mechanics one can consider the Hilbert space of the entire universe, in principle. This is a space of possible events, with a certain kinematic structure of probabilistic correlations between events, represented by the subspace structure or projective geometry of the space (different from the classical correlational structure represented by the subset structure of phase space). On the usual view, the quantum analogue of the classical state is a pure state represented by a ray or one-dimensional subspace in Hilbert space. There is, of course, no two-valued homomorphism on the non-Boolean algebra of subspaces of Hilbert space, but a pure state can be taken as distinguishing events that occur at a particular time (events represented by subspaces containing the state, and assigned probability 1 by the state) from events that don't occur (events represented by subspaces orthogonal to the state, and assigned probability 0 by the state). This leaves all remaining events represented by subspaces that neither contain the state nor are orthogonal to the state (i.e. events assigned a probability p by the state, where $0 < p < 1$) in limbo: neither occurring nor not occurring. The measurement problem then arises as the problem of accounting for the fact that an event that neither occurs not does not occur when the system is in a given quantum state can somehow occur when the system undergoes a measurement interaction with a macroscopic measurement device—giving measurement a very special status in the theory. Once the pure state is taken as the analog of the classical state in this sense, the only way out of this problem, without adding structure to the theory, is the Everettian maneuver.

On the information-theoretic interpretation, the quantum state is a derived entity, a credence function that assigns probabilities to events in alternative Boolean algebras associated with the outcomes of alternative measurement outcomes. The measurement outcomes are macroevents in a particular Boolean algebra, and the macroevents that actually occur, corresponding to a particular measurement outcome, define a two-valued homomorphism on this Boolean algebra. What has to be shown is how this occurrence of events in a particular Boolean algebra is consistent with the quantum dynamics.

It is a contingent feature of the dynamics of our particular quantum universe that events represented by subspaces of Hilbert space have a tensor product structure that reflects the division of the universe into microsystems (e.g. atomic nuclei), macrosystems (e.g. macroscopic measurement devices constructed from pieces of metal and other hardware), and the environment (e.g. air molecules, electromagnetic radiation). The Hamiltonians characterizing the interactions between microsystems and macrosystems, and the interactions between macrosystems and their environment, are such that a certain relative structural stability emerges at the macrolevel as the tensor-product structure of events in Hilbert space evolves under the unitary dynamics. Symbolically, an event represented by a one-dimensional projection operator like $P_{|\psi\rangle} = |\psi\rangle \langle \psi|$, where

$$|\psi\rangle = |s\rangle |M\rangle |\varepsilon\rangle \tag{15}$$

and s, M, ε represent respectively microsystem, macrosystem, and environment, evolves under the dynamics to $P_{|\psi(t)\rangle}$, where

$$|\psi(t)\rangle = \sum_k c_k |s_k\rangle |M_k\rangle |\varepsilon_k(t)\rangle, \tag{16}$$

and

$$|\varepsilon_k(t)\rangle = \sum_\nu \gamma_\nu e^{-ig_{k\nu}t} |e_\nu\rangle \tag{17}$$

if the interaction Hamiltonian $H_{M\varepsilon}$ between a macrosystem and the environment takes the form

$$H_{M\varepsilon} = \sum_{k\gamma} g_{k\nu} |M_k\rangle \langle M_k| \otimes |e_\nu\rangle \langle e_\nu| \tag{18}$$

with the $|M_k\rangle$ and the $|e_k\rangle$ orthogonal. That is, the 'pointer' observable $\sum_k m_k |M_k\rangle \langle M_k|$ commutes with $H_{M\varepsilon}$ and so is a constant of the motion induced by the Hamiltonian $H_{M\varepsilon}$.

Here $P_{|M_k\rangle}$ can be taken as representing, in principle, a configuration of the entire macroworld, and $P_{|s_k\rangle}$ a configuration of all the microevents correlated with macroevents. The dynamics preserves the correlation represented by the superposition $\sum_k c_k |s_k\rangle |M_k\rangle |\varepsilon_k(t)\rangle$ between microevents, macroevents, and the environment for the macroevents $P_{|M_k\rangle}$, even for non-orthogonal $|s_k\rangle$ and $|\varepsilon_k\rangle$, but not for macroevents $P_{|M'_l\rangle}$ where the $|M'_l\rangle$ are linear superpositions of the $|M_k\rangle$. Since the tri-decomposition $\sum_k c_k |s_k\rangle |M_k\rangle |\varepsilon_k(t)\rangle$ is unique (unlike the bi-orthogonal Schmidt decomposition; see Elby and Bub [1994]), a correlation of the form $|s\rangle |M\rangle |\varepsilon\rangle$ evolves to a linear superposition in which the macroevents $P_{|M'_l\rangle}$ become correlated with entangled system-environment events represented by subspaces (rays) spanned by linear superpositions of the form $\sum_k c_k d_{lk} |s_k\rangle |\varepsilon_k(t)\rangle$. (See Zurek [2005] pp.052105–14].)

It is characteristic of the dynamics that correlations represented by (16) evolve to similar correlations (similar in the sense of preserving the micro–macroenvironment division), and the macroevents represented by $P_{|M_k\rangle}$, at a sufficient level of coarse-graining, can be associated with structures at the macrolevel—the familiar macro-objects of our experience—that remain relatively stable under the dynamical evolution. So a Boolean algebra $\mathcal{B}_\mathcal{M}$ of macroevents $P_{|M_k\rangle}$ correlated with microevents $P_{|s_k\rangle}$ in (16) is emergent in the dynamics. Note that the emergent Boolean algebra is not the same Boolean algebra from moment to moment, because the correlation between microevents and macroevents changes under the dynamical evolution induced by the micro–macro interaction (e.g. corresponding to different measurement interactions). What remains relatively stable under the dynamical evolution are the *macrosystems* associated with macroevents in correlations of the form (16), even under a certain vagueness in the coarse-graining associated with these macroevents: macrosystems like grains of sand, tables and chairs, macroscopic measurement devices, cats and people, galaxies, etc.

It is further characteristic of the dynamics that the environmental events represented by $P_{|\varepsilon_k(t)\rangle}$ very rapidly approach orthogonality, i.e., the "decoherence factor"

$$\zeta_{kk'} = \langle \varepsilon_k | \varepsilon_{k'} \rangle = \sum_\nu |\gamma_\nu|^2 e^{i(g_{k'\nu} - g_{k\nu})t} \tag{19}$$

becomes negligibly small almost instantaneously. When the environmental events $P_{|\varepsilon_k(t)\rangle}$ correlated with the macroevents $P_{|M_k\rangle}$ are effectively orthogonal, the reduced density operator is effectively diagonal in the 'pointer' basis $|M_k\rangle$ and there is effectively no interference between elements of the emergent Boolean algebra $\mathcal{B}_\mathcal{M}$. That is, the conditional probabilities of events associated with a subsequent emergent Boolean algebra (a subsequent measurement) are additive on $\mathcal{B}_\mathcal{M}$. (See Zurek [2005] pp.052105–14], [2003].)

The Born probabilities are probabilities of events in the emergent Boolean algebra, i.e., the Born probabilities are probabilities of 'pointer' positions, the coarse-grained basis selected by the dynamics. Applying quantum mechanics kinematically, say in assigning probabilities to the possible outcomes of a measurement of some observable of a microsystem, we consider the Hilbert space of the relevant degrees of freedom of the microsystem and treat the measuring instrument as simply selecting a Boolean subalgebra in the non-Boolean event space of the microsystem to which the Born probabilities apply. In principle, we can include the measuring instrument in a dynamical analysis of the measurement process, but such a dynamical analysis—even though complete in terms of the quantum dynamics—does not provide a dynamical explanation of how individual outcomes come about. In such a dynamical analysis, the Born probabilities are probabilities of the occurrence of events in an emergent Boolean algebra. The information loss on conditionalization relative to classical conditionalization is a kinematic feature of the structure of quantum events, not accounted for by the unitary quantum dynamics, which conforms to the kinematic structure. This is analogous to the situation in special relativity, where Lorentz contraction is a kinematic effect of relative motion that is *consistent* with a dynamical account in terms of Lorentz covariant forces, but is not explained in Einstein's theory—by contrast with Lorentz's theory—as a dynamical effect in a Newtonian spacetime structure, in which this sort of contraction does not arise as a purely kinematic effect. That is, the dynamical explanation of Lorentz contraction in special relativity involves forces that are Lorentz covariant—in effect, the dynamics is assumed to have symmetries that respect Lorentz contraction as a kinematic effect of relative motion. In quantum mechanics, the possibility of a dynamical analysis of the measurement process conforming to the kinematic structure of Hilbert space provides a consistency proof that the familiar objects of our macroworld behave dynamically in accordance with the kinematic probabilistic constraints on correlations between events.

A physical theory of an indeterministic universe is primarily a theory of probability (or information). Probabilities are defined over an event structure, which in the quantum case is a family of Boolean algebras forming a particular sort of non-Boolean algebra. On the information-theoretic interpretation, no assumption is made about the fundamental 'stuff' of the universe. So, one might ask, what do tigers supervene on?[13] In the case of Bohm's theory or the GRW theory, the answer is relatively straightforward: tigers supervene on particle configurations in the case of Bohm's theory, and on mass density or 'flashes' in the case of the GRW theory, depending on whether one adopts the GRWm version or the GRWf version. In the Everett interpretation, tigers supervene on features of the quantum state, which describes an ontological entity. In the case of the information-theoretic interpretation, the 'supervenience base' is provided by the dynamical analysis: tigers supervene on events defining a two-valued homomorphism in the emergent Boolean algebra.

It might be supposed that this involves a contradiction. What is contradictory is to suppose that a correlational event represented by $P_{|\psi(t)\rangle}$ actually occurs, where $|\psi(t)\rangle$ is a linear superposition $\sum_k c_k |s_k\rangle |M_k\rangle |\varepsilon_k(t)\rangle$, as well as an event represented by $P_{|s_k\rangle |M_k\rangle |\varepsilon_k(t)\rangle}$ for some specific k. We do not suppose this. On the information-theoretic interpretation we propose, there is a kinematic structure of possible correlations (but no particular atomic correlational event is selected as the 'state' in a sense analogous to the pure classical state), and a particular dynamics that preserves certain sorts of correlations, i.e., correlational events of the sort represented by $P_{|\psi(t)\rangle}$ with $|\psi(t)\rangle = \sum_k c_k |s_k\rangle |M_k\rangle |\varepsilon_k(t)\rangle$ evolve to correlational events of the same form. What can be identified as emergent in this dynamics is an

---

[13] We thank Allen Stairs for raising the realism question in this form.

effectively classical probability space: a Boolean algebra with atomic correlational events of the sort represented by orthogonal one-dimensional subspaces $P_{|s_k\rangle|M_k\rangle}$, where the probabilities are generated by the reduced density operator obtained by tracing over the environment, when the correlated environmental events are effectively orthogonal.

The dynamics does not describe the (deterministic or stochastic) evolution of the two-valued homomorphism on which tigers supervene to a new two-valued homomorphism (as in the evolution of a classical state). Rather, the dynamics leads to the relative stability of certain event structures at the macrolevel associated with the familiar macrosystems of our experience, and to an emergent effectively classical probability space whose atomic events are correlations between events associated with these macrosystems and microevents.

It is part of the information-theoretic interpretation that events defining a two-valued homomorphism on the Boolean algebra of this classical probability space actually occur with the emergence of the Boolean algebra at the macrolevel. This selection of actually occurring events is only in conflict with the quantum pure state if the quantum pure state is assumed to have an ontological significance analogous to the ontological significance of the classical pure state as the 'truthmaker' for propositions about the occurrence and non-occurrence of events, and if the quantum pure state evolves unitarily—in particular, if it is assumed that the quantum pure state partitions all events into events that actually occur, events that do not occur, and events that neither occur nor do not occur, as on the usual interpretation. We argued that this assumption is one of the dogmas about quantum mechanics that should be rejected. Rather, we take the quantum state, pure or mixed, to represent a credence function: the credence function of a rational agent (an information-gathering entity 'in' the emergent Boolean algebra) who is updating probabilities on the basis of events that occur in the emergent Boolean algebra.

4 CONCLUDING REMARKS

We have argued that the 'big' measurement problem is like the problem for Newtonian physics raised by relativistic effects such as length contraction and time dilation, and that the solution to both problems involves the recognition of a fundamental change in the underlying *kinematics* of our physics, represented by the transition from a Newtonian spacetime to Minkowski spacetime in the case of special relativity, and from the set-theoretic structure of classical phase space to the subspace structure of Hilbert space in the case of quantum mechanics. So the two assumptions, about the ontological significance of the quantum state and about the dynamical account of how measurement outcomes come about, should be rejected as unwarranted dogmas about quantum mechanics.

The solutions to the 'big' measurement problem provided by Bohm's theory and the GRW theory are dynamical and involve adding structure to quantum mechanics. There is a sense in which adding structure to the theory to solve the measurement problem dynamically—insofar as the problem arises from a failure to recognize the significance of Hilbert space as the kinematic framework for the physics of an indeterministic universe—is like Lorentz's attempt to explain relativistic length contraction dynamically, taking the Newtonian spacetime structure as the underlying kinematics and invoking the ether as an additional structure for the propagation of electromagnetic effects. In this sense, Bohm's theory and the GRW theory are 'Lorentzian' interpretations of quantum mechanics.

The Everettian rejects the legitimacy of the problem by simply denying that measurements have definite outcomes, i.e., by denying that the pure states in a superposition describe alternative event complexes, only one of which actually occurs. This requires showing that a *particular* decomposition of the quantum state corresponding to our experience has a preferred significance, and that weights can be assigned to the individual terms in the preferred superposition that have the significance of probabilities, even though no one definite event complex is selected as actually occurring in contrast to the other event complexes in the superposition. The Everettian's solution to *this* problem is dynamical. So the Everettian, too, sees the underlying problem as dynamical.

We reject the legitimacy of the 'big' measurement problem on the basis of an information-theoretic interpretation of quantum mechanics, in terms of which the problem arises from the failure to see the significance of Hilbert space as the kinematic framework for an indeterministic physics. The dynamical analysis we provide is a solution to a consistency problem: the 'small' measurement problem. The analysis shows that a quantum dynamics, consistent with the kinematics of Hilbert space, suffices to underwrite the emergence of a classical probability space for the familiar macroevents of our experience, with the Born probabilities for macroevents associated with measurement outcomes derived from the quantum state as a credence function. The explanation for such non-classical effects as the loss of information on conditionalization is not provided by the dynamics, but by the kinematics, and given 'no cloning' as a fundamental principle, there can be no deeper explanation. In particular, there is no dynamical explanation for the definite occurrence of a particular measurement outcome, as opposed to other possible measurement outcomes in a quantum measurement process—the occurrence is constrained by the kinematic probabilistic correlations encoded in the projective geometry of Hilbert space, and only by these correlations.

Acknowledgments

Jeffrey Bub acknowledges support from the National Science Foundation under Grant No. 0522398. Itamar Pitowsky's research is supported by the Israel Science Foundation, Grant 744/07.

Appendix: The information loss theorem

We show that *it follows from the 'no cloning' principle* that information cannot be extracted from a non-classical source without changing the source irreversibly. (We prove this theorem for quantum information sources, but note that the proof does not depend on specific features of the Hilbert space formalism.)

We assume:

1 The 'no cloning' principle: there is no universal cloning machine.
2 Every (quantum) state ρ is specified by the probabilities of the measurement outcomes of a finite, informationally complete (or 'fiducial') set of observables.

Assumption (2) holds for a large class of theories, including quantum and classical theories. Note that an informationally complete set is not unique. For example, in the case of a qubit, the probabilities for spin-'up' and spin-'down' in three orthogonal directions suffice to define a direction on the Bloch sphere and hence to determine the state, so the spin observables σ_x, σ_y, σ_z form an informationally complete set. (For a classical system or a classical information source, an informationally complete set is given by of a single observable, with n possible outcomes, for some n.)

Let $\mathcal{F} = \{A, B, C, \ldots\}$ be an informationally complete set of observables represented by a finite set of Hermitian operators on an n-dimensional Hilbert space $\mathcal{H}_n$. A quantum state ρ assigns a probability distribution to every outcome of any measurement of an observable in $\mathcal{F}$. Measuring A yields one of the outcomes a_1, a_2, $\ldots$ with a probability distribution $P_\rho(a_1|A)$, $P_\rho(a_2|A)$, $\ldots$. Similarly, measuring B yields one of the outcomes b_1, b_2, $\ldots$ with a probability distribution $P_\rho(b_1|A)$, $P_\rho(b_2|A)$, $\ldots$, and so on. If $\mathcal{F}$ is informationally complete, the finite set of probabilities completely characterizes ρ as the state on $\mathcal{H}$.

Assuming that all measurement outcomes are independent and ignoring any algebraic relations among elements of $\mathcal{F}$, a classical probability measure on a classical (Kolmogorov) probability space can be constructed from these probabilities:

$$P_\rho(a, b, \ldots | A, B, \ldots) = P_\rho(a|A)P_\rho(b|B) \ldots \tag{20}$$

(cf. the 'trivial' hidden variable construction of Kochen and Specker [1967]). Note that the probability space is finite since $\mathcal{F}$ is finite and $\dim\mathcal{H} < \infty$. (The number of atoms in the probability space is at most $\dim \mathcal{H}^{|\mathcal{F}|}$.) The quantum state ρ can be reconstructed from P_ρ (given as a classical information source, or rationally approximated in the memory of a classical computer).

We now prove:

The information loss theorem. *Assumptions (1) and (2) entail that extracting information from a quantum information source given by a quantum state ρ, sufficient to generate the probabilities of an informationally complete set of observables, is either impossible or necessarily changes the state ρ irreversibly, i.e., there must be information loss in the extraction of such information.*

Proof Step 1: begin with a quantum source in the state ρ and measure $A, B, \ldots$ sufficiently many times to generate the classical probability measure P_ρ, to as good an approximation as required, without destroying ρ. Step 2: from P_ρ construct a copy of ρ.

$$\rho \overset{\text{measure}}{\longrightarrow} P_\rho \overset{\text{prepare}}{\longrightarrow} \rho \qquad (21)$$

This procedure defines a universal cloning machine, which we assume to be impossible. Since Step 2 is possible by assumption (2), the 'no cloning' assumption (1) entails that Step 1 is blocked.

We are left with two options: either there is no way to generate P_ρ from ρ (which is the case in quantum mechanics if we have only one copy of ρ, or too few copies of ρ), or else, if we can generate P_ρ from ρ, assumption (1) entails that the original 'blueprint' ρ must have been changed irreversibly by the process of extracting the information to generate P_ρ (if not, the change in ρ could be reversed dynamically and cloning would be possible):

$$\cancel{\rho} \overset{\text{measure}}{\longrightarrow} P_\rho \overset{\text{prepare}}{\longrightarrow} \rho \qquad (22)$$

$\square$

Since we can prepare multiple copies of the state ρ from P_ρ, one might think that even if the original state is destroyed in generating P_ρ, we still end up with multiple copies of ρ:

$$\cancel{\rho} \overset{\text{measure}}{\longrightarrow} P_\rho \overset{\text{prepare}}{\longrightarrow} \rho$$
$$\overset{\text{prepare}}{\searrow} \rho$$
$$\vdots \qquad (23)$$

But note that to generate P_ρ, we need to begin with multiple copies of ρ, i.e., we need to begin with a state $\rho \otimes \rho \cdots$, so what we really have is:

$$\cancel{\rho} \otimes \cancel{\rho} \cdots \overset{\text{measure}}{\longrightarrow} P_\rho \overset{\text{prepare}}{\longrightarrow} \rho \otimes \rho \cdots \qquad (24)$$

which simply restates (22).

Corollary. *No complete dynamical (i.e., unitary) account of the state transition in a measurement process is possible in quantum mechanics, in general.*

Proof. Any measurement can be part of an informationally complete set, so any measurement must lead to an irreversible (hence non-unitary) change in the quantum state of the measured system.□

We conclude—essentially from the 'no cloning' principle—that there can be no measurement device that functions dynamically in such a way as to identify with certainty the output of an arbitrary quantum information source without altering the source irreversibly or 'uncontrollably', to use Bohr's term—no device can distinguish a given output from every other possible output by undergoing a dynamical (unitary) transformation that results in a state that represents a distinguishable record of the output, without an irreversible transformation of the source.

References

Allori, V., S Goldstein, R. Tumulka, and N. Zanghì [2007], 'On the common structure of Bohmian Mechanics and the Ghirardi–Rimini–Weber Theory', *British Journal for Philosophy of Science* 59, 353–89

Balashov, Y. and M. Janssen [2003], 'Critical notice: Presentism and relativity', *British Journal for the Philosophy of Science* 54, 327–46.

Barnum, H., J. Barrett, M. Leifer, and A. Wilce [2006], 'Cloning and broadcasting in generic probabilistic models'. Available online at http://arXiv.org/pdf/quant-ph/0611295.

—— [2007], 'A generalized no-broadcasting theorem'. Available online at arXiv.org/pdf/quant-ph/0707.0620.

Barnum, H., C.M. Caves, D. Finkelstein, C.A. Fuchs, and R. Schack [2000], 'Quantum probability from decision theory?', *Proceedings of the Royal Society of London* A456, 1175–82.

Bell, J.S. [1987], 'How to teach special relativity', in *Speakable and Unspeakable in Quantum Mechanics*, pp.67–80, Cambridge University Press, Cambridge.

—— [1990], 'Against measurement', *Physics World* 8, 33–40. Reprinted in *Sixty-Two Years of Uncertainty: Historical, Philosophical and Physical Inquiries into the Foundations of Quantum Mechanics*, A. Miller (ed.), pp.17–31, Plenum, New York (1990).

Brown, H. [2005], *Physical relativity: spacetime structure from a dynamical perspective*, Oxford University Press, Oxford.

Brown, H.R. and C.G. Timpson [2007], 'Why special relativity should not be a template for a fundamental reformulation of quantum mechanics', in *Physical Theory and its Interpretation: Essays in Honor of Jeffrey Bub*, W. Demopoulos and I. Pitowsky (eds), Springer, Berlin.

Bub, J. [1977], 'Von Neumann's projection postulate as a probability conditionalization rule in quantum mechanics', *Journal of Philosophical Logic* 6, 381–90.

—— [2006], 'Quantum information and computation', in *Handbook of Philosophy of Physics*, North-Holland. Available online at arXiv.org/pdf/quant-ph/0512125.

—— [2007], 'Quantum probabilities as degrees of belief', *Studies in the History and Philosophy of Modern Physics* 38, 232–54.

Demopoulos, W. [2009], 'Effects and propositions', *Foundations of Physics* **40**, 368–89.

Deutsch, D. [1999] 'Quantum theory of probability and decisions', *Proceedings of the Royal Society of London* **A455**, 3129–37.

Elby, A. and J. Bub. [1994], 'Triorthogonal uniqueness theorem and its relevance to the interpretation of quantum mechanics', *Physical Review* **A 49**, 4213–16.

Fuchs, C.A. [2002], 'Quantum mechanics as quantum information (and only a little more)'. Available online at arXiv.org/pdf/quant-ph/0205039.

Ghirardi, G.C. [2002], 'Collapse theories', in *The Stanford Encyclopedia of Philosophy*, E.N. Zalta (ed.). Available online at http://plato.stanford.edu/entries/qm-collapse/.

Gleason, A.N. [1957], 'Measures on the closed sub-spaces of Hilbert spaces', *Journal of Mathematics and Mechanics* **6**, 885–93.

Goldstein, S. [2001], 'Bohmian mechanics', in *The Stanford Encyclopedia of Philosophy*, E.N. Zalta (ed.). Available online at http://plato.stanford.edu/entries/qm-bohm.

Greaves, H. [2004], 'Understanding Deutsch's probability in a deterministic multiverse', *Studies in History and Philosophy of Modern Physics* **35**, 423–56.

—— [2007a], 'On the Everettian epistemic problem', *Studies in History and Philosophy of Modern Physics* **38**, 120–52.

—— [2007b], 'Probability in the Everett interpretation', *Philosophy Compass* **2**, 109–28.

Hemmo, M. and I. Pitowsky [2007], 'Quantum probability and many worlds', *Studies in History and Philosophy of Modern Physics* **38**, 333–50.

Hughston, L.P., R. Jozsa, and W.K. Wootters [1993], 'A complete classification of quantum ensembles having a given density matrix', *Physics Letters* **A 183**, 14–18.

Janssen, M. [2007], 'Drawing the line between kinematics and dynamics', *Symposium on Time and Relativity*, Institute for Advanced Study, University of Minnesota, October 26, 2007. Available online at http://www.tc.umn.edu/˜janss011/pdf%20files/IAS-symposiumF07.pdf.

Kochen, S. and E.P. Specker [1967], 'On the problem of hidden variables in quantum mechanics', *Journal of Mathematics and Mechanics* **17**, 59–87.

Pitowsky, I. [2003], 'Betting on the outcomes of measurements: a Bayesian theory of quantum probability', *Studies in History and Philosophy of Modern Physics* **34**, 395–414.

—— [2007], 'Quantum mechanics as a theory of probability', in *Physical Theory and its Interpretation: Essays in Honor of Jeffrey Bub*, W. Demopoulos and I. Pitowsky (eds), Springer, Berlin. Available online at arXiv.org/pdf/quant-ph/0510095.

Saunders, S. [1995], 'Time, decoherence and quantum mechanics', *Synthese* **102**, 235–66.

—— [1998], 'Time, quantum mechanics, and probability', *Synthese* **114**, 373–404.

—— [2004], 'Derivation of the born rule from operational assumptions', *Proceedings of the Royal Society of London* **A460**, 1–18.

Uhlhorn, U. [1963], 'Representation of symmetry transformations in quantum mechanics', *Arkiv Fysik* **23**, 307.

von Neumann, J. [2001], 'Unsolved problems in mathematics', in *John von Neumann and the Foundations of Quantum Physics*, M. Redei and M. Stoltzner (eds), pp.231–45, Kluwer Academic Publishers: Dordrecht.

Wallace, D. [2003a], 'Everett and structure', *Studies in History and Philosophy of Modern Physics* **34**, 86–105.

—— [2003b], 'Everettian rationality: defending Deutsch's approach to probability in the Everett interpretation', *Studies in the History and Philosophy of Modern Physics*, **34**, 415–38. Available online at arXiv.org/pdf/quant-ph/0303050.

—— [2006], 'Epistemology quantized: circumstances in which we should come to believe in the Everett interpretation', *British Journal for the Philosophy of Science* **57**, 655–89. Available online at philsci-archive.pitt.edu/archive/00002839.

—— [2007], 'Quantum probability from subjective likelihood: improving on Deutsch's proof of the probability rule', *Studies in the History and Philosophy of Modern Physics* **38**, 311–32. Available online at arXiv.org/abs/quant-ph/0312157.

Wigner, E. [1959], *Group Theory and its Applications to Quantum Mechanics of Atomic Spectra*, Academic Press, New York.

Zurek, W.H. [2003], 'Decoherence, einselection, and the quantum origins of the classical', *Reviews of Modern Physics* **75**, 715.

—— [2005], 'Probabilities from entanglement, Born's rule $p_k = |\psi_k|^2$ from envariance', *Physical Review* A **71**, 052105–1–0525105–29.

Rabid Dogma? Comments on Bub and Pitowsky

Christopher Timpson

1 THE THEORY

Bub and Pitowsky present an interpretation of quantum mechanics which treats the theory as realist and as universal in the traditional sense: it is deemed apt to describe everything in the universe, all in one go. (They call their view an *information-theoretic* interpretation: we shall return to the question of how apposite this label is presently.) In outline their proposal is this: Begin with the Hilbert space for the universe; subspaces of this Hilbert space represent possible events. Heisenberg evolution of the projectors onto these subspaces tells us how the structure of possible events evolves over time. But the theory is realist, descriptive: one needs to say more than simply how what is possible is structured: one also needs to say how things *are*. Accordingly, some one of the possible events (a one-dimensional subspace or projector) is actual at each time. Moreover, dynamical decoherence prefers events corresponding to determinate macroscopic goings on: superpositions of macroscopically distinct objects are not available as options.[1] Thus the (approximately) classical world emerges as a consequence of the actual being selected from amongst the decohering alternatives.

This view has affinities with modal interpretations and with decoherent histories approaches (perhaps stronger affinities than Bub and Pitowsky would wish to aver). Let us borrow a piece of terminology from the former: *value state*. The value state is the item in the theory which tells us how things are at any given time (what, from amongst the possible, is actual). Although Bub and Pitowsky have much to say about the structure of the possible (the Heisenberg-evolving Hilbert space), they have precious little to say about the value state (beyond reminding us that in their theory it is not the *quantum state*—which plays no

---

[1] Why? They don't say. It may be that given the structure of their position there is nothing they can say. Arguably the Bub–Pitowsky proposal faces a problem here which Everett does not; but I shall not pursue this question further here.

role in describing reality, for them). In particular, they refuse to countenance providing any dynamics for the value state. This seems a troubling omission. Surely it is natural to desire in one's fundamental physical theory not only statements about how things are at a given time, but also (precise) rules governing how things will change, even if (perhaps) one can only arrive at stochastic rules. Much as they might like to turn us away from dwelling on it, Bub and Pitowsky cannot do without a value state (forgo it and they forgo their realism) and this state does, moreover, change. When what is actual is in danger of evolving into a macrosuperposition (e.g., when a measurement is being performed) then the value state jumps to one or other of the classically acceptable alternatives.[2] But exactly when and where does this jump take place? With no dynamics for the value state we have no answer to this question in Bub and Pitowsky's setting. Usually, putative interpretations of quantum mechanics which do not address this question are deemed woefully inadequate and incomplete. To say in effect only: 'jumps occur when they are needed to save the appearances'—while possibly true—is to provide no kind of acceptable theory.

How do Bub and Pitowsky defend themselves from this kind of challenge? By going on the offensive. They suggest that the simple-minded desire for laws describing how the actual will change over time (a dynamics for the value state) is atavistic and misguided, once we come to quantum mechanics. What are their arguments in support of this surprising claim?

2 THE ARGUMENT

Bub and Pitowsky bring two sets of considerations to bear. On the one hand they identify, and urge us to reject, what they deem two dogmas about quantum mechanics:

- **D1** (Bell) Measurement should not be introduced as a primitive in a fundamental theory; one should always be able to provide an analysis of how *individual outcomes* come about dynamically;
- **D2** The quantum state has a role to play in representing reality.

On the other hand, they encourage us to recognize the significance of Hilbert space as providing a new kinematical structure for events. It is doing *this* which is supposed to justify rejecting (D1) and (D2). (And with (D1) and (D2) out of the way, the measurement problem in one of its standard forms does not arise.)

---

[2] When a macrosuperposition is not looming then there is no need for the value state to change: it can be left as it is and the (unitary) Heisenberg evolution of the subspaces is adequate to describe change in the actual.

The actual argument seems to go like this. (Main argument): If Hilbert space is duly recognized as a new non-classical kinematic structure for possible events (a 'non-Boolean probability structure') then no explanation for the discontinuous change on measurement is required (nor, maybe, possible). So we should reject (D1). (Supporting argument—by analogy): In fact, trying to provide such an explanation is like trying to provide a dynamical account of length contraction in special relativity à la Lorentz—unnecessary: at most one can get a consistency argument.

Note that neither the main nor the supporting argument indicate why we should be unhappy with (D2). Perhaps the thought is just that the presentation of Bub and Pitowsky's own interpretation *shows* that one need not take the quantum state to represent reality, that one can take it, rather, as a credence function which tracks the objective probability relations provided by the Hilbert space structure. Perhaps; but showing that one *can* reject a propostion is far from establishing that one *should*. But let us leave this difficulty to one side.[3]

Is the main argument convincing? I shall suggest not. Notice first its conditional form. It may well be that those naturally inclined towards (D1) or something similar would also be inclined towards rejecting the advertised construal of Hilbert space; and reasonably so, perhaps. We are not, after all, given any persuasive reason why one *ought to* or *has to* understand Hilbert space in this way. So as it stands, the argument is unpersuasive. But more importantly, even if we grant the antecedent, the conditional itself is false. Bub and Pitowsky are guilty, I think, of an equivocation. What they require is an argument that the demand for a dynamics for the value state is misbegotten, but this is not at all what their considerations deliver.

When unpacked, the main argument turns on the links Bub and Pitowsky emphasize between information loss (disturbance) on measurement, the probabilistic structure built into Hilbert space and the no-broadcasting ('no-cloning') theorem (Barnum et al. [1996]). Given a system prepared in a certain basis, measurement in any other basis will, in the usual Hilbert space setting, *inevitably* involve a disturbance or information loss, in the sense of a loss of predictivity for the former observables. Things couldn't be otherwise given the structure of probabilistic relations encoded in Hilbert space. Moreover this follows from the no-broadcasting theorem, which holds in quantum mechanics and in many other non-classical probabilistic theories. So again, the information loss and disturbance is *inevitable* given no-broadcasting; it couldn't be otherwise when the structure of the possibilities is as it is. Therefore it is redundant to try

---

[3] In fact, I have doubts that Bub and Pitowsky can sustain the claim that in their setting the quantum state has no role to play in representing reality. If one looks closely, the value state will be seen to behave suspiciously like a standardly construed quantum state would.

and *explain* how the information loss comes about: it is redundant to try and describe how the non-unitary jump to the particular measurement outcome takes place.

The equivocation takes place in the very final step. What is crucial in a single universe (i.e. non-Everett) account of quantum mechanics is a satisfactory story about the transition from the possible to the actual. Faced with a choice, the world needs to jump one way in a measurement setting; it is this which we (simple-mindedly?) desire a dynamics for. But notice that Bub and Pitowsky have shifted the question quite *away* from the possible-to-actual transition. Instead they dwell on comparative features of the probability distributions associated with the before and after measurement states of affairs. But it is of no matter *at all* to the question of the appropriateness of a dynamics for the value state that in quantum mechanics (and any other no-broadcasting theory) the post-measurement probabilities will be sharper for some observables and more diffuse for others than the pre-measurement probabilities. Yes, the jumps in the value state will (inevitably) be (information) lossy; why should that relieve us of the duty of providing a dynamics for them? *Of course* the dynamics will take me from kinematically allowed to kinematically allowed state, and in the measurement scenarios adumbrated, that will involve differently peaked probability distributions; but one still needs the dynamics. Without it one does not have a theory. So it seems that Bub and Pitowsky's interpretation of quantum mechanics is, after all, seriously lacking; they still face a form of the 'big' measurement problem; and (D1) is unshaken by their arguments.[4]

With the main argument malfunctioning, there is little left for the supporting argument by analogy to get a grip on. Elsewhere, as the authors note, Harvey Brown and I have registered our doubts about what can be drawn from this kind of analogy regarding principle and constructive theory approaches to quantum mechanics (Brown and Timpson [2006]). Suffice it here to say two things. Drawing the analogy with Lorentz (one might also pertinently add the name of Fitzgerald, perhaps) has a tendency to backfire. Einstein himself came to doubt the appropriateness of the purely principle theory approach in special relativity, referring to the 'sin' of treating rods and clocks as unanalysed rather than as moving atomic configurations subject to dynamical laws; and he always emphasized the greater explanatory power of constructive over principle theory approaches. Lorentz was not wrong to seek a dynamical explanation of length contraction; if the right kind of dynamics did not hold, rods would not contract and clocks would not slow down. The second point is that it is part of the

---

[4] Distinguish two tasks: (i) explaining why the change in measurement is one which will be information lossy; (ii) explaining (providing a dynamics for) how the change takes place. Reflections on the kinematic structure of Hilbert space and the no-broadcasting theorem only engage with (i), while it is (ii) with which Bub and Pitowsky need to be concerned.

argument of Brown [2005] that dynamics and kinematics are not wholly distinct fields of enquiry: they are mutually interdependent.

3 RABID DOGMA?

As Carnap might have said to Quine, one man's dogma is another man's axiom. We have seen that Bub and Pitowsky have not done enough to shake faith in (D1) and (D2) (if faith there be). But are these propositions aptly named dogma? Some remarks.

1. Neither (D1) nor (D2) seem to me to be held on the basis of appeal to authority, nor truly themselves to be a starting point. (So neither dogma nor axiom.) Rather they can be derived or *justified* from more general principles: broadly speaking, realism about fundamental physical theories. The reluctance to countenance measurement as a primitive simply follows from realism and an unassuming physicalism: measuring devices and observers are just complicated physical systems amongst many, with no especially distinguishing features; measurement interactions are just one more physical interaction, on a par with any other; they are only of particular interest to us *qua* epistemic agents; they are not physically special in any significant way. Taking the state to represent (at least some part of) reality is even more directly arrived at: it just follows from a familiar literalism about the physical formalism. Often the *simplest* reading of the formalism is the right one. (Compare a literal reading of Maxwell's equations delivering us the electromagnetic field as an entity.)

2. (D1) seems misattributed to Bell. In Bub and Pitowsky's formulation the (putative) dogma involves essential reference to *individual outcomes* occurring, i.e., not a pluralism of outcomes à la Everett. This plays an important role in their argument; they see Everett also as rejecting (D1). But as I read him (Bell [1987, 1990]) Bell's insistence that measurement should not be a primitive does not involve commitment to single outcomes of measurement, rather it is just the claim that measurement should not be treated as primitive because it is not primitive (see remark 1).[5]

3. It follows that what is justified by realism and (unassuming) physicalism is not quite (D1), but the weaker claim that measurement should not be introduced as a primitive but rather should be apt for a dynamical analysis. To reach (D1) one needs to add the further clause of 'only one world!'. (Many might be happy to add this.) Notice, though, that in a sense, Bub and Pitowsky's

---

[5] His complaints against the Everett interpretation lie elsewhere, in what he sees as the lack of precision in the Everettian story about branching—fundamental physics should not be imprecise. But branching in Everett is not fundamental physics, so this challenge can, ultimately, be discounted. See, e.g., Wallace (this volume, Chapter 1).

own account is consistent with the weaker measurement requirement—the true Bellian requirement: they do give a fully dynamical treatment of measurement interaction (the unitary one, giving rise to decoherence); they only get into trouble because they then in addition insist on only one world and provide no dynamics for the entailed value state.

4 WHAT'S INFORMATION GOT TO DO WITH IT?

It is of considerable current interest to determine exactly what lessons information-theoretic principles have to teach us about the interpretation of quantum mechanics (cf. Fuchs [2003], Timpson [2008]). Is Bub and Pitowsky's interpretation aptly called an information-theoretic one? I can't see that it is. This label is motivated on two counts: (i) that their interpretation concerns itself much with probabilities, as information theories do; (ii) that their interpretation is the *constructive* theory to associate with a no-broadcasting, no-signalling (hence information-based) *principle* theory. (i), I think, is quickly recognized to be a spurious association; (ii) does not seem to hold much water either. What determines that Bub and Pitowsky's approach (neglecting for the moment its internal difficulties) should be *the* constructive theory to go along with the information-theoretic principles they identify? *Any* of the standard (realist) interpretations would do as well (Everett, GRW, Bohm, . . .). No-broadcasting (etc.) is entailed by the quantum formalism so will hold appropriately in any of the standard interpretations (cf. Timpson [2004 section 9.2.3]). Each of these interpretations is suitably consistent with the information-theoretic principles; could each be an information-theoretic interpretation? That would seem to render the label rather vacuous.

References

Barnum, H., C.M. Caves, C.A. Fuchs, R. Jozsa, and B. Schumacher [1996], 'Noncommuting mixed states cannot be broadcast', *Phys. Rev. Lett*, 76, 2818.
Bell, J.S. [1987], *Speakable and Unspeakable in Quantum Mechanics*, Cambridge University Press, second edition 2004.
——[1990], 'Against "measurement" ', *Physics World*, August, 33–40. Repr. in Bell [1987].
Brown, H.R. [2005], *Physical Relativity: Spacetime Structure from a Dynamical Perspective*, Oxford University Press, Oxford.
Brown, H.R. and C.G. Timpson [2006], 'Why special relativity should not be a template for a fundamental reformulation of quantum mechanics', in W. Demopoulos and I. Pitowsky (eds), *Physical Theory and Its Interpretation: Essays in Honor of Jeffrey Bub*, Springer. Available online at arXiv:quant-ph/0601182.

Fuchs, C.A. [2003], *Notes on a Paulian Idea: Foundational, Historical, Anecdotal and Forward Looking Thoughts on the Quantum (Selected Correspondence)*, Växjö University Press, Oxford. Available online at arXiv:quant-ph/0105039.

Timpson, C.G. [2004], 'Quantum Information Theory and the Foundations of Quantum Mechanics', PhD thesis, University of Oxford, Oxford. Available online at arXiv:quant-ph/0412063.

—— [2010]. *Quantum Information Theory and the Foundations of Quantum Mechanics*. Oxford University Press, Oxford. Forthcoming.

15

The Principal Principle and Probability in the Many-Worlds Interpretation

Rüdiger Schack

In the decision-theoretic approach to the many-worlds interpretation due to Deutsch and Wallace, probabilities are taken to be Bayesian degrees of belief. Their connection to the quantum-mechanical wavefunction, derived from a set of decision-theoretic axioms, can be regarded as a quantum version of Lewis's principal principle. In this chapter we show that applying the principal principle in quantum mechanics is problematical because it is impossible to give precise criteria for what constitutes a repeated quantum-mechanical trial. This difficulty is resolved in a full quantum Bayesian theory where quantum states, as well as probabilities, are interpreted as expressing an agent's degrees of belief, rather than corresponding to objective properties of physical systems. A similar resolution exists in the many-worlds interpretation; as we will show below, however, this resolution leads to a new question for the latter.

Probability is often regarded as a problem for the many-worlds interpretation: if all branches of the splitting wavefunction are equally real, what sense does it make to say that the branches have different probabilities? To address this problem, Deutsch [1999] and Wallace [2003, 2006, 2007] have proposed a decision-theoretic approach to the many-worlds interpretation. The starting point of Deutsch and Wallace is the decision-theoretic approach to probability pioneered by Savage [1972], where probabilities acquire meaning through the preferences of a rational agent.

In Savage's theory, all probabilities are expressions of an agent's Bayesian degrees of belief. Bayesian probabilities are a function of the agent in that they always depend on the agent's prior; in this sense degrees of belief are subjective. The agent updates his probabilities using observations and experimental data. Bayesian theory is conceptually straightforward. It provides simple and compelling accounts of the analysis of repeated trials in science (Savage

[1972]), statistical mechanics and thermodynamics (Jaynes [1957a] and [1957b]), general statistical practice (Bernardo and Smith et al. [1994]), and quantum mechanics (Caves et al. [2002]).

In the many-worlds interpretation, the quantum state or wavefunction is regarded as a property of the world, independent of any agent's belief. The problem addressed by Deutsch and Wallace is how to connect the agent's decision-theoretic preferences, i.e., his probabilities, to the quantum state. Wallace's solution to this problem takes the form of a set of decision-theoretic axioms in the spirit of Savage's seminal work. From his axioms Wallace can derive that an agent who happens to know the state $|\psi\rangle$ of a quantum system must assign his decision-theoretic probabilities according to the Born rule, i.e., the usual quantum probability rule. This way of connecting the wavefunction (a property of the world) with probability (an agent's degree of belief) can be viewed (Wallace [2006]) as a quantum realization of Lewis's principal principle (Lewis [1986]).

The principal principle distinguishes between *chance* and probability. Chance is supposed to be objective; its numerical value is a property of the world, independent of any agent's belief. Chance is often introduced in the form of a physical parameter. Probabilities, on the other hand, are an agent's Bayesian degrees of belief. In this paper we will always make this distinction. The term 'probability' will thus always refer to a Bayesian degree of belief.

If E is an event, and $0 \leq q \leq 1$, the statement 'the chance of E is q' is a proposition. Denote this proposition by C_q. The principal principle links chance and probability by requiring that an agent's conditional probability of E, given C_q, must be q, irrespective of any observed data. More precisely, if D refers to some other compatible event, e.g., frequency data, then the principal principle states that the Bayesian probability must satisfy

$$\Pr(E|C_q \& D) = q. \tag{1}$$

The statement C_q is a proposition in the sense that it is true or false, i.e., it has a definite truth value. The statement C_q differs from an ordinary proposition, however, in that its truth value cannot be determined in general. The only exception is the trivial case when q is 0 or 1. For instance, the observation that E has occurred makes the statement C_0, 'the chance of E is 0', false.

To establish the truth of a non-trivial statement about chance at least with high confidence, one usually invokes repeated trials. In the simplest case, a trial determines whether the event E occurs or not. If this trial is repeated n times, it gives rise to n instances of the event E, which we may denote $E_1, \ldots, E_n$. The events $E_1, \ldots, E_n$ are assumed to be independent, and for each trial ($j = 1, \ldots, n$), the chance that E_j occurs is assumed to be equal to the chance of E. In order to say that E_j occurs in the j-th trial, we will also use the terminology 'success in the j-th trial'. Now suppose an agent whose prior for the chance of E is $f_{\text{prior}}(q)$ performs the n repeated trials. Using the principal principle, the agent's probability for observing E_j in the j-th trial is

$$
\Pr(E_j) = \int_0^1 \Pr(E_j|C_q)\Pr(C_q)\,dq
$$
$$
= \int_0^1 q\,f_{\text{prior}}(q)\,dq, \tag{2}
$$

and the agent's probability that exactly k of the events $E_1,\ldots,E_n$ occur is

$$
\Pr(k\text{ successes}) = \int_0^1 \Pr(k\text{ successes}\mid C_q)\Pr(C_q)\,dq
$$
$$
= \int_0^1 \binom{n}{k} q^k (1-q)^{n-k} f_{\text{prior}}(q)\,dq. \tag{3}
$$

Conversely, if the agent observes exactly k of the events $E_1,\ldots,E_n$, his updated (posterior) probability for the chance of E is

$$
f_{\text{posterior}}(q) = \Pr(C_q \mid k\text{ successes}), \tag{4}
$$

where the right-hand side can be found by applying Bayes's rule. Typically, when n is large, the posterior will be strongly peaked near the 'true' chance of E.

The above procedure is sound, but it relies on the concept of repeated trials. In particular, these trials must be performed under conditions that guarantee that the chance of a success is the same for every trial. In a classical deterministic theory, any attempt to define such a repeated chance set-up must lead to an infinite regress (see, e.g., Jaynes [2003]). This can be seen easily as follows. If one specifies the initial conditions of a trial precisely, the outcome of the trial will be determined, thus leading to the trivial case where the chance is 0 or 1. If, on the other hand, the initial conditions are not given precisely, the chance of a success depends on the probability (or chance) distribution of the initial conditions. In order to guarantee that each trial has the same chance of a success, one has to guarantee certain properties of the distribution of the initial conditions. This only pushes the problem up one level and therefore leads to a regress.

Bayesian theory provides a simple solution to this problem (Savage [1972]). The problem exists only because the repeated-trial scenario has been phrased in terms of objective chance. A fully Bayesian analysis of repeated trials, without referring to chance, proceeds as follows. The full space of outcomes of our n

trials has size 2^n. The Bayesian prior is a probability distribution, representing an agent's degrees of belief, on this multi-trial space. A repeated trial corresponds to a prior that is *exchangeable* (de Finetti, 1990), i.e., which is a member of a sequence of distributions for m trials, $P^{(m)}$, such that for all integers m, $P^{(m)}$ is symmetric under permutations of the m trials and is the marginal of $P^{(m+1)}$. Exchangeability is not a property of the world, rather it depends on the agent's prior judgement. Of course, this judgement will often be informed by relevant facts such as observations and measurement outcomes.

According to de Finetti's representation theorem (de Finetti [1990]), any exchangeable prior can be expressed as a convex sum of i.i.d. distributions (i.e. a convex sum of products of independent and identical distributions). In the simple binary case described above, exchangeability therefore implies that the probability of k successes in n trials can be written as

$$\Pr(k \text{ successes}) = \int_0^1 \binom{n}{k} q^k (1-q)^{n-k} f(q)\, dq, \qquad (5)$$

where $f(q)$ has the mathematical form of a probability distribution. This is exactly the same equation as in our discussion of the principal principle above, except that $f(q)$ is no longer interpreted as the probability of C_q (the statement that the chance of a success is q). To analyse our repeated trials, we can now use the well-developed apparatus of Bayesian statistics (Bernardo and Smith [1994]). The concept of chance is therefore simply not needed in a classical deterministic theory.

At first, the situation in quantum mechanics seems to be fundamentally different. In the quantum formalism, we can represent an event and its negation by a pair of positive operators E and $1 - E$ (these form a POVM) defined on a Hilbert space $\mathcal{H}$. If the state of a quantum system is ρ (a density operator), the probability of a 'success', i.e., the probability of obtaining the measurement outcome E, is given by the Born rule,

$$\Pr(\text{success}) = \operatorname{tr}(\rho E). \qquad (6)$$

In the case of a projective measurement, $E = |\phi\rangle\langle\phi|$, and a pure state, $\rho = |\psi\rangle\langle\psi|$, the rule takes the better-known form $\Pr(\text{success}) = |\langle\phi|\psi\rangle|^2$.

In Wallace's version of the many-worlds interpretation, the probability on the left-hand side of Eq. (6) is an agent's degree of belief, whereas the state ρ on the right-hand side is a property of the world, which is independent of the agent. The state ρ could be pure or mixed; it is mixed if the system in question is entangled with other parts of the world. The Born rule (6) is thus precisely of the form of the principal principle, with the state ρ playing the role of chance. In this interpretation, the Born rule says that an agent who happens to know that the state of the system is ρ must assign the probability $\mathrm{tr}(\rho E)$ to the outcome E. We can write

$$\mathrm{Pr}(E|C_\rho) = \mathrm{tr}(\rho E), \tag{7}$$

where C_ρ is the statement 'the state of the system is ρ'. As in the general case, C_ρ is regarded as a proposition with a definite truth value, but, again as in the general case, the truth of C_ρ cannot in general be established by any measurement.

To establish the truth of a statement about the quantum state of a system, at least with high confidence, one usually invokes repeated trials, often in the form known as quantum tomography (see, e.g., Vogel and Risken [1989]). Suppose an agent whose prior for the state of the system is $f_{\mathrm{prior}}(\rho)$ performs n repeated trials. Using the principal principle, which is now the Born rule, the agent's probability for observing E in the j-th trial (event E_j) is

$$\mathrm{Pr}(E_j) = \int \mathrm{Pr}(E|C_\rho)\,\mathrm{Pr}(C_\rho)\,d\rho$$

$$= \int \mathrm{tr}(\rho E)f_{\mathrm{prior}}(\rho)\,d\rho, \tag{8}$$

and the agent's probability that exactly k of the events $E_1, \ldots, E_n$ occur is

$$\mathrm{Pr}(k \text{ successes}) = \int \mathrm{Pr}(k \text{ successes} \mid C_\rho)\,\mathrm{Pr}(C_\rho)\,d\rho$$

$$= \binom{n}{k} \int tr[\rho E]^k tr[\rho(1-E)]^{n-k} f_{\mathrm{prior}}(\rho)\,d\rho$$

$$= \binom{n}{k} \int \mathrm{tr}[\rho^{\otimes n} E^{\otimes k} \otimes (1-E)^{\otimes(n-k)}] f_{\mathrm{prior}}(\rho)\,d\rho, \tag{9}$$

where, e.g., $\rho^{\otimes n} = \rho \otimes \cdots \otimes \rho$ is the n-fold tensor product of ρ with itself.

Conversely, when the agent observes k successes in n trials, he can use a quantum Bayes rule (Schack et al. [2001]) to update his probability distribution for the state of the system, i.e., to find

$$f_{\mathrm{posterior}}(\rho) = \mathrm{Pr}(C_\rho \mid k \text{ successes}). \tag{10}$$

In quantum tomography, one normally uses measurements with more than two possible outcomes, for instance informationally complete POVMs (Caves et al. [2007]). In any case, the idea is to use frequency data to obtain information about the truth of the proposition C_ρ.

The procedure for quantum tomography described above depends on the ability to conduct the trials in a way that guarantees that the system state is the same each time. In other words, the above procedure presupposes that one can specify a procedure to prepare the same, possibly unknown, quantum state n times, i.e., to prepare the total n-trial system in the state

$$\rho^{(n)} = \int \rho^{\otimes n} f_{\text{prior}}(\rho) \, d\rho = \int \rho \otimes \ldots \otimes \rho \, f_{\text{prior}}(\rho) \, d\rho. \qquad (11)$$

The operator $\rho^{(n)}$ is defined on the n-fold tensor product Hilbert space $\mathcal{H}^{\otimes n}$.

The existence of such a preparation procedure is a central assumption of the Copenhagen interpretation (or at least some versions of it—see the discussion and further references in Caves et al. [2007]. A preparation device in the spirit of the Copenhagen interpretation can be given a complete description in classical terms. It can therefore be used to prepare the same state many times.

In most modern interpretations of quantum mechanics, including many worlds, however, it is assumed that the quantum laws apply to all physical systems, and therefore also to the preparation device. This has the consequence that the attempt to specify a procedure preparing the same state multiple times leads to the same regress as in the classical case. Any deterministic preparation procedure can be realized by first entangling the system and the apparatus, then performing a measurement on the apparatus, and finally changing the system state conditional on the measurement outcome. This can be realized in turn by a purely unitary interaction between apparatus and system (see, e.g., Caves et al. [2007]). The final quantum state of the system depends therefore on the initial (or prior) quantum state of the apparatus. To guarantee that the same system state is prepared in each trial, one has to make assumptions about the quantum state of the preparation device, which means that one has simply moved the problem up one level. This leads to a regress as claimed.

Notice the close similarity between the classical and quantum cases. In order to specify a (classical) probabilistic situation, one has to specify a prior. This is a problem for an analysis in terms of chance, but not for a fully Bayesian analysis, where the prior represents an agent's decision-theoretic preferences. To specify the conditions for a quantum experiment, one has to specify a prior state for the apparatus. This is a problem if the state is regarded as an objective property of system and apparatus, independent of any agent's belief. The problem is resolved in the full quantum Bayesian approach (Caves et al. [2002, 2007], Fuchs [2002]), where all quantum states, including the prior, represent an agent's decision-theoretic preferences.

Within the many-worlds interpretation, the regress is not an infinite regress, however. The regress stops with the wavefunction of the universe, from which, ultimately, the state of any subsystem can be derived. This implies that, in the many-worlds interpretation, the following solution exists for the repeated-trial problem. The n copies of the system that make up our n trials have an objective quantum state which in principle can be derived from the wavefunction of the universe. This state is an (unknown) density operator defined on $\mathcal{H}^{\otimes n}$. In Wallace's approach, only the unknown quantum state $\rho^{(n)}$ is a property of the world, whereas probabilities represent an agent's decision-theoretic preferences or Bayesian degrees of belief. The Bayesian analysis of the n trials now starts from a prior, which is a probability density on the space of all density operators on $\mathcal{H}^{\otimes n}$, i.e. a probability density on a d^{2n}-dimensional space, where d is the dimension of $\mathcal{H}$. This density represents the agent's degrees of belief about the unknown state of the n copies of the system.

In this approach, the purpose of the repeated trial is for the agent to update his probabilities. The agent's updated (posterior) probabilities depend on data or measurement outcomes, but they also depend on the agent's prior. If the agent judges that the quantum state is the same for each of the n trials, his prior will be non-zero for product states of the form $\rho^{\otimes n}$ and zero for all other states. This prior is equivalent to specifying a probability distribution $f_{\text{prior}}(\rho)$ on the space of single-system density operators on $\mathcal{H}$. This means that the assumptions of the simple Bayesian analysis of a repeated trial starting from Eq. (8) above are fulfilled.

To put it simply, to do quantum tomography in Wallace's decision-theoretic approach, one does not require that the true quantum state of the n systems is of the form $\rho^{\otimes n}$. One requires instead that the agent's subjective prior is non-zero only for states of this form. By making clear that the probabilities that feature in the analysis of the measurement outcomes express an agent's degrees of belief rather than objective properties of the physical system, one eliminates the regress and puts the analysis of repeated trials on a firm footing.

The resolution of the problem of repeated trials sketched in the last paragraphs shifts the emphasis away from the objective properties of the physical system towards the agent's degrees of belief. It de-emphasizes the objective quantum state and puts the agent's belief in the centre. The role of the objective quantum state is diminished to the point that it can be eliminated from the analysis without any other changes. This raises the question of whether the concept of an objective quantum state has any useful role to play at all. From the perspective of a full quantum Bayesian theory (Caves et al. [2002, 2007], Fuchs [2002]) the answer is that objective quantum states, and therefore the objective wavefunction of the universe, have no useful place in quantum theory.

To show that nothing is lost, and much can be gained when objective states are eliminated from the theory, we will conclude this paper by giving a brief account of repeated trials in a full quantum Bayesian setting. From this perspective, quantum states have exactly the same status as Bayesian probabilities. A quantum state represents an agent's degrees of belief, i.e., his decision-theoretic preferences. The Born rule, rather than functioning as a quantum version of the principal principle, is now a rule for transforming degrees of belief and for relating the

Hilbert-space formalism to the classical decision-theoretic framework. Faced with n trials as above, a quantum Bayesian agent starts by assigning a prior, which is now a quantum state, $\rho^{(n)}$, on the tensor-product Hilbert space $\mathcal{H}^{\otimes n}$. We are in the situation of a repeated trial if the agent's prior is exchangeable, i.e., if it is a member of a sequence of states $\rho^{(m)}$ defined on $\mathcal{H}^{\otimes m}$ such that for all integers m, $\rho^{(m)}$ is symmetric under permutations of the m subsystems and can be obtained from $\rho^{(m+1)}$ by tracing over one subsystem. According to the quantum de Finetti theorem (Hudson and Moody [1976], Caves et al. [2002]), exchangeability implies that the prior state is of the form (11). We have thus reached exactly the same point as before, enabling us to do a simple Bayesian analysis of our measurement outcomes.

Instead of two fundamental concepts—objective quantum states and decision-theoretic preferences—and their arguably awkward connection via the principal principle, we now have a single fundamental concept. This translates into much greater conceptual and mathematical simplicity. For instance, consider the prior for n trials. In the special case of a repeated trial, the quantum Bayesian description is isomorphic to the description in a many-worlds context by virtue of de Finetti's representation theorem. In the general case of n trials, without the exchangeability assumption, the quantum Bayesian prior is a single density operator on $\mathcal{H}^{\otimes n}$. Even though this is an operator on a space of large dimension, it is a much simpler mathematical object than the most general prior in the many-worlds approach, which is a probability distribution over *all* density operators on $\mathcal{H}^{\otimes n}$.

In conclusion, to interpret probability in quantum mechanics as representing an agent's decision-theoretic preferences appears to be the right move. In Wallace's approach to the many-worlds interpretation, where a distinction between objective quantum states and subjective probabilities is maintained, a coherent analysis of repeated trials is possible. This analysis, however, suggests that the concept of objective quantum states can be eliminated in favour of a fully quantum Bayesian approach where quantum states as well as probabilities represent an agent's decision-theoretic preferences.

References

Bernardo, J.M. and A.F.M. Smith [1994], *Bayesian Theory*, Wiley, Chichester.

Caves, C.M., C.A. Fuchs, and R. Schack [2002], 'Unknown quantum states: The quantum de Finetti representation', *Journal of Mathematical Physics* **43**, 4537.

—— [2007],'Subjective probability and quantum certainty', *Studies in History and Philosophy of Modern Physics* **38**, 255.

de Finetti, B. [1990], *Theory of Probability*, Wiley New York.

Deutsch, D. [1999], 'Quantum theory of probability and decisions', *Proceedings of the Royal Society of London* **A455**, 3129–37.

Fuchs, C.A. [2002], 'Quantum mechanics as quantum information (and only a little more)', e-print quant-ph/0205039.

Hudson, R.L. and G.R. Moody [1976], 'Locally normal symmetric states and an analog of de Finetti's theorem', *Z. Wahrscheinlichkeitstheorie verw.* 33, 343.

Jaynes, E.T. [1957a], 'Information theory and statistical mechanics', *Physical Review* 106, 620.

—— [1957b], 'Information theory and statistical mechanics II', *Physical Review* 108, 171.

—— [2003], *Probability Theory. The Logic of Science*, G. Larry Bretthorst (ed.), p.324, Cambridge University Press, Cambridge.

Lewis, D. [1986], 'A subjectivist's guide to objective chance,' in *Philosophical Papers, Vol. II*, p.83, Oxford University Press, Oxford.

Savage, L. [1972], *The Foundations of Statistics*, Dover, New York.

Schack, R., T.A. Brun, and C.M. Caves [2001], 'Quantum Bayes rule', *Physical Review* A 64, 014305.

Vogel, K. and H. Risken [1989], 'Determination of quasiprobability distributions in terms of probability distributions for the rotated quadrature phase', *Physical Review* A 40, 2847.

Wallace, D. [2003], 'Everettian rationality: defending Deutsch's approach to probability in the Everett interpretation', *Studies in the History and Philosophy of Modern Physics* 34, 415–39. Available online at arXiv.org/abs/quant-ph/0303050.

—— [2006], 'Epistemology quantized: circumstances in which we should come to believe in the Everett interpretation', *British Journal for the Philosophy of Science* 57, 655–89. Available online at philsci-archive.pitt.edu/archive/00002839.

—— [2007], 'Quantum probability from subjective likelihood: improving on Deutsch's proof of the probability rule', *Studies in the History and Philosophy of Modern Physics* 38, 311–32. Available online at arXiv.org/abs/quant-ph/0312157.

16

De Broglie–Bohm Pilot-Wave Theory: Many Worlds in Denial?

Antony Valentini

ABSTRACT

We reply to claims (by Deutsch, Zeh, Brown, and Wallace) that the pilot-wave theory of de Broglie and Bohm is really a many-worlds theory with a superfluous configuration appended to one of the worlds. Assuming that pilot-wave theory does contain an ontological pilot wave (a complex-valued field in configuration space), we show that such claims arise from not interpreting pilot-wave theory on its own terms. Specifically, the theory has its own ('subquantum') theory of measurement, and in general describes a 'non-equilibrium' state that violates the Born rule. Furthermore, in realistic models of the classical limit, one does not obtain localized pieces of an ontological pilot wave following alternative macroscopic trajectories: from a de Broglie–Bohm viewpoint, alternative trajectories are merely mathematical and not ontological. Thus, from the perspective of pilot-wave theory itself, many worlds are an illusion. It is further argued that, even leaving pilot-wave theory aside, the theory of many worlds is rooted in the intrinsically unlikely assumption that quantum measurements should be modelled on classical measurements, and is therefore unlikely to be true.

1 INTRODUCTION

It used to be widely believed that the pilot-wave theory of de Broglie [1928] and Bohm [1952a,b] had been ruled out by experiments demonstrating violations of Bell's inequality. Such misunderstandings have largely been overcome, and in recent times the theory has come to be widely accepted by physicists as an alternative (and explicitly non-local) formulation of quantum theory. Even so, some workers claim that pilot-wave theory is not really a physically distinct formulation of quantum theory, that instead it is actually a theory of Everettian many worlds. The principal aim of this paper is to refute that claim. We shall also end with a counter-claim, to the effect that Everett's theory of many worlds is unlikely to be true, as it is rooted in an intrinsically unlikely assumption about measurement.

Pilot-wave theory is a first-order, non-classical theory of dynamics, grounded in configuration space. It was first proposed by de Broglie, at the 1927 Solvay conference (Bacciagaluppi and Valentini [2009]). From de Broglie's dynamics, together with an assumption about initial conditions, it is possible to derive the full phenomenology of quantum theory, as was first shown by Bohm in 1952.

In pilot-wave dynamics, a closed system with configuration q has a wave-function $\Psi(q,t)$—a complex-valued field on configuration space obeying the Schrödinger equation $i\partial\Psi/\partial t = \hat{H}\Psi$. The system has an actual configuration $q(t)$ evolving in time, with a velocity $\dot{q} \equiv dq/dt$ determined by the gradient ∇S of the phase S of Ψ (for systems with standard Hamiltonians $\hat{H}$).[1] In principle, the configuration q includes all those things that we normally call 'systems' (particles, atoms, fields) as well as pieces of equipment, recording devices, experimenters, the environment, and so on.

Let us explicitly write down the dynamical equations for the case of a non-relativistic many-body system, as they were given by de Broglie [1928]. For N spinless particles with positions $\mathbf{x}_i(t)$ and masses m_i $(i = 1, 2, \ldots, N)$, in an external potential V, the total configuration $q = (\mathbf{x}_1, \mathbf{x}_2, \ldots, \mathbf{x}_N)$ evolves in accordance with the de Broglie guidance equation

$$m_i \frac{d\mathbf{x}_i}{dt} = \nabla_i S \tag{1}$$

---

[1] More generally, $\dot{q} = j/|\Psi|^2$ where j is the current associated with the Schrödinger equation (Struyve and Valentini [2009]).

(where $\hbar = 1$ and $\Psi = |\Psi| e^{iS}$), while the 'pilot wave' Ψ (as it was originally called by de Broglie) satisfies the Schrödinger equation

$$i\frac{\partial \Psi}{\partial t} = \sum_{i=1}^{N} -\frac{1}{2m_i} \nabla_i^2 \Psi + V\Psi \ . \tag{2}$$

Mathematically, these two equations define de Broglie's dynamics—just as, for example, Maxwell's equations and the Lorentz force law may be said to define classical electrodynamics.

The theory was revived by Bohm in 1952, though in a pseudo-Newtonian form. Bohm regarded the equation

$$m_i\frac{d^2\mathbf{x}_i}{dt^2} = -\nabla_i(V + Q) \tag{3}$$

as the true law of motion, with a 'quantum potential'

$$Q \equiv -\sum_{i=1}^{N} \frac{1}{2m_i} \frac{\nabla_i^2 |\Psi|}{|\Psi|}$$

acting on the particles. (Taking the time derivative of (1) and using (2) yields (3).) On Bohm's view, (1) is not a law of motion but a condition $\mathbf{p}_i = \nabla_i S$ on the initial momenta—a condition that happens to be preserved in time by (3), and which could in principle be relaxed (leading to corrections to quantum theory) (Bohm [1952a pp.170–71]). One should therefore distinguish between de Broglie's first-order dynamics of 1927, defined by (1) and (2), and Bohm's second-order dynamics of 1952, defined by (3) and (2). In particular, Bohm's rewriting of de Broglie's theory had the unfortunate effect of making it seem much more like classical physics than it really was. De Broglie's original intention had been to depart from classical dynamics at a fundamental level, and indeed the resulting theory is highly non-Newtonian. As we shall see, it is crucial to avoid making classical assumptions when interpreting the theory.

Over an ensemble of quantum experiments, beginning at time $t = 0$ with the same initial wavefunction $\Psi(q,0)$ and with a Born-rule or 'quantum equilibrium' distribution

$$P(q, 0) = |\Psi(q, 0)|^2 \tag{4}$$

of initial configurations $q(0)$, it follows from de Broglie's dynamics that the distribution of final outcomes is given by the usual Born rule (Bohm [1952a,b]). On the other hand, for an ensemble with a 'quantum non-equilibrium' distribution

$$P(q, 0) \neq |\Psi(q, 0)|^2, \tag{5}$$

in general, one obtains a distribution of final outcomes that *disagrees* with quantum theory (for as long as P has not yet relaxed to $|\Psi|^2$, see below) (Valentini [1991a,b, 1992, 1996, 2001, 2002, 2004a]; Pearle and Valentini [2006]).

The initial distribution (4) was assumed by both de Broglie and Bohm, and subsequently most workers have regarded it as one of the axioms of the theory. As we shall see, this is a serious mistake that has led to numerous misunderstandings, and is partially responsible for the erroneous claim that pilot-wave theory is really a theory of many worlds.

We shall not attempt to provide an overall assessment of the relative merits of de Broglie–Bohm pilot-wave theory and Everettian many-worlds theory. Instead, here we focus on evaluating the following claim—hereafter referred to as 'the Claim'—which has more or less appeared in several places in the literature (Deutsch [1996], Zeh [1999], Brown and Wallace [2005]) (author's paraphrase):

- Claim: *If one takes pilot-wave theory seriously as a possible theory of the world, and if one thinks about it properly and carefully, one ought to see that it really contains many worlds—with a superfluous configuration q appended to one of those worlds.*

Were the Claim correct, one could reasonably add a corollary to the effect that one should then drop the superfluous configuration q, and arrive at (some form of) many-worlds theory.

Deutsch's way of expressing the Claim has inspired the title of this paper (Deutsch [1996 p.225]):

In short, pilot-wave theories are parallel-universes theories in a state of chronic denial.

We should emphasize that here we shall interpret pilot-wave theory (for a given closed system) as containing an ontological—that is, physically real—complex-valued field $\Psi(q,t)$ on configuration space, where this field drives the motion of an actual configuration $q(t)$. The Claim asserts that, if the theory is regarded in these terms, then proper consideration shows that Ψ contains many worlds, with q amounting to a superfluous appendage to one of the worlds. One might try to side-step the Claim by asserting that Ψ has no ontological status in pilot-wave theory, that it merely provides a mathematical account of the motion $q(t)$. In this case, one could not even begin to make the Claim, for the complete ontology would be defined by the configuration q. For all we currently know, this view might turn out to be true in some future derivation of pilot-wave theory from a deeper theory. But in pilot-wave theory as we know it today—the subject of this paper—such a view seems implausible and physically unsatisfactory (see below). In any case, even if only for the sake of argument, let us here assume that the pilot wave Ψ is ontological, and let us show how the Claim may still be refuted.

It will be helpful first to review the distinction between ontological and mathematical structure in current physical theory, and then to give a brief overview of pilot-wave theory interpreted on its own terms.

Generally speaking, theories should be evaluated on their own terms, *without assumptions that make sense only in rival theories*. We shall see that, in essence, the Claim in fact arises from not interpreting and understanding pilot-wave theory on its own terms.

2   ONTOLOGY VERSUS MATHEMATICS

Physics provides many examples of the distinction between ontological and mathematical structure. Let us consider three.

(1) *Classical mechanics.* This may be formulated in terms of a Hamiltonian trajectory $(q(t),p(t))$ in phase space. For a given individual system, there is only one real trajectory. The other trajectories, corresponding to alternative initial conditions $(q(0),p(0))$, have a purely mathematical existence. Similarly, in the Hamilton–Jacobi formulation, the Hamilton–Jacobi function $S(q,t)$ is associated with a whole family of trajectories (with $\dot{q}$ determined by ∇S), only one of which is realized.

(2) *A test particle in an external field.* This provides a particularly good parallel with pilot-wave theory. A charged test particle, placed in an external electromagnetic field $\mathbf{E}(\mathbf{x},t)$, $\mathbf{B}(\mathbf{x},t)$, will follow a trajectory $\mathbf{x}(t)$. One would normally say that the field is real, and that the realized particle trajectory is real; while the alternative particle trajectories (associated with alternative initial positions $\mathbf{x}(0)$) are not real, even if they might be said to be contained in the mathematical structure of the electromagnetic field. Similarly, if a test particle moves along a geodesic in a background spacetime geometry, one can think of the geometry as ontological, and the mathematical structure of the geometry contains alternative geodesic motions—but again, only one particle trajectory is realized, and the other geodesics have a purely mathematical existence.

(3) *A classical vibrating string.* Consider a string held fixed at the endpoints, $x = 0$, L. (This example will also prove relevant to the quantum case.) A small vertical displacement $\psi(x,t)$ obeys the partial differential equation

$$\frac{\partial^2 \psi}{\partial t^2} = \frac{\partial^2 \psi}{\partial x^2}$$

(setting the wave speed $c = 1$). This is conveniently solved using the standard methods of linear functional analysis. One may define a Hilbert space of functions ψ, with a Hermitian operator $\hat{\Omega} = -\partial^2/\partial x^2$ acting thereon. Solutions of the wave equation may then be expanded in terms of a complete set of eigenfunctions $\phi_m(x) = \sqrt{2/L}\sin(m\pi x/L)$, where $\hat{\Omega}\phi_m = \omega_m^2 \phi_m$ with $\omega_m^2 = (m\pi/L)^2$ $(m = 1, 2, 3, \ldots)$. Assuming for simplicity that $\dot{\psi}(x, 0) = 0$, we have the general solution

$$\psi(x, t) = \sum_{m=1}^{\infty} c_m \phi_m(x) \cos \omega_m t \quad \left(c_m \equiv \int_0^L dx\, \phi_m(x)\psi(x, 0) \right)$$

or (in bra-ket vector notation)

$$|\psi(t)\rangle = \sum_{m=1}^{\infty} |m\rangle\langle m|\psi(0)\rangle \cos \omega_m t$$

(where $\hat{\Omega}|m\rangle = \omega_m^2|m\rangle$). Any solution may be written as a superposition of oscillating 'modes'. Even so, the true ontology consists essentially of the total displacement $\psi(x,t)$ of the string (perhaps also including its velocity and energy). Individual modes in the sum would not normally be regarded as physically real. One would certainly not assert that ψ is composed of an ontological multiplicity of strings, with each string vibrating in a single mode. Instead one would say that, in general, the eigenfunctions and eigenvalues have a mathematical significance only.

All this is not to say that the question of ontology in physical theories is trivial or always obvious. On the contrary, it is not always self-evident as to whether mathematical objects in our physical theories should be assigned ontological status or not. For example, classical electrodynamics may be viewed in terms of a field theory (with an ontological electromagnetic field), or in terms of direct action-at-a-distance between charges (where the electromagnetic field is merely an auxiliary field, if it appears at all). Most physicists today prefer the first view, probably because the field seems to contain a lot of independent and contingent structure (see below).

The question to be addressed here is: in the pilot-wave theory of de Broglie and Bohm, if one regards the pilot wave Ψ as ontological (which seems the most natural view at present), does this amount to an ontology of many worlds?

3 PILOT-WAVE THEORY ON ITS OWN TERMS

In the author's view, pilot-wave theory continues to be widely misinterpreted and misrepresented, even by some of its keenest supporters. Here, for illustration, we confine ourselves to de Broglie's original dynamics for a system of non-relativistic (and spinless) particles, defined by (1) and (2).

Basic History

Let us begin by setting the historical record straight,[2] as historical arguments sometimes play a role in evaluations of pilot-wave theory.

Pilot-wave dynamics was constructed by de Broglie in the period 1923–27. His motivations were grounded in experiment. He wished to explain the quantization of atomic energy levels and the interference or diffraction of single photons. To this end, he proposed a unification of the physics of particles with the physics of waves. De Broglie argued that Newton's first law of motion had to be abandoned, because a particle diffracted by a screen does not touch the screen and yet does not move in a straight line. During 1923–24, de Broglie then proposed a new, non-Newtonian form of dynamics in which the *velocity* of a particle is determined by the phase of a guiding wave. As a theoretical guide, de Broglie sought to unify the classical variational principles of Maupertuis ($\delta \int m\mathbf{v} \cdot d\mathbf{x} = 0$, for a particle with velocity $\mathbf{v}$) and of Fermat ($\delta \int dS = 0$, for a wave with phase S). The result was the guidance equation (1) (at first applied to a single particle and later generalized), which de Broglie regarded as the basis of a new form of dynamics.

At the end of a rather complicated development in the period 1925–27 (including a crucial contribution by Schrödinger, who found the correct wave equation for de Broglie's waves), de Broglie proposed the many-body dynamics defined by (1) and (2). De Broglie regarded his theory as provisional, much as Newton regarded his own theory of gravity as provisional. And de Broglie regarded the observation of electron diffraction, by Davisson and Germer in 1927, as a vindication of his prediction (first made in 1923), and as clear evidence for his new (first-order) dynamics of particle motion.

Clearly, de Broglie's construction of pilot-wave dynamics was motivated by experimental puzzles and had its own internal logic. Note in particular that de Broglie did not construct his theory to 'solve the measurement problem', nor did he construct it to provide a (deterministic or realistic) 'completion of quantum theory': for in 1923, there was no measurement problem and there was no quantum theory.

Getting the history right is important, for its own sake and also because some criticisms of pilot-wave theory are based on a mistaken appraisal of history. For example, Deutsch [1986 pp.102–3] has said the following about the theory:

... to append to the quantum formalism an additional structure ... solely for the purpose of interpretation, is I think a very dangerous thing to do in physics. These

---

[2] For a detailed account, see chapter 2 of Bacciagaluppi and Valentini [2009].

structures are being introduced solely to solve the interpretational problems, without any physical motivation. . . . the chances of a theory which was formulated for such a reason being right are extremely remote.

But there is no sense in which de Broglie 'appended' something to quantum theory, for quantum theory did not yet exist. And de Broglie had ample physical motivation, grounded in experimental puzzles and in a compelling analogy between the principles of Maupertuis and Fermat.

A proper historical account also undermines discussions in which pilot-wave theory is presented as being motivated by the desire to 'solve the measurement problem'. For example, Brown and Wallace [2005]—who discuss Bohm's motivations but ignore de Broglie's—argue that many-worlds theory provides a more natural solution to the measurement problem than does pilot-wave theory. The discussion is framed as if the measurement problem were the prime motivation for considering pilot-wave theory in the first place. As a matter of historical fact, this is false.

The widespread misleading historical perspective has been exacerbated by some workers who present de Broglie's 1927 dynamics as a way to 'complete' quantum theory by adding trajectories to the wavefunction (Dürr et al. [1992, 1996]), an approach that furthers the mistaken impression that the theory is a belated reformulation of an already-existing theory. Matters are further confused by some workers who refer to de Broglie's first-order dynamics by the misnomer 'Bohmian mechanics', a term that should properly be applied to Bohm's second-order dynamics. De Broglie's dynamics pre-dates quantum theory; and it was given in final form in 1927, not as an afterthought (or reformulation of quantum theory) in 1952.

We may then leave aside certain spurious objections that are grounded in a mistaken version of historical events. In the author's view, the proper way to pose the question addressed in this paper is: *given* de Broglie's dynamics (as it was in 1927), if we examine it carefully on its own terms, does it turn out to contain many worlds?

Basic Ontology

As stated in the introduction, we regard the theory as having a dual ontology: the configuration $q(t)$ together with the pilot wave $\Psi[q,t]$. We need to give the relation between this ontology and what we normally think of as physical reality.

De Broglie constructed the theory as a new dynamics of particles: specifically, the basic constituents of matter and radiation (as understood at the time). It is then natural to assume that physical systems, apparatus, people, and so on, are 'built from' the configuration q. (In extensions of the theory, q may of course include configurations of fields, the geometry of 3-space, strings, or whatever may be thought of as the modern fundamental constituents. Further,

macroscopic systems—such as experimenters—will usually supervene on q
under some coarse-graining.) This view has been explicitly stated in the literature
by several workers—for example Bell [1987 p.128], Valentini [1992 p.26],
Holland [1993 pp.337, 350], and others—though perhaps it is not clearly
stated in some of the de Broglie–Bohm literature (as Brown and Wallace
[2005] suggest). In any case, we shall take this to be the correct and natural
viewpoint.

That Ψ is also to be regarded as ontological is often not explicitly stated. A
notable exception was Bell [1987 p.128, original italics]:

> . . . the wave is supposed to be just as 'real' and 'objective' as say the fields of classical
> Maxwell theory . . . . *No one can understand this theory until he is willing to think of ψ as
> a real objective field . . . . Even though it propagates not in 3-space but in 3N-space.*

Could Ψ instead be regarded as 'fictitious', that is, as a merely mathematical
field appearing in the law of motion for q? As already mentioned, this does not
seem reasonable, at least not for the theory in its present form, where—like the
electromagnetic field—Ψ contains a lot of independent and contingent structure,
and is therefore best regarded as part of the state of the world (Valentini [1992
p.17]; Brown and Wallace [2005 p.532]).

Valentini [1992 p.13] considered the possibility that Ψ might merely provide
a convenient mathematical summary of the motion $q(t)$; to this end, he drew
an analogy between Ψ and physical laws such as Maxwell's equations, which
also provide a convenient mathematical summary of the behaviour of physical
systems. On this view, 'the world consists purely of the evolving variables $X(t)$,
whose time evolution may be summarised mathematically by Ψ' (ibid. p.13).
But Valentini argued further (p.17) that such a view did not do justice to
the physical information stored in Ψ, and he concluded instead that Ψ was
a new kind of causal agent acting in configuration space (a view that the
author still takes today). The former view, that Ψ is law-like, was adopted
by Dürr et al. [1997].[3] They proposed further that the time dependence
and contingency of Ψ—properties that argue for it to be ontological (see
Brown and Wallace [2005 p.532])—may be illusions, as the wavefunction
for the whole universe is (so they claim) expected to be static and unique.
However, the present situation in quantum gravity indicates that solutions
for Ψ (satisfying the Wheeler–DeWitt equation and other constraints) are
far from unique, and display the same kind of contingency (for example in
cosmological models) that we are used to for quantum states elsewhere in
physics (Rovelli [2004]). Should the universal wavefunction be static—and
the notorious 'problem of time' in quantum gravity urges caution here—this
alone is not enough to establish that it should be law-like: contingency, or

---

[3] ' . . . the wave function is a component of physical law rather than of the reality described by
the law' (Dürr et al. [1997 p.33]).

under-determination by physical law, is the more important feature.[4] Therefore, current theoretical evidence speaks against the idea. And in any case, our task here is to consider the theory we have now, not ideas for theories that we may have in the future: in the present form of pilot-wave theory, the time-dependence and (especially) the contingency of Ψ make it best regarded as ontological.

Note that in 1927 de Broglie regarded Ψ as providing—as a temporary measure—a mathematically convenient and phenomenological summary of motions generated from a deeper theory, in which particles were singular regions of 3-space waves (Bacciagaluppi and Valentini [2009, section 2.3.2]). De Broglie hoped the theory would later be derived from something deeper (as Newton believed of gravitational attraction at a distance). Should this eventually happen, ontological questions will have to be addressed anew. Alternatively, perhaps de Broglie's 'deeper theory' (the theory of the double solution) should be regarded merely as a conceptual scaffolding which he used to arrive at pilot-wave theory, and the scaffolding should now be forgotten.[5] In any case, the theory has come to be regarded as a theory in its own right, and the question at hand is whether *this* theory contains many worlds or not.

Equilibrium and Non-equilibrium

Many workers take the quantum equilibrium distribution (4) as an axiom, alongside the laws of motion (1) and (2). It has been argued at length that this is incorrect and deeply misleading (Valentini [1991a,b, 1992, 1996, 2001, 2002]; Valentini and Westman [2005]; Pearle and Valentini [2006]). A postulate concerning the distribution of initial conditions has no fundamental status in a theory of dynamics. Instead, quantum equilibrium is to pilot-wave dynamics as thermal equilibrium is to classical dynamics. In both cases, equilibrium may be understood as arising from a process of relaxation. And in both cases, the equilibrium distributions are mere contingencies, not laws: the underlying theories allow for more general distributions, that violate quantum physics in the first case and thermal-equilibrium physics in the second.

Taken on its own terms, then, pilot-wave theory is *not* a mere alternative formulation of quantum theory. Instead, the theory itself tells us that quantum physics is a special case of a much wider 'non-equilibrium' physics (with $P \neq |\Psi|^2$), which may exist for example in the early inflationary universe, or for relic

---

[4] One should also guard against the idea—sometimes expressed in this context—that the existence of 'only one universe' somehow suggests that the universal wavefunction cannot be contingent. Equally, in non-Everettian cosmology, there is only one intergalactic magnetic field, and yet it would be generally agreed that the precise form of this field is a contingency (not determined by physical law).

[5] Cf. the role played by the ether in electromagnetism, or in Newton's thinking about gravitation. For a discussion of this parallel, see section 2.3.2 of Bacciagaluppi and Valentini [2009].

particles that decoupled soon after the big bang, or for particles emitted by black holes (Valentini [2004b, 2007, 2008a,b]).

True (Subquantum) Measurements

The wider physics of non-equilibrium has its own theory of measurement— 'subquantum measurement' (Valentini [1992, 2002]; Pearle and Valentini [2006]). This is to be expected, since measurement is theory-laden: given a (perhaps tentative) theory, one should look to the theory itself to tell us how to perform correct measurements (cf. Section 8).

In pilot-wave theory, an 'ideal subquantum measurement' (analogous to the ideal, non-disturbing measurement familiar from classical physics) enables an experimenter to measure a de Broglie–Bohm system trajectory without disturbing the wavefunction. This is possible if the experimenter possesses an apparatus whose 'pointer' has an arbitrarily narrow non-equilibrium distribution (Valentini [2002], Pearle and Valentini [2006]). Essentially, the system and apparatus are allowed to interact so weakly that the joint wavefunction hardly changes; yet, the displacement of the pointer contains information about the system configuration, information that is visible if the pointer distribution is sufficiently narrow. A sequence of such operations allows the experimenter to determine the system trajectory without disturbing the wavefunction, to arbitrary accuracy.

Generally False Quantum 'Measurements' (Formal Analogues of Classical Measurements)

We are currently unable to perform such true measurements, because we are trapped in a state of quantum equilibrium. Instead, today we generally carry out procedures that are known as 'quantum measurements'. This terminology is misleading, because such procedures are—at least according to pilot-wave theory—generally *not* correct measurements: they are merely experiments of a certain kind, designed to respect a formal analogy with *classical* measurements (cf. Valentini [1996 pp.50–1]).

Thus, in classical physics, to measure a system variable ω using an apparatus pointer y, Hamilton's equations tell us that we should switch on a Hamiltonian $H = a\omega p_y$ (where a is a coupling constant and p_y is the momentum conjugate to y). One obtains trajectories $\omega(t) = \omega_0$ and $y(t) = y_0 + a\omega_0 t$. From the displacement $a\omega_0 t$ of the pointer, one may infer the value of ω_0. An experimental operation represented by $H = a\omega p_y$ then indeed realizes a correct measurement of ω (according to classical physics). But there is no reason to expect the same experimental operation to constitute a correct measurement of ω for a non-classical system. Even so, remarkably, so-called quantum 'measurements' are in general designed using classical measurements as a guide. Specifically, in quantum theory, to measure an observable ω using an apparatus pointer y, one switches on a Hamiltonian operator $\hat{H} = a\hat{\omega}\hat{p}_y$. The quantum procedure is obtained, in effect, by 'quantizing' the classical procedure.

But what does this analogous quantum procedure actually accomplish? According to pilot-wave theory, it merely generates a branching of the total wavefunction, with branches labelled by eigenvalues ω_n of the linear operator $\hat{\omega}$, and with the total configuration $q(t)$ ending in the support of one of the (non-overlapping) branches. Thus, for example, if the system is a particle with position x, the initial wavefunction

$$\Psi_0(x, y) = \left(\sum_n c_n \phi_n(x) \right) g_0(y)$$

(where $\hat{\omega}\phi_n = \omega_n \phi_n$ and g_0 is the initial (narrow) pointer wavefunction) evolves into

$$\Psi(x, y, t) = \sum_n c_n \phi_n(x) g_0(y - a\omega_n t).$$

The effect of the experiment is simply to create this branching.[6]

From a pilot-wave perspective, the eigenvalues ω_n have no particular ontological status: we simply have a complex-valued field Ψ on configuration space, obeying a linear wave equation, whose time evolution may in some situations be conveniently analysed using the methods of linear functional analysis (as we saw for the classical vibrating string).

It cannot be sufficiently stressed that, generally speaking, by means of this procedure one has *not measured anything* (so pilot-wave theory tells us). In quantum theory, if the pointer is found to occupy the nth branch, it is common to assert that therefore 'the observable ω has the value ω_n'. But in pilot-wave theory, all that has happened is that, at the end of the experiment, the system trajectory $x(t)$ is guided by the (effectively) reduced wavefunction $\phi_n(x)$.[7] This does not usually imply that the system has or had some property with value ω_n (at the end of the experiment or at the beginning), because in pilot-wave theory there is no general relation between eigenvalues and ontology.[8]

---

[6] Over an ensemble, if x and y have an initial distribution $P_0(x, y) = |\Psi_0(x, y)|^2$, one of course finds that a fraction $|c_n|^2$ of trajectories $q(t) = (x(t), y(t))$ end in the (support of) the nth branch $\phi_n(x)g_0(y - a\omega_n t)$.

[7] Because the branches have separated in configuration space, it follows from de Broglie's equation of motion that the 'empty' branches no longer affect the trajectory.

[8] For example, the eigenfunction $\phi_E(x) \propto (e^{ipx} + e^{-ipx})$ of the kinetic-energy operator $\hat{p}^2/2m$ has eigenvalue $E = p^2/2m \neq 0$; and yet, the actual de Broglie–Bohm kinetic energy vanishes, $\frac{1}{2}m\dot{x}^2 = 0$ (since $\partial S/\partial x = 0$). If the system had this initial wavefunction, and we performed a so-called 'quantum measurement of kinetic energy' using a pointer y, then the initial joint

Thus, a so-called 'ideal quantum measurement of ω' is not a true measurement (a notable exception being the case $\omega = x$). And in general, it is usually incorrect to identify eigenvalues with values of real physical quantities: one must beware of 'eigenvalue realism'.

4 SOME EXAMPLES OF THE CLAIM

Before evaluating the Claim, let us quote some examples of it from the literature.
    First, Deutsch [1996 p.225] argues that parallel universes are

... a logical consequence of Bohm's 'pilot-wave' theory (Bohm [1952]) and its variants (Bell [1986]). . . . The idea is that the 'pilot-wave' . . . guides Bohm's single universe along its trajectory. This trajectory occupies one of the 'grooves' in that immensely complicated multidimensional wavefunction. The question that pilot-wave theorists must therefore address, and over which they invariably equivocate, is what are the *unoccupied* grooves? It is no good saying that they are merely a theoretical construct and do not exist physically, for they continually jostle both each other and the 'occupied' groove, affecting its trajectory (Tipler [1987], p.189). . . . So the 'unoccupied grooves' must be physically real. Moreover they obey the same laws of physics as the 'occupied groove' that is supposed to be 'the' universe. But that is just another way of saying that they are universes too . . . . In short, pilot-wave theories are parallel-universes theories in a state of chronic denial.

Zeh [1999 p.200] puts the matter thus:

It is usually overlooked that Bohm's theory contains the *same* 'many worlds' of dynamically separate branches as the Everett interpretation (now regarded as 'empty' wave components), since it is based on precisely the same ('*absolutely* real') global wavefunction. . . . Only the 'occupied' wavepacket itself is thus meaningful, while the assumed classical trajectory would merely point at it: 'This is where *we* are in the quantum world.'

Similarly, Brown and Wallace [2005 p.527] write the following:

. . . the corpuscle's role is minimal indeed: it is in danger of being relegated to the role of a mere epiphenomenal 'pointer', irrelevantly picking out one of the many branches defined by decoherence, while the real story—dynamically and ontologically—is being told by the unfolding evolution of those branches. The 'empty wavepackets' in the configuration space which the corpuscles do not point at are none the worse for its absence: they still contain cells, dust motes, cats, people, wars and the like.

In the case of Zeh, and of Brown and Wallace, the key assertion is that pilot-wave theory and many-worlds theory contain the same multitude of wavefunction

wavefunction $\phi_E(x)g_0(y)$ would evolve into $\phi_E(x)g_0(y - aEt)$ and the pointer would indicate the value E—even though the particle kinetic energy was and would remain equal to zero. The experiment has not really measured anything.

branches, and that in pilot-wave theory the 'empty' branches nevertheless constitute parallel worlds (which 'still contain cells, dust motes, cats, people, wars and the like').

Deutsch's argument leads to the same assertion—if one interprets his word 'grooves' to mean what are normally called 'branches'. However, Deutsch may in fact have used 'grooves' to mean the set of de Broglie–Bohm trajectories, in which case his version of the Claim states that pilot-wave theory is really a theory of "many de Broglie–Bohm worlds".[9] (This version of the Claim is addressed in Section 7.) In any case, in essence Deutsch argues that the unoccupied grooves are real, and that they 'obey the same laws of physics' as the occupied groove, thereby constituting a 'multiverse'.

Today, it is often said that in Everettian quantum theory the notion of parallel 'worlds' or 'universes' applies only to the macroscopic worlds defined (approximately) by decoherence. Formerly, it was common to assert the existence of many worlds at the microscopic level as well. Without entering into any controversy that might still remain about this, here for completeness we shall address the Claim for both 'microscopic' and 'macroscopic' cases.

5 'MICROSCOPIC' MANY WORLDS?

In pilot-wave theory, is there a multiplicity of parallel worlds at the microscopic level? To see that there is not, let us consider some examples.

(1) *Superposition of eigenvalues.* Let a single particle moving in one dimension have the wavefunction $\psi(x,t) \propto e^{-iEt}(e^{ipx} + e^{-ipx})$, which is a mathematical superposition of two distinct eigenfunctions of the momentum operator $\hat{p} = -i\partial/\partial x$. Are there in any sense two particles, with two different momenta $+p$ and $-p$? Clearly not. While the field $\psi \propto \cos px$ has two Fourier components e^{ipx} and e^{-ipx}, there is only one single-valued field ψ (as in our example of the classical vibrating string). And a true (subquantum) measurement of the particle trajectory $x(t)$ would reveal that the particle is at rest (since $S = -Et$ and $\partial S/\partial x = 0$). In a so-called 'quantum measurement of momentum', at the end of the experiment $x(t)$ is guided by e^{ipx} or e^{-ipx}: during the experiment the particle is accelerated and *acquires* a momentum $+p$ or $-p$, as could be confirmed by a true subquantum measurement. Any impression that there may be two particles present arises from a mistaken belief in eigenvalue realism.

(2) *Double-slit experiment.* Let a single particle be fired at a screen with two slits, where the incident wavefunction ψ passes through both slits, leading to

---

[9] Deutsch cites the rather confused paper by Tipler [1987], which argues among other things that de Broglie–Bohm trajectories must affect each other in unphysical ways. Tipler's critique is mostly aimed at a certain stochastic version of pilot-wave theory. While it is not really relevant to Deutsch's argument, for completeness we note that, as regards conventional (deterministic) pilot-wave theory, Tipler's critique stems from an elementary misunderstanding of the role of probability in the theory.

an interference pattern on the far side of the screen. Are there in any sense two particles, one passing through each slit? Again, clearly not. There is a single-valued field ψ passing through both slits, and there is one particle trajectory $\mathbf{x}(t)$ in 3-space, passing through one slit only (as again could be tracked by a true subquantum measurement).

(3) *Superposition of 'Ehrenfest' packets for a hydrogen atom.* Finally, consider a single hydrogen atom, with a centre-of-mass trajectory $\mathbf{x}(t)$ and with a wavefunction that is a superposition

$$\psi = \frac{1}{\sqrt{2}} (\psi_1 + \psi_2)$$

of two localized and spatially separated 'Ehrenfest' packets ψ_1 and ψ_2. Each packet, with centroid $\langle \mathbf{x} \rangle_1$ or $\langle \mathbf{x} \rangle_2$, follows an approximately classical trajectory, and let us suppose that the actual trajectory $\mathbf{x}(t)$ lies in ψ_2 only. Is there any sense in which we have *two* hydrogen atoms? The answer is no, because, once again, a true subquantum measurement could track the unique atomic trajectory $\mathbf{x}(t)$ (without affecting ψ).

This last example has a parallel in the macroscopic domain, to be discussed in the next section. Before proceeding, it will prove useful to consider the present example further. In particular, one might argue that each packet ψ_1 and ψ_2 *behaves like* a hydrogen atom, under operations defined by changes in the external potential V. Specifically, the motion of the empty packet ψ_1 will respond to changes in V, in exactly the same way as will the motion of the occupied packet ψ_2. One might then claim that, if one regards each packet as physically real, one may as well conclude that there really are two hydrogen atoms present. But this argument fails, because the similarity of behaviour of the two packets holds only under the said restricted class of operations (that is, modifying the classical potential V). In pilot-wave theory, in principle, other experimental operations are possible, under which the behaviours of ψ_1 and ψ_2 will be quite different.

For example, suppose one first carries out an ideal subquantum measurement, which shows that the particle is in the packet ψ_2. One may then carry out an additional experiment—say an ordinary quantum experiment, using a piece of macroscopic apparatus—designed to find out whether or not a given packet is occupied. One may *predict* that, in the second experiment, if the operation is performed on packet ψ_1 the apparatus pointer will point to 'unoccupied', while if the operation is performed on ψ_2 the pointer will point to 'occupied'.[10] It will then become operationally apparent that ψ_1 consists solely of a bundle

---

[10] In quantum theory too, of course, the second experiment will always give different results for the two packets. But the outcome will be random, making the operational difference between the packets less clear.

of the complex-valued ψ-field, whose centroid happens to be *simulating* the approximately classical motion of a hydrogen atom in an external field (under the said restricted class of operations).

It is of course hardly mysterious that in some circumstances one may have an ontological but empty ψ-packet whose motion approximately traces out the trajectory of a classical body—just as, in some circumstances, a localized classical electromagnetic pulse travelling through an appropriate medium (with variable refractive index) might trace out a trajectory similar to that of a moving body. In both cases, it would be clear from other experiments that the moving pulse is not really a moving body.

6 'MACROSCOPIC' MANY WORLDS?

Let us now ask if there is any sense in which pilot-wave theory contains many worlds at the 'macroscopic' level.

We shall begin with an utterly unrealistic example, involving a superposition of two 'Ehrenfest' packets each (supposedly) representing a classical macroscopic world. This example has the virtue of illustrating the Claim in what we believe to be its strongest possible form. We shall see that, even for this example, the Claim may be straightforwardly refuted, along the lines given in the last section for the case of the hydrogen atom.

We then turn to a further unrealistic example, involving a superposition of two delocalized 'WKB' packets which, again, are each supposed to represent a classical macroscopic world. This example has the virtue of showing that, if one cannot point to some piece of localized 'Ψ-stuff' following an alternative classical trajectory, then the Claim simply cannot be formulated. The lesson learned from this example is then readily applied to realistic cases with decoherence, for which the wavefunctions involved are also generally delocalized, and for which, therefore, the Claim again cannot be formulated.

The Claim in a 'Strong Form'

Let us again consider an 'Ehrenfest' superposition

$$\Psi(q, t) = \frac{1}{\sqrt{2}} \left(\Psi_1(q, t) + \Psi_2(q, t) \right),$$

where now the configuration q represents not just a single hydrogen atom but all the contents of a macroscopic region—for example, a region including the Earth, with human experimenters, apparatus, and so on. We shall imagine that the centroids $\langle q \rangle_1$, $\langle q \rangle_2$ of the respective packets Ψ_1, Ψ_2 follow approximately classical trajectories, corresponding to alternative histories of events on Earth.

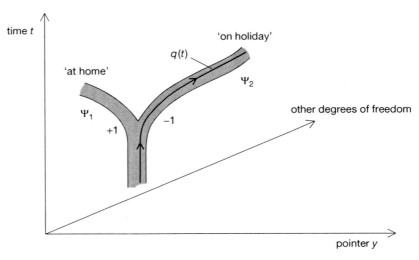

Figure 1. The Claim in a 'strong form'.

This is of course not at all a realistic formulation of the classical limit for a complex macroscopic system: wavepackets spread, and they do so particularly rapidly for chaotic systems. But we shall ignore this for a moment, because the example is nevertheless instructive.

Let us assume that Ψ consists initially of a single narrow packet, and that the subsequent splitting of Ψ into the (non-overlapping) branches Ψ_1, Ψ_2 occurs as a result of a 'quantum measurement' with two possible outcomes $+1$ and -1. (See Fig. 1.) One might imagine that, at first, the branches Ψ_1, Ψ_2 develop a non-overlap with respect to the apparatus pointer coordinate y, which then generates a non-overlap with respect to other (macroscopic) degrees of freedom—beginning, perhaps, with variables in the eye and brain of the experimenter who looks at the pointer. We may imagine that it had been decided in advance that if the outcome were $+1$, the experimenter would stay at home; while if the outcome were -1, the experimenter would go on holiday. These alternative histories for the experimenter are supposed to be described by the trajectories of the narrow packets Ψ_1 and Ψ_2 (whose arguments include all the relevant variables, constituting the centre-of-mass of the experimenter, his immediate environment, the plane he may or may not catch, and so on). Let us assume that the actual de Broglie–Bohm trajectory $q(t)$ ends in the support of Ψ_2, as shown in Fig. 1.

One could of course extend the example to superpositions of the form $\Psi = \Psi_1 + \Psi_2 + \Psi_3 + \ldots$, where Ψ_1, Ψ_2, $\Psi_3, \ldots$ are non-overlapping narrow packets that trace out—in configuration space—approximately classical motions corresponding to alternative macroscopic histories of the world, with each history containing, in the words of Brown and Wallace, 'cells, dust motes, cats, people, wars and the like'.

Now, with these completely unrealistic assumptions, the Claim seems to be at its strongest. For if Ψ is ontological, then in the example of Fig. 1 the narrow packets Ψ_1 and Ψ_2 are both real objects moving along approximately classical paths in configuration space. There is certainly *something real* moving along each path. One of the paths has an extra component too—the actual configuration $q(t)$—but even so the fact remains that something real is moving along the other path as well.

This situation seems to be the strongest possible realization of the Claim. One might say, for example with Brown and Wallace (Section 4 above), that '[t]he "empty wavepackets" in the configuration space which the corpuscles do not point at are none the worse for its absence'.[11] One might assert that here there really are two macroscopic worlds, one built from Ψ_1 alone, and one built from Ψ_2 together with q. And again, as in the case of the hydrogen atom discussed in Section 5, one might argue that there is no difference in the behaviour of these two worlds, and that the motion of Ψ_1 represents a world every bit as bona fide as the world represented by Ψ_2 (together with q, which one might assert is superfluous).

But again, as in the case of the hydrogen atom, pilot-wave theory tells us that a remote experimenter with access to non-equilibrium particles could in principle track the true history $q(t)$, without affecting Ψ. Further, once it is known which packet is empty and which not, the experimenter could perform additional experiments showing that Ψ_1 and Ψ_2 (predictably) behave *differently* under certain operations. Again, the empty packet is merely simulating a classical world, and the simulation holds only under a class of operations more restrictive than those allowed in pilot-wave theory. The situation is conceptually the same as in the case of the single hydrogen atom.[12]

We conclude that the Claim fails, even in a 'strong form'.

The Claim in a 'Weak Form'

Before considering more realistic approaches (with decoherence), it is instructive to reconsider the above scenario in terms of a different—and equally unrealistic—approach to the classical limit, namely the WKB approach, in which the amplitude of Ψ is taken to vary slowly over relevant lengthscales. It is often said that the resulting wavefunction may be 'associated with' a family of classical trajectories, defined by the equation $p = \nabla S$ giving the classical momentum p in

[11] This is not to suggest that Brown and Wallace, or other proponents of the Claim, actually make the Claim in the 'strong' form given here. We consider this form first, because it seems to us to be the strongest possible version of the argument.

[12] Except, one might argue, if one is talking about the 'whole universe'. One could restrict the argument to approximately independent regions; this does not seem an essential point.

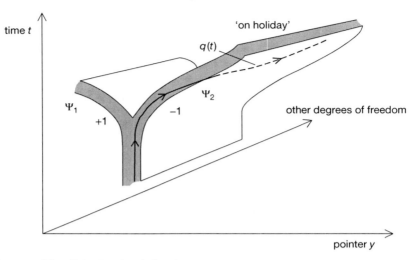

Figure 2. The Claim in a 'weak form'.

terms of the phase gradient. (This approach is frequently used, for example, in quantum cosmology.) Where such trajectories come from is not clear in standard quantum theory, but in pilot-wave theory it is clear enough: in the WKB regime, the de Broglie–Bohm trajectory $q(t)$ (within the extended wave) will indeed follow a classical trajectory defined by $p = \nabla S$.

Now let the superposition

$$\Psi(q, t) = \frac{1}{\sqrt{2}}\left(\Psi_1(q, t) + \Psi_2(q, t)\right)$$

be composed of two non-overlapping 'WKB packets' Ψ_1, Ψ_2, that formed from the division of a single WKB packet Ψ, where again q represents the contents of a macroscopic region including the Earth. As in the earlier example, we imagine that the division occurred because a quantum experiment was performed, with two possible outcomes indicated by a pointer coordinate y: and again, Ψ_1 corresponds to the outcome $+1$, while Ψ_2 corresponds to the outcome -1, and the actual $q(t)$ ends in the support of Ψ_2. Unlike the earlier example, though, in this case the packets Ψ_1, Ψ_2 are narrow with respect to y but broad with respect to the other (relevant) degrees of freedom—so broad, in fact, that with respect to these other degrees of freedom the packets are effectively plane waves. The only really significant difference between Ψ_1 and Ψ_2 is in their support with respect to y. (See Fig. 2.)

To be sure, this is not a realistic model of the macroscopic world, any more than the Ehrenfest model was. But it is instructive to see the effect that this alternative approach has on the Claim.

Under the above assumptions, the actual trajectory $q(t)$ will be approximately classical (except in the small branching region), and might be taken to correctly model the macroscopic history with outcome -1 and the experimenter going on holiday. But is there now any other discernible realization of an alternative classical macroscopic motion, such as the experimenter staying at home? Clearly not. While the empty branch Ψ_1 is ontological, it is spread out over all degrees of freedom except y, so that its time evolution does *not* trace out a trajectory corresponding to an approximately classical alternative motion. The experimenter 'staying at home' is nowhere to be seen. Unlike in the Ehrenfest case, one cannot point to some piece of localized 'Ψ-stuff' following an alternative classical trajectory.

Of course, different initial configurations $q(0)$ (with the same initial Ψ) would yield different trajectories $q(t)$. And the 'information' about these alternative paths certainly exists in a mathematical sense, in the structure of the complex field Ψ. But there is no reason to ascribe anything other than mathematical status to these alternative trajectories—just as we saw in Section 2, for the analogous classical case of a test particle moving in an external electromagnetic field or in a background spacetime geometry. The alternative trajectories are mathematical, not ontological.

Realistic Models (with Environmental Decoherence)

A more realistic account of the macroscopic, approximately classical realm may be obtained from models with environmental decoherence. (For a review, see Zurek [2003].)

Consider a system with configuration q, coupled to environmental degrees of freedom $y = (x_1, x_2, \ldots, x_N)$. For a pure state the wavefunction is $\Psi(q,y,t)$, and one often considers mixtures with a density operator

$$\hat{\rho}(t) = \sum_\alpha p_\alpha |\Psi_\alpha(t)\rangle \langle \Psi_\alpha(t)|.$$

(For example, in 'quantum Brownian motion', the system is a single particle in a potential and the environment consists of a large number of harmonic oscillators in a thermal state.) By tracing over y one obtains a reduced density operator for the system, with matrix elements

$$\rho_{\text{red}}(q, q', t) \equiv \sum_\alpha p_\alpha \int dy \Psi_\alpha(q, y, t) \Psi_\alpha^*(q', y, t),$$

from which one may define a quasi-probability distribution in phase space for the system:

$$W_{\text{red}}(q, p, t) \equiv \frac{1}{2\pi} \int dz\, e^{ipz} \rho_{\text{red}}(q - z/2, q + z/2, t)$$

(the reduced Wigner function). In certain conditions, one obtains an approximately non-negative function $W_{red}(q,p,t)$ whose time evolution approximates that of a classical phase-space distribution.

For some elementary systems, such as a harmonic oscillator, the motion of a narrowly localized packet $W_{red}(q,p,t)$ can trace out a thin 'tube' approximating a classical trajectory in phase space (Zurek et al. [1993]). However, such simple quantum-classical correspondence breaks down for chaotic systems, because of the rapid spreading of the packet: even an initial minimum-uncertainty packet spreads over macroscopic regions of phase space within experimentally accessible timescales (Zurek [1998]). On the other hand, at least for some examples it can be shown that, even in the chaotic case, the evolution of $W_{red}(q,p,t)$ approximates the evolution of a classical phase-space distribution $W_{class}(q,p,t)$ (a Liouville flow with a diffusive contribution from the environment), where both distributions *rapidly delocalize* (Habib et al. [1998]; Zurek [2003 pp.745–7]).[13]

In pilot-wave theory, a mixed quantum state is described by a preferred decomposition of $\hat{\rho}$ into a statistical mixture (with weights p_a) of ontological pilot waves Ψ_a (Bohm and Hiley [1996]). For a given element of the ensemble, the de Broglian velocity of the configuration is determined by the actual pilot wave Ψ_a. (A different decomposition generally yields different velocities, and so is physically distinct at the fundamental level.) Now, the pilot-wave theory of quantum Brownian motion has been studied by Appleby [1999]. Under certain conditions it was found that, as a result of decoherence, the de Broglie–Bohm trajectories of the system become approximately classical (as one might have expected). While Appleby made some simplifying assumptions in his analysis, pending further studies of this kind it is reasonable to assume that Appleby's conclusions hold more generally.

We may now evaluate the Claim in the context of realistic models. First of all, as in the unrealistic examples considered above, the Claim fails because an ideal subquantum measurement will always show that there is just one trajectory $q(t)$; and further experiments will show that empty wavepackets (predictably) behave differently from packets containing the actual configuration. This alone suffices to refute the Claim. Even so, it is interesting to ask if it is possible to have localized ontological packets ('built out of Ψ') whose motions execute alternative classical histories: that is, it is interesting to ask if the 'strong form' of the Claim discussed above—which in any case fails, but is still rather intriguing—could ever occur in practice in realistic models. The answer, again, is no.

For an elementary non-chaotic system, one can obtain a narrow 'Wigner packet' $W_{red}(q,p,t)$ approximating a classical trajectory, and one could also have a superposition of two or more such packets (with macroscopic separations).

---

[13] The examples are based on the weak-coupling, high-temperature limit of quantum Brownian motion. The system consists of a single particle moving in one dimension in a classically chaotic potential.

One might then argue that, since W_{red} is built out of Ψ, we have (in a realistic setting, with decoherence) something like the 'strong form' of the Claim discussed above. However, the models usually involve a mixture of Ψs, of which W_{red} is not a local functional. So the ontological status of a narrow packet W_{red} is far from clear. But even glossing over this, having a narrow packet W_{red} following an approximately classical path is in any case unrealistic in a world containing chaos, where, as we have already stated, one can show only that W_{red}—an approximately non-negative function, with a large spread over phase space—has a time evolution that approximately agrees with the time evolution of a classical (delocalized) phase-space distribution; that is, W_{red} follows an approximately Hamiltonian or Liouville flow (with a diffusive contribution). Again, one cannot obtain anything like 'localized ontological Ψ-stuff' (or something locally derived therefrom) executing an approximately classical trajectory—not even for one particle in a chaotic potential, and certainly not for a realistic world containing turbulent fluid flow, double pendulums, people, wars, and so on.

One *can* obtain localized trajectories from a quantum description of a chaotic system, if the system is continuously measured—which in practice involves an experimenter continuously monitoring an apparatus or environment that is interacting with the system (Bhattacharya et al. [2000]). Such trajectories for the Earth and its contents might in principle be obtained by monitoring the environment (the interstellar medium, the cosmic microwave background, etc.), but in the absence of an experimenter performing the required measurements it is difficult to see how this could be relevant to our discussion. And in any case, in a pilot-wave treatment, there is no reason why such a procedure would yield 'localized ontological Ψ-stuff' executing the said trajectories.

In a realistic quantum-theoretical model, then, the outcome is a highly delocalized distribution $W_{red}(q,p,t)$ on phase space, obeying an approximately Hamiltonian or Liouville evolution (with a diffusive contribution). As in the unrealistic WKB example above, in pilot-wave theory there will be one trajectory for each system. And, while different initial conditions will yield different trajectories, there is no reason to ascribe anything other than mathematical status to these alternatives—just as in the analogous classical case of a test particle moving in an external field or background geometry. Once again, the alternative trajectories are mathematical, not ontological.

Of course, given such a distribution $W_{red}(q,p,t)$, *if one wishes* one may identify the flow with a set of trajectories representing parallel (approximately classical) worlds, as in the decoherence-based approach to many worlds of Saunders and Wallace. This is fair enough from a many-worlds point of view. But if we start from pilot-wave theory understood on its own terms, there is no motivation for doing so: such a step would amount to a reification of mathematical structure (assigning reality to all the trajectories associated with the velocity field at all points in phase space). If one does so reify, one has constructed a different physical theory, with a different ontology; one may do so if one wishes, but from a pilot-wave perspective there is no special reason to take this step.

Other Approaches to Decoherence

Finally, decoherence and the emergence of the classical limit has also been studied using the decoherent histories formulation of quantum theory.[14] In these treatments, there will still be no discernible 'localized ontological Ψ-stuff' following alternative classical trajectories, for realistic models containing chaos. Therefore, again, the 'strong form' of the Claim (which in any case fails by virtue of subquantum measurement) could never occur in practice.

7 FURTHER REMARKS

Many de Broglie–Bohm Worlds?

In the Saunders–Wallace approach to many worlds, one ascribes reality to the full set of trajectories associated with the reduced Wigner function $W_{\text{red}}(q,p,t)$ in the classical limit (for some appropriately defined macrosystem with configuration q). This raises a question. Why not *also* ascribe reality to the full set of de Broglie–Bohm trajectories outside the classical limit, for arbitrary (pure) quantum states, resulting in a theory of 'many de Broglie–Bohm worlds'?[15]

After all, just as W_{red} has a natural velocity field associated with it (on phase space), so an arbitrary wavefunction Ψ has a natural velocity field associated with it (on configuration space)—namely, de Broglie's velocity field derived from the phase gradient ∇S (or more generally, from the quantum current). In both cases, the velocity fields generate a set of trajectories, and one may ascribe reality to them all if one wishes. Why do so in the first case, but not in the second?

Furthermore, if the results due to Appleby [1999] (mentioned in Section 6) for quantum Brownian motion hold more generally, the parallel de Broglie–Bohm trajectories will reduce to the parallel classical trajectories in an appropriate limit; in which case, the theory of 'many de Broglie–Bohm worlds' will reproduce

---

[14] See, for example, Gell-Mann and Hartle [1993] and Halliwell [1998], as well as the reviews in this volume.

[15] Such a theory has, in effect, been considered by Tipler [2006].

the Saunders–Wallace multiverse in the classical limit, and will provide a simple and natural extension of it outside that limit—that is, one will have a notion of parallel worlds that is defined generally, even at the microscopic level, and not just in the classical-decoherent limit.[16]

However, since the de Broglie velocity field is single-valued, trajectories $q(t)$ cannot cross. There can be no splitting or fusion of worlds. The above 'de Broglie–Bohm multiverse' then has the same kind of 'trivial' structure that would be obtained if one reified all the possible trajectories for a classical test particle in an external field: the parallel worlds evolve independently, side by side. Given such a theory, on the grounds of Occam's razor alone, there would be a conclusive case for taking only one of the worlds as real.

On this point we remark that, in Deutsch's version of the Claim, if his word 'grooves' is interpreted as referring to the set of de Broglie–Bohm trajectories, then the Claim amounts to asserting that pilot-wave theory implies the de Broglie–Bohm multiverse. But again, because the parallel worlds never branch or fuse, it would be natural to reduce the theory to a single-world theory with only one trajectory.

A theory of many de Broglie–Bohm worlds, then, can only be a mere curiosity—a foil, perhaps, against which to test conventional Everettian ideas, but not a serious candidate for a physical theory. On the other hand, it appears to provide the basis for an argument against the Saunders–Wallace multiverse. For as we have seen, it is natural to extend the Saunders–Wallace multiverse to a deeper and more general de Broglie–Bohm multiverse.[17] And this, in turn, reduces naturally to a single-universe theory—that is, to standard de Broglie–Bohm theory. Thus, we have an argument that begins by extending the Saunders–Wallace worlds to the microscopic level, and ends by declaring only one of the resulting worlds to be real.

Quantum Non-equilibrium and Many Worlds

Since pilot-wave theory generally violates the Born rule, while conventional many-worlds theory (apparently) does not, on this ground alone any attempt to argue that the two theories are really the same must fail. Further, if such violations were discovered,[18] then Everett's theory would be disproved and that of de Broglie and Bohm vindicated.

---

[16] One need not think of this as 'adding' trajectories to the wavefunction; one could think of it as an alternative reading of physical structure already existing in the 'bare' wavefunction.

[17] It might be claimed that, outside the non-relativistic domain, such an extension is neither simple nor natural. However, the (deterministic) pilot-wave theory of high-energy physics has achieved a rather complete (if not necessarily final) state of development—for recent progress see Colin [2003], Colin and Struyve [2007], Struyve [2007], and Struyve and Westman [2007].

[18] See Valentini [2007, 2008a,b] for recent discussions of possible experimental evidence.

On the other hand, it might be suggested that violations of the Born rule could be incorporated into an Everett-type framework, by adopting the theory of 'many de Broglie–Bohm worlds' sketched above. Restricting ourselves for simplicity to the pure case, if one assumes a non-equilibrium probability measure $P_0 \neq |\Psi_0|^2$ on the set of (parallel) initial configurations $q(0)$, then for as long as relaxation to quantum equilibrium has yet to occur completely, one will obtain a non-equilibrium set of parallel trajectories $q(t)$, and one expects (in general) to find violations of the Born rule within individual parallel worlds.[19] If one accepts this, then observation of quantum non-equilibrium would not suffice to disprove many worlds (though of course conventional Everettian quantum theory *would* be disproved). On the other hand, however, as stated above it is natural to reduce the theory of many de Broglie–Bohm worlds to a single-world theory, and this is equally true in the non-equilibrium case. Therefore, the de Broglie–Bohm multiverse would not provide a plausible refuge for the Everettian faced with non-equilibrium phenomena.

Even so, it might be worth exploring the theory of many de Broglie–Bohm worlds with a non-equilibrium measure, in particular to highlight the assumptions made in the Deutsch–Wallace derivation of the Born rule (Deutsch [1999], Wallace [2003a]).

On Arguments Concerning 'Structure'

One might argue that the mathematical structure in the quantum state that is reified by many-worlds theorists plays such an explanatory and predictive role that it should indeed be regarded as real. To quote Wallace [2003b p.93]:

A tiger is any pattern which behaves as a tiger. . . . the existence of a pattern as a real thing depends on the usefulness—in particular, the explanatory power and predictive reliability—of theories which admit that pattern in their ontology.

However, the behaviour of a system depends on the allowed set of experimental operations. If one considers subquantum measurements, the patterns reified by many-worlds theorists will cease to be explanatory and predictive. From a pilot-wave perspective, then, such mathematical patterns are explanatory and predictive only in the confines of quantum equilibrium: outside that limited domain, subquantum measurement theory would provide a more explanatory and predictive framework.

At best, it can only be argued that, if approximately classical experimenters are confined to the quantum equilibrium state, so that they are unable to perform subquantum measurements, then they will encounter a phenomenological *appearance* of many worlds—just as they will encounter a

---

[19] On the other hand, quantum equilibrium for a multicomponent closed system implies the Born rule for measurements performed on subsystems (Valentini [1991a], Dürr et al. [1992]).

phenomenological appearance of locality, uncertainty, and of quantum physics generally.

On Arguments Concerning Computation

It might be argued that quantum computation provides evidence for the existence of many worlds (Deutsch [1985, 1997]). Deutsch asks 'how' and 'where' the supposedly huge number of parallel computations are performed, and has challenged those who doubt the existence of parallel universes to provide an explanation for quantum-computational processes such as Shor's factorization algorithm (Deutsch [1997 p.217]).

However, while it often used to be asserted that the advantages of quantum computation originated from quantum superposition, the matter has become controversial. Some workers, such as Jozsa [1998] and Steane [2003], claim that entanglement is the truly crucial feature. Further, the ability to find periods seems to be the mechanism underlying Shor's algorithm, and this is arguably more related to the 'wave-like' aspect of quantum physics than it is to superposition (Mermin [2007]).

Leaving such controversies aside, we know in any case that, in quantum equilibrium, pilot-wave theory yields the same predictions as ordinary quantum theory, including for quantum algorithms. In an assessment of precisely how pilot-wave theory provides an explanation for a specific quantum algorithm, it should be borne in mind that: (a) the theory contains an ontological pilot wave propagating in many-dimensional configuration space; (b) the theory is non-local; and (c) with respect to quantum 'measurements', the theory is contextual. There is then ample scope for exploring the pilot-wave-theoretical account of quantum-computational processes, if one wishes to do so, just as there is for any other type of quantum process.

8 COUNTER-CLAIM: A GENERAL ARGUMENT AGAINST MANY WORLDS

We have refuted the Claim, that pilot-wave theory is 'many worlds in denial'. Here, we put forward a Counter-Claim:

- Counter-Claim: *The theory of many worlds is unlikely to be true, because it is ultimately motivated by the puzzle of quantum superposition, which arises from a belief in eigenvalue realism, which is in turn based (ultimately) on the intrinsically unlikely assumption that quantum measurements should be modelled on classical measurements.*

We saw in Section 3 that quantum theorists call an experiment 'a measurement of ω' only because it formally resembles what *would have been* a correct measurement of ω had the system been classical. Thus, the system–apparatus interaction Hamiltonian is chosen by means of (for example) the mapping

$$H = a\omega p_y \longrightarrow \hat{H} = a\hat{\omega}\hat{p}_y, \qquad (6)$$

so that quantum 'measurements' are in effect modelled on classical measurements. That this is a mistake is clear from a pilot-wave perspective.[20] But the key point is more general, and does not depend on pilot-wave theory. In fact, it was made by Einstein in 1926 (see below).

The Argument

Everett's initial motivation for introducing many worlds was the puzzle of quantum superposition, in particular the apparent transfer of superposition from microscopic to macroscopic scales during a quantum measurement (Everett [1973 pp.4–6]). While our understanding of the theory today differs in many respects from Everett's, it is highly doubtful that the theory would ever have been proposed, were it not for the puzzle of quantum superposition.

Now, the puzzle of superposition stems from what we have called 'eigenvalue realism': the assignment of an ontological status to the eigenvalues of linear operators acting on the wavefunction. For if an initial wavefunction

$$\psi_0(x) = \sum_n c_n \phi_n(x)$$

is a superposition of different eigenfunctions $\phi_n(x)$ of $\hat{\omega}$ with different eigenvalues ω_n, then if one takes eigenvalue realism literally it appears as if all the values ω_n should somehow be regarded as simultaneous ontological attributes of a single system.

Why do so many physicists believe in eigenvalue realism? The answer lies, ultimately, in their belief in the quantum theory of measurement. For example, it is widely thought that an experimental operation described by the Hamiltonian operator $\hat{H} = a\hat{\omega}\hat{p}_y$ constitutes a correct measurement of an observable ω, as indicated by the value of the pointer coordinate y. To see that this leads to a belief in eigenvalue realism, consider a system with wavefunction $\phi_n(x)$. Under such an operation, the pointer y will indicate the value ω_n.

---

[20] In the classical limit of pilot-wave theory, emergent effective degrees of freedom have a purely mathematical correspondence with linear operators acting on the wavefunction. Physicists trapped in quantum equilibrium have made the mistake of taking this correspondence literally (Valentini [1992 pp.14–16, 19–29]; [1996 pp.50–1]).

Because the experimenter *believes* that this pointer reading provides a correct measurement, the experimenter will then believe that the system must have a property ω with value ω_n—that is, the experimenter will believe in eigenvalue realism.

Now, why do so many physicists believe that an operation described by (for example) $\hat{H} = a\hat{\omega}\hat{p}_y$ constitutes a correct measurement of ω, for any observable ω? The answer, as we have seen, is that the said operation formally resembles a classical measurement of ω, via the mapping (6).

We claim that this is the heart of the matter: it is widely assumed, in effect, that classical physics provides a reliable guide to measurement for non-classical systems. We claim further that this assumption is intrinsically unlikely, so that the conclusions stemming from it—eigenvalue realism, superposition of properties, multiplicity of worlds—are in turn intrinsically unlikely (Valentini [1992 pp.14–16, 19–29]; [1996 pp.50–1]).

The assumption is unlikely because, generally speaking, one cannot use a theory as an accurate guide to measurement outside the domain of validity of the theory. For experiment is theory-laden, and correct measurement procedures must be laden with the correct theory. As an example, consider what might happen if one used Newton's theory of gravity to interpret observations close to a black hole: one would encounter numerous puzzles and paradoxes that would be resolved only when the observations were interpreted using general relativity. It is intrinsically improbable that measurement operations taken from an older, superseded physics will remain valid in a fundamentally new domain for all possible observables. It is much more likely that a new domain will be better understood in terms of a new theory based on new concepts, with its own new theory of measurement—as shown by the example of general relativity, and indeed by the example of de Broglie's non-classical dynamics.[21]

'Einstein's Hot Water'

This very point was made by Einstein in 1926, in a well-known conversation with Heisenberg (Heisenberg [1971 pp.62–9]). This conversation is often cited as evidence of Einstein's view that observation is theory-laden. But a crucial element is usually missed: Einstein also warned Heisenberg that his treatment of observation was unduly laden with the superseded theory of classical physics, and that this would eventually cause trouble (Valentini [1992 p.15]; [1996 p.51]).

---

[21] In contrast with Bohr's unwarranted claim: 'The unambiguous interpretation of any measurement must be essentially framed in terms of the classical physical theories, and we may say that in this sense the language of Newton and Maxwell will remain the language of physicists for all time' (Bohr [1931]).

During the conversation, Heisenberg made the (at the time fashionable) claim that 'a good theory must be based on directly observable magnitudes' (p.63). Einstein replied that, on the contrary (p.63):

... it is quite wrong to try founding a theory on observable magnitudes alone. In reality the very opposite happens. *It is the theory which decides what we can observe.* [Italics added.]

Einstein added that there is a long, complicated path underlying any observation, which runs from the phenomenon, to the production of events in our apparatus, and from there to the production of sense impressions. And theory is required to make sense of this process:

Along this whole path ... we must be able to tell how nature functions ... before we can claim to have observed anything at all. Only theory, that is, knowledge of natural laws, enables us to deduce the underlying phenomena from our sense impressions.

Einstein's key point so far is that, as we have said, there is no a priori notion of how to perform a correct measurement: one requires some knowledge of physics to do so. If we wish to design a piece of apparatus that will correctly measure some property ω of a system, then we need to know the correct laws governing the interaction between the system and the apparatus, to ensure that the apparatus pointer will finish up pointing to the correct reading. (One cannot, for example, design an ammeter to measure electric current without some knowledge of electromagnetic forces.)

Now, Einstein went on to note that, when new experimental phenomena are discovered—phenomena that require the formulation of a new theory—in practice the old theory is at first assumed to provide a reliable guide to interpreting the observations (pp.63–4):

When we claim that we can observe something new, we ought really to be saying that, although we are about to formulate new natural laws that do not agree with the old ones, we nevertheless assume that the existing laws—covering the whole path from the phenomenon to our consciousness—function in such a way that we can rely upon them and hence speak of 'observations'.

Note that this is a practical necessity, for the new theory has yet to be formulated. However—and here is the crucial point—once the new theory *has* been formulated, one ought to be careful to use the new theory to design and interpret measurements, and not continue to rely on the old theory to do so. For one may well find that consistency is obtained only when the new laws are found *and applied to the process of observation.* If one fails to do this, one is likely to cause difficulties. That Einstein saw this very point is clear from a subsequent passage (p.66):

I have a strong suspicion that, precisely because of the problems we have just been discussing, your theory will one day get you into hot water.... When it comes to observation, you behave as if everything can be left as it was, that is, as if you could use the old descriptive language.

Here, then, is Einstein's warning to Heisenberg: not to interpret observations of quantum systems using the 'old descriptive language' of classical physics. The point, again, is that while observation is in general theory-laden, in quantum theory observations are incorrectly laden with a *superseded* theory (classical physics), and this will surely lead to trouble.

We claim that the theory of many worlds is precisely an example of what one might call 'Einstein's hot water'. Specifically, the apparent multiplicity of the quantum domain is an illusion, caused by an over-reliance on a superseded (classical) physics as a guide to observation and measurement—a mistake that is the ultimate basis of the belief in eigenvalue realism, which in turn led to the puzzle of superposition and to Everett's valiant attempt to resolve that puzzle.

9 CONCLUSION

Pilot-wave theory is intrinsically non-classical, with its own ('subquantum') theory of measurement, and it is in general a 'non-equilibrium' theory that violates the quantum Born rule. From the point of view of pilot-wave theory itself, an apparent multiplicity of worlds at the microscopic level (envisaged by some theorists) stems from the generally mistaken assumption that eigenvalues have an ontological status ('eigenvalue realism'), which in turn ultimately derives from the generally mistaken assumption that "quantum measurements" are true and proper measurements.

At the macroscopic level, it might be thought that the universal (and ontological) pilot wave can develop non-overlapping and localized branches that evolve just like parallel classical worlds. But in fact, such localized branches are unrealistic (especially over long periods of time, and even for short periods of time in a world containing chaos). And in any case, subquantum measurements could track the actual de Broglie–Bohm trajectory, so that in principle one could distinguish the branch containing the configuration from the empty ones, where the latter would be regarded merely as concentrations of a complex-valued configuration-space field.

In realistic models of decoherence, the pilot wave is delocalized, and the identification of a set of parallel (approximately) classical worlds does not arise in terms of localized pieces of actual 'Ψ-stuff' executing approximately classical motions. Instead, such identification amounts to a reification of purely mathematical trajectories—a move that is fair enough from a many-worlds perspective, but which is unnecessary and unjustified from a pilot-wave perspective because according to pilot-wave theory there is nothing actually moving along any of the trajectories except one (just as in the classical theory of a test particle in an external field or background spacetime geometry). In addition to being unmotivated, such reification begs the question of why the mathematical trajectories should not also be reified outside the classical limit for general wavefunctions, resulting in a theory of 'many de Broglie–Bohm worlds' (which in turn naturally reduces to a single-world theory).

Properly understood, pilot-wave theory is not 'many worlds in denial': it is a different physical theory. Furthermore, from the perspective of pilot-wave theory itself, many worlds are an illusion. And indeed, even leaving pilot-wave theory aside, we have seen that the theory of many worlds is rooted in the intrinsically unlikely assumption that quantum measurements should be modelled on classical measurements, and is therefore in any case unlikely to be true.

Acknowledgements

This work was partly supported by grant RFP1-06-13A from the Foundational Questions Institute (fqxi.org). For their hospitality, I am grateful to Carlo Rovelli and Marc Knecht at the Centre de Physique Théorique (Luminy), to Susan and Steffen Kyhl in Cassis, and to Jonathan Halliwell at Imperial College London.

References

Appleby, D.M. [1999], 'Bohmian trajectories post-decoherence', *Foundations of Physics*, 29, 1885–916.

Bacciagaluppi, G. and A. Valentini, [2009], *Quantum Theory at the Crossroads: Reconsidering the 1927 Solvay Conference*, Cambridge University Press, Cambridge. Available online at quant-ph/0609184.

Bell, J.S. [1987], *Speakable and Unspeakable in Quantum Mechanics*, Cambridge University Press, Cambridge.

Bhattacharya, T., S. Habib, and, K. Jacobs [2000], 'Continuous quantum measurement and the emergence of classical chaos', *Physical Review Letters* 85, 4852–5.

Bohm, D. [1952a], 'A suggested interpretation of the quantum theory in terms of "hidden" variables, I', *Physical Review* 85, 166–79.

——— [1952b], 'A suggested interpretation of the quantum theory in terms of "hidden" variables, II', *Physical Review* 85, 180–93.

Bohm, D. and, B.J. Hiley [1996], 'Statistical mechanics and the ontological interpretation' *Foundations of Physics* 26, 823–46.

Bohr, N. [1931], 'Maxwell and modern theoretical physics', *Nature* 128, 691–2. Reprinted in *Niels Bohr: Collected Works*, vol. 6, ed. J. Kalckar. North-Holland, Amsterdam 1985, p.357.

Brown, H.R. and D. Wallace, [2005], 'Solving the measurement problem: de Broglie–Bohm loses out to Everett', *Foundations of Physics* 35, 517–40.

Colin, S. [2003], 'A deterministic Bell model', *Physics Letters* A 317, 349–58. Available online at quant-ph/0310055.

Colin, S. and W. Struyve, [2007], 'A Dirac sea pilot-wave model for quantum field theory', *Journal of Physics A: Mathematical and Theoretical* **40**, 7309–41. Available online at quant-ph/0701085.

de Broglie, L. [1928], 'La nouvelle dynamique des quanta', in *Électrons et Photons: Rapports et Discussions du Cinquième Conseil de Physique'*, Gauthier-Villars, Paris, pp.105–32. [English translation: Bacciagaluppi, G. and A. Valentini, (2009).]

Deutsch, D. [1985], 'Quantum theory, the Church-Turing principle and the universal quantum computer', *Proceedings of the Royal Society of London* **A 400**, 97–117.

—— [1986], 'Interview, in *The Ghost in the Atom*, P.C.W. Davies and J.R. Brown (eds). Cambridge, University Press, Cambridge, pp.83–105.

—— [1996], 'Comment on Lockwood', *British Journal for the Philosophy of Science* **47**, 222–8.

—— [1997], *The Fabric of Reality*, London, Penguin.

—— [1999], 'Quantum theory of probability and decisions', *Proceedings of the Royal Society of London* **A 455**, 3129–37.

—— [1996], 'Bohmian mechanics as the foundation of quantum mechanics', in *Bohmian Mechanics and Quantum Theory: an Appraisal*, J.T. Cushing et al. (eds), Kluwer, Dordrecht, pp.21–44.

Dürr, D., Goldstein, S., and Zanghì, N. [1992], 'Quantum equilibrium and the origin of absolute uncertainty', *Journal of Statistical Physics*, **67**, 843–907.

—— 'Bohmian mechanics and the meaning of the wavefunction', in *Experimental Metaphysics: Quantum Mechanical Studies for Abner Shimony*, R.S. Cohen et al. (eds), Kluwer, Dordrecht, pp.25–38.

Everett, H. [1973], 'The theory of the universal wave function', in *The Many-Worlds Interpretation of Quantum Mechanics*, B.S. DeWitt and N. Graham (eds), Princeton University Press, Princeton, pp.3–140.

Gell-Mann, M. and, J.B. Hartle [1993], 'Classical equations for quantum systems', *Physical Review* **D 47**, 3345–82.

Habib, S., K. Shizume, and, W.H. Zurek [1998], 'Decoherence, chaos, and the correspondence principle', *Physical Review Letters* **80**, 4361–5.

Halliwell, J.J. [1998], 'Decoherent histories and hydrodynamic equations', *Physical Review* **D 58**, 105015.

Heisenberg, W. [1971], *Physics and Beyond*, Harper & Row, New York.

Holland, P.R. [1993], *The Quantum Theory of Motion: An Account of the de Broglie–Bohm Causal Interpretation of Quantum Mechanics*, Cambridge University Press, Cambridge.

Jozsa, R. [1998], 'Entanglement and quantum computation', in *The Geometric Universe: Science, Geometry, and the Work of Roger Penrose*, Oxford University Press, Oxford, pp.369–79.

Mermin, N.D. [2007], 'What has quantum mechanics to do with factoring?' *Physics Today* **60** (April 2007), 8–9.

Pearle, P. and, A. Valentini [2006], 'Quantum mechanics: generalizations', in *Encyclopaedia of Mathematical Physics*, J.-P. Françoise et al. (eds), Elsevier, Amsterdam, pp.265–76. Available online at quant-ph/0506115.

Rovelli, C. [2004], *Quantum Gravity*, Cambridge University Press, Cambridge.

Steane, A.M. [2003], 'A quantum computer only needs one universe'. Available online at arXiv:quant-ph/0003084v3 (24 March 2003).

Struyve, W. [2007], 'De Broglie–Bohm Field beables for quantum field theory'. Available online at arXiv:0707.3685.

Struyve, W. and, A. Valentini [2009], 'De Broglie–Bohm guidance equations for arbitrary Hamiltonians', *Journal of Physics A: Mathematical and Theoretical* 42, 035301. Available online at arXiv:0808.0290.

Struyve, W. and, H. Westman [2007], 'A minimalist pilot-wave model for quantum electrodynamics', *Proceedings of the Royal Society of London* A 463, 3115–29 Available online at arXiv:0707.3487.

Tipler, F.J. [1987], 'Non-Schrödinger forces and pilot waves in quantum cosmology', *Classical and Quantum Gravity* 4, L189–L195.

—— [2006], 'What about quantum theory? Bayes and the Born interpretation'. Available online at arXiv:quant-ph/0611245.

Valentini, A. [1991a], 'Signal-locality, uncertainty, and the subquantum *H*-theorem, I', *Physics Letters* A 156, 5–11.

—— [1991b], 'Signal-locality, uncertainty, and the subquantum *H*-theorem, II', *Physics Letters* A 158, 1–8.

—— [1992], 'On the pilot-wave theory of classical, quantum and subquantum physics', PhD thesis, International School for Advanced Studies, Trieste, Italy. Available online at http://www.sissa.it/ap/PhD/Theses/valentini.pdf.

—— [1996], 'Pilot-wave theory of fields, gravitation and cosmology', in *Bohmian Mechanics and Quantum Theory: an Appraisal*, J.T. Cushing et al. (eds), Kluwer, Dordrecht, pp.45–66.

—— [2001], 'Hidden variables, statistical mechanics and the early universe', in *Chance in Physics: Foundations and Perspectives*, J. Bricmont et al. (eds), Springer-Verlag, Berlin, pp.165–81. Available online at quant-ph/0104067.

—— [2002b], 'Subquantum information and computation', *Pramana—Journal of Physics* 59, 269–77. Available online at quant-ph/0203049.

—— [2004a], 'Universal signature of non-quantum systems', *Physics Letters* A 332, 187–93. Available online at quant-ph/0309107.

—— [2004b], 'Black holes, information loss, and hidden variables', Available online at arXiv:hep-th/0407032.

—— [2007], 'Astrophysical and cosmological tests of quantum theory', *Journal of Physics A: Mathematical and Theoretical* 40, 3285–303. Available online at arXiv:hep-th/0610032.

—— [2008a], 'Inflationary cosmology as a probe of primordial quantum mechanics'. Available online at arXiv:0805.0163.

—— [2008b], 'De Broglie–Bohm prediction of quantum violations for cosmological super-Hubble modes'. Available online at arXiv:0804.4656.

Valentini, A. and, H. Westman [2005], 'Dynamical origin of quantum probabilities', *Proceedings of the Royal Society of London* A 461, 253–72.

Wallace, D. [2003a], 'Everettian rationality: defending Deutsch's approach to probability in the Everett interpretation', *Studies in History and Philosophy of Modern Physics* 34, 415–39.

—— [2003b], 'Everett and structure,' *Studies in History and Philosophy of Modern Physics* 34, 87–105.

Zeh, H.D. [1999], 'Why Bohm's quantum theory?' *Foundations of Physics Letters* 12, 197–200.

Zurek, W.H. [1998], 'Decoherence, chaos, quantum-classical correspondence, and the algorithmic arrow of time', *Physica Scripta* **T76**, 186–98.

——[2003], 'Decoherence, einselection, and the quantum origins of the classical', *Reviews of Modern Physics*, **75**, 715–75.

Zurek, W.H., S. Habib, and, J.P. Paz [1993], 'Coherent states via decoherence', *Physical Review Letters* **70**, 1187–90.

Reply to Valentini, 'De Broglie–Bohm Pilot-Wave Theory: Many Worlds in Denial?'

Harvey Brown

1 INTRODUCTION

What a privilege it is to be invited to reply to Antony Valentini's paper. If anyone is capable of persuading me of the plausibility of the pilot-wave picture of quantum reality, it is he. But I am not convinced that his defence of pilot-wave theory from the accusation that it is really Everett theory encumbered with otiose ontology (which Valentini calls 'the Claim') is successful. In the space available, I cannot do justice to all of his arguments, so I will restrict myself to what I take to be the central ones.

On a number of occasions in his paper, Valentini stresses the philosophical point that theories should be assessed on their own terms—that it is unfair to criticize a theory for failing to concur with assumptions that 'make sense only in rival theories'. Valentini argues that the Everett and pilot-wave pictures differ on their own terms for several reasons.

First, the 'correct and natural viewpoint' about pilot-wave ontology is that which Valentini attributes to de Broglie, according to which physical systems, apparatuses, people, etc., are built from the configuration variable q. In particular, all macroscopic, observable phenomena, including the very stuff of our mental sensations, supervene on the configurations of the punctiform corpuscles hypothesized to coexist with the wavefunction (pilot-wave). Although it is not clear that this viewpoint—let us call it *the matter assumption*—is common to all variants of the de Broglie–Bohm approach, as Valentini admits in Section 3,[1] modern disagreements within the camp do seem to concentrate on distinct issues such as the reality or otherwise of the pilot-wave, or whether the appropriate formulation of the corpuscle dynamics is first- or second-order. Second, and

---

[1] In Bohm's 1952 work it would seem that the role of the corpuscles in the measurement context is indexical, picking out the relevant component of the wavefunction that itself describes macroscopic physical systems; see Brown and Wallace [2005].

more significantly, Valentini stresses the role of non-equilibrium statistics in (his own version of) pilot-wave theory. This allows him to assert that the theory is '*not* a mere alternative formulation of quantum theory'.

Finally, Valentini provides a 'counter-claim', to the effect that the Everett picture is motivated by erroneous reasoning, and thus is unlikely to be true. In what follows, I will discuss each of these arguments in turn.

2 ASSESSING PILOT-WAVE THEORY ON ITS OWN TERMS

A key passage in the paper occurs at the end of the discussion in Section 6 of the role of decoherence in the physics of the measurement process.

> Of course, . . . , *if one wishes* one may identify the flow with a set of trajectories representing parallel (approximately classical) worlds, as in the decoherence-based approach to many worlds of Saunders and Wallace. This is fair enough from a many-worlds point of view. But if we start from pilot-wave theory understood on its own terms, there is no motivation for doing so: such a step would amount to a reification of mathematical structure (assigning reality to all the trajectories asociated with the velocity field at all points in phase space). If one does so reify, one has constructed a different physical theory, with a different ontology: one may do so if one wishes, but from a pilot-wave perspective there is no special reason to take this step.

The trouble is that this argument looks more like a restatement of the rival positions than a critical comparison of them, or at any rate a defence of the pilot-wave from the Claim. At the risk of belabouring this point, let me tell a story.

Prof. X has just published the latest version of his dualist philosophy of mind, which lies somewhere between solipsism and scepticism concerning other minds. Prof. X hypothesizes the existence of a mental substance attached to his own person—he sees no other way of accounting for his own consciousness, and qualia in particular. But he rejects solipsism and Berkeleian idealism, believing in the existence of an external material world including other persons. And through a questionable application of Occam's razor, Prof. X argues that he can save appearances by denying mental substances, and hence (in his view) consciousness, to persons other than himself. Others may act as if they were conscious, but there was no need to go so far as to postulate that they actually are conscious.[2] In her response, Prof. X's arch critic, Prof. Y, reiterates a point made widely in the literature, namely that to account for his own consciousness, Prof. X need not appeal to dualism and the existence of mental substance; he could avail himself

---

[2] This story may not be as contrived as it might seem. If one adopts van Fraassen's constructive empiricism, it is not entirely clear, to me at least, how any agent is supposed to avoid agnosticism about the existence of other minds.

of a materialist, functionalist theory of mind. Were he to do this, Prof. X would of course have to conclude that he lives in a world with many minds. But Prof. X rebuts Prof. Y as follows:

Of course, *if one wishes* one may take the view that the behaviour and physical constitution of other persons are jointly evidence for the existence of other minds. This is fair enough from the point of view of a functionalist philosophy of mind. But if we start from my dualist theory understood on its own terms, there is no motivation for doing so: such a step would amount to postulating unnecessary entities. If one does so postulate, one has constructed a different theory, with a different ontology: one may do so if one wishes, but from my dualist perspective there is no special reason to take this step.

Well, functionalists like Prof. Y would be forgiven for feeling a degree of frustration at this reply. Their basic claim is that dualism is unnecessary for consciousness, and therefore that Prof. X's argument, rather than exploiting Occam's razor, violates it *ab initio*. If they are reasonable, functionalists should expect debate on their basic claim (which cannot be regarded as obviously true), but they will naturally regard Prof. X's assertion that they are failing to assess his theory on its own terms as beside the point.

The analogy in pilot-wave theory to dualism, and in particular to mental substance, in this story is obviously the matter assumption. Why impose it? Why is it necessary within quantum mechanics to understand the nature of physical systems, apparatuses, people, etc., in terms of configurations of hypothetical point corpuscles? If it can be shown that the wavefunction or pilot-wave is structured enough to do the job, why go further?

For many workers in quantum mechanics, the answer is clear: because without further ado unitary quantum theory faces the measurement problem. Quantum mechanics generically predicts a superposed state widely interpreted as a bizarre schizophrenia of distinct measurement outcomes. Pilot-wave theory by way of the matter assumption restores sanity, or at any rate a single definite measurement outcome.[3] But is it necessary to follow this route?

If the object of the exercise is to save the appearances, it is not obviously so. Everettians plausibly claim that the multiplicity of outcomes (in the sense defined by the Saunders–Wallace decoherence analysis of the superposed wavefunction) is not actually schizophrenia in any observable sense; it is consistent with experience.

If both Everett and pilot-wave theories save the appearances, it might seem that choosing between them is a question of taste. But this would be misleading.

---

[3] Valentini complains in Section 3 that the recent critique of pilot-wave theory by Brown and Wallace [2005] is 'framed as if the measurement problem were the prime motivation for considering pilot-wave theory in the first place. As a matter of historical fact, this is false.' Indeed, much of the historical discussion in [2005] purports to show that Bohm was unaware of the measurement problem in 1952. But it is hard to imagine a more significant selling-point for pilot-wave theory than its supposed ability to solve the measurement problem, whatever de Broglie's and Bohm's original motivations were.

Let us consider the case of advocates of the view that the pilot-wave is physically real (such as Valentini himself). The onus is on such advocates not just to justify the introduction of structure over and above the wavefunction on configuration space. The further onus is to explain how the matter assumption *even makes sense* in the light of the possibility that the wavefunction is sufficiently highly structured on its own to account for physical systems, apparatuses, people, etc. After all, it is hard to see how the process of adding further degrees of freedom, hidden or otherwise, in the theory does anything to detract from the wavefunction's potency in this sense. What is really needed is an argument to the effect that the wavefunction does not have such potency as the Everettians attribute to it.[4]

Valentini goes some way to addressing this crucial matter at two points in his paper.

At the end of Section 5 of his paper, a hydrogen atom in a superposition of two 'Ehrenfest' packets is discussed, and appeal is made to the possibility of a subquantum measurement designed to establish which component in the superposition is 'unoccupied'. For present purposes, the likelihood of such a measurement being possible (see below) is largely irrelevant, the question being what the status of such a component is in itself. According to Valentini, the unoccupied component is merely 'simulating' the approximately classical motion of the atom. Valentini further claims in Section 6 that the treatment of the analogous, and more pressing, case of a superposition of non-overlapping packets representing distinct *macroscopic* arrangements is conceptually just the same. But in both cases, this notion of simulation is hard to reconcile with the plausible claim that, even in pilot-wave theory taken on its own terms, the intrinsic properties of quantum systems such as mass (both inertial and gravitational), charge and magnetic moment pertain to (at least) the pilot wave.[5] If in the second case the macroscopic systems involved contain human observers, and the superposition is defined relative to the appropriate decoherence basis, it is hard to see why phenomenologically the unoccupied component does not have the same status as it does in the Everett picture.[6]

Further clarification is offered by Valentini in his Section 7, where he question's Wallace's account [2003] of the phenomenology of the wavefunction in terms of Dennett's notion of macro-objects as patterns. Valentini admits

---

[4] In Holland [1993], it is merely claimed that the wavefunction fails to have both 'form and substance' (see the discussion in Brown and Wallace [2005]). A more sustained argument was offered recently by Maudlin [2007] in the context of any 'bare' GRW-type theory of spontaneous collapse, which of course is just as vulnerable as Everett theory to this kind of objection—such as it is. See his chapter in this volume.

[5] See Brown et al. [1995] and Brown et al. [1996].

[6] A further plausibility argument to the effect that such an unoccupied component of the superposition has the 'credentials' to represent a bona fide measurement outcome in standard pilot-wave theory is found in Brown and Wallace [2005]; the generalization of the argument to pilot-wave theory with regimes of non-equilibrium statistics is, I believe, straightforward. A critique of the credentials argument is found in Lewis [2007].

that in the quantum equilibrium regime, approximately classical experimenters 'will encounter a phenomenological *appearance* of many worlds just as they will encounter a phenomenological appearance of locality, uncertainty, and of quantum physics generally'.[7] He again appeals to the possibility of subquantum measurements in the non-equilibrium regime to question the explanatory and predictive role of such patterns reified by Everettians. But the reality of these patterns is *not* like locality and uncertainty, which are ultimately statistical notions and are supposed to depend on whether equilibrium holds. The patterns, on the other hand, are features of the wavefunction and are either there or they are not, regardless of the equilibrium condition.

3 NON-EQUILIBRIUM STATISTICS

A theme running throughout Valentini's paper is that pilot-wave theory cannot be a mere alternative formulation of quantum theory, or a sort of Everett theory in denial, because it allows for non-equilibrium physics. Indeed, we have just seen that Valentini effectively concedes that equilibrium pilot-wave theory is not a serious rival to Everett theory. But I argued in the last section that what is essential in the Everett picture, namely the analysis of the structural properties of the wavefunction and its ramifications for the measurement problem—what Wallace is striving to articulate in his metaphor of patterns in the context of decoherence—is untouched by the possibility of an additional ontology of corpuscles whether distributed in equilibrium or not. If the analysis is correct, it has implications for pilot-wave theory (or that version in which the pilot wave is real) and bare GRW-type theories just as much as for the Everett picture. Valentini is, I think, not justified in ignoring the potency of the wavefunction in the non-equilibrium regime.

In fact, it is hard to avoid the conclusion that the very notion of non-equilibrium quantum physics is problematic. If the wavefunction is indeed potent in the relevant sense, there are strong decision-theoretic arguments to the effect that rational observers should expect Born statistics.[8] It is not just that de

---

[7] The phrase 'phenomenological *appearance* of many worlds' is perhaps unhappy; the whole point of the Everett account of measurement is to demonstrate that the multiplicity of worlds is dynamically unavoidable but effectively unobservable, thus saving appearances! I take it, however, that what Valentini means here is that in the equilibrium regime, whatever one says about the significance of the wavefunction in Everett theory, one can say about the pilot-wave in pilot-wave theory. Valentini's position is quite different from Lewis's recent defence [2007] of pilot-wave theory, which involves simply rejecting Dennett's treatment of macro-objects as patterns (even in the equilibrium regime), but for no better reason than that it saves the day.

[8] See Chapter 8 by Wallace, in this volume; and strong arguments showing that rational observers will empirically confirm the Born rule are found in Chapter 9 by Greaves and Myrvold, in this volume. Such arguments suggest that Everett theory is more Popperian than pilot-wave theory à la Valentini: it is more falsifiable because it rules out non-equilibrium physics in the very special regimes where Valentini thinks it is likely, but not certain.

Broglie–Bohm corpuscles are surplus to requirement. Their irrelevance to the issue of defining what measurement outcomes are means that their contingent distribution should pose no threat to the Born rule. Needless to say, if, as Valentini hopes, we were eventually to observe strange non-local phenomena associated with, say, relic particles that decoupled soon after the big bang, Everettians would have to throw in the towel. But pilot-wave theorists who treat the wavefunction as part of physical, and not just nomological, reality, should, it seems to me, be doubtful about this possibility—unless they can show where the Everettians have gone wrong.

4 THE COUNTER-CLAIM

Valentini claims that Everett theory, indeed Everett's own 1957 thinking, is motivated by the 'puzzle of superposition', which in turn stems from the notion of 'eigenvalue realism'—itself allegedly based on classical reasoning.

The argument has several strands and is not easy to summarize concisely. But let us first consider the issue of what a measurement is in quantum mechanics. Valentini is surely right: the choice of an interaction Hamiltonian *qua* measurement must ultimately be legitimized by quantum, not classical, considerations. But how does one begin? What does one mean by an observable in the first place? What makes a given interaction a 'measurement' of that observable? And how does treatment of the measurement process tie in both with the dynamical principles in the theory and the rules governing stochastic behaviour, if any? The matter is intricate, and depends on diverse aspects of the theory.

At the beginning of Section 8, Valentini stresses that his critique of eigenvalue realism 'does not depend on pilot-wave theory', which leaves the matter somewhat ill-defined. Exactly what version of quantum theory is in play? In the middle of Section 8, Valentini states that it is 'much more likely that a new domain will be better understood in terms of a new theory based on new concepts, with its own new theory of measurement—as shown by the example of general relativity, and indeed by the example of de Broglie's nonclassical dynamics'. Yet recall that Bohm in his 1952 version of pilot-wave theory availed himself of *standard* quantum measurement theory, at least as regards the choice of interaction Hamiltonians and the definition of observables.

Valentini, however, emphasizes in Section 3 that the semantics of measurement in pilot-wave theory (equilibrium or otherwise) is quite different from that in classical mechanics. But that is also largely true for orthodox quantum theory. It is widely accepted in quantum theory that generically one is not measuring what is already there, one is not revealing a pre-existing element of reality—unless, perhaps, when prior to measurement the system is in an eigenstate of the observable in question. In that special case, eigenvalue realism is very close to what in the philosophical literature is called the 'eigenstate eigenvalue link', which in turn is very close in spirit to the 1935 Einstein–Podolsky–Rosen sufficiency criterion for the existence of an element of physical reality. The common notion here is that if theory predicts that measurement of some observable will yield a certain value with probability one, then the measurement must be revealing a property of the system that was already there. If this is right (which I doubt), then what one is supposed to infer in the generic case where the pre-measurement state is not an eigenstate is to some extent open to discussion. But it is not obvious how to avoid value-fuzziness in this case and hence a version of the puzzlement of superposition in the context of the measurement process.

Like Valentini, I think that eigenvalue realism *is* questionable—even in the absence of hidden variables. But this does not remove the need to make sense of the superposition in the context of measurement, and Everettians do not appeal to any classical prejudices in doing so. Valentini's further argument is that Everett theory is unlikely to be true because its followers are first led to the puzzlement of superposition on the basis of eigenvalue realism.[9] Even if he were right about Everettians, the argument strikes me as unconvincing. First, successful theories are sometimes developed partly on the basis of misguided or questionable motivations. (Amongst the principal ideas driving the development of general relativity were a dubious version of Mach's principle, and the erroneous notion that the principle of general covariance represents a generalization of the relativity principle.) Second, and perhaps more pertinently, a successful theory may be developed in part to solve a long-standing conceptual problem, and in the process of doing so show precisely how the the original assumptions leading to the problem are ill-founded.

5 ACKNOWLEDGEMENTS

My thanks go to the editors of this volume for the kind invitation to contribute in this way, and to Chris Timpson, Steve Weinstein, and especially Simon Saunders for helpful discussions. I gratefully acknowledge the support of the Perimeter Institute for Theoretical Physics, where part of this work was undertaken. Research at Perimeter Institute is supported by the Government of Canada through Industry Canada and by the Province of Ontario through the Ministry of Research & Innovation.

[9] To be fair, Valentini is, consciously or otherwise, turning on its head the 1986 Deutsch argument cited in (Valentini's) Section 3 against the plausibility of pilot-wave theory based on *its* allegedly suspect motivation.

References

Brown, H.R., C. Dewdney, and G. Horton, [1995], 'Bohm particles and their detection in the light of neutron interferometry', *Foundations of Physics* **25**, 329–47.

Brown, H.R., A. Elby, and R. Weingard, [1996], Cause and effect in the pilot-wave interpetation of quantum mechanics', in J.T. Cushing, A. Fine, and S. Goldstein (eds), *Bohmian Mechanics and Quantum Theory: An Appraisal*, Kluwer Academic Publishers, Netherlands, (1996); pp.309–19.

Brown, H.R. and D. Wallace, [2005], 'Solving the measurement problem: de Broglie–Bohm loses out to Everett', *Foundations of Physics* **35**, 517–40. Available online at quant-ph/0403094; PITT-PHIL-SCI 1659.

Holland, P. [1993], *The Quantum Theory of Motion: an Account of the de Broglie–Bohm Causal Interpretation of Quantum Mechanics*, Cambridge University Press, Cambridge.

Lewis, P.J [2007], 'Empty waves in Bohmian quantum mechanics', *British Journal for the Philosophy of Science* **58**, 787–803.

Maudlin, T. [2007], 'Completeness, supervenience, and ontology', *Journal of Physics A: Math. Theor.* **40**, 3151–171.

Wallace, D. [2003], 'Everett and structure', *Studies in the History and Philosophy of Modern Physics* **34**, 87–105.

PART VI

NOT ONLY MANY WORLDS

17

Everett and Wheeler: the Untold Story

Peter Byrne

In the beginning, John Archibald Wheeler was Hugh Everett III's champion. In July 1957, *Reviews of Modern Physics* published Everett's doctoral dissertation, ' "Relative state" formulation of quantum mechanics.' Accompanying it into print was an assessment of Everett's work by Wheeler, who was his thesis advisor at Princeton University. The physics professor enthused: 'It is difficult to make clear how decisively the "relative state" formulation drops classical concepts. One's initial unhappiness at this step can be matched but few times in history: when Newton described gravity by anything so preposterous as action at a distance; when Maxwell described anything as natural as action at a distance in terms as unnatural as field theory; when Einstein denied a privileged character to any coordinate system, and the whole foundations of physical measurement at first sight seemed to collapse. . . . No escape seems possible from this relative state formulation. . . . [It] does demand a totally new view of the foundational character of physics' (Wheeler [1957 p.464]).

If Wheeler's assessment rang true to the physics world at large, Everett's career in theoretical physics was assured. But this was not to be—Everett never published another word of quantum mechanics. Partly, this was because he was unhappy with the final version of his thesis, which, Everett thought, failed to fully explain his theory. And, partly, it had to do with the distaste for academic discourse he felt after his theory was shot down by Niels Bohr and his circle in Copenhagen. And, partly, it was because Everett enjoyed applying his genius to military operations research, which provided him with access to state secrets and state-of-the-art computers—not to mention a competitive salary.

The paper printed in *Reviews of Modern Physics* was drastically abridged from the doctoral thesis that Everett had originally submitted to Wheeler. Upset by the colorful language in which the dissertation was couched, the professor had insisted that Everett cut and rewrite the bulk of his work. In Wheeler's view, the logical consequence of what he called Everett's 'impeccable formalism' was troubling enough without creating metaphors of human observers and cannonballs splitting into countless versions of themselves inside a tangle of branching universes. Nor was Wheeler happy that Everett had dismissed Bohr's interpretation of quantum mechanics as 'mathematical artifice'.

Wheeler's support for Everett's theory was born of an agenda: quantizing gravity. And for this project, Everett's formulation of a universal wavefunction was useful, provided that its baggage—a non-denumerable infinity of branching worlds—could be, somehow, lightened. He told Everett to tone down his language, and he threatened to reject the dissertation should he fail to do so. Under Wheeler's close supervision, Everett reluctantly complied, and three-quarters of the original thesis was excised or condensed. Mission accomplished, Wheeler publicly compared his student's work to the achievements of Newton, Maxwell, and Einstein. Not everybody agreed with him, to say the least. And, eventually, Wheeler ceased advocating for, and then attacked the 'many-worlds' interpretation of quantum mechanics.

A paper trail detailing the stages of a protracted struggle between Everett and Wheeler and members of Bohr's inner circle over the content of the dissertation has emerged from files at the Niels Bohr Archive in Copenhagen, the American Philosophical Society in Philadelphia, and at the American Institute of Physics in College Park, Maryland. Some of these records were unearthed by Professor Olival Freire Jr. of the Universidade Federal da Bahia, Brazil and Anja Jacobsen of the Niels Bohr Archive. Additional letters and related materials surfaced in May 2007 when Everett's son, Mark Everett, and I began opening cardboard boxes of Everett's personal papers that had been stored for many years in a dusty Los Angeles basement.

Because Everett's thesis evolved through multiple versions it had several different titles, a situation which has (understandably) confused Everett scholars. To clarify: in Mark Everett's basement are the original sheaves of yellow legal paper upon which Everett wrote, in pencil, his (untitled) thesis, which he began working on in late 1954, during his third semester of graduate school. Wheeler was aware of the essence of Everett's theory in January 1955, when he wrote a laudatory report on Everett for the National Science Foundation. Sections of the thesis were typed during the summer of 1955 by Nancy Gore (who later became Mrs Hugh Everett III). These sections were shown to Wheeler for comment in the fall. In January 1956, Everett submitted a 137-page dissertation to Wheeler: 'Quantum Mechanics by the Method of the Universal WaveFunction'. In bound copies distributed in April to select physicists, including Bohr and Petersen, it was retitled, 'Wave Mechanics Without Probability'.

In a April 24, 1956 letter to Bohr, Wheeler wrote, 'I would be appreciative of comments by you and Aage Petersen about the work of Everett. . . . The title itself, "Wave Mechanics Without Probability", like so many of the ideas in it, need further analysis and rephrasing, as I know Everett would be the first to say. But I am more concerned with you[r] reaction to the more fundamental question, whether there is any escape from a formalism like Everett's when one wants to deal with a situation where several observers are at work, and wants to include the observers themselves on the system that is to receive mathematical analysis.'

After the edit by Wheeler and Everett, the dissertation was retitled, 'On the Foundations of Quantum Mechanics', and that is the title of the doctoral thesis that was officially accepted by Princeton on April 15, 1957. For publication in *Reviews of Modern Physics*, it was renamed, at Wheeler's insistence, ' "Relative State" Formulation of Quantum Mechanics'. Then, in 1973, the *unedited* thesis was published for the first time, by Princeton University Press, in *The Many Worlds Interpretation of Quantum Mechanics,* edited by Bryce DeWitt and Neill Graham. The manuscript of 'Wave Mechanics Without Probability' that Everett sent to DeWitt for typesetting was, once again, retitled, 'The Theory of the Universal WaveFunction.' (DeWitt coined the term 'many worlds').

1 GENESIS OF THE THEORY

After graduating with a degree in chemical engineering from Catholic University, Everett spent his first year as a physics grad student at Princeton (1953–54) concentrating on game theory—he was a regular at the now-legendary teas and game theory conferences at Fine Hall attended by such icons of the field as John Nash, Lloyd Shapley, John von Neumann, Oskar Morgenstern, Harold Kuhn, and Albert Tucker. He wrote an influential paper, 'Recursive Games', which was printed in *Annals of Mathematics Studies* (Princeton University Press, 1957) and reprinted by Kuhn in 'Classics in Game Theory' (Princeton University Press, 1997). He also studied quantum mechanics with Robert Dicke and Eugene Wigner, and gravitated toward Wheeler's circle of graduate students, which included his friend, Charles Misner.

In the fall of 1954, Bohr was in residence at the Institute for Advanced Study in Princeton and, according to Abraham Pais, lectured on why 'he thinks that the "quantum theory of measurement" is wrongly put' (Pais [1991 p.435]). Bohr's philosophy of 'complementarity' did not recognize the existence of a measurement problem, per se. (The measurement problem occurs because the Schrödinger wave equation shows superposed quantum states evolving linearly through time, whereas, upon interaction with a macroscopic entity, or scientific measuring device, only one of the possible states emerges or 'collapses' from the superposition. The 'problem' is to explain why we only experience one state out of all possible states.)

Many years later, Everett laughingly recounted to Misner, in a tape-recorded conversation at a cocktail party in May 1977, that he came up with his many-worlds idea in 1954 'after a slosh or two of sherry', when he, Misner, and Aage Petersen (Bohr's assistant) were thinking up 'ridiculous things about the implications of quantum mechanics'.

Inspirational flashes aside, the theory developed in a controlled fashion. In the taped conversation, Misner reminded Everett that Wheeler, 'was preaching this idea that you ought to just look at the equations and if there were the fundamentals of physics, why you followed their conclusions and give them a serious hearing. He was doing that on these solutions of Einstein's equations like Wormholes and Geons.' Everett replied, 'I've got to admit that is right, and might very well have been totally instrumental in what happened.'

In the early and mid 1950s, Wheeler and a few of his graduate students were exploring the possibility of uniting quantum mechanics and general relativity using his former student Richard Feynman's path integral formulation as a guide. Misner applied himself to the task, which was the focus of the 'Conference on the Role of Gravitation in Physics', held in January 1957 at the University of North Carolina.

To jump ahead of the narrative for a moment: the Chapel Hill meeting was attended by Wheeler, Feynman, Misner, Leon Rosenfeld, and other prominent physicists, including conference organizers, Bryce S. DeWitt and Cecile M. DeWitt. Everett did not attend, but according to the official conference report, the measurement problem, and the need for an Everett-type universal wavefunction in cosmology were subjects of discussion. His theory proposed a solution by positing a non-collapsing wavefunction describing the whole universe. Since it is not possible to observe the universe from outside the universe, a non-collapse theory along the lines of what Everett was proposing was viewed by Wheeler as a necessary step toward quantizing a universe filled with gravitational fields. At the conference, Everett's theory was not well received. Feynman commented: '[T]he concept of a "universal wavefunction" has serious difficulties. This is so since the function must contain amplitudes for all possible worlds depending upon all quantum mechanical possibilities in the past and thus one is forced to believe in the equal reality of an infinity of possible worlds' (DeWitt [1957 p.149]).

2 FIRST DRAFTS

In an unpublished section of his draft thesis, Everett outlined the argument of what later became known as the Many-Worlds Interpretation, including his claim that a mathematical equivalent to Born's Rule emerges from his formalism. This nine-page work, called 'Probability in Wave Mechanics', is essentially an abstract—light on mathematical notation, heavy on metaphor.

The section begins by delineating the contradiction in the 'orthodox' (John von Neumann's) interpretation of quantum mechanics in which the evolution of the wave equation proceeds linearly, continuously, until it mysteriously collapses, apparently defying special relativity and logic. Everett questions what happens to the observer of a quantum mechanical measurement: 'Why doesn't our observer see a smeared out needle? The answer is quite simple. He behaves just like the apparatus did. When he looks at the needle (interacts) he himself becomes smeared out, but at the same time correlated to the apparatus, and hence to the system. . . . [T]he observer himself has split into a number of observers, each of which sees a definite result of the measurement. . . . As an analogy one can imagine an intelligent amoeba with a good memory. As time progresses the amoeba is constantly splitting.' Everett observed of his own theory, 'It can lay claim to a certain completeness, since it applies to all systems, of whatever size, and is still capable of explaining the appearance of the macroscopic world. The price, however, is the abandonment of the concept of uniqueness of the observer, with its somewhat disconcerting philosophical implications.'

In 'Probability in Wave Mechanics' (much of which was dropped in subsequent drafts) Everett presaged elements of the decoherence model of quantum mechanics. 'In fact, . . . whenever any two systems interact some degree of correlation is always produced. . . . Consider a large number of interacting particles. If we suppose them to be initially independent, then throughout the course of time the position amplitude of any single particle spreads further and further, approaching uniformity over the whole universe, while at the same time, due to the interactions, strong correlations will be built up, so that we might say that the particles have coalesced to form a solid object. That is, even though the position amplitude of any single object would be "smeared out" over a vast region, if we consider a "cross-section" of the total wavefunction for which one particle has a definite position, then we immediately find all the rest of the particles nearby, forming a solid object. *It is this phenomenon which accounts for the classical appearance of the macroscopic world, the existence of solid bodies, etc. since we ourselves are strongly correlated to our environment.* Even though it is possible for a macroscopic object to "smear out", . . . we would never be aware of it due to the fact that the interactions between the object and our senses are so strong that we become correlated to almost instantly.' (Emphasis added.)

Everett concluded, 'The physical "reality" is assumed to be the wavefunction of the whole universe itself. By properly interpreting the internal correlations in this wavefunction it is possible to explain the appearance of the macroscopic world to us, as well as the apparent probabilistic aspects.' On the margins of Everett's mini-paper, Wheeler wrote, 'Have to discuss questions of know-ability of the universal ψ function—and latitude with which we can ever determine it. . . . Question of whether new view has any practical consequence.'

'Probability in Wave Mechanics', was a summary of the longer work-in-progress. In September 1955, Wheeler wrote to Everett, 'I am frankly bashful about showing it to Bohr in its present form, valuable and important as I consider it to be, because of parts subject to mystical misinterpretations by too many unskilled readers.' Wheeler was particularly disturbed by Everett's use of the verb 'split' to describe what happens to an observer correlated to a superposed system.

Wheeler was more positive, however, about two related sections of the thesis draft. 'Quantitative Measure of Correlation' utilized 'the concepts of information theory' to measure the amount of correlation between two quantum variables in a probability distribution. This was cutting-edge material for the day. And 'Objective vs. Subjective Probability' argued that a continually branching observer will subjectively experience quantum determinism (everything happens) as indeterminism (chance rules) because of possessing incomplete information about the quantum environment. Regarding the prevailing notion, 'that even in principle quantum mechanics cannot describe the process of measurement itself', Everett wrote: 'This is somewhat repugnant, since it leads to an artificial dichotomy of the universe into ordinary phenomena and measurements.'

By January 1956, Everett had abandoned the amoeba metaphor in the evolving dissertation, but he did not eliminate 'split', nor did he shy away from painting pictures of multiple, disconnecting universes stocked with armies of bifurcating humans and superposed cannonballs. Nor was Everett the least bit bashful about criticizing the prevailing interpretations of quantum mechanics. He said that the 'popular' (von Neumann) interpretation, including its postulate of wavefunction collapse, was 'untenable'. Speaking directly of the Copenhagen Interpretation, 'developed by Bohr', Everett declared, 'While undoubtedly safe from contradiction, due to its extreme conservatism, it is perhaps overcautious. We do not believe that the primary purpose of theoretical physics is to construct "safe" theories at severe cost in the applicability of their concepts, which is a sterile occupation, but to make useful models which serve for a time and are replaced as they are outworn' (DeWitt and Graham [1973 p.111]).

Everett concluded, 'Our theory in a certain sense bridges the positions of Einstein and Bohr, since the complete theory is quite objective and deterministic . . . and yet on the subjective level . . . it is probabilistic in the *strong sense* that there is no way for observers to make any predictions better than the limitations imposed by the uncertainty principle.' He added, 'The constructs of classical physics are just as much fictions of our own minds as those of any other theory; we simply have more confidence in them' (DeWitt and Graham [1973 pp.117–34]).

Lest there be any misunderstanding about the depth of Everett's disenchantment with Bohr, here is what he wrote to Bryce DeWitt in May 1957. '[T]he Copenhagen Interpretation is hopelessly incomplete because of its apriori reliance on classical physics (excluding *in principle* any deduction of classical physics from quantum theory, or any adequate investigation of the measuring process), as well as a philosophical monstrosity with a "reality" concept for the macroscopic world and denial of the same for the microcosm' (Everett [May 31, 1957]).

Decades later, in an unpublished referee report, DeWitt commented, 'I know that John Wheeler admires brevity and probably urged Everett to try and "sum up in a nutshell" the essential points of his new interpretation of quantum mechanics. It is also possible that Wheeler was reluctant to support a more blatant statement because it would mean setting himself into direct opposition to his hero, Niels Bohr. What is sure is that Wheeler long ago abandoned his support for Everett. What is equally sure is that if the Urwerk [the original, unedited 137-page thesis that DeWitt published in 1973] had been published [in 1957], Everett would not have been ignored for so long' (DeWitt [1988]).

3 THE BATTLE WITH COPENHAGEN

In 2006, I submitted several questions to the ailing Wheeler, via his biographer, Kenneth Ford. Asked if he had 'distanced' himself from Everett's theory, Wheeler replied, ' "Not embracing his theory" would be better than "distancing myself from it." How I wish I had kept up the sessions with Everett. The questions that he brought up were important. Maybe I did not have my radar operating.'

But in January 1956, in regard to Everett's theory, Wheeler's radar was sharp. Before agreeing to approve the dissertation, he sent a bound copy of Everett's long thesis ('Wave Mechanics Without Probability') to Copenhagen for review by Bohr. Misner explains Wheeler's dilemma: 'John Wheeler got along with everybody, but in Hugh's case Wheeler had a very difficult time applying his usual tactics because he couldn't just encourage Hugh to follow his ideas and present them as powerfully as possible since they ran contrary to Bohr's ideas. And Wheeler regarded Bohr as his most important mentor. So he was really torn and I think he kept trying to play both sides of that tension by trying to get Hugh to tone down the thesis so it wouldn't be quite so needling to people, and then writing a comment on it himself to publish along side of it to try and smooth things over a bit.'

May 1956 found Wheeler in Copenhagen discussing Everett's work with Bohr and Petersen. On May 26, Wheeler wrote to Everett, 'After my arrival the three of us had three long and strong discussions about [the thesis]. . . . Stating conclusions briefly, your beautiful wavefunction formalism of course remains unshaken; but all three of us feel that the real issue is the words that are to be attached to the quantities of the formalism. We feel that complete mis-interpretation of what physics is about will result unless the words that go with the formalism are drastically revised.' Wheeler urged Everett to strug-gle it out in Copenhagen directly with Bohr. And he warned Everett that he would not schedule his final exam, 'until this whole issue of words is straightened out'.

A few hours later, Wheeler wrote Everett another letter, this time enclosing a copy of the notes he had taken of his meetings with Bohr and Petersen. He told Everett, 'Much of what is said in objection to your work is irrelevant. Much is relevant: The difficulty of expressing in everyday words the goings on in a mathematical scheme that is about as far removed as it could be from the everyday description; the contradictions and misunderstandings that will arise;

the very very heavy burden and responsibility you have to state everything in such a way that these misunderstandings can't arise. . . . The combination of qualities, to accept corrections in a humble spirit, but to insist on the soundness of certain fundamental principles, is one that is rare but indispensable; and you have it. But it won't do much good unless you go and fight with the greatest fighter. Frankly, I feel about two more months of nearly solid day by day argument are needed to get the bugs out of the words, not out of the formalism.' Wheeler offered to pay half of Everett's steamship fare to Denmark and said Bohr would cover the rest.

Everett scribbled several caustic remarks on his copy of Wheeler's notes. Next to Petersen's assertion that 'the wavefunction for [the] electron doesn't make sense until we get something like a probability distribution of spots', Everett wrote: 'Nonsense!' Then Petersen argued, 'Math can never be used in physics until [we] have words. [We] aren't comparing [our]selves with servomechanisms. What [we] mean by physics is what can be expressed unambiguously in ordinary language.' Everett penciled, 'Obviously hasn't completed reading of thesis! It does just that.'

Also in May, Alexander Stern, an American physicist visiting Bohr's institute, wrote to Wheeler saying he had just given a seminar on Everett's paper, and Bohr had opened the discussion. Reflecting the tenor of that discussion, Stern commented that Everett, 'lack[s] an adequate understanding of the measuring process. Everett does not seem to appreciate the Fundamentally irreversible character and the Finality of a macroscopic measurement. . . . It is an Indefinable interaction.' Stern complained that Everett excluded probability from wave mechanics and did not understand the concept of 'observer'.

And failing to appreciate that Everett was *totally* eliminating the role of external observation (real or ideal) so crucial to Bohr's interpretation, Stern concluded, 'If Everett's universal wave equation demands a universal observer, an idealized observer, then this becomes a matter of theology . . . The subjective aspect of physics, which some scholars and philosophers have claimed to detect but have not understood, has its origin in the fact that physics must make contact with reality which is, after all, the way the world appears to us, and can be understood by us.'

Like many of Everett's critics, past and present, Stern was troubled by Everett's treatment of probability: 'I do not follow him when he claims that . . . one can view the accepted probabilistic interpretation of quantum theory as representing the subjective appearances of observers.' After commenting that probability distributions provide physicists with a 'meaningful information Pattern', Stern remarked, 'Wave mechanics without probability excludes physicists' (Stern to Wheeler [May 20, 1956]).

In a conciliatory letter, Wheeler immediately replied to Stern: 'I would not have imposed upon my friends the burden of analyzing Everett's ideas . . . if I did not feel that the concept of "universal wavefunction" offers an illuminating and satisfactory way to present the content of quantum theory. I do not in any way question the self consistency and correctness of the present quantum mechanical formalism when I say this. On the contrary, I have vigorously supported and expect to support in the future the current and inescapable approach to the measurement problem. To be sure, Everett may have felt some questions on this point in the past, but I do not. Moreover, I think I may say that this very fine and able and independently thinking young man has gradually come to accept the present approach to the measurement problem as correct and self consistent, despite a few traces that remain in the present thesis draft of a past dubious attitude. So, to avoid any possible misunderstanding, let me say that Everett's thesis is not meant to *question* the present approach to the measurement problem, but to accept and *generalize* it' (Wheeler to Stern [May 25, 1956]).

Wheeler went on to mount a spirited defense of Everett's formalism, with the caveat that the relative state theory applies to a possible 'model' of the world, and not to the real world, per se. He copied Everett on the letter, attaching another pedagogical warning, 'I have no escape from one sad but important conclusion: that your thesis must receive heavy revision of words and discussion, very little of mathematics, before I can rightfully take the responsibility to recommend it for acceptance. . . . I feel that your work is most interesting and am sure that it will receive discussion of a scope comparable to that which has attended Bohm's publications. But in your case I must ask that the bugs be got out and the sources of misunderstanding be clarified *before* the job is made public, not *afterwards*. I hope you will realize that I mean this as what is called here your "promoter," and one actively interested in your reputation and promising future' (Wheeler to Everett [May 25, 1956]).

Wheeler wrote to Allen Shenstone, chairman of the physics department at Princeton, 'I think [Everett's] very original ideas are going to receive wide discussion. . . . Since the strongest present opposition to some parts of it comes from Bohr, I feel that acceptance in the Danish Academy would be the best public proof of having passed the necessary tests' (Wheeler to Shenstone [May 28, 1956]).

In one of the boxes in Mark's basement, we found Everett's handwritten comments on Stern's letter. He wrote, 'Technically, "observer" can be applied to any physical system capable of changing its state to a new state with some fairly permanent characteristics which depend upon the object system (with which it interacts) . . . Stern's remarks about [my] misunderstanding of [the] fundamental irreversibility of [the] measurement process indicate rather clearly that he has had insufficient time to read the entire work. Several rereadings on his part seem to be called for. Also, Stern is quite guilty in these remarks of begging the question—one of the fundamental motivations of the paper is the question of *how can it be* that [many] measurements are "irreversible", the answer to which *is* contained in my theory, but is a serious lacuna in the other theory.'

Everett's notes were attached to a letter from Petersen to Wheeler, in which Petersen sent Bohr's copy of Everett's thesis back to Wheeler. Petersen also sent a note to Everett inviting him to Copenhagen. Everett replied that he'd like to visit and that, by the way, he was enclosing another copy of his thesis since, 'Judging from Stern's letter to Wheeler, which was forwarded to me, there has not been a copy in Copenhagen long enough for anyone to have read it thoroughly, a situation which this copy may rectify. I believe that a number of misunderstandings will evaporate when it has been read more carefully (say two or three times)' (Everett to Petersen [July, 1956]).

4 JAVELIN PROOF

Everett was not to get to Copenhagen for three years. In the summer of 1956, he took a job with the top-secret Weapons Systems Evaluation Group at the Pentagon and moved in with Gore, who soon became his wife, and then a mother. During the next half year, Everett and Wheeler negotiated the objectionable language out of the thesis to make it, in Wheeler's phrase, 'javelin proof'. In March 1957, preprints of the truncated dissertation, now entitled 'On the Foundations of Quantum Mechanics,' and Wheeler's supporting article, which attempted to frame Everett's (anti-complementarity) argument as in accord with complementarity, were sent to a score of prominent physicists, including Bohr, Erwin Schrödinger, Robert Oppenheimer, Leon Rosenfeld, and Norbert Wiener. DeWitt later told Wheeler: 'It always amused me to read in your Assessment of Everett's Theory how highly you praised Bohr, when the whole purpose of the theory was to undermine the stand which he had for so long taken!' (DeWitt to Wheeler [April 20, 1967]).

A few weeks later, Bohr dropped a note to Wheeler, saying, 'I have not found time to write to you and Everett about the papers you kindly sent me. It appears that the argumentation contains some confusion as regards the observational problem and . . . Aage Petersen will write to Everett about our discussions' (Bohr to Wheeler [April 12, 1957]).

Days later, Petersen wrote to Everett, 'As you can imagine, the papers have given rise to much discussion at the Institute. . . . We cannot agree with you and Wheeler that the relative state formulation entails a further clarification of the foundations of quantum mechanics. . . . There can on this view be no special observational problem in quantum mechanics—in accordance with the fact that the very idea of observation belongs to the frame of classical concepts. . . . There is no arbitrary distinction between the use of classical concepts and the formalism since the large mass of the apparatus compared with that of the individual atomic objects permits the neglect of quantum effects which is demanded for the account of the experimental arrangement. . . . Of course, I am aware that from the point of view of your model-philosophy most of these remarks are besides the point. However, to my mind this philosophy is not suitable for approaching the measuring problem. I would not like to make it a universal principle that ordinary language is indispensable for definition or communication of physical experience, but for the elucidation of the measuring problems hitherto met with in physics the correspondence approach has been quite successful' (Petersen to Everett [April 24, 1957]).

So now we have the curious situation in which Wheeler and Everett had stripped away much (but not all) of the explanatory language while cutting nearly three-quarters of the original paper. For example, an entire chapter on information theory, probability, and the measurement problem was eliminated. (Stern had thought this chapter to be the 'best in the book'.) Much of the colorful language that Everett used to bring his theory alive in 'ordinary' terms was excised, as was his criticism of Bohr. It must be noted, however, that the editing did clarify arguments on the significance of applying a universal wavefunction to gravitation. In fact, the revised dissertation was now reframed in its first sentence as 'the task of quantizing general relativity', which had not been Everett's primary goal. He had been primarily concerned, in the long thesis, with deriving an interpretation of quantum mechanics directly from its formalism, without inserting wavefunction collapse or postulating Born's rule. In fact, Everett had consciously upended Bohr's complementary approach to physics by choosing to describe the world as fundamentally quantum mechanical, *not* classical, and by treating the Schrödinger equation as universally valid. The branching universes were a consequence, not a predicate of Everett's interpretation. But, after Everett had allowed Wheeler to basically dictate what was to remain intact of his original thesis partially in order to minimize the impact of its language, Bohr, through Petersen, complained that the formalism was not explained in terms of ordinary language, classical language.

Everett replied to Petersen angrily, 'Lest the discussion of my paper die completely, let me add some fuel to the fire with . . . criticisms of the "Copenhagen interpretation." . . . I do not think you can dismiss my viewpoint as simply a misunderstanding of Bohr's position. . . . I believe that basing quantum mechanics upon classical physics was a necessary provisional step, but that the time has come . . . to treat [quantum mechanics] in its own right as a fundamental theory without any dependence on classical physics, and to derive classical physics from it. . . . Let me mention a few more irritating features of the Copenhagen Interpretation. You talk of the massiveness of macro systems allowing one to neglect further quantum effects (in discussions of breaking the measuring chain), but never give any justification for this flatly asserted dogma. [And] there is nowhere to be found any consistent explanation for this "irreversibility" of the measuring process. It is again certainly not implied by wave mechanics, nor classical mechanics either. Another independent postulate?' (Everett to Petersen [May 31, 1957]).

In April 1957, H.J. Groenewold of Natuurkundig Laboratorium der Rijks-Universiteit te Groningen wrote a long critique of the edited thesis ('relative states' preprint) in which he 'profoundly disagree[d]' with the premise and conclusion of Everett's theory. Groenewold began by saying that in the summer of 1956 he had 'borrowed' a copy of 'Wave Mechanics Without Probability', and that the preprint of ' "Relative State" Formulation of Quantum Mechanics' was 'much improved'. (Believing that the preprint was abstracted from an *improved* longer version, he asked to read the latter!) (Groenewold to Everett [April 11, 1957]).

Groenewold wrote, 'I fully sympathize with the idea of describing the measuring process on purely physical systems without including living observers. So the "measuring chain" has to be cut off. But it is extremely fundamental that the [cut] off is made after the measuring result has been recorded [in a] permanent way, so that it no longer can be essentially changed if it is observed on its turn. . . . This recording has to be more or less irreversible and can only take place in a macrophysical (recording) system.' Everett penciled in the margin, 'Nonsense. Whole idea not to cut off till after final observ[ation] QM says it effected just like microsystem. Whence this magic irrevers[ibility]?'

Groenewold continued: 'Because all observable quantities may ultimately be expressed in statistical relations between measuring results and the latter are represented by essentially macrophysical recordings, the former ones may ultimately be expressed in macrophysical language.' In the margin Everett scribbled, 'Epistemologically garbage. Lack of understanding of the nature of physical theory. Why base concept of reality on classical macrophysical realms?'

When Groenewold complained that Everett's theory could not avoid introducing the 'cat' and EPR paradoxes, Everett exclaimed, 'Didn't even read my paper . . . the paradoxes [are] more easily explained than usual.' In a subsequent letter, Groenewold insinuated that Wheeler and Everett had 'abandoned the idea of interaction at a distance'. And, perhaps, they had—since Everett believed he had accounted for what he described, in the edited version, as the 'fictitious' EPR paradox.

Not all of the reactions to Everett's and Wheeler's preprints were negative. Henry Margenau, a professor of physics and natural philosophy at Yale University wrote, 'The problem with which you deal has irritated many minds. I, for one, find your disposal quite acceptable.' Norbert Wiener (of Massachusetts Institute of Technology) weighed in: '[T]he inclusion of the observer as an intrinsic part of the observed system is absolutely sound.' But Wiener remarked that Everett was wrong to introduce a Lebesque measure in Hilbert space. He concluded, '[Y]our paper should be published, but more as comments on the present intellectual situation than as a definitive result' (Margenau to Everett [April 8, 1957]; Wiener to Wheeler and Everett [April 9, 1957]).

Everett replied to Wiener, 'I would like to correct any impression that my theory requires a Lebesque measure on Hilbert space. The only measure which I introduced was a measure on the orthogonal states which are superposed to form another state . . . and not a measure of Hilbert space itself, the difficulties of which I am fully aware' (Everett to Wiener [May 31, 1957]). In fact, Everett believed that a probabilistic calculus emerged from his theory—subjectively. He opined that, subjectively, for each of the splitting observers, an apparent collapse of the wavefunction makes phenomenological sense as a probabilistic event, even though, objectively, the wavefunction does not collapse and all branchings are equally 'real' in a quantum universe.

5 PHILOSOPHICAL DEBATE

In May 1957, after more than a year of battling unsuccessfully with Wheeler, Bohr, and others over the fate of his interpretation, Everett sent a copy of the abridged thesis to Professor Philipp Frank, a philosopher of science at Harvard who had recently edited a collection of essays, 'The Validation of Scientific Theories'. Everett wrote to Frank, 'I have received several of your works on the philosophy of science. I have found them extremely stimulating and valuable. I find that you have expressed a viewpoint which is very nearly identical with the one I have developed independently over the last few years, concerning the nature of physical reality' (Everett to Frank [May 31, 1957]). A former member of the 'operationalist' Vienna Circle, Frank wrote often of the interplay between sociology and science.

In his essay, 'The Variety of Reasons for the Acceptance of Scientific Theories', Frank examined the furor around Nicolas Copernicus, a 16th-century scientist whose heliocentric theory was not fully recognized as true until Isaac Newton substantiated it a century after it was proposed. As an example of scientific rigidity toward the counter-intuitive, Frank cited Francis Bacon—who had rejected the Copernican view because it did not accord to common sense. Frank elaborated, 'Looking at the historical record, we notice that the requirement of compatibility with common sense and the rejection of "unnatural theories" have been advocated with a highly emotional undertone, and it is reasonable to raise the question: What was the source of heat in those fights against new and absurd theories? Surveying those battles, we easily find one common feature, the apprehension that a disagreement with common sense may deprive scientific theories of their value as incentives for a desirable human behavior. In other words, by becoming incompatible with common sense, scientific theories lose their fitness to support desirable attitudes in the domain of ethics, politics, and religion' (Frank [1954 p.9]).

Also in May, Everett corresponded with Bryce DeWitt, who was guest editing the issue of *Reviews of Modern Physics* in which Everett and Wheeler's 'relative state' papers were slated to appear, along with the other Chapel Hill gravitational conference papers. DeWitt had written to Wheeler that Everett's paper was 'valuable' and 'beautifully constructed', but 'the real world does not branch'. Everett rejoined in a letter to DeWitt that the same sort of objection was raised by Copernicus' critics: when he asserted that the earth revolved around the sun, they said that was impossible because they could not feel it move. Everett poked DeWitt: 'I can't resist asking: Do you feel the motion of the earth?' He then remarked, 'It is impossible to do full justice to the subject in so brief an article as the one you read' (DeWitt to Wheeler [May 7, 1957]); Everett to DeWitt [May 31, 1957]).

DeWitt recalled years later, 'His reference to the anti-Copernicans left me with nothing to say but "Touché!" ' DeWitt did not read the unexpurgated thesis until the early 1970s, but he said he put Everett's paper in *Reviews of Modern Physics* because, 'Although Everett had not been a conference participant and I had never met him, his paper was accompanied by (1) a strong letter from John Wheeler and (2) a paper by Wheeler assessing Everett's ideas. Since Wheeler had been a very active conference participant [he was a main organizer of the conference] and since Everett's paper seemed to be relevant to the themes of the conference, I agreed to include it' (DeWitt [*c.* 1988]).

In 1995, Ken Ford interviewed DeWitt about Everett and Wheeler. Regarding the editing of Everett's paper, DeWitt remarked, 'I asked [Wheeler] why the original article, I mean the [Urwerk], wasn't ever published. Wheeler said, "Because I sat down with Everett and told him what to say." ' DeWitt said, 'The funny thing is, you have to read the *Reviews of Modern Physics* article very carefully, as I did, to see what's really there. Whereas in the Urwerk it's quite well spelled out, to me' (Ford [1995]).

In the end, after the rebellious, anti-Bohr comments in the original work largely vanished, along with much of the explanatory language, and much of the formal argument, Everett's dissertation was accepted by Wheeler. In April 1957, Everett became Dr. Everett. One of his classmates congratulated him on finally having his thesis posted for reading in the physics department. He commented, "Incidentally, did you know that there was a rumor here that there were no faculty members willing to be second and third readers on it? On checking, this was scotched by Charlie [Misner] who claimed it to be a sort of ploy by Wheeler who wanted you to keep rewriting until it was in shape to convince the world. How do you figure the odds on that?' (Rockman to Everett [March 2, 1957]).

In mid April 1957, Wheeler wrote a memo to Everett's student file: 'This work is almost completely original with Mr Everett both as to the formulation of the problem and its solution. It is too early to assess its final contribution to physics, but there is a distinct possibility that Everett's work may be a significant contribution to our understanding of the foundations of quantum theory' (Wheeler to Everett student file [April 15, 1957], Seeley G. Mudd Library, Princeton University).

When Everett's paper appeared in July, it included 'split', inserted by Everett, without Wheeler's approval, in a footnote during the proof process, here excerpted: 'From the viewpoint of the theory *all* elements of a superposition (all "branches") are "actual," none any more "real" than the rest. It is unnecessary to suppose that all but one are somehow destroyed, since all the separate elements of a superposition individually obey the wave equation with complete indifference to the presence or absence ("actuality" or not) of any other elements. This total lack of effect of one branch on another also implies that no observer will ever be aware of any "splitting" process' (Everett [1957 p.459]).

6 THE THEORY PERCOLATES

After the abbreviated thesis was published as ' "Relative state" Formulation of Quantum Mechanics', the physics community was largely silent about Everett's idea, which many thought to be crazy. Disappointed by the lack of response, Everett occupied himself in operations research—calculating nuclear bomb kill ratios at the Pentagon.

For a while, Wheeler kept up a correspondence with his protégé, urging him to visit Copenhagen and 'fight' it out with Bohr. He encouraged him to get a job in academia and continue his theoretical work in quantum mechanics. He invited him to make presentations at several seminars, and occasionally copied him on correspondence with eminent physicists of the day. In 1959, Everett visited Bohr in Copenhagen; they had several discussions, but neither man budged from his position.

During the ensuing years, the argument over the paper began slowly percolating into the consciousnesses of physicists around the world. Everett received requests for reprints, and gave a few talks on his theory at universities. And in October, 1962 he made a semi-private presentation at Xavier University in Cincinnati, Ohio to P.A.M. Dirac, Eugene Wigner, Nathan Rosen, Yakir Aharonov, Abner Shimony, Wendell Furry, Boris Podolsky, and several others. There is an unpublished transcript of this 'Conference on the Foundations of Quantum Mechanics', archived at the American Institute of Physics. Many of the conferees were not happy with the 'orthodox' (wave reduction) interpretation of quantum mechanics, and were willing to entertain such strange notions as hidden variables and multiple universes.

Everett, who loathed public speaking, was introduced by Rosen and asked to describe his non-collapse theory. He began, 'My position is simply that I think you can make a tenable theory out of allowing the superposition to continue forever, even for a single observer.' Panel members grilled him about the number of branching universes, and how probabilities emerged from his formalism, and the mentality of a splitting consciousness.

Everett agreed with Podolsky that the worlds were 'non-denumerably infinite'. He also agreed with Podolsky's statement, 'Every time a decision is made, the observer proceeds along one particular time while the other possibilities still exist and have physical reality.'

Furry summed up a common feeling amongst panelists, 'To me, the hard thing about it is that one must picture the world, oneself, and everybody else as consisting not in just a countable number of copies but somehow or another in an non-denumerable number of alternative Furrys.' The meeting adjourned without settling the debate (Werner [1962]).

Bohr's close colleague, Leon Rosenfeld, on the other hand, actively campaigned against the propagation of Everett's many-worlds interpretation for many years.

His anti-Everett letters are archived at the Niels Bohr Institute in Copenhagen. In 1959, Rosenfeld wrote to a colleague, Saul Bergmann, who had inquired about Everett's work: 'This work suffers from the fundamental misunderstanding which affects all attempts at "axiomatizing" any part of physics. The "axiomatizers" do not realize that every physical theory must necessarily make use of concepts which *cannot* in principle be further analyzed. . . . The fact, emphasized by Everett, that it is actually possible to set-up a wavefunction for the experimental apparatus and a Hamiltonian for the interaction between system and apparatus is perfectly trivial, but also terribly treacherous; in fact, it did mislead Everett to the conception that it might be possible to describe apparatus + atomic object as a closed system. . . . This, however, is an illusion' (Rosenfeld to Bergmann [December 21, 1959]).

And in November 1971, Rosenfeld wrote to John Bell in the theoretical division at CERN. Bell, who was famous for his theorem of quantum non-locality, was a firm non-Copenhagenist. He was soon to publish several papers taking Everett's universal wavefunction seriously by tying it to the wave theories of de Broglie and Bohm. Rosenfeld, never one to mince words, wrote: 'My dear Bell, Many thanks for the preprint of your last paper which I *did* read because you are one of the very few heretics from whom I always expect to learn something, and, indeed, I found this new paper of yours exceedingly instructive. To begin with, it is no mean achievement to have given Everett's damned nonsense an air of respectability by presenting it as a refurbishing of the idea of preestablished harmony. . . . [I]s it not complacent of you to think that you can contemplate the world from the point of view of God?' (Rosenfeld to Bell [November 30, 1971]).

In letters to Prof. H.J. Belifante of Purdue University in June 1972, Rosenfeld called Everett's theory a 'heresy' and a 'muddle', commenting, 'With regard to Everett neither I nor even Niels Bohr could have any patience with him, when he visited us in Copenhagen more than 12 years ago in order to sell the hopelessly wrong ideas he had been encouraged, most unwisely, by Wheeler to develop. He was undescribably [sic] stupid and could not understand the simplest things in quantum mechanics. . . . I would suggest that Occam's Razor could be most profitably used to rid us of Everett or at least his writings' (Rosenfeld to Belinfante [June 22, 1972]).

Rosenfeld's calumny was, no doubt, brought on by an evolving appreciation of Everett's theory by his peers. In 1967, DeWitt wrote an article for *Physical Review* presenting the Wheeler–DeWitt equation: a universal wavefunction that a theory of quantum gravity should satisfy. In the paper, he credited Everett's groundbreaking analysis of the need for a universal wavefunction. In September 1970, *Physics Today* published DeWitt's 'Quantum Mechanics and Reality'. In this article, DeWitt broadly attacked the 'conventionalist' Copenhagen Interpretation as 'external a priori metaphysics', and promoted, in its stead, the 'Everett–Wheeler–Graham metatheorem'. (Neill Graham was DeWitt's graduate student whose doctoral thesis was on Everett's derivation of probability, which he found lacking.) A lively debate about the merits and demerits of the once-ignored theory ensued in subsequent issues of *Physics Today*, and Everett followed it from afar (DeWitt [September 1970]; DeWitt et al. [1971]).

In 1973, DeWitt and Graham published Everett's unedited 'urwerk', along with favorable commentaries as *The Many Worlds Interpretation of Quantum Mechanics*. Quantum cosmologists in need of a universal wavefunction were paying serious attention to Everett's theory, and they were eventually joined by quantum computationists and philosophers. In Max Jammer's widely read book, *The Philosophy of Quantum Mechanics*, the author called Everett's theory, '[O]ne of the most daring and most ambitious theories ever constructed in the history of science' (Jammer [1974 p.509]).

7 PROBLEMS WITH PROBABILITY

Among Everett's effects, is a pre-print written in 1971 by John Bell, 'On the Hypothesis that the Schrödinger Equation is Exact' (probably the same paper that provoked Rosenfeld's ire). In a section on Everett, Bell wrote, '[T]his multiplication of universes is extravagant, and serves no real purpose in the theory and can simply be dropped without repercussions. So I see no reason to insist on this particular difference between the Everett theory and the pilot-wave theory . . . [T]he Everett theory provides a resting place for those who do not like the pilot wave trajectories but who would regard the Schrödinger equation as exact. But a heavy price has to be paid. We would live in a present which had no particular past, not indeed any particular (even if predictable) future. If such a theory were taken seriously it would hardly be possible to take anything else seriously. So much for the social implication.' (A version of this pre-print was later published as 'Quantum mechanics for cosmologists', in Bell [1987]). In the margin, Everett scrawled 'Ha' and a mostly illegible sentence that ends with 'probabilities also no unique past!'

Nor was Everett happy with how his strongest supporter, DeWitt, viewed his claim to have derived probability. In Everett's copy of DeWitt's lecture on 'The Many Universes Interpretation of Quantum Mechanics' at Varenna, Italy in July 1970, he furiously penciled, 'Goddamn it, you don't see it' next to DeWitt's assertion that Everett's derivation of probability is 'rather too brief to be entirely satisfying'.

In a September 1973 letter to Jammer (found in Mark's basement), Everett commented, 'I was somewhat surprised, and a little amused, that none of the physicists [in Copenhagen in 1959 and at Xavier University in 1962] had grasped one of what I considered to be the major accomplishment of the theory—the "rigorous" deduction of the probability interpretation of quantum mechanics from wave mechanics alone . . . . That this point was essentially completely overlooked at that time I can now only ascribe to my failure in writing the paper' (Everett to Jammer [September 19, 1973]).

DeWitt's 'Many Universes' essay was reprinted in *The Many Worlds Interpretation of Quantum Mechanics*. The sentence that Everett had found so offensive three years previously reappeared. DeWitt claimed that the Born rule could be derived from the formalism of quantum mechanics, thanks to work by Graham and James Hartle, but that, 'Everett's original derivation of this result invokes the formal equivalence of measure theory and probability theory, and is rather too brief to be entirely satisfying.' In Everett's personal hardcover copy of the book, he scribbled, next to DeWitt's sentence, 'only to you!' And Graham asserted, 'In short we criticize Everett's interpretation on the grounds of insufficient motivation. Everett gives no connection between his measure and the actual operations involved in determining a relative frequency . . . Furthermore, it is extremely difficult to see what significance such a measure can have when its implications are completely contradicted by a simple count of the worlds involved, worlds that Everett's own work assures us must all be on the same footing' (Graham [1973 p. 236]). Next to Graham's paragraph, Everett scrawled a single word: 'bullshit'.

8 WHEELER'S MANY MINDS

By the early 1970s, Wheeler was fascinated by the quantum chaos underlying the seeming order of the universe: 'law without law'. In a 1973 article, 'From Relativity to Mutability', he credited Everett's 'many universes formulation' as a contribution to quantum mechanics equal to those of Heisenberg, Dirac, and Feynman (Wheeler [1973]).

In April 1977, J.A. Wheeler and DeWitt invited Everett to give a seminar at the University of Texas in Austin. David Deutsch was there and he recalls that Everett was up on the latest in quantum theory. He was also addicted to smoking several packs of Kent cigarettes a day and was afflicted by alcoholism and obesity. But he was invigorated by the attention being paid to his theory. He told Charles Misner a few weeks later (on the tape-recording) that Wheeler had recently 'confessed, he actually now believes it [Everett's theory], except on Tuesdays, once a month.'

Wheeler's oscillations settled into a ground state. In July 1977, he publicly disavowed the many-worlds interpretation, writing, 'Imaginative Everett's thesis is, and instructive, we agree. We once subscribed to it. In retrospect, however, it looks like the wrong track. First, this formulation of quantum mechanics denigrates the quantum. It denies from the start that the quantum character of Nature is any clue to the plan of physics. Take this Hamiltonian for the world, that Hamiltonian, or any other Hamiltonian, this formulation says. I am in principle too lordly to care which, or why there should be any Hamiltonian at all. You give me whatever world you please, and in return I give you back many worlds. Don't look to me for help in understanding this universe.

'Second, its infinitely many unobservable worlds make a heavy load of metaphysical baggage. They would seem to defy Mendeleev's demand of any proper scientific theory, that it should "expose itself to destruction." ' (Wheeler [1979 p.396]).

And in another paper published in 1977, 'Include the Observer in the Wave-Function?', Wheeler endorsed the concept of an 'idealized observer' that Stern had accused him of promoting two decades earlier. Retracting his endorsement of Everett's theory of a universal wavefunction that includes the observer, Wheeler appealed to the authority of *both* Wigner and Bohr (who were not in agreement on interpretive matters): 'The "consciousness of the observer" is outside the wavefunction. An observation is only an observation then when it is recorded in the consciousness (Wigner 1974). An observation is only then an observation when one observer can tell another the result of the observation "in plain language" (Bohr 1962)' (Wheeler [1977 p.14]).

We do not know if Everett was aware of these attacks by his former mentor, but two years after the Austin trip, Nancy Everett typed an odd letter to Wheeler, after she and her husband had seen him on a television show. The letter was signed by Everett, but it is written in the third person. It reads, in part, 'There are two things about Hugh that perhaps need clearing up. One is, tho' it appears he plays hard-to-get by refusing to correspond, the truth is, he feels the written word is totally inadequate in comparison to a one-to-one conversation. This is why the meeting in Austin two years ago with Bryce DeWitt, the young Britishers, and others was such a great thing for him to participate in. . . . Far from being totally unconcerned, Hugh may even feel some gratification to be receiving a small measure of recognition for his work done under your counsel. . . . Now we read in *Physics Today* that even more is being done to expedite the flow and exchange of ideas what with the Institute forming in Santa Barbara. (Hugh always thought Santa Barbara a lovely spot.)' (Everett to Wheeler [March 21, 1979]).

In July 1979, in what may very well have been their last communication, Wheeler wrote to Everett: 'Thank you for your letter of too many weeks back. I think you got a great subject going and I am overjoyed at the thought of your getting back and going to bat for it!' On that same day, Wheeler wrote to Douglas Scalapino at the Institute for Theoretical Physics at Santa Barbara: 'Hugh Everett who did that fascinating Everett interpretation of quantum mechanics and who ought to be got back into it to go on with it has written to me indicating that he might conceivably get free to spend a period at the Institute. I have written Bryce DeWitt about this and believe that it has real possibilities for quite fruitful interac-tions' (Wheeler to Everett [July 12, 1979]; Wheeler to Scalapino [July 12, 1979]).

Despite Wheeler's enthusiasm, Everett made no move to reignite his theoretical career. Regardless, DeWitt wrote back to Wheeler, 'Everett suggests (and I believe) that it is a mistake to transform the wonderful lessons that Bohr has taught us into points of dogma. The history of physics has taught 1. that one should never be dogmatic, 2. that one should never hesitate to push a formalism to its ultimate logical conclusions however absurd. In the case of the formalism of quantum mechanics one cannot say that the interferences are there at one moment but gone the next. All that Everett is really trying to say is that the interferences are in principle always there. As David Deutsch so aptly puts it: "Quantum theory *is* the Everett interpretation." The theory may ultimately be proved wrong, but at the present time you cannot have one without the other' (DeWitt to Wheeler [September 25, 1979]).

9 END GAME

Everett, 51, died of a heart attack in July 1982, a quarter century after his dissertation was published. And on April 2008, Wheeler, 96, died in Princeton, New Jersey. According to his obituary in *The New York Times*, his two most prominent students were Richard Feynman and Hugh Everett III. Had he read that, Everett would have, surely, grinned.

In 1980, two years before his death, Everett wrote a letter to physics enthusiast L. David Raub: 'I certainly still support all of the conclusions of my thesis. . . . Dr Wheeler's position on these matters has never been completely clear to me (perhaps not to John either). He is, of course, heavily influenced by Bohr's position . . . It is equally clear to me that, at least sometimes, he wonders very much about that mysterious process, "the collapse of the wavefunction". The last time we discussed such subjects at a meeting in Austin several years ago he was even wondering if somehow human consciousness was a distinguished process and played some sort of critical role in the laws of physics.

'I, of course, do not believe any such special processes are necessary, and that my formulation is satisfactory in all respects. The difficulties in finding wider acceptance, I believe, are purely psychological. It is abhorrent to many individuals that there should not be a single unique state for them (in the world view), even though my interpretation explains all subjective feelings quite adequately and is consistent with all observations' (Everett to Raub [April 7, 1980]).

Inside a box of Everett's papers, is a list he made after a professional organization asked him to prioritize his top five scientific capabilities. At the bottom, Everett put 'servomechanisms', followed by 'operations research'. Skill number three was 'relativity and gravity', and two was 'decision game theory'. And listed in pride of first place was 'quantum mechanics'.

Postscript: In July 2007, just in time for the Oxford conference on Everett, *Nature* featured Everett's many-worlds theory on the cover, with several explanatory articles celebrating its 50th anniversary. In December 2007, *Scientific American* ran a profile of Everett by Peter Byrne. Oxford University Press commissioned a full-length biography of Everett by Byrne to appear in 2009,

The Many Worlds of Hugh Everett III: Multiple Universes, Mutually Assured Destruction, and the Meltdown of a Nuclear Family. And in November 2007, BBC4 premiered its film on Everett and his son, *Parallel Worlds, Parallel Lives*.

References

Bell, J.S. [1987], *Speakable and Unspeakable in Quantum Mechanics*, Cambridge University Press, Cambridge, 2004.

DeWitt, B. [1970], 'Quantum mechanics and reality', *Physics Today* 23(9): Sept.

DeWitt, B. et al. [1971] 'Letters to editor', *Physics Today* 24(10): Oct.

DeWitt, B. and N. Graham (eds) [1973], *The Many Worlds Interpretation of Quantum Mechanics*, Princeton University Press.

DeWitt, B. [*c.* 1988], Unpublished referee report for *Am. Jour. Ph.* on Yoav BenDov's 'Everett's theory and the "many worlds" interpretation', courtesy C. DeWitt-Morette.

DeWitt, C. (ed.) [1957], *Conference on the Role of Gravitation in Physics*, National Technical Information Service, UCI.

Everett III, H. [1957], ' "Relative state" formulation of quantum mechanics', *Reviews of Modern Physics* 29(3), 454–62.

Ford, K. [1995], 'Interview' with B. DeWitt and C. DeWitt, Wheeler archive, American Institute of Physics.

Frank, P. G. [1954], *The Validation of Scientific Theories*, The Beacon Press.

Graham, N. [1970], 'The Everett Interpretation of Quantum Mechanics', PhD thesis, University of North Carolina, Chapel Hill, UCI.

Jammer, M. [1974], *The Philosophy of Quantum Mechanics*, Wiley.

Pais, A. [1991], *Niels Bohr's Times in Physics, Philosophy, and Politics*, Oxford University Press, Oxford.

Werner, F.G. [1962], *Conference of the Foundations of Quantum Mechanics* (transcript), Xavier University, AIP, UCI.

Wheeler, J.A. [1957], 'Assessment of Everett's "relative state" formulation of quantum mechanics', *Reviews of Modern Physics*, 29(3) July, 463–5.

—— [1973], 'From Relativity to Mutability', in *The Physicist's Conception of Nature*, J. Mehra (ed), D. Reidel, Dordrecht-Holland, 202–47.

—— [1977], 'Include the Observer in the Wave Function', in *Quantum Mechanics, a Half Century Later*, Jose Leite Lopes and Michel Paty (eds), D. Reidel Publishing Company, Dordrecht-Holland.

—— [1979], 'Law without law', in *Problems in the Foundation of Physics, Proceedings of the International School of Physics 'Enrico Fermi'*, Varenna on Lake Como, 25th July–6th August 1977, North-Holland, 1979.

Wheeler, J.A. and Ford, K. [1988], *Geons, Black Holes & Quantum Foam*, W.W. Norton.

Note: Everett's scientific papers, including the correspondence referenced herein, will be availables as of 2011 online at the Digital Archive of the University of California, Irvine (UCI): http://ucispace.lib.uci.edu/. The original materials will be (or already are) archived in the Everett Archive at the Niels Bohr Library at the American Institute of Physics. The Rosenfeld letters are at the Niels Bohr Archive, Copenhagen.

18

Apart from Universes

David Deutsch

I'll start with a simple fact: in this room, in some nearby universes, Hugh Everett is here with us, celebrating. Perhaps he's *there*, in that seat where Simon is. And therefore, in those universes, Simon is somewhere else.

It's customary to say how astonishing that is—that we know of other universes and can reason about them, and have evidence of their attributes. But actually, by this time, the only astonishing thing is that that's still controversial. After all, we know that no single-universe theory can explain even the Einstein–Podolsky–Rosen experiment, let alone, say, quantum computation. That is because *any* process (hidden variables, or whatever) that accounts for such phenomena must not only be exponentially more complex than everything that we see, but also contains many autonomous streams of information, each of which describes something resembling the universe as described by classical physics (Deutsch [2002]).

But I don't want to address that controversy today, for reasons that I'll explain, except for one aspect, which, on the face of it, may seem merely terminological: I'm not going to refer to Everett's discovery as an 'interpretation'—nor even as *the* interpretation of quantum theory, though that's what it is (no other is known). That is because the term 'interpretation' has come to have connotations that misrepresent not just quantum theory but science in general. We don't speak of the existence of dinosaurs millions of years ago as being 'an interpretation of our best theory of fossils'. We claim that it is *the explanation* of fossils. And the theory isn't primarily about fossils: it's about dinosaurs. We claim that there really were dinosaurs, even though there is an infinity of rival 'interpretations' of the same data which make all the same predictions and yet say that the dinosaurs weren't there.

For instance, there's the 'interpretation' that fossils only come into existence when they are consciously observed—in which case, among other things, they're no older than the human species. And under that 'interpretation', they are *not* evidence of dinosaurs, but only of those acts of observation. Or there's the 'interpretation' that dinosaurs were such weird animals that 'conventional logic' doesn't apply to them. Or there's the 'interpretation' that it's meaningless even to ask whether dinosaurs were real or just useful fiction. None of those 'interpretations' is empirically distinguishable from the rational explanation of fossils. But they are ruled out because all of them are general-purpose means of denying *anything*. You could even use them to deny that quantum theory is true.

So, insisting that parallel universes are 'only an interpretation' and not a—what? a scientifically established fact or something (as if there were such a thing)—has the same logic as those stickers that they paste in some American biology textbooks, saying that *evolution* is 'only a theory', by which they mean precisely that it's just an 'interpretation'. Or, in terms of the analogy that Everett used in his famous exchange of letters with Bryce DeWitt, it's like claiming that the motion of the Earth about its axis is only an 'interpretation' that we place on our observations of the sky.

A prime mistake in all those 'interpretations' is to conceive of scientific theories as being composed of separate: *formalism*, *predictions*, and *'interpretation'*. For that necessarily makes it look as though the 'interpretation', which then by definition doesn't make predictions, isn't testable, and therefore isn't scientific. Also, as I've just explained, 'interpretations' would then be infinitely variable. So would formalisms for that matter, because on that view, the only real content of a theory is its predictions. But that is equivalent to making *observation* an irreducible primitive in science—which is inconsistent with developing a proper theory of measurement that is itself scientific. So you can't go that way.

That is one reason why science can only be *explanation*: asserting what is there in reality. (For a discussion of other reasons, I refer you to my book *The Fabric of Reality* [1997].) The only purpose of formalism, predictions, and interpretation is to express explanatory theories about what is there in reality, not merely predictions about human perceptions. Restricting science to the latter would be arbitrary and intolerably parochial.

Regarded as explanations in this sense, all the so-called interpretations of quantum theory are (see Deutsch [1996]) either versions of Everett's theory, sometimes in heavy disguise and with heavy equivocation (like the Bohm theory), or are claims that no explanation of quantum phenomena exists—that is to say, no objective reality which accounts for the observations on the one hand, and obeys the equations on the other. But if you deny that there's an objective reality, then science stops being *about* anything beyond your own thoughts.

So, the idea that science is about interpreting mathematical formalisms is horribly misleading. Which is why I won't say 'Everett's interpretation', but 'Everett's *theory*'—which is quantum theory, and I'll use those terms more or less interchangeably.

So much for the terminological issue. Now, if Everett *were* here today (which is another way of saying 'in universes in which Everett *is* here today'), what would he make of the progress that's been made with his theory since its discovery?

I think it's fair to say that, averaged over the whole period, progress has been disappointingly slow. However, thanks in no small part to some of the people in this room, the rate of progress has been increasing, and I'll come to that in a moment. But first, why was progress slow?

A related issue is that *take-up* of Everett's theory was, and continues to be, disappointingly slow. But I think that that is an effect, not a cause. After all, ten per cent—one per cent—of a vigorous research community like the theoretical physics community is plenty to pursue any one topic. There are plenty of causes in physics that are pursued by only one per cent of the community, and in which progress is made. The real impediment to progress in our fundamental understanding of quantum physics, during those decades, was quite different.

But let me digress briefly to some historical speculation:

> 'When it comes to discussing the many-universes
>
> interpretation, the level of the debate falls to zero.'
>
> – Dennis Sciama

Figure 1.

That was hyperbole by Dennis Sciama (I quote it from memory). The phenomenon has been diminishing, but it was once an intense effect, and it's tempting to blame it. But no, you can't blame a theory's *opponents* for not developing it.

By the way, Schrödinger had the basic idea of multiple universes shortly before Everett, but he didn't publish anything. He mentioned it in a lecture in Dublin (Schrödinger [1996]), in which he predicted that his audience would think he was crazy. Isn't that a strange assertion coming from a Nobel Prize-winner—that he feared being considered crazy for claiming that his own equation, the very one that he won the prize for, might be *true*?

So: Schrödinger, and Everett—and DeWitt, and everyone else who realized that Everett was right—encountered that perplexing phenomenon of Fig. 1. And as a result they—the proponents of Everett's theory, to the extent that they worked on it at all—focused on defending it against criticism of rather poor quality. Initially it was against its overtly irrational predecessor (the Copenhagen 'interpretation'), and against the Bohm theory. By the way, future historians may regard the Bohm theory as just another pre-Everett speculation that almost got there. Bohm could have anticipated Everett if he had chosen not to equivocate about which of the things in his theory were real and which weren't. Schrödinger, too, might well have got there if he had chosen to pursue the matter. Then later, Everett proponents were defending against all the other disguises and denials of quantum theory that were being proposed in vain attempts to escape from the plain physics of the theory.

But that was no way to solve that problem (Fig. 1)—nor indeed any problem—because defending a theory against its predecessor is inherently backwards-looking. Some such defence is of course necessary, but preoccupation with looking backwards to refute bad theories makes it hard to move forwards and develop the right one. It's as if molecular biologists were to devote all their efforts to combating creationism.

Then, more sophisticated opponents arrived with better criticism. Everett's theory was indeed improved by meeting it. That was progress; but progress of the same kind as putting better tyres on your car. What's the point of doing that if you never drive the car anywhere? So, *that* (Fig. 1), I think, wasn't the reason for the slow progress.

This was the reason:

A historical speculation:

Backwards-looking defensiveness on the part of

proponents of Everett's theory was a cause of their

slowness to use the theory to make further progress.

Figure 2.

And that has palpably changed. But not yet enough. So today I want to advocate developing the sophisticated and far-reaching *consequences* of Everett's theory rather than endlessly defending and re-defending just the basic idea of it. Let's address the many perplexing problems to which quantum theory gives us access, each of which presumably promises new insight into reality. In most cases it's only Everett's theory that does that, because most of those problems are fatally obscured in other approaches by their vagueness or crudeness, or in many cases by their denial that there's anything real there to theorize about in the first place.

If someone opposes (say) realism, or insists on inherently vague or equivocal language, or relies on what Richard Dawkins calls the 'argument from personal incredulity' (which, by the way, I think you'll find in virtually every critique of Everett *and* in virtually every defence), you can't prove them wrong. But what you can do, with a bit of luck, is make progress. I think that most physicists and philosophers want to understand the world, no matter how loudly they may deny that.

Therefore it's going to be *progress that's difficult or impossible to achieve in other approaches* that will ultimately be the best and only possible effective defence of the rational quantum theory—just as progress in general is the only really effective argument for science and reason in general.

So I want to tell you what I think some of those unsolved problems are. They're all in one way or another about what is out there, in reality, *apart from universes*. Because the multiverse isn't just a collection of universes. In fact universes, the things that historically nearly all the fuss has been about, are really just classical physics. A collection of classical universes—even if they slightly interact—is still classical. It doesn't have entanglement; its elements don't have phases to their amplitudes; it doesn't have continuous motion of discrete observables; it doesn't support quantum computation, and so on. It doesn't even represent *observables* unless they're diagonal, or near-diagonal, in the decoherence basis. Those are some of the properties that the multiverse has in addition to universes.

Consider a single, free particle in empty space. It's described by a wave-packet—thanks to the uncertainty principle. Which means that, as far as universes go, it's at different positions in different universes. You might think that a non-interacting particle, at least, is something that happens in each universe independently of the others, so that we can forget about the multiverse when describing it. But no. Again because of the uncertainty principle, and because of the linearity of quantum mechanics, there is no region of the multiverse in which both the position and the velocity are behaving independently of what's happening elsewhere in the multiverse. That means that there are no autonomous information flows that would be universes. So in fine detail, even a free particle is an irreducibly multiversal object, not just a parallel-universes' one.

Furthermore, at a later time, the shape of the wavepacket will have changed. The instances of the particle in the multiverse will be at different positions. But none of them, individually, will have moved to where it is—because there is no such thing as one of them individually. When the universe approximation breaks down, the autonomy of the instances of a single particle in the multiverse breaks down too. They are then fungible.

All this is implicit in the equations, but we lack an explicit mathematical description of the multiverse at any level lower than the emergent level of universes. This is one of the most telling omissions from multiverse research to date: no one has yet done the spadework to construct a mathematical description of what the multiverse *is*. In contrast, in general relativity, Einstein's field equations are understood to apply to a physical object, spacetime, with a non-trivial structure: it has a geometry, a topology, a signature, singularities—all of which took a lot of careful mathematical and theoretical thinking to discover and to express. Whatever possesses people to think that, for instance, quantum gravity can be cracked before we've developed a similar understanding of the multiverse of quantum theory?

So, where should we look for that understanding? Even in general relativity, much of the theory of spacetime is the theory of *information flow*: the Lorentzian metric, event horizons, Cauchy surfaces, causality conditions, and so on. And at a more basic level, there's just the plain relativity of simultaneity as well. In the structure of the multiverse, information flow is even more important; it might even be the whole story. And it's studied in the quantum theory of computation. So that's a good place to look. (In fact, that's the place to look for most important things, I've found.)

At the crudest level, quantum computation is just quantum parallelism: many classical computations in parallel—which is the parallel-*universe* approximation, the thing we need to improve upon. Now, we know that the physical attributes of systems determine whether and how they can be used for information processing: to do Turing computations, you need discrete observables, such as bits, that can interact in certain ways, and you need stable information storage. The next step, say, quantum cryptography, requires certain operations on qubits rather than bits, but only individual ones. To build general quantum gates you need to be able to do more, and for universal quantum computation still more, such as error correction. And then it stops; then we have something universal. So, different kinds of information processing are possible in different kinds of multiverse region. Therefore, the understanding that's currently being developed about these different kinds of computational resources is telling us about the structure of the multiverse.

But the sophistication of these new results about computation contrasts sharply with the woolly and archaic *explanations* that still accompany them. In quantum teleportation, for instance, we're still routinely told that information passes from A to B without passing through the space in between, or that it travels backwards in time to the source of the entanglement and then forwards again to the destination, or that it 'jumps instantaneously across the gap' (whatever that means) and so on. Again, there is no substance accompanying these descriptions, no mathematical object that is purported to represent these alleged processes of jumping and so on. Only their observed outcomes are described mathematically, because, again, the quantum formalism is being used purely instrumentally, together with the same old single-universe apologetics.

The only way of understanding what's really happening in quantum teleportation, and in the newly discovered, important processes of cluster/graph-quantum computation, is to apply Everett's theory. Patrick Hayden and I analysed information flow in general entangled systems—using the quantum theory of computation—and showed that there's quantum information *inside* decoherent observables (Deutsch and Hayden [1999]). There's no way even to express that fact, other than in Everett's theory, because it means that there is no 'classical level'. Even physical variables such as those in an observer's brain, which behave autonomously in each universe, also carry this hidden quantum information that's not confined to universes and can later participate in quantum computations. By the way, in that work, we used the Heisenberg Picture, which I don't have time to go into today, but suffice it to say that working in the Heisenberg Picture is the key to eliminating many of the stubbornest misconceptions in quantum theory, including those about non-locality in processes like quantum teleportation.

There's a related and even worse spanner in the works of elementary particle physics: particles (or fields, strings, or whatever) are supposed to be fully quantum-mechanical entities. But the people who work on them only ever construct classical, single-universe theories. Why? Because they think that the quantum part of the theory necessarily has to be trivial. It is assumed that in order to discover the true quantum-dynamical equations of the world, you have to enact a certain ritual. First you have to invent a theory that you know to be false, using a traditional formalism and laws that were refuted a century ago. Then you subject this theory to a formal process known as quantization (which for these purposes includes renormalization). And that's supposed to be your quantum theory: a classical ghost in a tacked-on quantum shell.

In other words, the true explanation of the world is supposed to be obtained by the mechanical transformation of a false theory, without any new explanation being added. This is almost magical thinking. How far could Newtonian physics have been developed if everyone had taken for granted that there had to be a ghost of Kepler in every Newtonian theory—that the only valid solutions of Newtonian equations were those based on conic sections, because Kepler's Laws had those. And because the early successes of Newtonian theory had them too?

The 'quantization' recipe did successfully discover new quantum systems—what?—three times, four times? It was worth trying for one or two more times. Lots of things are worth *trying*. *String theory* was worth trying. But there was no justification for relying exclusively on nature being a classical ghost in a quantum shell. I think it's no coincidence that the decades of not taking seriously what quantum theory says in its own right (by following up on Everett) was also the time when discoveries of new kinds of quantum systems frustratingly slowed down as well.

It's surely a mistake to stick to the historical single-universe worldview even while using quantum-mechanical equations. We should be seeking explanations, not in the confines of one spacetime, or even multiple spacetimes, but in the multiverse. (Such theories are certainly possible. I even constructed one a while ago for unrelated reasons: *Qubit Field Theory* (Deutsch [2004]).)

Another problem is the nature of time. In 1983 Don Page and Bill Wootters discovered that *times are Everett universes*. They were addressing, again, an apparently unrelated theoretical problem, namely: since the total energy is conserved, how can we tell that the world isn't in an eigenstate of it? They discovered that we *can't* tell; we can assume without loss of generality that the world is in an eigenstate of its Hamiltonian: a *stationary* state. But then, what's all this motion that we see? By analysing closed physical systems that included clocks, they showed that motion is an entanglement effect: clock states are correlated with states of other systems. This important result should be the basis of everyone's conception of time. Yet it has barely, and rarely, been taken on board. (Julian Barbour [1999] is an honourable exception, as is Andreas Albrecht's talk at this conference.) Why has it not been taken on board? Because, again, this is an issue that one can address, and solve, and understand, only in Everett's theory. All other approaches just command us to talk nonsense, or assume a classical conception of time by fiat—or if they were being consistent, would say that the present is the only real time and the rest just possibilities.

I should also mention time *travel*. I think that more work needs to be done on information flow round closed time-like curves. Because even if it turns out that we have to rule them out, how can we do that without understanding what causality means in the multiverse? And one thing we do know is that it means something different from what it means in spacetime.

Then there's the arrow of time. Or arrows of time. Entropy is a measure of the 'branchingness' of the multiverse; and the so-called subjective or 'knowledge' arrow of time is also related to multiverse structure, because knowledge is associated with complexity that extends over a multiversal region (Deutsch [1997]). So it seems likely that the entropy and knowledge arrows of time are connected via the structure of the multiverse.

An existing success along the lines that I'm advocating is decoherence theory. This solved the 'preferred basis problem'—but only in (Everett) quantum theory, not in any of the 'other interpretations': in those, one still had to say by fiat which states the wavefunction collapsed to, and so on. Only in the unmodified quantum theory does nothing unusual happen at measurements, and so only it can accommodate the fact that universes turn out to be approximate, emergent structures in the multiverse. Decoherence theory opened up the study of the structure of the multiverse: not just how the quasiclassical universes emerge, but also how what is happening exactly when universes are present emergently.

Another such success has been in the understanding of *probability*: what do probability statements mean (without circularity or nonsense)? Many of the other presentations at this conference discuss that work. What I want to stress is that Everett's theory provides the solution to a philosophical puzzle that people had worried about quite independently of quantum theory. The problem is a bit like deriving an ought from an is—it's deriving a tends to from a does.

You try to explain what a probability statement means—say, that heads will probably come up about half the time when you toss coins—and you find yourself reducing it to other probability statements—such as that if you bet on some other proportion you are 'likely' to lose. It seems that you can never reduce it to 'so and so *will* happen'. Consequently, people have been reduced to talking nonsense about this, such as the 'frequency' or 'ensemble' interpretations of probability. They say that something is 'probable' if it would happen in the majority of cases, if the experiment were repeated infinitely many times. But then they are not talking about reality, because real experiments aren't performed infinitely many times. And even if they were, we want to know what to do on particular occasions. And in any case, the frequency interpretation, whether with an infinite or a finite number of instances, logically doesn't reproduce probabilities unless the instances are equally likely, which gives a circular definition of 'likely'.

Other people have tried to say that probability is a measure of subjective belief. But then you have to explain why my subjective belief about the coin should be a property of the coin rather than just of me—and the answer has to be that it's not any old belief, it's the belief that I *should* have, if I'm rational. But then, why should rational people have that subjective belief? Because if they were to adopt some other belief they would be *likely* to lose bets. There's the circularity again.

David Papineau once called this state of affairs at the foundations of probability theory a scandal. Well, as I said, it has been solved: Everett's theory, which is a completely deterministic theory of the multiverse without stochastic processes, determines what rational behaviour is for observers who bifurcate in the multiverse (Deutsch [1999], Wallace [2003]). Consequently, anyone who sets out to deny Everett now, has the embarrassing problem that the philosophy of probability is a scandal without it. I know that some people here disagree. But some agree! And this which goes to illustrate the merit of seeking progress instead of ever more devastating demolitions of what were always bad ideas in the first place.

However, there still exist some outstanding and promising problems in regard to probability, because there's more to probability than just decision-making when observers bifurcate: there are also important situations where an observer only exists in some regions of the multiverse and not in others.

One example is in anthropic-principle arguments, where one tries to explain the values of observed quantities by calculating their expectation values *given* that someone exists to ask what the values should be. In particular, one is usually concerned with ensembles of universes, or often ensembles of multiverses, with many different sets of laws of physics. We exist in some of them and not in others, which, it is hoped, will allow predictions about what we shall observe, conditional on our existing. But one problem that arises is 'who is we?'

This isn't just the problem of personal identity in the philosophy of mind (though that too is an issue that one can't hope to understand without the full quantum context, as Michael Lockwood [1991] and others have pointed out). For instance, the universe may be spatially infinite, in which case, again, there may be infinitely many instances of the whole universe that are observable from here. If we identify all identical instances of us as 'the same person' (which we surely have to, if we are physicalists), then should we not also be identifying identical observable universes across space as being the same universe? But in that case, space would really be finite after all, even in cosmologies where space is infinite in the classical, general-relativity approximation.

Then there's the so-called 'quantum suicide' argument, to the effect that if you want to win the lottery, all you have to do is buy yourself a ticket and set up a machine that will automatically kill you in your sleep if you lose. Then, in all the universes in which you do wake up, you're a winner.

Then there are various versions of what's called the 'simulation argument', for instance by Nick Bostrom [2003]. Its premise is that in the distant future the whole universe as we know it is going to be simulated in computers, many times. Therefore the overwhelming majority of the instances of whatever we have observed will also be observed in those simulations. And therefore we probably are in one of them. So the argument goes.

At present, little if any progress can be made in any of those controversies. Ingenious arguments have been devised for them, but at present, they all reduce to hand-waving, because at the root of each of them there is some assumption about probability: all of them are guilty of using the frequency interpretation, or the subjective-belief interpretation, or both. They are therefore all in a state of sin, at present.

What is needed is to express such arguments in the framework of a theory of a multiverse—sometimes it has to be a bigger multiverse than Everett's. One has to derive the premises of the argument from laws of physics, as Everett did. One has to determine, as was done with the Everett multiverse, whether the theory forces a unique meaning on the numbers that those arguments want to regard as probabilities—a meaning such that those numbers can be used to make predictions and place rational bets, make Bayesian arguments about the state of the universe and so on. If it doesn't force such a meaning, then no progress can be made with it and the argument in question isn't really valid—isn't really scientific.

My guess is that when that's done we'll find that the quantum suicide argument and some versions of the simulation argument are actually invalid. The anthropic argument, I suspect, is right as far as it goes, but insufficient in itself to explain anything. But those are just my opinions: just hand-waving. Hand-waving, as I said, can't resolve those issues, only future developments in multiverse theory can.

The open problems I've mentioned here are just the tip of the iceberg, I'm sure. Everett's theory is very far-reaching because quantum theory is deeper than all the other theories in physics that are considered fundamental. The implication of that for physicists is: we've got to understand this thing, the multiverse, because otherwise in every fundamental branch of physics it's as if we were planning an expedition to the moon while still thinking that the earth is flat.

To philosophers, the implication is: some of our basic intuitions about the physical world are plain wrong, not just in regard to how many universes there are (as I said, that's the least of it; that's really just classical). It's what's there *apart* from universes. And hence, since philosophy is about understanding things—like probability, and time, and the nature of existence and non-existence, and the self, causation, laws of nature, the relationship of mathematics to physical reality—for any of those, you've got to understand the multiverse. Learn to think in terms of it, build on that understanding, and apply it to learning about everything else. And discover its successor.

As I said, there are people here already doing those things. But to the rest, and to the physics and philosophy communities at large: don't let it be another 50 years before you too become serious about building on Everett's discovery.

References

Barbour, J. [1999], *The End of Time*, Penguin Books.

Bostrom, N. [2003], 'Are you living in a computer simulation?', *Philosophical Quarterly* **53**, 211, 243–55.

Deutsch, D. [1996], 'Comment on Lockwood,' *British Journal for the Philosophy of Science* **47**(2), 222–8.

—— [1997], *The Fabric of Reality*, Penguin Books.

—— [1999], 'Quantum theory of probability and decisions', *Proceedings of the Royal Society of London* **A 455**, 3129–37.

—— [2002], 'The Structure of the multiverse', *Proceedings of the Royal Society of London* **A 458**, 2028, 2911–23.

—— [2004], *Qubit Field Theory*. Available online at arXiv.org/pdf/quant-ph/0401024.

Deutsch, D. and P. Hayden [1999], 'Information flow in entangled quantum systems', *Proceedings of the Royal Society of London* **A 456**, 1759.

Lockwood, M. [1991], *Mind, Brain and the Quantum*, Blackwell.

Page, D.N. and W.K. Wootters [1983], 'Evolution without evolution: Dynamics described by stationary observables', *Physical Review* **D 27**, 2885.

Schrödinger, E. [1996], *The Interpretation of Quantum Mechanics: Dublin Seminars (1949–1955) and other unpublished essays*, M. Bitbol (ed.), OxBow Press.

Wallace, D. [2003], 'Everettian rationality: defending Deutsch's approach to probability in the Everett interpretation', *Studies in the History and Philosophy of Modern Physics* **34**, 415–38.

19

Many Worlds in Context

Max Tegmark

ABSTRACT

Everett's many-worlds interpretation of quantum mechanics is discussed in the context of other physics disputes and other proposed kinds of parallel universes. We find that only a small fraction of the usual objections to Everett's theory are specific to quantum mechanics, and that all of the most controversial issues crop up also in settings that have nothing to do with quantum mechanics.

1 INTRODUCTION

There is now great interest in Everett's many-worlds interpretation of quantum mechanics and the controversy surrounding it.[1] A key reason for this is undoubtedly that it connects with some of our deepest questions about the nature of reality. How large is physical reality? Are there parallel universes? Is there fundamental randomness in nature?

---

[1] The controversy shows no sign of abating, as evidenced by the results of the following highly unscientific poll carried out by the author at the Perimeter Institute Everett@50 Conference 9/22-07:

1. *Do you believe that new physics violating the Schrödinger equation will make large quantum computers impossible?* (4 Yes; 29 No; 11 Undecided)
2. *Do you believe that all isolated systems obey the Schrödinger equation (evolve unitarily)?* (17 Yes; 10 No; 20 Undecided)
3. *Which interpretation of quantum mechanics is closest to your own?*
   - 2 Copenhagen or consistent histories (including postulate of explicit collapse)
   - 5 Modified dynamics (Schrödinger equation modified to give explicit collapse)
   - 19 Many worlds/consistent histories (no collapse)
   - 2 Bohm
   - 1.5 Modal
   - 22.5 None of the above/undecided
4. *Do you feel comfortable saying that Everettian parallel universes are as real as our universe?* (14 Yes; 26 No; 8 Undecided)

Table 1. Most Common Worries about Everett's Many-Worlds Interpretation are not Specific to Quantum Mechanics.

Worry	QM-specific?	Counterexamples/resolution
1 Popper worry: falsifiable?	No	General relativity, inflation
2 Occam worry: parallel universes feel wasteful	No	Level I & II multiverses
3 Aristotle worry: math mere approximation	No	Linked to external reality hypothesis
4 Uncertainty worry: How can omniscience allow uncertainty?	No	Occurs whenever observer ensemble
5 How derive probabilities from deterministic theory?	No	Occurs whenever observer cloning
6 Unequal probability worry: Why square the amplitudes?	Yes	Comes from Hilbert space structure
7 ρ worry: describes world or my knowledge of it?	Partly	Can describe both
8 How judge evidence for/against such a theory?	No	Classical statistical mechanics
9 Word worry: What do we mean by 'exist', 'real', 'is', etc?	No	Level I & II multiverses
10 Invisibility worry: Why can't we detect the parallel worlds?	Yes	Solved by decoherence
11 Basis worry: What selects preferred basis?	Yes	Solved by decoherence
12 Weirdness worry	No	Electric fields, black holes, Levels I & II

The goal of this article is to place both Everett's theory and the standard objections to it in context. We will review how Everett's many worlds may constitute merely one out of four different levels of parallel universes, the rest of which have little to do with quantum mechanics. We will also analyze the many objections to Everett's theory listed in Table 1, concluding that most of them are not specific to quantum mechanics. By better understanding this context, quantum physicists can hopefully avoid reinventing many wheels that have been analyzed in detail in other areas of physics or philosophy, and focus their efforts on those remaining aspects of Everett's theory that are uniquely quantum-mechanical. This is not to say that the issues in Table 1 with a 'No' in the *QM-specific* column are necessarily unimportant—merely that it is unfair to blame Hugh Everett for them or to use them as evidence against his theory alone.

Rather than discuss these objections one by one in the order they appear in Table 1, this article is structured as a survey of multiverse theories, addressing the objections in their natural context. We then return to Table 1 and summarize our conclusions in Section 6.

1.1 The MWI: What it is and What it isn't

Let us first spell out what we mean by the many-worlds interpretation (MWI). Much of the early criticism of the MWI was based on confusion as to what it meant. Here we grant Everett the final say in how the MWI is defined, since he did after all invent it (Everett [1957], [1973]), and take it to consist of the following postulate alone:

- Everett postulate:
  All isolated systems evolve according to the Schrödinger equation $\frac{d}{dt}|\psi\rangle = -\frac{i}{\hbar}H|\psi\rangle$.

More succinctly, 'physics is unitary'. Although this postulate sounds rather innocent, it has far-reaching implications:

1. **Corollary 1:** the entire universe evolves according to the Schrödinger equation, since it is by definition an isolated system.
2. **Corollary 2:** when a superposition state is observed, there can be no definite outcome (wavefunction collapse), since this would violate the Everett postulate.

Because of corollary 1, 'universally valid quantum mechanics' is often used as a synonym for the MWI. What is to be considered 'classical' is therefore *not* specified axiomatically (put in by hand) in the MWI—rather, it can be derived from the Hamiltonian dynamics, by computing decoherence rates.

How does corollary 2 follow? Consider a measurement of a spin 1/2 system (a silver atom, say) where the states 'up' and 'down' along the z axis are denoted $|\uparrow\rangle$ and $|\downarrow\rangle$. Assuming that the observer will get happy if she measures spin up, we let $|\overset{..}{-}\rangle$, $|\overset{..}{\smile}\rangle$ and $|\overset{..}{\frown}\rangle$ denote the states of the observer before the measurement, after perceiving spin-up and after perceiving spin-down, respectively. If the measurement is to be described by a unitary Schrödinger time evolution operator $U = e^{-iH\tau/\hbar}$ applied to the total system, then U must clearly satisfy

$$U|\uparrow\rangle \otimes |\overset{..}{-}\rangle = |\uparrow\rangle \otimes |\overset{..}{\smile}\rangle \quad \text{and} \quad U|\downarrow\rangle \otimes |\overset{..}{-}\rangle = |\downarrow\rangle \otimes |\overset{..}{\frown}\rangle. \tag{1}$$

Therefore if the atom is originally in a superposition $\alpha|\uparrow\rangle + \beta|\downarrow\rangle$, then the Everett postulate implies that the state resulting after the observer has interacted with the atom is

$$U(\alpha|\uparrow\rangle + \beta|\downarrow\rangle)) \otimes |\overset{..}{-}\rangle = \alpha|\uparrow\rangle \otimes |\overset{..}{\smile}\rangle + \beta|\downarrow\rangle \otimes |\overset{..}{\frown}\rangle. \tag{2}$$

In other words, the outcome is not $|\uparrow\rangle \otimes |\smile\rangle$ or $|\downarrow\rangle \otimes |\frown\rangle$ with some probabilities, merely these two states in superposition. Very few physicists have actually read Everett's original 137-page PhD thesis (reprinted in Everett: [1973]), which has led to a common misconception that it contains a second postulate along the following lines:

- What Everett does **NOT** postulate:
  At certain magic instances, the world undergoes some sort of metaphysical 'split' into two branches that subsequently never interact.

This is not only a misrepresentation of the MWI, but also inconsistent with the Everett postulate, since the subsequent time evolution could in principle make the two terms in Eq. (2) interfere. According to the MWI, there is, was and always will be only one wavefunction, and only decoherence calculations, not postulates, can tell us when it is a good approximation to treat two terms as non-interacting.

1.2 Many Worlds Galore

Parallel universes are now all the rage, cropping up in books, movies and even jokes: 'You passed your exam in many parallel universes—but not this one.' However, they are as controversial as they are popular, and it is important to ask whether they are within the purview of science, or merely silly speculation. They are also a source of confusion, since many forget to distinguish between different types of parallel universes that have been proposed (Tegmark [1998]).

The farthest you can observe is the distance that light has been able to travel during the 14 billion years since the big-bang expansion began. The most distant visible objects are now about 4×10^{26} meters away,[2] and a sphere of this radius defines our observable universe, also called our *Hubble volume*, our *horizon volume*, or simply our universe. Below I survey physics theories involving parallel universes, which form a natural four-level hierarchy of multiverses (Fig. 1) allowing progressively greater diversity.

- **Level I:** A generic prediction of cosmological inflation is an infinite 'ergodic' space, which contains Hubble volumes realizing all initial conditions—including an identical copy of you about $10^{10^{29}}$ m away.
- **Level II:** Given the *fundamental* laws of physics that physicists one day hope to capture with equations on a T-shirt, different regions of space can exhibit different *effective* laws of physics (physical constants, dimensionality, particle content, etc.) corresponding to different local minima in a landscape of possibilities.

---

[2] After emitting the light that is now reaching us, the most distant things we can see have receded because of the cosmic expansion, and are now about 40 billion light years away.

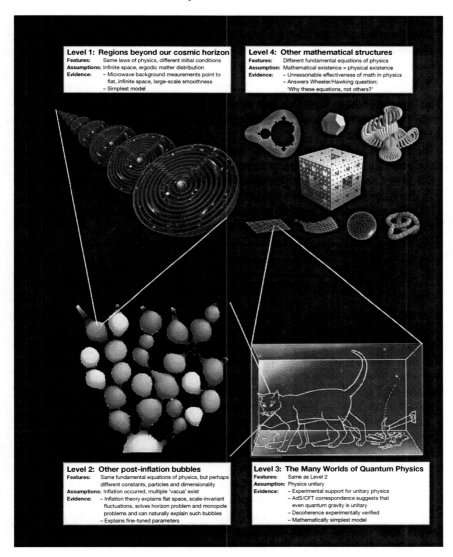

Figure 1. The four parallel universe levels.

- **Level III:** In Everett's unitary quantum mechanics, other branches of the wavefunction add nothing qualitatively new, which is ironic given that this level has historically been the most controversial.
- **Level IV:** Other mathematical structures give different fundamental equations of physics for that T-shirt.

The key question is therefore not whether there is a multiverse (since Level I is the rather uncontroversial cosmological concordance model), but rather how many levels it has.

Below we will discuss at length the issue of evidence and whether this is science or philosophy. For now, the key point to remember is that *parallel universes are not a theory, but a prediction of certain theories*. The Popper worry listed in Table 1 is the question of whether Everett's theory is falsifiable. For a theory to be falsifiable, we need not be able to observe and test all its predictions, merely at least one of them. Consider the following analogy:

General relativity	Black hole interiors
Inflation	Level I parallel universes
Unitary quantum mechanics	Level III parallel universes

Because Einstein's theory of general relativity has successfully predicted many things that we *can* observe, we also take seriously its predictions for things we cannot observe, e.g., that space continues inside black hole event horizons and that (contrary to early misconceptions) nothing funny happens right at the horizon. Likewise, successful predictions of the theories of cosmological inflation and unitary[3] quantum mechanics have made some scientists take more seriously their other predictions, including various types of parallel universes.

Let us conclude with two cautionary remarks before delving into the details. Hubris and lack of imagination have repeatedly caused us humans to underestimate the vastness of the physical world, and dismissing things merely because we cannot observe them from our vantage point is reminiscent of the ostrich with its head in the sand. Moreover, recent theoretical insights have indicated that nature may be tricking us. Einstein taught us that space is not merely a boring static void, but a dynamic entity that can stretch (the expanding universe), vibrate (gravitational waves) and curve (gravity). Searches for a unified theory also suggest that space can 'freeze', transitioning between different phases in a landscape of possibilities just like water can be solid, liquid, or gas. In different phases, effective laws of physics (particles, symmetries, etc.) could differ. A fish never leaving the ocean might mistakenly conclude that the properties of water are universal, not realizing that there is also ice and steam. We may be smarter than fish, but could be similarly fooled: cosmological inflation has the deceptive

---

[3] As described below, the mathematically simplest version of quantum mechanics is 'unitary', lacking the controversial process known as wavefunction collapse.

property of stretching a small patch of space in a particular phase so that it fills our entire observable universe, potentially tricking us into misinterpreting our local conditions for the universal laws that should go on that T-shirt.

2 LEVEL I: REGIONS BEYOND OUR COSMIC HORIZON

Let us return to your distant twin. If space is infinite and the distribution of matter is sufficiently uniform on large scales, then even the most unlikely events must take place somewhere. In particular, there are infinitely many other inhabited planets, including not just one but infinitely many with people with the same appearance, name and memories as you. Indeed, there are infinitely many other regions the size of our observable universe, where every possible cosmic history is played out. This is the Level I multiverse.

2.1 Evidence for Level I Parallel Universes

Although the implications may seem crazy and counter-intuitive, this spatially infinite cosmological model is in fact the simplest and most popular one on the market today. It is part of the cosmological concordance model, which agrees with all current observational evidence and is used as the basis for most calculations and simulations presented at cosmology conferences. In contrast, alternatives such as a fractal universe, a closed universe, and a multiply connected universe have been seriously challenged by observations. Yet the Level I multiverse idea has been controversial (indeed, an assertion along these lines was one of the heresies for which the Vatican had Giordano Bruno burned at the stake in 1600[4]), so let us review the status of the two assumptions (infinite space and 'sufficiently uniform' distribution).

How large is space? Observationally, the lower bound has grown dramatically (Fig. 1) with no indication of an upper bound. We all accept the existence of things that we cannot see but could see if we moved or waited, like ships beyond the horizon. Objects beyond cosmic horizon have similar status, since the observable universe grows by one light-year every year as light from further away has time to reach us.[5] If anything, the Level I multiverse sounds trivially obvious. How could space not be infinite? Is there a sign somewhere saying 'Space ends here—mind the gap'? If so, what lies beyond it? In fact, Einstein's theory of gravity calls this intuition into question. Space could be finite if it has a convex curvature or an unusual topology (that is, interconnectedness). A spherical,

---

[4] Bruno's ideas have since been elaborated by, for example, Brundrit [1979], Garriga and Vilenkin [2001], all of whom have thus far avoided the stake.

[5] If the cosmic expansion continues to accelerate (currently an open question), the observable universe will eventually stop growing.

doughnut-shaped or pretzel-shaped universe would have a limited volume and no edges. The cosmic microwave background radiation allows sensitive tests of such scenarios. So far, however, the evidence is against them. Infinite models fit the data, and strong limits have been placed on the alternatives (de Oliveira-Costa et al. [2003], Shapiro et al. [2007]). In addition, a spatially infinite universe is a generic prediction of the cosmological theory of inflation (Garriga and Vilenkin [2001]), so the striking successes of inflation listed below therefore lend further support to the idea that space is after all infinite just as we learned in school.

Another loophole is that space is infinite but matter is confined to a finite region around us—the historically popular 'island universe' model. In a variant on this model, matter thins out on large scales in a fractal pattern. In both cases, almost all universes in the Level I multiverse would be empty and dead. But recent observations of the three-dimensional galaxy distribution and the microwave background have shown that the arrangement of matter gives way to dull uniformity on large scales, with no coherent structures larger than about 10^{24} meters. Assuming that this pattern continues, space beyond our observable universe teems with galaxies, stars, and planets.

2.2 What Are Level I Parallel Universes Like?

The physics description of the world is traditionally split into two parts: initial conditions and laws of physics specifying how the initial conditions evolve. Observers living in parallel universes at Level I observe the exact same laws of physics as we do, but with different initial conditions than those in our Hubble volume. The currently favored theory is that the initial conditions (the densities and motions of different types of matter early on) were created by quantum fluctuations during the inflation epoch (see Section 3). This quantum mechanism generates initial conditions that are for all practical purposes random, producing density fluctuations described by what mathematicians call an ergodic random field. *Ergodic* means that if you imagine generating an ensemble of universes, each with its own random initial conditions, then the probability distribution of outcomes in a given volume is identical to the distribution that you get by sampling different volumes in a single universe. In other words, it means that everything that could in principle have happened here did in fact happen somewhere else.

Inflation in fact generates all possible initial conditions with non-zero probability, the most likely ones being almost uniform with fluctuations at the 10^{-5} level that are amplified by gravitational clustering to form galaxies, stars, planets, and other structures. This means both that pretty much all imaginable matter configurations occur in some Hubble volume far away, and also that we should expect our own Hubble volume to be a fairly typical one—at least typical among those that contain observers. A crude estimate suggests that the closest identical copy of you is about $\sim 10^{10^{29}}$ m away. About $\sim 10^{10^{91}}$ m away, there should be a sphere of radius 100 light-years identical to the one centered here, so all perceptions that we have during the next century will be identical to those of our counterparts over there. About $\sim 10^{10^{115}}$ m away, there should be an entire Hubble volume identical to ours.[6]

2.3 How to Derive Probabilities From a Causal Theory?

Let us now turn to worry 4 and worry 5 in Table 1. The Level I multiverse raises an interesting philosophical point: you would not be able to compute your own future even if you had complete knowledge of the entire state of the cosmos! The reason is that there is no way for you to determine which of these copies is 'you' (they all feel that they are). Yet their lives will typically begin to differ eventually, so the best you can do is predict probabilities for what you will observe, corresponding to the fractions of these observers that experience different things. This kills the traditional notion of determinism even without invoking quantum mechanics.

However, perhaps it is a uniquely quantum-mechanical phenomenon that you can end up with subjective indeterminism even if only a single you exists to start with? No, because we can create the same phenomenon in the following simple gedanken experiment involving only classical physics, without even requiring any sort of multiverse (not even Level I). You are told that you will be sedated, that a perfect clone of you will be constructed (including your memories), and that the two yous will be woken up by a bell at the same time the next morning in two identical-looking rooms. The rooms are numbered 0 and 1, and these numbers are printed on a sign outside the door. When asked by the anesthesiologist to place a bet on where you will wake up, you realize that you have to give her 50-50 odds, because someone feeling that they are you will wake up in both. When you awaken, you realize that you'd still give 50-50 odds, because even if you knew the position of every atom in the universe, you still couldn't know which of the two yous is the one having your current subjective experience. When you go outside, the room number you read will therefore feel just like a random number to you.

---

[6] This is an extremely conservative estimate, simply counting all possible quantum states that a Hubble volume can have that are no hotter than 10^8 K. 10^{115} is roughly the number of protons that the Pauli exclusion principle would allow you to pack into a Hubble volume at this temperature (our own Hubble volume contains only about 10^{80} protons). Each of these 10^{115} slots can be either occupied or unoccupied, giving $N = 2^{10^{115}} \sim 10^{10^{115}}$ possibilities, so the expected distance to the nearest identical Hubble volume is $N^{1/3} \sim 10^{10^{115}}$ Hubble radii $\sim 10^{10^{115}}$ meters. Your nearest copy is likely to be much closer than $10^{10^{29}}$ meters, since the planet formation and evolutionary processes that have tipped the odds in your favor are at work everywhere. There are probably at least 10^{20} habitable planets in our own Hubble volume alone.

Now suppose instead that you were told that this experiment would be repeated 10 more times, resulting in a total of 2^{10} clones in 1024 identical rooms which have their numbers written out in binary. When asked to place bets on your room number, you assign an equal probability for all of them. However, you can give more interesting odds on what fraction of the ten binary digits on your door will be zeros, knowing from the binomial theorem that it's 50% for $\binom{10}{5} = 254$ yous, 20% for $\binom{10}{1} = 45$ yous, etc. You can therefore say that you will probably see a random-looking string of zeros and ones on your door, with an 89% chance that the fraction of ones will be between 30% and 70%. This conclusion is exactly the same as you would draw if you instead assumed that there was only one of you, and that you would be placed in a random room. Or that there was only one you and one room, whose 10 digits were each generated randomly with 50% probability for both 0 and 1.

In Everett's MWI, probability appears from randomness in exactly the same way if the branches have equal amplitude: one you evolves into more than one through deterministic Hamiltonian dynamics as in Eq. (2) with $\alpha = \beta = 2^{-1/2}$. The only difference is what the physical nature of the cloning process is. In our example above, another difference is that the hospital guests can meet and verify the existence of their clones, whereas quantum clones cannot because of decoherence—however, the hospital experiment could easily be modified to have this property too, say by keeping the rooms locked forever or shipping the clones off into deep space without radios. It is therefore observer cloning that is the crux, not what physics is involved in the cloning process. You need to end up with more than one post-experiment you with different recent experiences, but having identical memories from before the experiment.

In summary, although these classical parallels have not ended the debates over probability in the Everett picture, as evidenced by the continuing controversy over whether probability makes sense in the many-worlds interpretation (this volume, Parts 3 and 4), they do show that worries 4 and 5 appear already in classical physics. That is, whenever there are multiple observers with identical memories of what happened before a certain point but differing afterwards, these observers will perceive apparent randomness even if the evolution of their universe is completely deterministic. Whenever an observer is cloned, she will perceive something completely indistinguishable from true randomness. Since both of these phenomena can be realized without quantum mechanics, apparent causality breakdown and randomness are therefore not quantum-specific. Unitary quantum mechanics has these attributes simply because it routinely creates observer cloning when an instability rapidly amplifies microsuperpositions into macrosuperpositions, while decoherence ensures—effectively—that the doors between the clones are kept locked forever. Examples of such instabilities include most quantum measurements, Schrödinger's cat experiment and, probably, certain snap decision processes in the brain.

2.4 How a Multiverse Theory Can be Tested and Falsified

Is a multiverse theory one of metaphysics rather than physics? This is the concern listed as worry 1 in Table 1. As emphasized by Karl Popper, the distinction between the two is whether the theory is empirically testable and falsifiable. Containing unobservable entities does clearly *not* per se make a theory non-testable. For instance, a theory stating that there are 666 parallel universes, all of which are devoid of oxygen makes the testable prediction that we should observe no oxygen here, and is therefore ruled out by observation.

As a more serious example, the Level I multiverse framework is routinely used to rule out theories in modern cosmology, although this is rarely spelled out explicitly. For instance, cosmic microwave background (CMB) observations have recently shown that space has almost no curvature. Hot and cold spots in CMB maps have a characteristic size that depends on the curvature of space, and the observed spots appear too large to be consistent with the previously popular 'open universe' model. However, the average spot size randomly varies slightly from one Hubble volume to another, so it is important to be statistically rigorous. When cosmologists say that the open universe model is ruled out at 99.9% confidence, they really mean that if the open universe model were true, then fewer than one out of every thousand Hubble volumes would show CMB spots as large as those we observe—therefore the entire model with its entire Level I multiverse of infinitely many Hubble volumes is ruled out, even though we have of course only mapped the CMB in our own particular Hubble volume.

A related issue is worry 8 in Table 1: how does one judge evidence for/against a multiverse theory, if some small fraction of the observers get fooled by unusual data? For example, if a Stern–Gerlach apparatus is used to measure the spin in the z-direction of 10,000 particles prepared with their spin in the x-direction, most of the 2^{10000} resulting observers will observe a random-looking sequence with about 50% spin-up, but one of them will be unlucky enough to measure spin-up every time and mistakenly conclude that quantum mechanics is incorrect.

This issue clearly has nothing to do with quantum mechanics per se, since it also occurs in our last hospital example from Section 2.3. Suppose the 1024 clones are all considering the hypothesis that what happened to them is indeed the cloning experiment as described in Section 2.3, trying to decide whether to believe it or not. They all observe their room numbers, and most of them find it looking like a random sequence of zeros and ones, consistent with the hypothesis. However, one of the clones observes the room number '0000000000', and declares that the hypothesis has been ruled out at 99.9% confidence, because if the hypothesis were true, the probability of finding oneself in the very first room is only 1/1024. Similar issues also tormented some of the pioneers of classical statistic mechanics: in the grand ensemble at the heart of the theory, there would always be some totally confused observers who repeatedly saw eggs unbreak and cups of water spontaneously separate into steam and ice.

When they occur in examples not involving quantum mechanics, these issues are generally considered resolved, merely exemplifying what confidence levels are all about. If anybody in any context says that she has ruled something out at 99.9% confidence, she means that there is a 1 in 1000 chance that she has been fooled. Whenever there is any form of randomness, either ontologically fundamental as in the Copenhagen interpretation, apparent as in the MWI, or merely epistemological (reflecting our inability to predict detector noise, say), there is a risk that we get fooled by fluke data. In most cases, we can reduce this risk as much as we want by performing more measurements, but in some cases we cannot, say when measuring the large-scale power spectrum in the cosmic microwave background, where further measurements would only help if we could perform them outside of our cosmic horizon volume, i.e., in Level I parallel universes.

The take-home message from this section is that the MWI and indeed any multiverse theories *can* be tested and falsified, but only if they predict what the ensemble of parallel universes is and specify a probability distribution (or more generally what mathematicians call a *measure*) over it. This measure problem can be quite serious and is still unsolved for some multiverse theories (see Tegmark [2005], Garriga et al. [2005], Easther et al. [2006], Aguirre et al. [2007], Bousso [2006], Page [2008] for recent reviews), but is solved for both statistical mechanics and quantum mechanics in a finite space.

3 LEVEL II: OTHER POST-INFLATION BUBBLES

If you felt that the Level I multiverse was large and hard to stomach, try imagining an infinite set of distinct ones (each symbolized by a bubble in Fig. 1), some perhaps with different dimensionality and different physical constants. This is what is predicted by most currently popular models of inflation, and we will refer to it as the Level II multiverse. These other domains are more than infinitely far away in the sense that you would never get there even if you traveled at the speed of light forever. The reason is that the space between our Level I multiverse and its neighbors is still undergoing inflation, which keeps stretching it out and creating more volume faster than you can travel through it. In contrast, you could travel to an arbitrarily distant Level I universe if you were patient and the cosmic expansion decelerates.[7]

---

[7] Astronomical evidence suggests that the cosmic expansion is currently accelerating. If this acceleration continues, then even the Level I parallel universes will remain forever separate, with the intervening space stretching faster than light can travel through it. The jury is still out, however, with popular models predicting that the universe will eventually stop accelerating and perhaps even recollapse.

3.1 Evidence for Level II Parallel Universes

Inflation is an extension of the big bang theory and ties up many of its loose ends, such as why the universe is so big, so uniform, and so flat. An almost exponentially rapid stretching of space long ago can explain all these and other attributes in one fell swoop (see reviews Linde [1994], Guth and Kaiser [2005]). Such stretching is predicted by a wide class of theories of elementary particles, and all available evidence bears it out. Much of space is stretching and will continue doing so forever, but some regions of space stop inflating and form distinct bubbles, like gas pockets in a loaf of rising bread. Infinitely many such bubbles emerge (Fig. 1, lower left, with time increasing upwards). Each is an embryonic Level I multiverse: infinite in size[8] and filled with matter deposited by the energy field that drove inflation. Recent cosmological measurements have confirmed two key predictions of inflation: that space has negligible curvature and that the clumpiness in the cosmic matter distribution used to be approximately scale invariant.

The prevailing view is that the physics we observe today is merely a low-energy limit of a more general theory that manifests itself at extremely high temperatures. For example, this underlying fundamental theory may be 10-dimensional, supersymmetric and involving a grand unification of the four fundamental forces of nature. A common feature in such theories is that the potential energy of the field(s) relevant to inflation has many different minima (sometimes called 'metastable vacuum states'), and ending up in different minima corresponds to different effective laws of physics for our low-energy world. For instance, all but three spatial dimensions could be curled up ('compactified') on a tiny scale, resulting in an effectively three-dimensional space like ours, or fewer could curl up leaving a five-dimensional space. Quantum fluctuations during inflation can therefore cause different post-inflation bubbles in the Level II multiverse to end up with different effective laws of physics in different bubbles—say different dimensionality or different types of elementary particles, like two rather than three generations of quarks.

In addition to such discrete properties as dimensionality and particle content, our universe is characterized by a set of dimensionless numbers known as *physical constants*. Examples include the electron/proton mass ratio $m_p/m_e \approx 1836$ and the cosmological constant, which appears to be about 10^{-123} in so-called Planck units. There are models where also such non-integer parameters can vary from one post-inflationary bubble to another.[9] In summary, the Level II multiverse is

---

[8] Surprisingly, it has been shown that inflation can produce an infinite Level I multiverse even in a bubble of finite spatial volume, thanks to an effect whereby the spatial directions of spacetime curve towards the (infinite) time direction (Bucher and Spergel [1999]).

[9] Although the fundamental equations of physics are the same throughout the Level II multiverse, the approximate effective equations governing the low-energy world that we observe will differ.

likely to be more diverse than the Level I multiverse, containing domains where not only the initial conditions differ, but perhaps also the dimensionality, the elementary particles, and the physical constants.

This is currently a very active research area. The possibility of a string theory 'landscape' (Bousso and Polchinski [2000], Susskind [2003]), where the above-mentioned potential has perhaps 10^{500} different minima, may offer a specific realization of the Level II multiverse which would in turn have four sub-levels of increasing diversity: **IId**: different ways in which space can be compactified, which can allow both different effective dimensionality and different symmetries/elementary articles (corresponding to different topology of the curled-up extra dimensions). **IIc**: different 'fluxes' (generalized magnetic fields) that stabilize the extra dimensions (this sublevel is where the largest number of choices enter, perhaps 10^{500}). **IIb**: once these two choices have been made, there may be a handful of different minima in the effective supergravity potential. **IIa**: the same minimum and effective laws of physics can be realized in a many different post-inflationary bubbles, each constituting a Level I multiverse.

Before moving on, let us briefly comment on a few closely related multiverse notions. First of all, if one Level II multiverse can exist, eternally self-reproducing in a fractal pattern, then there may well be infinitely many other Level II multiverses that are completely disconnected. However, this variant appears to be untestable, since it would neither add any qualitatively different worlds nor alter the probability distribution for their properties. All possible initial conditions and symmetry breakings are already realized within each one.

An idea proposed by Tolman and Wheeler and recently elaborated by Steinhardt and Turok [2002] is that the (Level I) multiverse is cyclic, going through an infinite series of big bangs. If it exists, the ensemble of such incarnations would also form a multiverse, arguably with a diversity similar to that of Level II.

An idea proposed by Smolin [1997] involves an ensemble similar in diversity to that of Level II, but mutating and sprouting new universes through black holes rather than during inflation. This predicts a form of natural selection favoring universes with maximal black hole production.

In braneworld scenarios, another three-dimensional world could be quite literally parallel to ours, merely offset in a higher dimension. However, it is unclear whether such a world ('brane') deserves to be called a parallel universe separate from our own, since we may be able to interact with it gravitationally much as we do with dark matter.

For instance, moving from a three-dimensional to a four-dimensional (non-compactified) space changes the observed gravitational force equation from an inverse square law to an inverse cube law. Likewise, breaking the underlying symmetries of particle physics differently will change the line-up of elementary particles and the effective equations that describe them. However, we will reserve the terms 'different equations' and 'different laws of physics' for the Level IV multiverse, where it is the fundamental rather than the effective equations that change.

3.2 Fine-tuning and Selection Effects

Although we cannot interact with other Level II parallel universes, cosmologists can infer their presence indirectly, because their existence can account for unexplained coincidences in our universe. To give an analogy, suppose you check into a hotel, are assigned room 1967 and note that this is the year you were born. What a coincidence, you say. After a moment of reflection, however, you conclude that this is not so surprising after all. The hotel has hundreds of rooms, and you would not have been having these thoughts in the first place if you had been assigned one with a number that meant nothing to you. The lesson is that even if you knew nothing about hotels, you could infer the existence of other hotel rooms to explain the coincidence.

As a more pertinent example, consider the mass of the sun. The mass of a star determines its luminosity, and using basic physics, one can compute that life as we know it on Earth is possible only if the sun's mass falls into the narrow range between 1.6×10^{30}kg and 2.4×10^{30}kg. Otherwise Earth's climate would be colder than that of present-day Mars or hotter than that of present-day Venus. The measured solar mass is $M \sim 2.0 \times 10^{30}$kg. At first glance, this apparent coincidence of the habitable and observed mass values appears to be a wild stroke of luck. Stellar masses run from 10^{29} to 10^{32}kg, so if the sun acquired its mass at random, it had only a small chance of falling into the habitable range. But just as in the hotel example, one can explain this apparent coincidence by postulating an ensemble (in this case, a number of planetary systems) and a selection effect (the fact that we must find ourselves living on a habitable planet). Such observer-related selection effects are referred to as 'anthropic' (Carter [1974]), and although the 'A-word' is notorious for triggering controversy, physicists broadly agree that these selection effects cannot be neglected when testing fundamental theories. In this weak sense, the anthropic principle is not optional.

What applies to hotel rooms and planetary systems applies to parallel universes. Most, if not all, of the attributes set by symmetry breaking appear to be fine-tuned. Changing their values by modest amounts would have resulted in a qualitatively different universe—one in which we probably would not exist. If protons were 0.2% heavier, they could decay into neutrons, destabilizing atoms. If the electromagnetic force were 4% weaker, there would be no hydrogen and no normal stars. If the weak interaction were much weaker, hydrogen would not exist; if it were much stronger, supernovae would fail to seed interstellar space with heavy elements. If the cosmological constant were much larger, the universe would have blown itself apart before galaxies could form. Indeed, most if not all the parameters affecting low-energy physics appear fine-tuned at some level, in the sense that changing them by modest amounts results in a qualitatively different universe.

Although the degree of fine-tuning is still debated (see Barrow and Tipler [1986], Hogan [2000], Tegmark et al. [2006], and Tegmark [1998] for more technical reviews), these examples suggest the existence of parallel universes with other values of some physical constants. The existence of a Level II multiverse implies that physicists will never be able to determine the values of all physical constants from first principles. Rather, they will merely compute probability distributions for what they should expect to find, taking selection effects into account. The result should be as generic as is consistent with our existence.

4 LEVEL III: THE MANY WORLDS OF QUANTUM PHYSICS

If Everett was correct and physics is unitary, then there is a third type of parallel worlds that are not far away but in a sense right here. The universe keeps branching into parallel universes as in the cartoon (Fig. 2, bottom): whenever a quantum event appears to have a random outcome, all outcomes in fact occur, one in each branch. This is the Level III multiverse. Although it is more debated and controversial than Level I and Level II, we will see that, surprisingly, this level adds no new types of universes.

Since the volume to which this chapter belongs discusses the MWI in great detail, we will summarize the key points only very briefly. Everett's MWI is simply standard quantum mechanics with the collapse postulate removed, so that the Schrödinger equation holds without exception (Section 1.1). From this, the following conclusions can be derived:

1. Microsuperpositions (say of an atom going through two slits at the same time) are inevitable (the Heisenberg Uncertainty principle).
2. Macrosuperpositions (say of a cat being dead and alive) are also perfectly legitimate quantum states.
3. Processes occur that amplify microsuperpositions into macrosuperpositions (spontaneous symmetry breaking, Schrödinger's cat, and quantum measurements being three examples).
4. The superposition of a single macroscopic object tends to spread to all other interacting objects, eventually engulfing our entire universe.
5. Decoherence makes most macrosuperpositions for all practical purposes unobservable.
6. Decoherence calculations can determine which quantities appear approximately classical.

There is consensus in the physics community that both double-slit interference and the process of decoherence have been experimentally observed, showing the predicted behavior. Conclusions 1, 2, 3, and 4 together imply that astronomically

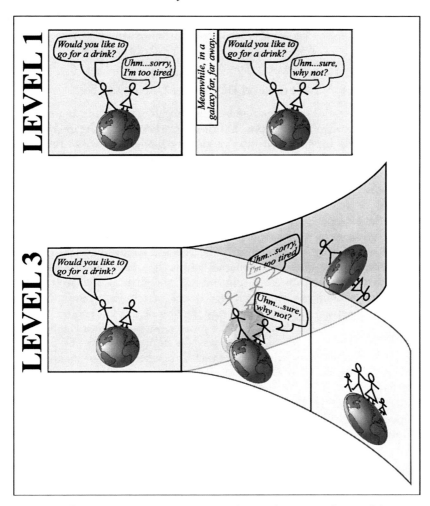

Figure 2. Difference between Level I and Level III. Whereas Level I parallel universes are far away in space, those of Level III are even right here, with quantum events causing classical reality to split and diverge into parallel storylines. Yet Level III adds no new storylines beyond Levels I or II.

large macrosuperpositions occur. These are Everett's parallel universes.[10] Worry 10 in Table 1 is addressed by 5, and worry 11 is addressed by 6 as reviewed in Giuline et al. [1996], Tegmark and Wheeler [2001], and Zurek [2009]. It should

[10] Note that to avoid creating macrosuperpositions, it is insufficient to abandon unitarity. Rather, it is *symmetry* that must be abandoned. For example, any theory where the wavefunction of

be borne in mind that these two worries remained serious open problems when Everett first published his work, since decoherence was only discovered in 1970 (Zeh [1970]).

4.1 What are Level III Parallel Universes Like?

Everett's many-worlds interpretation has been boggling minds inside and outside physics for more than four decades. But the theory becomes easier to grasp when one distinguishes between two ways of viewing a physical theory: the outside view of a physicist studying its mathematical equations, like a bird surveying a landscape from high above it, and the inside view of an observer living in the world described by the equations, like a frog living in the landscape surveyed by the bird.

From the bird perspective, the Level III multiverse is simple. There is only one wavefunction. It evolves smoothly and deterministically over time without any kind of splitting or parallelism. The abstract quantum world described by this evolving wavefunction contains within it a vast number of parallel classical storylines, continuously splitting and merging, as well as a number of quantum phenomena that lack a classical description. From their frog perspective, observers perceive only a tiny fraction of this full reality. They can view their own Level I universe, but the process of decoherence (Guiline et al. [1996], Zeh [1970])—which mimics wavefunction collapse while preserving unitarity—prevents them from seeing Level III parallel copies of themselves.

Whenever observers are asked a question, make a snap decision, and give an answer, quantum effects in their brains lead to a superposition of outcomes, such as 'Continue reading the article' and 'Put down the article'. From the bird perspective, the act of making a decision causes a person to split into multiple copies: one who keeps on reading and one who doesn't. From their frog

---

a system evolves deterministically (even if according to another rule than the Schrödinger equation) will evolve a perfectly sharp needle balanced on its tip into a superposition of needles pointing in macroscopically different directions unless the theory explicitly violates rotational symmetry. If the theory does violate this symmetry and 'collapses' the wavefunction, then there are two interesting possibilities: either the symmetry is broken early on while the superposition is still microscopic and unobservable (in which case the collapse process has nothing to do with measurement), or the symmetry is broken later on when the superposition is macroscopic (in which case local energy conservation is seriously violated by abruptly moving the center-of-mass by a macroscopic amount—even if the mass transfer is not superluminal, it would have to be fast enough to involve kinetic energy greatly exceeding the natural energy scale of the problem). If this experiment or Schrödinger's cat experiment were performed in a sealed free-falling box, the environment outside the box would learn how the needle had fallen or whether the cat had died from the altered gravitational field outside the box (and perhaps also from recoil motion of the box), causing decoherence. However, this complication can in principle be eliminated by keeping the moving parts spherically symmetric at all times. For example, if a metal sphere full of hydrogen contains a smaller sphere at its center full of oxygen at the same pressure which is opened if an atom decays (after which diffusion would mix the gases), the resulting superposition of two macroscopically different density distributions would leave all external fields unaffected.

perspective, however, each of these alter egos is unaware of the others and notices the branching merely as a slight randomness: a certain probability of continuing to read or not.

As strange as this may sound, the exact same situation occurs even in the Level I multiverse. You have evidently decided to keep on reading the article, but one of your alter egos in a distant galaxy put down the book after the first paragraph. The only difference between Level I and Level III is where your doppelgängers reside. In Level I they live elsewhere in good old three-dimensional space. In Level III they live on another quantum branch in infinite-dimensional Hilbert space (Fig. 2).

4.2 Level III Parallel Universes: Evidence and Implications

The existence of Level III depends on one crucial assumption: that the time evolution of the wavefunction is unitary. So far experimenters have encountered no departures from unitarity. In the past few decades they have confirmed unitarity for ever larger systems, including carbon 60 buckyball molecules and kilometer-long optical fibers. On the theoretical side, the case for unitarity has been bolstered by the discovery of decoherence (see Tegmark and Wheeler [2001] for a popular review). Some theorists who work on quantum gravity have questioned unitarity; one concern is that evaporating black holes might destroy information, which would be a non-unitary process. But a recent breakthrough in string theory known as AdS/CFT correspondence suggests that even quantum gravity is unitary. If so, black holes do not destroy information but merely transmit it elsewhere.

If physics is unitary, then the standard picture of how quantum fluctuations operated early in the big bang must change. These fluctuations did not generate initial conditions at random. Rather they generated a quantum superposition of all possible initial conditions, which coexisted simultaneously. Decoherence then caused these initial conditions to behave classically in separate quantum branches. Here is the crucial point: the distribution of outcomes on different quantum branches in a given Hubble volume (Level III) is identical to the distribution of outcomes in different Hubble volumes within a single quantum branch (Level I). This property of the quantum fluctuations is known in statistical mechanics as ergodicity.

The same reasoning applies to Level II. The process of symmetry breaking did not produce a unique outcome but rather a superposition of all outcomes, which rapidly went their separate ways. So if physical constants, spacetime dimensionality and so on can vary among parallel quantum branches at Level III, then they will also vary among parallel universes at Level II.

In other words, the Level III multiverse adds nothing new beyond Level I and Level II, just more indistinguishable copies of the same universes—the same old storylines playing out again and again in other quantum branches. The passionate debate about Everett's theory therefore seems to be ending in a grand anticlimax, with the discovery of less controversial multiverses (Levels I and II) that are equally large.

4.3 The Unequal Probability Worry

Let us now turn to worry 6 in Table 1: how to compute the apparent probabilities from the wavefunction amplitudes when they are not all equal and the wavefunction collapse postulate has been dropped from the theory. Since a single approximately classical state often evolves into a superposition of macroscopically different states that rapidly decohere as discussed above, it is obvious that observers will experience apparent randomness just as in our hospital examples from Section 2.3. However, why is it that these probabilities correspond to the square modulus of the wavefunction amplitudes (the so-called Born rule)? For example, in Eq. (2), why is the apparent probability for a happy observer equal to $|a^2|$ rather than some other real-valued function of a, say $|a|^4$?

There are a number of arguments that suggest that it must be this way. For example, one could argue that the sum of the probabilities should be conserved (so that it can be normalized to 1 once and for all), and $\int |\phi|^2$ is the only functional of ψ that is conserved under arbitrary unitary evolution, by the very definition of unitarity. In other words, the business about the squaring comes straight from the Hilbert-space structure of quantum mechanics, whereby the inner product defines an L_2 norm but no other norms.

Other arguments to this end have been proposed, based on information theory (Everett [1973]), decision theory (Deutsch [1999]) and other approaches (Zurek [2009], Caves et al. [2002], Saunders [2004]). But many authors have expressed a deeper concern about whether probability in the usual sense even makes sense in MWI (often focused around some combination of worries 4 and 5. To this end, arguments have been proposed based on Savage's approach: whatever intelligent observers actually believe, they will behave as though ascribing subjective probabilities to outcomes—probabilities which, as Deutsch [1999], Wallace [2002], and Wallace [2007] showed, match the Born rule. A rigorous mathematical treatment of this is given by Wallace in Chapter 8.

At the extensive debates about this issue at the Everett@50 conference at the Perimeter Institute in 2007, it was clear that these purported Born-rule derivations were still controversial. Interestingly, the entire controversy centered around the equal-probability case (say $a = \beta$ in Eq. (2)), *i.e.*, getting probabilities in the first place (worries 4 and 5 in Table 1). In contrast, the notion that this can be generalized to arbitrary amplitudes (worry 6 in Table 1) was fairly uncontroversial. In summary, worry 6 is the first one in Table 1 which is truly specific to quantum mechanics, but addressing it, if worries 4 and 5 have been settled, is arguably a solved problem.

4.4 Does the State Describe the World or my Knowledge of it?

A quantum state can be mathematically described by a density matrix. But what does this density matrix really describe? The state of the universe or your state of knowledge about it? This issue, listed as worry 7 in Table 1, is as old as quantum mechanics itself and still divides the physics community.

The Everett postulate implies a clear answer to it: both! On one hand, the entire universe has a quantum state which corresponds to a wavefunction, or to a density matrix if the state is mixed. Let us call this the *ontological* quantum state. On the other hand, our state of knowledge of the universe is described by a lower-dimensional density matrix for those degrees of freedom that we are interested in, both conditioned on what we already know (limiting to those branches that we could be on—what Everett termed the 'relative state') and partial-traced over those degrees of freedom that we know nothing about. I will refer to this as the *epistemological* quantum state, bearing in mind that it differs from one observer to the another—both from a colleague in this branch of the wavefunction and from yourself in another branch.[11] In other words, the epistemological quantum state is derivable from the ontological quantum state and your subjective observations. When quantum textbooks refer to 'the' state, they usually mean the epistemological state of a system according to you, after you have prepared it in a certain way. This is further elaborated in Tegmark [2003].

The density-matrix aspect of this issue is clearly quantum-specific. However, the dichotomy between objective and subjective descriptions appears in classical statistical mechanics as well: an ensemble of classical worlds can be completely described by a probability distribution in a high-dimensional phase space, whereas the knowledge of the world by an individual observer is described by a probability distribution in a lower-dimensional phase space, again computable by conditioning (the classical equivalent of computing a relative state) and marginalizing (the classical equivalent of partial tracing).

4.5 The Weirdness Worry

Despite all the elaborate technical and philosophical worries about the MWI listed in Table 1, many physicists probably find their strongest objection to the MWI not in their brain but in their gut: it simply feels too weird, crazy, counter-intuitive, and disturbing.

The complaint about weirdness is aesthetic rather than scientific, and it really makes sense only in the Aristotelian worldview. Yet what did we expect? When we ask a profound question about the nature of reality, do we not expect an

---

[11] Whereas the ontological state might be pure, and hence describably by a wavefunction, the epistemological state is generically mixed and cannot be described by a wavefunction, only by a density matrix. This has already been pointed out by Schrödinger [1935].

answer that sounds strange? I personally dismiss this weirdness worry as a failure to appreciate Darwinian evolution. Evolution endowed us with intuition only for those aspects of physics that had survival value for our distant ancestors, such as the parabolic trajectories of flying rocks. Darwin's theory thus makes the testable prediction that whenever we look beyond the human scale, our evolved intuition should break down.

We have repeatedly tested this prediction, and the results overwhelmingly support it: our intuition breaks down at high speeds where time slows down, on small scales where particles can be in two places at once, on large scales where we encounter black holes, and at high temperatures, where colliding particles change identity. To me, an electron colliding with a positron and turning into a Z-boson feels about as intuitive as two colliding cars turning into a cruise ship. The point is that if we dismiss seemingly weird theories out of hand, we risk dismissing the correct theory if we stumble across it.

4.6 Two Worldviews

The seemingly endless debate over the interpretation of quantum mechanics is in a sense the tip of an iceberg. In the sci-fi spoof 'Hitchhiker's Guide to the Galaxy', the answer is discovered to be '42', and the hard part is finding the real question. Questions about parallel universes may seem to be just about as deep as queries about reality can get. Yet there is a still deeper underlying question: there are two tenable but diametrically opposed paradigms regarding physical reality and the status of mathematics, a dichotomy that arguably goes as far back as Plato and Aristotle, and the question is which one is correct.

- **Aristotelian paradigm:** The subjectively perceived frog perspective is physically real, and the bird perspective and all its mathematical language is merely a useful approximation.
- **Platonic paradigm:** The bird perspective (the mathematical structure) is physically real, and the frog perspective and all the human language we use to describe it is merely a useful approximation for describing our subjective perceptions.

Which is more basic—the frog perspective or the bird perspective? Which is more basic—human language or mathematical language? Your answer will determine how you feel about parallel universes.

If you prefer the Aristotelian paradigm, you share worry 3 in Table 1. If you prefer the Platonic paradigm, you should find multiverses natural, since our feeling that say the Level III multiverse is 'weird' merely reflects that the frog and bird perspectives are extremely different. We break the symmetry by calling the latter weird because we were all indoctrinated with the Aristotelian paradigm as children, long before we even heard of mathematics—the Platonic view is an acquired taste!

In the second (Platonic) case, all of physics is ultimately a mathematics problem, since an infinitely intelligent mathematician given the fundamental equations of the cosmos could in principle *compute* the frog perspective, i.e., compute what self-aware observers the universe would contain, what they would perceive, and what language they would invent to describe their perceptions to one another. In other words, there is a 'Theory of Everything' (TOE) whose axioms are purely mathematical, since postulates in English regarding interpretation would be derivable and thus redundant. In the Aristotelian paradigm, on the other hand, there can never be a TOE, since one is ultimately just explaining certain verbal statements by other verbal statements—this is known as the infinite regress problem (Nozick [1981]).

In Tegmark [2007a, 2007b], I have argued that the Platonic paradigm follows logically from the innocuous-sounding *external reality hypothesis* (ERH) (Tegmark [2007a]): 'there exists an external physical reality completely independent of us humans'. More specifically, Tegmark [2007a] argues that the ERH implies the *mathematical universe hypothesis* (MUH) that our external physical reality is a mathematical structure. The detailed technical definition of a mathematical structure is not important here; just think of it as a set of abstract entities with relations between them—familiar examples of mathematical structures include the integers, a Riemannian manifold, and a Hilbert space.

5 LEVEL IV: OTHER MATHEMATICAL STRUCTURES

Suppose you buy the mathematical universe hypothesis and believe that we simply have not found the correct equations yet, or more rigorously, the correct mathematical structure. Then an embarrassing question remains, as emphasized by John Archibald Wheeler: *Why these particular equations, not others?* Tegmark [2007a] argues that, when pushed to its extreme, the MUH implies that all mathematical structures correspond to physical universes. Together, these structures form the Level IV multiverse, which includes all the other levels within it. If there is a particular mathematical structure that is our universe, and its properties correspond to our physical laws, then each mathematical structure with different properties is its own universe with different laws. The Level IV multiverse is compulsory, since mathematical structures are not 'created' and don't exist 'somewhere'—they just exist. Stephen Hawking once asked, 'What is it that breathes fire into the equations and makes a universe for them to describe?' In the case of the mathematical cosmos, there is no fire-breathing required, since the point is not that a mathematical structure describes a universe, but that it is a universe.

In a famous essay, Wigner [1967] argued that 'the enormous usefulness of mathematics in the natural sciences is something bordering on the mysterious', and that 'there is no rational explanation for it'. This argument can be taken as support for the MUH: here the utility of mathematics for describing the physical world is a natural consequence of the fact that the latter *is* a mathematical structure, and we are simply uncovering this bit by bit. The various approximations that constitute our current physics theories are successful because simple mathematical structures can provide good approximations of how an observer will perceive more complex mathematical structures. In other words, our successful theories are not mathematics approximating physics, but mathematics approximating mathematics. Wigner's observation is unlikely to be based on fluke coincidences, since far more mathematical regularity in nature has been discovered in the decades since he made it, including the standard model of particle physics. Detailed discussions of the Level IV multiverse, what it means and what it predicts, are given in Tegmark [1998, 2007a].

6 DISCUSSION

We have discussed Everett's many-worlds interpretation of quantum mechanics in the context of other physics disputes and the three other levels of parallel universes that have been proposed in the literature. We found that only a small fraction of the usual objections to Everett's theory (summarized in Table 1) are specific to quantum mechanics, and that all of the most controversial issues crop up also in settings that have nothing to do with quantum mechanics.

6.1 The Multiverse Hierarchy

We have seen that scientific theories of parallel universes form a four-level hierarchy, in which universes become progressively more different from ours. They might have different initial conditions (Level I), different effective physical laws, constants and particles (Level II), or different fundamental physical laws (Level IV). It is ironic that Everett's Level III is the one that has drawn the most fire in the past decades, because it is the only one that adds no qualitatively new types of universes.

Whereas the Level I universes join seamlessly, there are clear demarcations between those within levels II and III caused by inflating space and decoherence, respectively. The Level IV universes are completely disconnected and need to be considered together only for predicting your future, since 'you' may exist in more than one of them.

6.2 Are Parallel Universes Wasteful?

A common argument about all forms of parallel universes, including Everett's Level III ones, is that they feel wasteful. Specifically, the wastefulness worry (#2 in Table 1) is that multiverse theories are vulnerable to Occam's razor because they postulate the existence of other worlds that we can never observe. Why should nature be so wasteful and indulge in such opulence as an infinity of different worlds? Yet this argument can be turned around to argue for a multiverse. What precisely would nature be wasting? Certainly not space, mass, or atoms—the uncontroversial Level I multiverse already contains an infinite amount of all three, so who cares if nature wastes some more? The real issue here is the apparent reduction in simplicity. A skeptic worries about all the information necessary to specify all those unseen worlds.

But an entire ensemble is often much simpler than one of its members. This principle can be stated more formally using the notion of algorithmic information content. The algorithmic information content in a number is, roughly speaking, the length of the shortest computer program that will produce that number as output. For example, consider the set of all integers. Which is simpler, the whole set or just one number? Naively, you might think that a single number is simpler, but the entire set can be generated by quite a trivial computer program, whereas a single number can be hugely long. Therefore, the whole set is actually simpler.

Similarly, the set of all solutions to Einstein's field equations is simpler than a specific solution. The former is described by a few equations, whereas the latter requires the specification of vast amounts of initial data on some hypersurface. The lesson is that complexity increases when we restrict our attention to one particular element in an ensemble, thereby losing the symmetry and simplicity that were inherent in the totality of all the elements taken together.

In this sense, the higher-level multiverses are simpler. Going from our universe to the Level I multiverse eliminates the need to specify initial conditions, upgrading to Level II eliminates the need to specify physical constants, and the Level IV multiverse eliminates the need to specify anything at all. The opulence of complexity is all in the subjective perceptions of observers (Tegmark [1996])—the frog perspective. From the bird perspective, the multiverse could hardly be any simpler.

A common feature of all four multiverse levels is that the simplest and arguably most elegant theory involves parallel universes by default. To deny the existence of those universes, one needs to complicate the theory by adding experimentally unsupported processes and ad hoc postulates: finite space, wavefunction collapse, ontological asymmetry, etc. Our judgment therefore comes down to which we find more wasteful and inelegant: many worlds or many words.

6.3 Are Parallel Universes Testable?

We have discussed how multiverses are not theories but predictions of certain theories, and how such theories are falsifiable as long as they also predict something that we can test here in our own universe. There are ample future prospects for testing and perhaps ruling out these multiverse theories. In the coming decade, dramatically improved cosmological measurements of the microwave

background radiation, the large-scale matter distribution, etc., will test Level I by
further constraining the curvature and topology of space, and will test Level II
by providing stringent tests of inflation. Progress in both astrophysics and high-
energy physics should also clarify the extent to which various physical constants
are fine-tuned, thereby weakening or strengthening the case for Level II. If the
current worldwide effort to build quantum computers succeeds, it will provide
further evidence for Level III, since such computers are most easily explained
as, in essence, exploiting the parallelism of the Level III multiverse for parallel
computation (Deutsch [1997]). Conversely, experimental evidence of unitarity
violation would rule out Level III. Unifying general relativity and quantum
field theory, will shed more light on Level IV. Either we will eventually find
a mathematical structure matching our universe, or we will have to abandon
Level IV.

6.4 So was Everett Right?

Our conclusions regarding Table 1 do not per se argue either for or against the
MWI, merely clarify what assumptions about physics lead to what conclusions.
However, all the controversial issues arguably melt away if we accept the *external
reality hypothesis* (ERH) (Tegmark [2007a]): there exists an external physical
reality completely independent of us humans. Suppose that this hypothesis is
correct. Then the core MWI critique rests on some combination of the following
three dubious assumptions.

1. **Omnivision assumption:** physical reality must be such that at least one
   observer can in principle observe all of it.
2. **Pedagogical reality assumption:** physical reality must be such that all reasonably informed human observers feel they intuitively understand it.
3. **No-copy assumption:** no physical process can copy observers or create
   subjectively indistinguishable observers.

1 and 2 appear to be motivated by little more than human hubris. The omnivision
assumption effectively redefines the word 'exists' to be synonymous with what is
observable to us humans. Of course those who insist on the pedagogical reality
assumption will typically have rejected comfortingly familiar childhood notions
like Santa Claus, local realism, the Tooth Fairy, and creationism—but have they
really worked hard enough to free themselves from comfortingly familiar notions
that are more deeply rooted? In my personal opinion, our job as scientists is to
try to figure out how the world works, not to tell it how to work based on our
philosophical preconceptions.

   If the omnivision assumption is false, then there are unobservable things that
exist and we live in a multiverse. If the pedagogical reality assumption is false, then
the weirdness worry (#12 in Table 1) makes no sense. If the no-copy assumption
is false, then worries 4 and 5 from Table 1 are misguided: observers can perceive

apparent randomness even if physical reality is completely deterministic and known. In this case, these fundamental conceptual questions raised by the MWI will arise in physics anyway, independent of quantum mechanics, and will need to be solved—indeed, Everett, in providing a coherent and intelligible account of probability even in the face of massive copying, has blazed a trail in showing us how to solve them.

The ERH alone settles worry 9 in Table 1, since what is in the external reality defines what exists. In summary, if the ERH is correct, then the only outstanding question about the MWI is whether physics is unitary or not. So far, experiments have revealed no evidence of unitarity violation, and ongoing and upcoming experiments will test unitarity for dramatically larger systems.

My guess is that the only issues that worried Hugh Everett were #10 and #11 from Table 1, which are precisely those which were laid to rest by the subsequent discovery of decoherence. Perhaps we will gradually get more used to the weird ways of our cosmos, and even find its strangeness to be part of its charm. In fact, I met Hugh Everett the other day and he told me that he agrees—but alas not in this particular universe.

Acknowledgments

The author wishes to thank Anthony Aguirre, David Albert, Peter Byrne, Olaf Dreyer, Bryan Eastin, Mark Everett, Brian Greene, Alan Guth, Seth Lloyd, George Musser, David Raub, Martin Rees, Simon Saunders, Harold Shapiro, Lee Smolin, Alex Vilenkin, Frank Wilczek, Anton Zeilinger, and Wojciech Zurek for stimulating discussions, Simon Saunders for detailed feedback on this manuscript, and Adrian Kent, Jonathan Barrett, David Wallace, and the Perimeter Institute for hosting a very stimulating meeting on the MWI.

This work was supported by NSF grants AST-0071213 & AST-0134999, NASA grants NAG5-9194 & NAG5-11099, a grant from the John Templeton Foundation, a fellowship from the David and Lucile Packard Foundation, and a Cottrell Scholarship from Research Corporation.

References

Aguirre, A, S. Gratton, and M.C. Johnson [2007], 'Measures on transitions for cosmology from eternal inflation', *Physical Review Letters* **98**, 131301.

Barrow, J. D. and F. J. Tipler [1986], *The Anthropic Cosmological Principle*, Clarendon Press, Oxford.

Bousso, R. [2006], 'Holographic probabilities in eternal inflation', *Physical Review Letters* **97**, 191302.

Bousso, R. and J. Polchinski [2000], 'Quantization of four-form fluxes and dynamical neutralization of the cosmological constant', *Journal of High Energy Physics* **6**, 6. Available online at arXiv.org/abs/hep-th/0004134.

Brundrit G.B. [1979], 'Life in the infinite universe', *Journal of the Royal Astronomical Society* **20**, 37.

Bucher, M.A. and D.N. Spergel [1999], 'Inflation in a low-density universe', *Scientific American* **280**, 62.

Carter, B. [1974], 'Large number coincidences and the anthropic principle', *IAU Symposium 63, Longair S*, Reide, Dordrecht.

Caves, C.M., C.A. Fuchs, and R. Schack [2002], 'Quantum probabilities as Bayesian probabilities', *Physics Review* **A 65**, 022305. Available online at arXiv.org/pdf/quant-ph/0106133v2.

de Oliveira-Costa, A, M. Tegmark, M. Zaldarriaga, and A. J. S. Hamilton [2003], 'The significance of the largest scale CMB fluctuations in WMAP', available online at arXiv.org/pdf/astro-ph/0307282.

Deutsch, D. [1997], *The Fabric of Reality*, Allen Lane, New York.

—— [1999], 'Quantum theory of probability and decisions', *Proceedings of the Royal Society of London* **A455**, 3129–37.

Easther, R., E.A. Lim, and M.R. Martin [2006], 'Counting pockets with world lines in eternal inflation', available online at arXiv.org/pdf/astro-ph/0511233.

Everett, H. [1957], ' "Relative state" formulation of quantum mechanics', *Reviews of Modern Physics* **29**, 454–62. Reprinted in *The Many-Worlds Interpretation of Quantum Mechanics*, B. De Witt and N. Graham (eds), Princeton University Press (1973).

—— [1973], 'The theory of the universal wavefunction', in *The Many-Worlds Interpretation of Quantum Mechanics*, B. DeWitt and N. Graham (eds), Princeton University Press.

Garriga J, D. Perlov-Schwartz, A. Vilenkin, and S. Winitzki [2005], 'Probabilities in the inflationary multiverse', available online at arXiv.org/pdf/hep-th/0509184.

Garriga, J. and A. Vilenkin [2001], 'Manyworlds in one', *Physical Review* **D 64**, 043511.

Guth, A. and D. L. Kaiser [2005], 'Inflationary Cosmology: Exploring the Universe from the Smallest to the Largest Scales', *Science* **307**, 884.

Hogan, C.J. [2000], 'Why the universe is just so', *Reviews of Modern Physics* **72**, 1149. Available online at arXiv.org/pdf/astro-ph/9909295.

Joos, E.H., D. Zeh, C. Kiefer, D. Giulini, J. Kupsch, and I.O. Stamatescu [2003], *Decoherence and the Appearance of a Classical World in Quantum Theory*, Springer, Berlin.

Linde, A. [1994], 'The self-reproducing inflationary universe', *Scientific American* **271**; 32, Nov issue.

Nozick, R. [1981], *Philosophical Explanations*, Harvard University Press, Cambridge, Mass.

Page, D.N. [2008], 'Cosmological measures without volume weighting', JCAP;0810;025. Available online at arXiv.org/abs/0808.0351.

Saunders, S. [2004], 'Derivation of the Born rule from operational assumptions', *Proceedings of the Royal Society* **A 460**, 1–18. Available online at arXiv.org/pdf/quant-ph/0211138v2.

Schrödinger, E. [1935], 'Die gegenwärtige Situation in der Quantenmechanik', *Die Naturwissenschaften* **23**, 844–9.

Shapiro, J., N.J. Cornish, D. Spergel and G.D. Starkman [2007], 'Extending the WMAP bound on the size of the universe', *Physical Review* D 75, 084034.

Smolin, L. [1997], *The Life of the Cosmos*, Oxford University Press, Oxford.

Steinhardt, P.J. and N. Turok [2002], 'A cyclic model of the universe', *Science* 296, 1436.

Susskind, L. [2003], 'The anthropic landscape of string theory'. Available online at arXiv.org/abs/hepth/0302219.

Tegmark, M. [1996], 'Does the universe in fact contain almost no information?', *Foundations of Physics Letters* 9, 25.

—— [1998], 'Is "the theory of everything" merely the ultimate ensemble theory?', *Annals of Physics* 270, 1–51. Available online at arXiv.org/pdf/gr-qc/9704009.

—— [2000], 'The importance of quantum decoherence in brain processes', *Physical Review* E 61, 4194–206. Available online at arXiv.org/pdf/quant-ph/9907009.

—— [2003], 'Parallel universes', *Scientific American* 270, 40. Available online at arXiv.org/pdf/astro-ph/0302131.

—— [2005], 'What does inflation really predict?', JCAP 2005-4; 1. Available online at arXiv.org/abs/astro-ph/0410281.

—— [2007a], 'The mathematical universe', *Foundations of Physics* 11/07, 116. Available online at arXiv.org/pdf/gr-qc/0704.0646.

—— [2007b], 'Reality by numbers', *New Scientist*, 9/15/2007. Available online at arXiv.org/pdf/0709.4024.

Tegmark, M. and J.A. Wheeler [2001], '100 years of quantum mysteries', *Scientific American* 284, 68–75.

Tegmark, M., A. Aguirre, M. Rees, and F. Wilczek [2006], 'Dimensionless constants, cosmology and other dark matters', *Physical Review* D 73, 023505. Available online at arXiv.org/pdf/astroph/0511774.

Wallace D. [2002], 'Quantum probability and decision theory, revisited'. Available online at arXiv.org/pdf/quant-ph/0211104v1.

—— [2007], 'Quantum probability from subjective likelihood: improving on Deutsch's proof of the probability rule', *Studies in the History and Philosophy of Modern Physics* 38, 311–32. Available online at arXiv.org/abs/quant-ph/0312157.

Wigner, E. P. [1967], *Symmetries and Reflections*, MIT Press, Cambridge.

Zeh, D. [1970], 'On the interpretation of measurement in quantum theory', *Foundations of Physics* 1, 69–76.

Zurek, W.H. [2009], 'Quantum Darnwinism', *Nature* 5, 181–8. Available online at arXiv.org/pdf/quant-ph/0903.5082.

20

Time Symmetry and the Many-Worlds Interpretation

Lev Vaidman

ABSTRACT

An attempt to solve the collapse problem in the framework of a time-symmetric quantum formalism is reviewed. Although the proposal does not look very attractive, its concept—a world defined by two quantum states, one evolving forwards and one evolving backwards in time—is found to be useful in modifying the many-worlds picture of Everett's theory.

1 INTRODUCTION

Quantum mechanics is an almost unprecedented success as a physical theory, yielding precise predictions for the results of experiments. However, if quantum theory is viewed as a direct description of physical reality, there is a significant difficulty: in order to explain particular outcomes of quantum measurements, a *collapse* of the quantum state has to be introduced. The collapse, with its randomness, non-locality and the lack of a well-defined moment of occurrence, is such an ugly scar on quantum theory, that I, along with many others, am ready to follow Everett and deny its existence. The price is the many-worlds interpretation (MWI), i.e., the existence of numerous parallel worlds.

There are other attempts to avoid collapse. One of them is a proposal due to Aharonov [2005], according to which, in addition to the standard forwards evolving quantum state, there is a backwards evolving quantum state. Both evolve according to the Schrödinger equation without collapses, but a particular form of the backwards evolving state ensures that we do not experience multiple branches of the forwards evolving state, just one. The form of the backwards evolving state is chosen (somewhat artificially) exactly by this requirement: that each quantum measurement ends up with a single outcome.

There is a certain difference between the single world described by quantum mechanics with collapses at each measurement, and the single world which emerges with the backwards evolving quantum state of Aharonov. While the former, at each moment in time, is defined by the results of measurements in the past, the latter is defined, in addition, by the results of a complete set of measurements in the future. Although I am not ready to accept Aharonov's proposal, I think that his idea of a single world emerging from forwards- and backwards-evolving quantum states is useful. I will argue that there is an advantage in drawing the Everettian many-worlds picture using multiple Aharonov worlds, instead of multiple worlds generated solely by quantum measurements in the past.

I will start with a brief description of the time-symmetric two-state vector formalism (TSVF) (Aharonov and Vaidman [2008]), which provides the framework for Aharonov's proposal.

2 THE TWO-STATE VECTOR FORMALISM

The TSVF describes quantum systems at a given time by a backwards evolving quantum state (Vaidman [2007]), in addition to the standard, forwards evolving quantum state. An ideal (textbook) quantum measurement at time t of a variable A with an outcome a creates a quantum state $|A = a\rangle$ evolving forwards in time and, at the same time, creates a quantum state evolving backwards in time, towards the past, which we denote $\langle A = a|$. An ideal measurement also serves as a verification measurement: the outcome $A = a$ is obtained with probability 1 if the state $|A = a\rangle$ evolves towards time t from the past or (and) if the state $\langle A = a|$ evolves towards time t from the future.

In a real laboratory we usually have separate devices for the creation and verification of quantum states: emitters and detectors. We can place a single photon source (made today in a number of laboratories around the world) in front of a two-slit barrier. It creates quantum states of photons exhibiting an interference pattern, which can be tested by measuring the frequency of clicks of a photodetector as a function of its position on the plain located beyond the barrier (see Fig. 1a). The backwards evolving quantum state of the photon is created by the detector and it is tested by the source. We can observe the interference pattern of the backwards evolving state by fixing the position of the photodetector in front of the slits and moving the source along the plane parallel to the barrier. The frequency of emitted photons, post-selected by the condition of being observed by the detector, exhibits the familiar two-slits interference pattern, but this time it is interference of the backwards evolving state (see Fig. 1b).

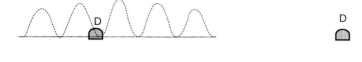

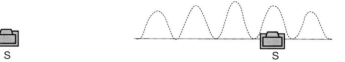

| (a) The number of absorbed particles as a function of the position of a detector shows the interference pattern of the forwards evolving quantum state. | (b) The number of emitted particles absorbed by the detector, as a function of the position of the source, shows the interference pattern of the backwards evolving quantum state. |

Figure 1. Interference patterns.

Somewhat surprisingly, we do have numerous realizations of essentially ideal von Neumann measurements in nature: the measuring device is the environment, and the measured systems are macroscopic objects or, sometimes, charged particles. Interactions with molecules, photons, phonons, etc. of the environment provide non-demolition localization measurements.

Given complete measurements preparing the state $|\Psi\rangle$ before time t and complete measurements verifying the state $|\Phi\rangle$ after time t, the quantum system at time t is described by the *two-state vector* (Aharonov and Vaidman [1990])

$$\langle\Phi| \; |\Psi\rangle. \tag{1}$$

The two-state vector provides maximal information regarding the way the quantum system can affect, at time t, any other system. In particular, the two-state vector describes the influence on a measuring device coupled with the system at time t. An ideal measurement of a variable O yields an eigenvalue o_n with probability given by the Aharonov–Bergman–Lebowitz (ABL) rule (Aharonov et al. [1964])

$$\text{Prob}(o_n) = \frac{|\langle\Phi|\mathbf{P}_{O=o_n}|\Psi\rangle|^2}{\sum_j |\langle\Phi|\mathbf{P}_{O=o_j}|\Psi\rangle|^2}. \tag{2}$$

An important case is when a particular measurement has an outcome that occurs with certainty. In this case I call it (somewhat misleadingly) an *element of reality* (Vaidman [1993]).

The most important result of the TSVF has been the discovery of *weak values* of physical variables (Aharonov et al. [1988]). When, at time t, another system couples weakly to a variable O of a pre- and post-selected system $\langle\Phi|$ $|\Psi\rangle$, the effective coupling is not to one of the eigenvalues, but to the weak value:

$$O_w \equiv \frac{\langle\Phi|O|\Psi\rangle}{\langle\Phi|\Psi\rangle}. \tag{3}$$

Since the quantum states remain effectively unchanged during the measurement, several weak measurements can be performed one after another and even simultaneously. *Weak-measurement elements of reality* (Vaidman [1996]), i.e., the weak values, provide a self-consistent but sometimes very unusual picture for pre- and post-selected quantum systems.

As a simple example, consider a particle emitted from a source S towards a beam splitter BS, and detected by a detector A (see Fig. 2). In our simplified model, the forwards-evolving state corresponds to the line from the source towards the beam splitter, and then to the two lines from the beam splitter towards the two detectors. The backwards-evolving state corresponds to the line from detector A towards the beam splitter, and then to two lines, one towards the source and one towards nowhere. If we denote the quantum state evolving from the beam splitter towards detector A as $|A\rangle$, and similarly, the state evolving from the beam splitter towards detector B as $|B\rangle$, then the two-state vector of

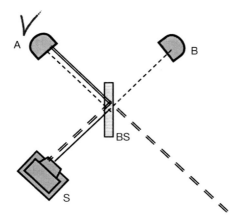

Figure 2. A particle is emitted by the source and absorbed by detector A. Single lines denote forwards, and double lines denote backwards evolving quantum states.

the particle, at a time when the particle has passed the beam splitter but has not reached the detectors, is

$$\langle A| \quad \frac{1}{\sqrt{2}}(|A\rangle + |B\rangle). \tag{4}$$

From the ABL rule and from the calculation of the weak values of projections on various locations, we can immediately see that both strong- and weak-measurement elements of reality show that the particle is well localized during all of its motion along the trajectory *S-BS-A*. The weak-measurement elements of reality for projections on the corresponding trajectories are: $(\mathbf{P}_A)_w = 1$ and $(\mathbf{P}_B)_w = 0$.

A more entertaining example is the *three-box paradox* (Aharonov and Vaidman [1991]). Consider a single particle in three boxes described by the two-state vector

$$\frac{1}{3}((\langle A| + \langle B| - \langle C|) \quad (|A\rangle + |B\rangle + |C\rangle), \tag{5}$$

where $|A\rangle$ is a quantum state of the particle located in box *A*, etc. For this particle, there is a set of *elements of reality*

$$\mathbf{P}_A = 1,$$
$$\mathbf{P}_B = 1. \tag{6}$$

Or, in words: if we open box *A*, we find the particle there for sure; if we open box *B* (instead), we also find the particle there for sure.

For this particle there are also corresponding weak-measurement elements of reality: $(\mathbf{P}_A)_w = 1$ and $(\mathbf{P}_B)_w = 1$. Any weak coupling to the particle in box *A* behaves as if there were a particle there. And, simultaneously, the same is true for box *B*. Note that these are properties of neither forwards nor backwards evolving states separately, but only of both together.

3 THE TWO-STATE VECTOR FORMALISM WITHOUT COLLAPSE

It is uncontroversial to apply the concepts of the TSVF, such as the two-state vector, weak values, etc., to a quantum system in the past, when both pre- and post-selection have already taken place. The revolutionary proposal of Aharonov (Aharonav and Gruss [2005]) is that the backwards evolving state exists at the present moment. Aharonov's backwards evolving state is a very special one. It ensures that all quantum measurements have definite outcomes. In particular, in our simple example with a beam splitter and a detector, the backwards evolving state has a component of triggered detector A and does not have a component of triggered detector B. Thus, although the forwards evolving state includes a component in which the particle arrives at detector B, it will leave no trace. Any weak coupling with the particle on the trajectory BS-B will show nothing.

I find Aharonov's proposal very problematic. It does remove action at a distance and randomness from basic physical interactions, two of the main difficulties with the collapse postulate. But it still has the third: it is not well defined. The backwards evolving quantum state needs to be tailored in such a way that all measurements will have a definite result, but what is the definition of a measurement?

The difficulty is increased by the fact that the backwards evolving state needs to be very specific. The backwards evolving state is specified by (vaguely defined) measurement events, whose probability of occurrence is given by the Born rule applied to the forwards evolving state. It will not be the case that the backwards evolving state alone describes well-defined results of measurements. For consider a longer history which includes our experiment (Fig. 3). Assume that the click of a detector A causes a lamp to be switched on. Then, a photon from the

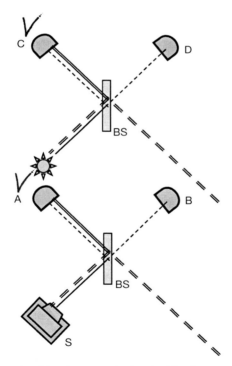

Figure 3. A particle emitted by the source is absorbed by detector A. It switches on the lamp and a photon from the lamp is absorbed by detector C.

lamp passes through another beam splitter and is detected by another detector *C*. A definite outcome for detector *C* implies that only part of the photon's backwards evolving wave reaches detector *A*. Therefore, the backwards evolving quantum state is not an eigenstate of the clicked detector *A*. Aharonov and Gruss [2005] claim that it is feasible to arrange the backwards evolving state in such a way that, together with the forwards evolving state, only a single history of 'macroscopic' events is significant. The weak value of the projection on the outcome *A* is exponentially larger (in the number of particles in the detector) than the weak value of the projection on the outcome *B*. However, it is hard to see how Aharonov's program can succeed for a long-lasting universe, and very many measurement events in parallel worlds.

The artificiality of the definition of the backwards evolving quantum state together with difficulties in making this program consistent make me very reluctant to accept Aharonov's hypothesis that at each moment in time, there is a quantum state evolving backwards. But in the context of Everettian quantum theory, Aharonov's idea is useful. It is possible to describe each individual world (i.e., branch of the universe) in terms of both a forwards and a backwards evolving state, with the outcomes of both past and future measurements fixed. I find this preferable to the standard Everettian approach.

I will begin my argument with a brief description of my understanding of the concept of a world in the MWI.

4 EVERETT 'WORLDS'

The basic concept in the MWI is a 'world'. It belongs not to the mathematical formalism of quantum mechanics, but to its interpretation. It helps to make a connection between the mathematical formalism and our experience.

One approach to the concept of a world (which seems to be very close to Everett's original proposal) is to define a subjective world for each observer—that is, a quantum state of the universe relative to a particular conscious state of the observer. This approach is certainly consistent. But it is not very effective, since it is hard to discuss objects which are not in a direct contact with the observer, e.g., objects in the far past, or the far future, or which are far away. We would like to discuss stars before life developed. Therefore, I prefer not to define concepts in terms of conscious observers. Instead, I define worlds in an objective way:

A **world** is the totality of (macroscopic) objects: stars, cities, people, grains of sand, etc. in a definite classically described state.

> The MWI in *Stanford Philosophy Encyclopedia* (Vaidman [2002])

I will now clarify (and slightly amend) this definition. First, a world is not a concept associated with a particular moment in time, but with all of time. It is a complete history in which all macroscopic objects have definite states. At the beginning, according to the standard MWI, there was one classical state, common to all worlds. At a later time there are multiple classically described states. Each one corresponds to numerous worlds with identical pasts and different futures.

The second clarification is that, apart from classical objects, the description of a world should include a description of some microscopic objects. A world is a sensible story, a causal connection between states of objects. We need not describe the quantum states of all the particles of a table, but the quantum state of a particle in an accelerator, which leaves a trace in a bubble chamber, or the state of a particle entering the beam splitter in one of the experiments described above, are certainly important parts of the 'story' of a world. Strictly speaking, the outcomes of quantum measurements, recorded in the definite classically described states of parts of the measuring devices, define the quantum states of the particle and we can tell the story replacing the particle's evolution by the set of outcomes of macroscopic measuring devices. But this is very artificial, so we adopt a description of a world as a classical state of macroscopic objects and, in addition, quantum states of (a few) relevant microscopic objects.

A good example of a quantum system which requires description in a world is a single particle passing through a beam splitter. If the particle passes the first beam splitter of a properly tuned Mach–Zehnder interferometer (MZI, see Fig. 4), its description as a superposition of being in two separate locations is necessary to explain why it will end up with certainty in one output of the interferometer and not in the other. If, after the first beam splitter, there are detectors, as in the experiment described in Fig. 2, then we may say that we need to describe the particle as being in a superposition for another reason. We might bet on one of the outcomes, and in order to place an intelligent bet we have to know the quantum state of the particle.

There is a big difference between strongly interacting quantum particles, like charged particles in the detector of an accelerator, and weakly interacting particles, like photons in an interferometer. Strongly interacting particles, in the same way as macroscopic bodies, are frequently measured by the environment, or by special detectors like bubble chambers. Therefore, when a strongly interacting particle is included as part of a world, it is nearly always well localized, with a definite classical description. Particles like photons are measured only occasionally, e.g., when they reach a detector. At the intermediate time, their quantum state becomes a superposition of localized states. Photons do interact with the environment between measurements, but only weakly, so that they are not localized in interferometers. Other particles, like neutrinos, interact only weakly most of the time.

In summary, a world consists of (a) macroscopic bodies and strongly interacting particles which are frequently measured by the environment, (b) microscopic particles which are measured, or interact strongly, only sometimes, and (c) microscopic particles which are almost always coupled only weakly with the

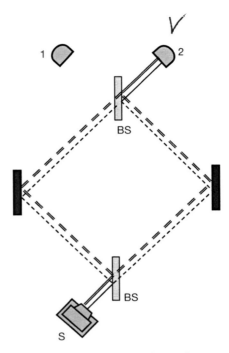

Figure 4. A Mach–Zehnder interferometer, tuned in such a way that all particles from *S* go to the port 2.

environment. Strongly interacting bodies (a) have a quantum state which is well localized at all times. The frequent measurements by the environment are non-demolition measurements and they create and measure both forwards and backwards evolving quantum states. Thus the backwards evolving state is identical to the time reversal of the forwards evolving state. Particles like photons (b) are measured occasionally, and mostly by demolition measurements. Strong measurements localize them, but in between measurements, their wavefunctions are typically spread out in space. Usually, their forwards and backwards evolving states are different: the backwards evolving state is not equal to the time reversal of the forwards evolving state, although it is never orthogonal to it. (Note that in some cases, several quantum particles are entangled and only their joint quantum state is pure.) Finally, weakly interacting particles (c) are spread out in space most of the time. The forwards and backwards evolving states need not be equal. Cosmology and experimental data tell us something about these states (mostly about the forwards evolving states).

Within a world, the objects of types (a), (b), and (c) influence one another, and their states are causally connected. In the standard MWI, only forwards evolving states are considered. I will argue that the story of a world is much clearer if both forwards and backwards evolving states are considered at a particular time. The forwards evolving state of quantum particles on its own (or the set of measurement outcomes in the past which defines it) does not describe correctly the influence of the particles upon other objects.

5 MODIFIED EVERETT WORLDS

Here is the main point I'd like to make. The standard quantum state of a particle, which is defined at a particular moment of time, which evolves forwards in time, and which depends on macroscopic records of measurements performed in the past, is not enough to define the particle's influence on other systems in the world. We also need to know the backwards evolving state, i.e., the outcomes of future measurements.

I suggest, therefore, the following change to the MWI picture of the universe. Instead of a tree of worlds, which starts from a single state common to all worlds and splits at every quantum measurement, the worlds are split by future measurements.

With this modification to the MWI, each Everett world is a possible Aharonov world. More precisely, it is a 'clean' Aharonov world because at each moment in time, both the forwards and the backwards evolving states have only the relevant components $|\psi\rangle$ and $\langle\phi|$ corresponding to the particular results of complete measurements before and after the time in question. This contrasts with Aharonov's original proposal, where both forwards and backwards evolving states $|\Psi\rangle$ and $\langle\Phi|$ are superpositions. The particular components $|\psi\rangle$ and $\langle\phi|$ in these superpositions have only tiny absolute weights which, however, ensure dominant contributions when we calculate weak values or (counterfactual) probabilities using the ABL rule.

Let us look again at our example with a single particle source, beam splitter and two detectors (Fig. 2). The world splits with the clicks of the detectors, creating a world in which the particle is reflected by the beam splitter and absorbed by detector A, and a world in which the particle passes through and is absorbed by detector B. In a standard approach to the MWI, at the time that the particle has passed the beam splitter, but has not yet reached a detector, there is only one world, with the particle in a superposition of two localized states. But nobody will ever observe any effect of this superposition (this is in contrast to an alternative experiment in which the beam splitter is part of an MZI). There will be two descendants of the experimenter who performs the experiment. One observes a click from detector A and might see traces of weak coupling on the route BS-A, and the other observes a click from detector B and might see traces of weak coupling on the route BS-B. The worlds, as I define them, are histories, one with the A outcome and another with the B outcome. In no world can a trace of the superposition of paths A and B be seen, i.e., in no world will we see something both on the route *BS-A* and on the route *BS-B*. In contrast, the traces of path *BS-A* in world *A* (and the traces of path *BS-B* in world *B*) can be seen. To this end we have to arrange an ensemble of identical experiments with weak measurements of projection on various paths at the appropriate time. Then, the results of the weak measurements on a pre- and post-selected ensemble will show the trajectories *BS-A* and *BS-B* respectively. (Especially effective is a weak measurement performed on a rare event pre- and post-selection ensemble when all particles in the ensemble end up with the desired outcome. Then, the same measuring device couples to all particles.)

The example with a single beam splitter is illustrative, but it is very simple, and although there is some inconsistency, we can also discuss it relatively well in the language of forwards evolving states only. Let us consider a more sophisticated example, in which the standard approach can lead to misleading conclusions (Hosten et al. [2006]). Consider a combination of beam splitters which create a MZI 'nested' in another MZI (see Fig. 5). The internal MZI is tuned in such a way that a particle entering the left input port always comes out in the right port. We consider a photon which enters the external interferometer from the source and ends up in detector 2. Since photons on the right arm of the

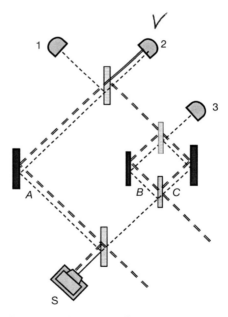

Figure 5. The internal Mach–Zehnder interferometer is tuned in such a way that the particle cannot go from source *S* to detector 2 through the right arm. Nevertheless, at B both forwards and backwards evolving states are present, and for any weak coupling to the particle at B, the effect is as if there were a single particle there.

external MZI leave the interferometer (due to the particular tuning of the internal interferometer), in the framework of the standard MWI we would conclude that the photon passed solely through arm A (and not through B and C). However, such a set-up of beam splitters can also be considered as the creation of the two-state vector (5) of the three box paradox, where $|A\rangle$, $|B\rangle$, and $|C\rangle$ are the states of a photon in arms A,B, and C respectively (Vaidman [2007]). Then, we know that the weak interaction influence at B is as if the photon were there. In any case, it is not less than the influence at A. This fact is immediately seen in the world described by the two-state vector and is very hard to understand in the standard approach.

Aharonov's proposal is to add a universal backwards evolving quantum state $\langle\Phi|$. My proposal is to add 'locally' to each world, in the period between complete measurements on quantum particles, a backwards evolving quantum state of the particles $\langle\phi|$. Aharonov's backwards evolving state has a fundamental ontological status, similar to that of the forwards evolving universal state in the standard MWI. My additional backwards evolving state is not of this kind (at least until I introduce a more speculative modification below). It is an explanatory concept for the inhabitants of a particular world.

Sometimes in discussions of the meaning of the quantum state, it is considered as a way to describe the information which belongs to a particular observer. In a particular world, at least for some periods in the past, forwards *and* backwards evolving states of particles describe correctly their influence on other objects. Note that the change is relevant only for objects of type (b); strongly interacting objects (a) have identical forwards and backwards evolving states and adding the backwards evolving state contributes nothing. Viewing the quantum state as information sometimes accompanies an attribution of ontological status to the outcomes of measurements. In this language, my proposal is to describe a quantum system at a particular time by the outcomes of measurement both before and after this time. Thus, the time of world splitting is the time of future measurements.

Let us clarify the concept of world splitting by returning to the experiment of Fig. 2. A world is a complete history of results of measurements. When we talk about 'splitting', we consider the following. We make, at consecutive times, snapshots of the universe and count the number of worlds. The increase of this number signifies splitting. During the time in which the photon has left the source and is flying towards the detectors, the snapshots count two worlds: in one, the photon is detected (in the future) by detector A, and in the other by detector B. The number of worlds continues to be two also for snapshots performed after the time at which the photon reaches the detectors (the time of splitting according to the standard approach). Regarding the detectors we can see a change: when the photon reaches the detectors, their states become different in the two worlds. Before the time of arrival of the photon, both detectors are in the 'ready state'. It is the photon which distinguishes the worlds via its different ways of weakly coupling with the environment. Note that when we consider the Mach–Zehnder interferometer experiment (Fig. 4), there are no different histories and there is a single world for snapshots at all times.

There is no paradox of backwards causality in this proposal, because the splitting of worlds is not a physical phenomenon: the concept of a 'world' is given by our own semantic definition which helps to explain our experiences. The fundamental ontological picture remains, as in standard MWI, that of a single forwards evolving quantum state. The forwards evolving state of measuring devices defines the outcomes of measurements which, in turn, define the forwards and backwards evolving states within a world.

The time asymmetry in this picture is even worse than in Aharonov's proposal. In Aharonov's theory, the asymmetry is due to a very big difference in the form and role of the forwards and backwards evolving states $|\Psi\rangle$ and $\langle\Phi|$. In my picture, there is only one, forwards evolving, fundamentally ontological state. It defines, in a more or less symmetric way, the forwards and backwards evolving states $|\psi\rangle$ and $\langle\phi|$ within various worlds, but on a global view there is an asymmetry: only about the forwards evolving states $|\psi\rangle$ we can say that their superposition is the universal quantum state.

Can we restore time symmetry? Can we apply the TSVF globally and accept, as Aharonov does, the fundamental ontological existence of both forwards and backwards evolving quantum states? In the next section I will discuss such a possibility.

6 TIME SYMMETRY

The most natural way to combine the backwards evolving states of all branches leads to a backwards evolving universal state which is just the time reversal of the forwards evolving universal state. All other components of the back-wards evolving states of different worlds interfere destructively. But it seems incorrect to assume that this is the fundamentally ontological backwards evolv-ing quantum state of the universe. The difficulty arises when we consider the issue of the probability of an outcome of a quantum experiment (Saunders [1998]).

In the MWI with a single ontological quantum state evolving forwards in time, we can postulate (or as David Deutsch [1999] claims, even derive) the observed Born rule for quantum mechanical probabilities. These are not 'real' probabilities, they are a 'measure of existence' or 'caring measure' (Vaidman [2002], Greaves [2004]), the main purpose of which is to advise a gambler how to bet on the outcomes of quantum measurements so as to gain a maximum reward for his multiple descendants. One can connect this measure to ignorance probabilities through a gedanken experiment involving a quantum gambler with a sleeping pill (Vaidman [1998]).

However, the assumption of identical ontological forwards and backwards evolving states leads to wrong probabilities: indeed, according to the ABL rule (2), the probability of an outcome $O = o_n$ is proportional to $|\langle\Psi|\mathbf{P}_{O=o_n}|\Psi\rangle|^2$, while according to the Born rule, the probability is proportional to $|\langle\Psi|\mathbf{P}_{O=o_n}|\Psi\rangle|$.

Apart from using the very special backwards evolving state proposed by Aharonov (in which the correct probabilities are fixed by fiat), we can obtain the correct probabilities if we assume that the backwards evolving state is a complete mixture, i.e., an equally weighted mixture of the states of an orthonormal basis.

This proposal might help us understand the main difficulty with the concept of probability in the MWI. In a quantum experiment all outcomes are realized, and so the standard concept—the probability that one outcome happens and not the others—evaporates. I had to introduce the 'sleeping pill' trick to associate the probability of an outcome with the concept of probability understood as representing ignorance of the experimenter (Vaidman [1998]). But now, when there is a fundamentally ontological mixture of states going backwards in time, corresponding to different outcomes of the experiment, we can associate the experimenter with both the forwards evolving state and a component of the backwards evolving mixture corresponding to one of the outcomes. For such an experimenter there is a matter of fact about the outcome of the experiment and the usual concept of probability applies.

Adding a backwards evolving quantum state which is a complete mixture is consistent with our observations, but it is only a small step towards time symmetry. The backwards and forwards evolving states are very different—one is a mixture and the other is a pure state.

The modification which I have advocated here, with the backwards evolving state either ontological or not, has no observable effects. But it clarifies our concept of 'world' and provides a step towards a better connection between the mathematical concepts of quantum mechanics and our experience.

It is a pleasure to thank Simon Saunders for suggesting the topic and for helping tremendously in writing this paper. I also thank Jon Barrett for numerous helpful comments. This work has been supported in part by grant 990/06 of the Israel Science Foundation.

References

Aharonov, Y., D.Z. Albert, and L. Vaidman [1988], 'How the result of measurement of a component of the spin of a spin-1/2 particle can turn out to be 100', *Physical Review Letters* **60**, 1351–54.

Aharonov, Y., P.G. Bergmann, and J.L. Lebowitz [1964], 'Time symmetry in the quantum process of measurement', *Physical Review* **134**, B1410–16.

Aharonov, Y. and E. Gruss [2005], 'Two-time interpretation of quantum mechanics', available online at arXiv:quant-ph/0507269v1.

Aharonov, Y. and L. Vaidman [1990], 'Properties of a quantum system during the time interval between two measurements', *Physical Review* A 41, 11–20.

—— [1991], 'Complete description of a quantum system at a given time', *Journal of Physics* A 24, 2315–28.

—— [2008], 'The two-state vector formalism, an updated review', *Lecture Notes in Physics* 734, 399–447.

Deutsch, D. [1999], 'Quantum theory of probability and decisions', *Proceedings of the Royal Society of London* A 455, 3129–37.

Greaves, H. [2004], 'Understanding Deutsch's probability in a deterministic multiverse' *Studies in History and Philosophy of Modern Physics* 35, 423–56. Available online at philsciarchive.pitt.edu/archive/00001742/.

Hosten, O. et al. [2006], 'Counterfactual quantum computation through quantum interrogation', *Nature* 439, 949–52.

Saunders, S. [1998], 'Time, quantum mechanics, and probability', *Synthese* 114, 405–44.

Vaidman, L. [1993], 'Lorentz-invariant "elements of reality" and the joint measurability of commuting observables', *Physical Review Letters* 70, 3369–72.

—— [1996], 'Weak-measurement elements of reality', *Foundations of Physics* 26, 895–966.

—— [1998], 'On schizophrenic experiences of the neutron or why we should believe in the many-worlds interpretation of quantum theory', *International Studies in the Philosophy of Science* 12, 245–61.

—— [2002], 'Many-words interpretation of quantum mechanics', *Stanford Encyclopaedia of Philosophy*, E. N. Zalta (ed.). Available online at http://plato.stanford.edu/entries/qm-manyworlds/.

—— [2007], 'Backward evolving quantum states', *Journal of Physics* A 40, 3275–84.

—— [2007], 'Impossibility of the counterfactual computation for all possible outcomes', *Physical Review Letters* 98, 160403.

Transcript: Not (Only) Many Worlds

Pitowsky (comment)

In our approach we begin with Hilbert space structure. You have the structure which is a realistic structure and this is what you should follow. It's not as if there's no alternative, our point is not a polemic point, and it's—this is the question that Simon had, is it a realistic interpretation?—it's a realistic interpretation in the following sense: you have events and you have measurement results. They have nothing to do with whether human beings are there or not because measurements are not privileged in the theory, events are. 'Event' is a primitive of probability theory—there's no analysis of what an event is in probability theory. You have a whole set of those that are actually occurring or ostensibly occurring, and you have a whole lot of very interesting structure *on* that, and this is the structure of the Hilbert space. So this is realism about the structure of the Hilbert space. It's realism about operators, not realism about a wavefunction, and in that sense it's completely orthogonal to the Everett interpretation. It's realism about events and operators, projection operators.

Wallace

OK, one person's dogma is another person's axiom is another person's empirical datum. I can imagine living in a world where measurement was a primitive. In this world there are indestructible black boxes dotted over the landscape; they have a knob on and a funnel and a dial. The world's *not* like that. The measurement devices we actually see in the world do not seem to be primitive, they seem to be built by very clever people, from principles that we basically think we can understand. One can look at them and deduce rather than postulate how they work.

I mean, this objection is utterly unoriginal, but if there is something to be said about it then I'm very interested to hear what that thing is.

(In this context, I think it's worth remembering the real reason why we adopt statistical mechanics rather than thermodynamics. It's not really about fluctuations, it's about the equation of state. Statistical mechanics doesn't require us to postulate that equation as primitive. it allows us to derive it from the underlying physics.)

Bub

Well, first of all the claim is certainly not that the no-cloning principle entails that there are measuring instruments around that are just black boxes which can't be analysed further. A quantum-mechanical analysis can be given, in principle, to any level of precision, so if somebody builds a measuring instrument, you can proceed to analyse any aspect of it down to its smallest microscopic constituents and keep going as far as you want. The claim is that it follows from the no-cloning principle that at the end of the day there will always be some aspect of the physical system that is left out of the analysis and is treated simply as a classical information source.

But I find it a bit surprising that there's some suggestion that if you say this then you're giving up on physics or something. It seems to me really implausible to try to perform some quantum-mechanical calculation from the point of view of the Bohm theory in order to get some new insight or new way of looking at things. I'm not talking about a new prediction derived from some non-equilibrium distribution, but some way of looking at things that is suggestive of something that supposedly, if we look at it from the Bohmian perspective, is really getting at the root of the way the world is put together, but that standard quantum mechanics can only reproduce in some artificial way.

Valentini

The discovery of Bell's theorem is a case where this actually happened. De Broglie–Bohm theory helped Bell discover non-locality.

Timpson

And contextuality.

Bub

Well, he knows that Bohm's theory is non-local, and he asked the question: must any hidden-variable theory have to be like Bohm's theory? But that seems to be something altogether different.

Valentini

It is an example of an insight obtained not just by looking at quantum mechanics.

Pitowsky

It wasn't *developed* on the basis of Bohm's theory, it was derived from Hilbert space. This is the point. Lead in the sense of inspire, yes; that I don't deny.

Wallace (to Valentini)

I thought that, perhaps unintentionally, perhaps not, Antony's closing comment, albeit that it was intended as a criticism of Everett, is quite a nice way of seeing how the ontological situation developed. I think it's entirely right that for a long time we (people doing quantum mechanics) thought much too much about quantum mechanics in terms of eigenvalues, eigenvectors, operators, definiteness, no-fact-of-the-matter-about, neither-one-nor-the-other. I think one of the things that's made it possible to get a real grip on what quantum mechanics is trying to tell us, at least Everett-style, has been the succession of moves away from that, to positive operator value measurements and Wigner function representations and consistent history frameworks and decoherence and pointer bases. It's led us to understand that if we do want to think about quantum mechanics realistically, thinking in the eigenvector-eigenvalue-operator way is simply flawed. So, I would say the Everett interpretation shows exactly how we should take the theory which we were bequeathed from half-baked classical considerations and read it properly. So I entirely agree with what you said but draw slightly different conclusions from it.

Bub (to Valentini)

Bohm always used to say that no one is interested in doing experiments just to find out the position of a particle. What we're really interested in is precisely explaining atomic spectra, for example—that is, measuring the eigenvalues of operators, and the story in terms of the position of a particle is very far removed from that. It's different in classical mechanics because the observables that you're looking for are just functions of position.

Valentini

Well, I agree, and look at what actually happened. The data that people were looking at were spectral lines, atomic energy levels—that was early on—and in the 1920s much of the experimental data that people were looking at were scattering experiments. You know, all kinds of ways that the differential cross-sectional area varies with angle. The Ramsauer effect stunned people: that a very low energy electron would pass almost freely through a gas. How do you explain that? It turned out that as Elsasser showed, if you have a long de Broglie wavelength, you could explain that the scattering cross section went almost to zero. Certainly they weren't measuring the positions of particles. But it was about electrons, atoms, then it became about photons, and I don't see how that really affects anything. Clearly the data you see, they're not just given directly in terms of particle positions, there's a complicated story about the experiments.

Saunders (to Valentini)

I share your view that given non-equilibrium, the accusation that pilot-wave theory is Everett in denial really takes on a different character. But sticking to the equilibrium case, I'd like a clear verdict from you. Sticking to the equilibrium case, as many Bohmians do, it seems to me that, to come back to the rhetorical issue of does one look at Bohm through Everettian spectacles or not, if it were the case that an Everettian is making some special postulate, alien to the formalism, alien to standard scientific methodology, then of course there would be something wrong with the accusation; it shouldn't anyway disturb the Bohmian. But if it is the case that the Everettian is not using any special postulates or assumptions, if the methodology of interpretation is standard scientific methodology, and therefore that that standard methodology applies to the Bohmian equilibrium case, then if all that goes through it seems to me that the argument is really compelling that the trajectories are epiphenomena.

Valentini

It may be difficult to come to a watertight conclusion about that. But I have an earlier problem with the equilibrium de Broglie–Bohm theory. It's a bit like if you only ever have thermal equilibrium in statistical mechanics, you never have access to some more detail about molecular motions, or more detail about the de Broglie–Bohm trajectories. Look, here we have a theory where there are two fundamental equations of motion, the Schrödinger equation and the equation of motion for the trajectory. Now, the details of the trajectories are almost completely washed out in equilibrium. All you see is that the velocity law must preserve the Born distribution with time, a kind of very crude average. Now, if that was the world, if someone says to me this is my theory of the world, these two equations of motion plus this distribution which wipes out almost all the details of one of them, then to me, scientifically, I'm very unhappy with that. To me, if you take the de Broglie–Bohm theory seriously, it is crying out to you that, look, quantum theory is just for a special state, there must be non-equilibrium somewhere or at some time; otherwise it doesn't seem at all like a reasonable way to make the world for a scientist. It's logically possible but I wouldn't accept it. So I would say that if in 100 years we've measured everything in the early universe, everything coming out of black holes, everything that one might conceivably think of in a place where there might be non-equilibrium, and there's still no sign of non-equilibria, then I will rethink.

Albert (to Saunders)

Just relevant to this, sort of intended I guess as an answer to Simon. Look, I don't think the two options you laid out are the only two options. I think one of the things that was enormously helpful about John Hawthorne's talk today was precisely the description of a very reasonable middle ground. If we have very good scientific reasons for believing that such and such is the basic ontology of the world, OK, then we're going to look around in the world and say, well, I guess those must be the tables and those must be the chairs because that's all we've got. But if we have reasons for entertaining a different ontology all of a sudden, and ontology with extra things there that aren't in the first ontology I was considering, of course it may be eminently reasonable to say, oh, no, *those* are the tables and chairs, we don't have to think of *these* as the tables and chairs any more and it just doesn't follow at all, and that's what seems to be the basis of all these arguments, that once you've granted, subject to the constraint that this is the basic ontology, well I guess those are going to be the tables and chairs in this ontology, that that somehow commits you to sticking with that in alternate circumstances where you're considering different ontologies where there are much more plausible, much more credible candidates for what the tables and chairs are. So, you're pointing to a dichotomy: either you have to reject that in a pure wavefunction monist theory and a collapse theory, that is, in the simple case, where it's an eigenstate, either you have to reject that that's a table or a chair or you're going to have to grant that inventing the corpuscles does no good. That seems to me an absolutely false dichotomy.

Brown (to Hawthorne)

I'm going to appeal to what I consider to be the best principle of metaphysics to pit against your metasemantics: Occam's razor. You have a formalism which is shared by two theories, so there's a dynamics associated with some fundamental element of reality, but one of the theories adds something else. Now the question arises: can you save the appearances in both theories? There is no consensus on this but most commentators would say yes.

Albert

Not most of us at this conference.

Brown

So it's not an obvious issue. Now, there is no law as far as I know that says that spacetime or indeed even space has to exist in the theory at some primordial level. Why does it? We would be perfectly happy in physics if everything we see around us, in a particular the three-dimensional and maybe even four-dimensional aspects of reality, was an emergent property out of something that is more fundamental. So, given that saving the appearances, well-defined dynamics, that all these things are a property of the first theory, why in your right mind would you introduce extra structure?

Hawthorne

There are two questions. One question is: is Tim's ontology too profligate? Maybe yes. But there's another claim that you have definitely been making in the literature which is the conditional: even if his ontology's true, grooves in configuration space are deserving of the name four-dimensional spacetime. I am challenging that.

Albert

Things are being presented as if they were exactly two options whereas there are in fact a continuum of them it seems to me. It doesn't follow from my saying—if I'm agreeing to play the game, OK, there's only the wavefunction. Tim is just going to refuse to play that game; that's not my position. My position would be, OK, I can see how to make sense of that. I guess, if there are good reasons for believing that, these must be the tables and chairs. But if I'm now entertaining a different ontology with, say, corpuscles in it, of course I'm going to withdraw that claim because there are much better-credentialed candidates around for being the tables and chairs. So I'm not going to say there are tons of tables and chairs here, I'm going to say the tables and chairs are made up by the corpuscles.

Loewer

In order to argue that Bohm is Everett in denial, you're thinking that whatever it is that the mathematical apparatus is describing in Bohm, is the same sort of thing as you think of it as describing in many worlds. But at least some Bohmians think of it really differently from that. They think of it as describing something that's like a law. That is, determining how particles are moving. Now, I don't know exactly what the content of defending that it is a law is, but I guess whatever the structure of a law is it's not the stuff that material objects are made up out of. So this really looks like an alternative theory. And not one which has ghost particles moving around.

Valentini

I would agree with the first bit of what you said. But as for the second bit—about some people who claim that the wavefunction is just like a law, I think that's just a bad idea. It's as bad as saying that the electromagnetic field is like a law and that, you know, the ontology's just charged particles. I think that view is bad.

Deutsch

I'd say the reason it's bad is that this law contains *people* walking around and interacting with each other and affecting each other. I'm surprised that no one has yet mentioned quantum computers in this connection. I don't think that measurement theory today is based on this classical assumption that if P1, Q2. It's more based on the quantum theory of computation. You imagine a computer that can be programmed to simulate a tiger that's going to do something like eat food, with classical computation, but it's a quantum computer. And then you start it off in a superposition of different initial conditions with the food in different places; then if you say that the first computation contained the simulation of a tiger eating food in position A (which you're committed to, with the wavefunction being ontological), then I don't see how you can avoid saying that the second place contains lots of simulations of tigers eating in all different places simultaneously.

Valentini

As I said, from the pilot-wave perspective you can point to mathematical structure in the pilot wave that you could identify with alternative evolutions. But you're not driven to give them an ontological status.

Deutsch

Are you saying it does not contain those evolutions?

Valentini

No, it contains a complex-valued field on configuration space which contains mathematical structure.

Deutsch

No, in reality.

Valentini

The wave is a reality. And if you look at the mathematical structure of that wave, yes, it does contain mathematical simulations.

Deutsch

So, in reality it contains running programs simulating tigers.

Valentini

In reality it consists of a complex field evolving on configuration space which you can decompose in certain ways and regard as, well, here's one piece of this.

Deutsch

There's your 'Everett in denial'.

Valentini

No.

Deutsch

It exists on the one hand . . .

Valentini

No, it exists as a complex-valued field. I mean, like the string exists as a displacement; it does not exist as a superposition of this mode plus that mode plus that; that's mathematical structure. And this is an important point because in your book you made a challenge: how do you explain quantum computing in a theory that doesn't have parallel worlds? Well, look, first of all, even within the standard quantum theory community people have questioned the degree to which the explanation is superposition. Some people like Richard Jozsa say really it's entanglement. It is controversial.

But second of all, here, I agree that there's an ontological complex-valued field in configuration space. And if you look at this theory, it's a non-local theory, it's a contextual theory and there's no doubt that it reproduces the empirical predictions of quantum theory for experiments, for quantum computers, and I'm sure that if you look at specific algorithms and you take into account that there is an ontological pilot wave in a higher-dimensional space, that it's non-local, there's entanglement, there's contextuality, then there's plenty of room for understanding how quantum computations work just as there's plenty of room for understanding how many other quantum experiments work. If there were no ontological pilot wave in configuration space then one might worry.

Vaidman to Deutsch

I'm certainly very much with you that Everett's is by far the best interpretation, that it's correct and it's a miracle why it's not understandable until now. I'm trying to think why, and I think you may be partly responsible. The first reason is the biggest; it's many worlds and you said they're really separate and it looks like your multiverse, which had before been given different names; you consider theories of the multiverse. And I think Everett means, and especially Schrödinger, it's a *single* physical universe. I don't think multiverse, it's not a theory of multiverse, the important thing is no collapse, that's the main point. And the other thing, and you made a big contribution which put it forward, but you also didn't ask which things can be achieved and which might not, and I think Everett was the first to say that you can get probability out of many worlds. If he didn't say so I think now it would be accepted, because this was not proven at that time, it's not proven until today. And when others show that this doesn't work they kill the whole theory. But the main thing is there's no collapse and that does not contradict our experience. This is Everett's theory. If you would not add some other things which are maybe not achievable then it would be accepted.

Deutsch

OK, two things. First of all, in regard to what you say about universe/multiverse, I don't care about the terminology, I'm quite happy to call the multiverse the universe and the universe a branch. In fact I did that in one paper. I don't care. Certainly the point of Everett's theory is that we don't need collapse and that quantum theory can be regarded as a complete theory of the universe. As for probability, I agree with you that the objection to Everett that it didn't describe probability properly was always a mistake. What was happening was that there is a problem at the foundations of probability theory which was obscured by silly interpretations. But in Everett's interpretation you could actually see that problem. It wasn't a problem with Everett's interpretation, it's something that you could see via Everett's interpretation and then, we believe, solve. But if we hadn't solved it, it would still be no argument against Everett.

Wallace to Deutsch

This is a sort of friendly amendment, I suppose. You considered a single particle localized in space, and in this situation the classical multiverse approximation, I think, breaks down. It seems to me that it's a retrograde move to describe that particle as living in different universes at different points. Those sorts of position-basis eigenstates are there and we can talk about them by all means, but calling them universes doesn't seem to me to be necessarily that helpful. They not autonomous, they're just components of the quantum state.

Deutsch

Yes, I entirely agree with that. And if only we had a proper mathematical description of the multiverse, I'd be very happy to say that what's *actually* happening in a certain situation is so-and-so and you can see you can't then make the 'universes' approximation because it wouldn't be accurate. Yes, I agree.

Saunders

When you say we don't have a proper understanding of the multiverse what sort of understanding do you want? Is it mathematical, is it . . .?

Deutsch

There's a mathematical physics issue and also a philosophical issue. The mathematical physics issue is that we don't know what mathematical object the equations of quantum theory are supposed to apply to. We know how to get answers out of them and we know how to describe *some* things about this object in emergent approximations, but we don't have an exact theory of it. For instance, in that work I did on time travel—going back in time and coming back in a different universe—I had a diagram there where a loop in spacetime gets unfolded into a causal spacetime with two sheets—I don't know if any of you are familiar with that work, but anyway the point is that that two-sheet manifold, which is a sort of hybrid conception of a relativity thing with a quantum thing—there's no mathematical object that I can say that's an approximation to. There ought to be. So that's the mathematical physics part of it; I think we need that in order to make progress in certain directions. There's also the philosophical issue that we don't have a vocabulary and a language. Again, we have a vocabulary to talk about situations where an observable is sharp, or to talk about parallel universes, but most of the multiverse isn't like that and we don't have a good way of talking about what that is like, even in the case of a single particle. So, those are the two things that are lacking.

Ladyman

I thought you overstated the argument against anti-realism, and also that you ran together two distinct things. An anti-realist about science doesn't have to deny that there's an objective reality; they can just deny that we're finding out about it. You shouldn't run those things together because obviously one's much less plausible than the other. I'm not personally an anti-realist but it is an intelligible view to take about science, as is exemplified by the attitude a lot of people take towards Newtonian mechanics: you can think it's a wonderful theory, use it every day, predict lots of things with it, and yet not believe that there really are Newtonian action-at-a-distance forces. That's not a crazy view to have of the world, and someone might think about quantum mechanics just like that: it's a good theory to use, but it doesn't compel anyone to believe that it is telling the truth about unobservables.

Deutsch

First of all, I'm glad you're not an anti-realist because it's very hard to argue with someone who denies that they exist. About whether I was too hard on anti-realism—all I said about it was this: you say it could be that there is a reality but our science doesn't have access to it. I said that if you take that view then science isn't about anything, apart from your own mind, and I think that would be true of the person that you outlined as well. Now, as for explanation, the trouble is, if you regard a theory as being *purely* a set of observable predictions, then you must, logically, be thinking of observations as unanalysable primitive things. And that is incompatible with having a universal scientific theory. Also there's the fact that there are infinitely many interpretations of those observed events which make different predictions for the future, then you'll run into the problem of induction and so on. I'd have to refer you to 'The Fabric of Reality' for why science has to be explanatory to solve that.

Index

Lightning Source UK Ltd.
Milton Keynes UK
UKOW042335240712

196491UK00001B/15/P